| | | |
|---|---|---|
| **CONTINUITY EQUATION FOR ANY FLUID** | $\rho_1 A_1 v_1 = \rho_2 A_2 v_2$ | |
| **CONTINUITY EQUATION FOR LIQUIDS** | $A_1 v_1 = A_2 v_2$ | (6–5) |
| **BERNOULLI'S EQUATION** | $\dfrac{p_1}{\gamma} + z_1 + \dfrac{v_1^2}{2g} = \dfrac{p_2}{\gamma} + z_2 + \dfrac{v_2^2}{2g}$ | (6–9) |
| **TORRICELLI'S THEOREM** | $v_2 = \sqrt{2gh}$ | (6–16) |
| **TIME REQUIRED TO DRAIN A TANK** | $t_2 - t_1 = \dfrac{2(A_t/A_j)}{\sqrt{2g}}\left(h_1^{1/2} - h_2^{1/2}\right)$ | (6–26) |
| **GENERAL ENERGY EQUATION** | $\dfrac{p_1}{\gamma} + z_1 + \dfrac{v_1^2}{2g} + h_A - h_R - h_L = \dfrac{p_2}{\gamma} + z_2 + \dfrac{v_2^2}{2g}$ | (7–3) |
| **POWER ADDED TO A FLUID BY A PUMP** | $P_A = h_A W = h_A \gamma Q$ | (7–5) |
| **PUMP EFFICIENCY** | $e_M = \dfrac{\text{Power delivered to fluid}}{\text{Power put into pump}} = \dfrac{P_A}{P_I}$ | (7–6) |
| **POWER REMOVED FROM A FLUID BY A MOTOR** | $P_R = h_R W = h_R \gamma Q$ | (7–8) |
| **MOTOR EFFICIENCY** | $e_M = \dfrac{\text{Power output from motor}}{\text{Power delivered by fluid}} = \dfrac{P_O}{P_R}$ | (7–9) |
| **REYNOLDS NUMBER—CIRCULAR SECTIONS** | $N_R = \dfrac{v D \rho}{\mu} = \dfrac{v D}{\nu}$ | (8–1) |
| **HYDRAULIC RADIUS** | $R = \dfrac{A}{WP} = \dfrac{\text{area}}{\text{wetted perimeter}}$ | (8–5) |
| **REYNOLDS NUMBER— NONCIRCULAR SECTIONS** | $N_R = \dfrac{v(4R)\rho}{\mu} = \dfrac{v(4R)}{\nu}$ | (8–6) |
| **DARCY'S EQUATION FOR ENERGY LOSS** | $h_L = f \times \dfrac{L}{D} \times \dfrac{v^2}{2g}$ | (9–1) |
| **HAGEN-POISEUILLE EQUATION** | $h_L = \dfrac{32 \mu L v}{\gamma D^2}$ | (9–2) |
| **FRICTION FACTOR FOR LAMINAR FLOW** | $f = \dfrac{64}{N_R}$ | (9–3) |

# Applied Fluid Mechanics

## Fifth Edition

# Applied Fluid Mechanics

## Fifth Edition

## Robert L. Mott
University of Dayton

Prentice Hall
*Upper Saddle River, New Jersey    Columbus, Ohio*

**Library of Congress Cataloging-in-Publication Data**

Mott, Robert L.
   Applied fluid mechanics / Robert L. Mott.—5th ed.
     p. cm
   Includes bibliographical references and index.
   ISBN 0-13-023120-7
   1. Fluid mechanics.   I. Title.

   TA357.M67 2000
   620.1'06—dc21           99-046580

Editor: Stephen Helba
Production Editor: Louise N. Sette
Production Supervision: York Production Services
Design Coordinator: Karrie Converse-Jones
Cover Designer: Michael Osadciw
Cover art: Rendering artist Larry Hewitt, Modern Medium, Inc., Portland, OR, for EDA,
   Berkeley, CA
Production Manager: Matthew Ottenweller
Marketing Manager: Chris Bracken

This book was set in Times Roman by York Graphic Services, Inc., and was printed and
bound by Courier/Westford, Inc. The cover was printed by Phoenix Color Corp.

© 2000, 1994 by Prentice-Hall, Inc.
Pearson Education
Upper Saddle River, New Jersey 07458

Printed in the United States of America

10  9  8  7  6  5  4

ISBN: 0-13-023120-7

Prentice-Hall International (UK) Limited, *London*
Prentice-Hall of Australia Pty. Limited, *Sydney*
Prentice-Hall of Canada, Inc., *Toronto*
Prentice-Hall Hispanoamericana, S. A., *Mexico*
Prentice-Hall of India Private Limited, *New Delhi*
Prentice-Hall of Japan, Inc., *Tokyo*
Prentice-Hall (Singapore) Pte. Ltd., *Singapore*
Editora Prentice-Hall do Brasil, Ltda., *Rio de Janeiro*

*To my wife Marge;*
*our children Lynné, Robert Jr., and Stephen;*
*and my Mother and Father.*

# Preface

**Introduction**  The objective of this book is to present the principles of fluid mechanics and the application of these principles to practical, applied problems. Primary emphasis is on fluid properties, the measurement of pressure, density, viscosity, and flow, fluid statics, flow of fluids in pipes and noncircular conduits, pump selection and application, open channel flow, forces developed by fluids in motion, the design and analysis of HVAC ducts, and the flow of gases.

Applications are shown in the mechanical field including fluid power, heating, ventilation, and air conditioning (HVAC); the chemical field including flow in materials processing systems; and the civil and environmental fields as applied to water and waste water systems, fluid storage and distribution systems, and open channel flow. This book is directed to anyone in an engineering field where the ability to apply the principles of fluid mechanics is the primary goal.

Those using this book are expected to have an understanding of algebra, trigonometry, and physics mechanics. After completing the book, the student should have the ability to design and analyze practical fluid flow systems and to continue learning in the field. Other applied courses, such as fluid power, HVAC, and civil hydraulics, could be taught following this course. Alternatively, this book could be used to teach selected fluid mechanics topics within such courses.

**Approach**  The approach used in this book encourages the student to become involved in the learning of the principles of fluid mechanics at six levels:

1. The understanding of concepts.
2. The recognition of the logical approach to problem solutions.
3. The ability to perform the details required in the solutions.
4. The ability to critique the design of a given system and recommend improvements.
5. The ability to design practical, efficient fluid systems.
6. The ability to use computer-assisted approaches for design and analysis of fluid flow systems.

This multilevel approach has been successful in building students' confidence in their ability to analyze and design fluid systems.

Concepts are presented in clear language and illustrated by reference to physical systems with which the reader should be familiar. An intuitive justification as well as a mathematical basis is given for each concept. The methods of solution to many types of complex problems are presented in step-by-step procedures. The importance of recognizing the relationships among what is known, what is to be found, and the choice of a solution procedure is emphasized.

Many practical problems in fluid mechanics require relatively long solution procedures. It has been my experience that students often have difficulty in accomplishing the details of the solution. For this reason, each example problem is worked in complete detail including the manipulation of units in equations. In the more complex examples, a programmed instruction format is used in which the student is asked to perform a small segment of the solution before being shown the correct result. The programs are of the linear type in which one panel presents a concept and then either poses a question or asks that a certain operation be performed. The following panel gives the correct result and the details of how it was obtained. The program then continues.

The International System (Système International) of units (SI) and the U.S. Customary System units are used approximately equally. The SI notation in this book follows the guidelines set forth by the American National Metric Council. Other references used are

> International Standard ISO 1000, *SI Units and Recommendations for the Use of their Multiples and of Certain Other Units,* International Organization for Standardization.
>
> *ASME Orientation and Guide for the Use of SI (Metric) Units,* 5th ed., The American Society of Mechanical Engineers.
>
> *SI Units in Fluid Mechanics,* The American Society of Mechanical Engineers.

**Computer-Assisted Problem Solving and Design**

Computer-assisted approaches to solving fluid flow problems are recommended after the student has demonstrated competence in solving problems manually. Computer-based assignments are included at the end of many chapters. These can be solved by a variety of techniques such as:

- programming in a technical language such as C, C++, BASIC, PASCAL, or FORTRAN
- the use of a spreadsheet such as Microsoft Excel
- the use of technical computing software such as MATLAB®
- the use of commercially available software for fluid flow analysis

Chapter 11, "Series Pipe Line Systems," and Chapter 13, "Pump Selection and Application," include example Excel spreadsheet aids to solve fairly complex system design and analysis problems. A CD-ROM is included with the solutions manual for this book that contains all of the spreadsheets used in the book and several more. They may be copied and distributed to students at the discretion of the instructor.

Also on the CD-ROM are student versions of two powerful programs commercially available from Tahoe Design Software of Truckee, California. HYDROFLO™ is a Windows-based tool for the design and analysis of fluid flow systems that have a single source and a single point of discharge. Additionally, parallel branches can be modeled. Pipe elements, pumps, valves, and fittings can be easily inserted from a large library of standard components or with unique data. The systems can be solved for steady-state flows, pressures, head losses, total dynamic head, and net positive suction head (NPSH) values. This software can be used to solve most of the problems of the types shown in Chapters 11, 12, and 13 of this book.

PumpBase™ is an extensive data base of performance curves for real, commercially available pumps that can be used in conjunction with HYDROFLO™ to select an appropriate pump for the system being designed. PumpBase™ automatically generates the flow versus head curve for the pump, superimposes the system curve, and determines the operating point of the pump in the system you have designed. Actual efficiency, power required, and NPSH required are given for the pump. This tool is very useful to supplement the pump performance curves in Chapter 13 of this book.

Both HYDROFLO™ and PumpBase™ can be placed on computers available to students or shared on a college network.

**Features New to the Fifth Edition**

The fifth edition continues the pattern of earlier editions in refining the presentation of several topics, enhancing the visual attractiveness and usability of the book, updating data and analysis techniques, and adding selective new material. The following list highlights some of the changes in this edition:

- Each chapter begins with a new section called *The Big Picture.* The student is asked to recall or discover real products or systems where the principles of fluid mechanics are used and to consider what kind of fluid is used, what the fluid is used for, how it behaves, and what conditions exist in the system that affect its behavior. Then the relationship between those systems and the concepts to be experienced in the given chapter will be explained. Instructors are urged to use *The Big Picture* to engage students in dialog and to help them draw from their own experiences to enhance their understanding of the topics presented in each chapter.
- Several illustrations have been upgraded for attractiveness and technical clarity.
- Surface tension has been added in the discussion of fluid properties in Chapter 1.
- Additional data for properties of air have been added to the appendix and the graphs for pressure and temperature versus altitude in the atmosphere have been upgraded.
- Data for ISO and SAE viscosity grades of lubricating oils in Chapter 2 have been updated to current standards.
- The title of Chapter 4 has been changed to "Forces Due to Static Fluids." Drawings of tanks and other vessels containing static fluids have improved realism.
- In Chapter 6 on fluid flow rate, extensive new tables for recommended ranges of velocity in suction and discharge lines of fluid flow systems have been added. The accompanying discussion gives students much better guidance for selecting a suitable pipe size to carry a given volume flow rate with a reasonable velocity of flow. This should aid students' understanding of the importance of velocity in relation to energy losses. The data are useful in any design problem where the specification of pipe size is required.
- In Chapter 9 on energy losses due to friction, data for pipe wall roughness have been adjusted to meet current recommendations of national organizations dealing with ductile iron and concrete piping. Also, the discussion of the computational approach to determining the friction factor has been streamlined.
- Chapter 10 on minor losses includes new data for foot valves with strainers that are very useful when designing systems that draw water or other fluids from open tanks or streams. Also, the resistance coefficient, $K$, is used for all minor losses for general purpose fluid distribution systems. Additional discussion and data are included for energy losses in typical valves used in fluid power systems.
- Approaches to the solution of Class II and Class III series pipe-line problems in Chapter 11 have been reworked to take advantage of computer-assisted techniques. This supplements or replaces the manual, computationally intensive approach used in earlier editions. With the spreadsheets, multi-stepped iterative solutions are completed very rapidly. Friction factors are computed using the technique discussed in Chapter 9 on energy losses due to friction.

- The chapter on pump selection and application has been moved to Chapter 13 and immediately follows the chapters on series and parallel pipe-line systems. This encourages the use of design problems keyed around the detailed design of a piping system and the selection of a suitable pump for the system.
- The performance of positive displacement pumps such as those used in fluid power systems has been discussed in more detail in Chapters 7 and 13.
- Two pump performance curves for centrifugal pumps have been added in Chapter 13. These additional curves cover lower flow rates than the seven curves that have been carried over from the fourth edition.
- The discussion of the operating point of a pump in a given system has been dramatically expanded in Chapter 13. An extensive example problem with the computer-assisted determination of the operating point has been added. This is then linked to the pump performance data in the chapter to select a suitable pump for the system.
- Data for the vapor pressure of water have been expanded to include tabular as well as graphical presentation of the data in both SI and U.S. Customary Units. This enhances the discussion of net positive suction head (NPSH) in Chapter 13, "Pump Selection and Application."
- The appendix table of conversion factors has been extensively expanded and reformatted. It should assist students to apply the factors correctly.
- The solutions manual for the book includes a CD-ROM containing all spreadsheets that are used in the book along with student versions of two powerful programs commercially available from Tahoe Design Software of Truckee, California, called HYDROFLO™ and PumpBase™.

**Acknowledgments**   I would like to thank all who helped and encouraged me in the writing of this book, including users of earlier editions and the several reviewers who provided detailed suggestions: William E. Cole, Northeastern University; Mark S. Frisina, Wentworth Institute of Technology; Dr. Roy A. Hartman, P. E., Texas A & M University; Dr. Greg E. Maksi, State Technical Institute at Memphis; Paul Ricketts, New Mexico State University; Mohammad E. Taslim, Northeastern University at Boston; Paolien Wang, University of North Carolina at Charlotte; and Steve Wells, Old Dominion University. Special thanks go to my colleagues, Jesse H. Wilder, David H. Myszka, and Joseph A. Untener of the University of Dayton, who used earlier editions of this book in class many times and offered helpful suggestions. Robert L. Wolff, also of the University of Dayton, has provided much help in the use of the SI system of units, based on his long experience in metrication through the American Society for Engineering Education. Professor Wolff also consulted on fluid power applications. Comments from students who have used the book are also appreciated, since the book is written for them.

*Robert L. Mott*

# Contents

# 1 The Nature of Fluids and the Study of Fluid Mechanics

## 1.1 The Big Picture

### Discussion Map

- *Fluid mechanics* is the study of the behavior of fluids, either at rest (fluid statics) or in motion (fluid dynamics).

- Fluids can be either *liquids* or *gases*.

- Liquids tend to flow freely and take the shape of their container. Examples are water, oil, gasoline, glycerine, and alcohol.

- Gases tend to expand to completely fill their container. Their molecules are in constant motion. Examples are air, water vapor, oxygen, nitrogen, or helium.

- You must learn to recognize common liquids and to characterize them in terms of their physical properties, particularly *density*, *specific weight*, *specific gravity*, and *viscosity*.

- As you complete the study of this book, you should also learn how to analyze the behavior of fluids as they flow through circular pipes and tubes, and through conduits with other shapes.

- You should learn how to consider energy contained by the fluid because of its pressure, velocity, and position.

- You should learn how to account for energy losses and additions that occur as the fluid flows through many types of systems.

### Discover

*Think, now about any system that you are familiar with that contains a fluid. For that system, describe the following:*

- The basic function or purpose of the system.
- The kind of fluid or fluids that are in the system.
- The kinds of containers for the fluid or the conduits through which it flows.
- If the fluid flows, what causes the flow to occur? Describe the flow path.
- What, if anything, resists the flow of the fluid.
- What characteristics of the fluid are important to its proper performance in the system?

*Consider fluid systems related to your home, commercial buildings, your car, other vehicles, consumer products, toys, amusement park devices, recreation equipment, construction equipment, or manufacturing operations.*

*Discuss these systems among your colleagues with whom you are studying and with the course instructor or facilitator.*

Here are some examples of fluid systems and discussions of how they relate to the material in this book:

1. In your home, you use water for many different purposes such as drinking, cooking, bathing, cleaning, and watering plants. Water also eliminates wastes from the home through sinks, drains, and toilets. Rain water, melting snow, and water in the ground must be managed to conduct it away from the home using gutters, downspouts, ditches, and sump pumps. Consider how the water is delivered to your home. What is the ultimate source of the water—a river, a reservoir, or natural groundwater? Is the water stored in tanks at some points in the process of getting it to your home? Notice that the water system needs to be at a fairly high pressure to be effective for its uses and to flow reliably through the system. How is that pressure created? Are there pumps in the system? Describe their function and how they operate. From where does each pump draw the water? To what places is the water delivered? What quantities of fluid are needed at the delivery points? What pressures are required? How is the flow of water controlled? What materials are used for the pipes, tubes, tanks, and other containers or conduits? As you study Chapters 6–13, you will learn how to analyze and design systems in which the water flows in a pipe or tube. Chapter 14 discusses the cases of open channel flow such as that in the gutters on your home.

2. In your car, describe the system that stores gasoline and then delivers it to the car's engine. How is the windshield washer fluid managed? Describe the cooling system and the nature of the coolant. Describe what happens when you apply the brakes, particularly as it relates to the hydraulic fluid in the braking system. The concepts in Chapters 6–13 will help you describe and analyze these kinds of systems.

3. Consider the performance of an automated manufacturing system that is actuated by fluid power systems such as that shown in Figure 1.1. Describe the fluids, pumps, tubes, valves, and other components of the system. What is the function of the system? How does the fluid accomplish that function? How is energy introduced to the system and how is it dissipated away from the system?

4. Consider the kinds of objects that must float in fluids such as boats, rafts, barges, and buoys. Why do they float? In what position or orientation do they float? Why do they maintain that position and orientation? The principles of buoyancy and stability are discussed in Chapter 5.

5. What examples can you think of where fluids at rest or in motion exert forces on an object? Any vessel containing a fluid under pressure should yield examples. Consider a swimming pool, a hydraulic cylinder, a dam or retaining wall holding a fluid, a high-pressure washer system, a fire hose, wind during a tornado or hurricane, and water flowing through a turbine to generate power. What other examples can you think of? Chapters 4, 16, and 17 discuss these cases.

6. Think of the many situations in which it is important to measure the flow rate of fluid in a system or the total quantity of fluid delivered. Consider measuring the gasoline that goes

**FIGURE 1.1**   Typical piping system for fluid power.

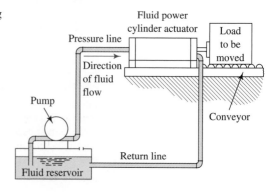

into your car so you can pay for just what you get. The water company wants to know how much water you use in a given month. Fluids often must be metered carefully into production processes in a factory. Liquid medicines and oxygen delivered to a patient in a hospital must be measured continuously for patient safety. Chapter 15 covers flow measurement.

There are many other ways in which fluids affect your life. Completion of this course in fluid mechanics will help you understand how those fluids can be controlled.

These are just a few of the many practical problems you are likely to encounter that require an understanding of the principles of fluid mechanics for their solution. The objective of this book is to help you solve these kinds of problems. Included in each chapter are problems representing situations from many fields of technology. Your ability to solve these problems will be a measure of how well the objective of this book has been accomplished.

In order to understand the behavior of fluids, it is necessary to understand the nature of the fluids themselves. This chapter defines the properties of fluids, introduces the symbols and units involved, and discusses the kinds of calculations required in the study of fluid mechanics.

**1.2**
**OBJECTIVES OF**
**THIS CHAPTER**

After completing this chapter, you should be able to:

1. Differentiate between a gas and a liquid.
2. Define the relationship between force and mass.
3. Identify the units for the basic quantities of time, length, force, and mass in the SI system (metric unit system).
4. Identify the units for the basic quantities of time, length, force, and mass in the U.S. Customary System.
5. Properly set up equations to ensure consistency of units.
6. Define *compressibility* and *bulk modulus*.
7. Define *pressure*.
8. Define *density*.
9. Define *specific weight*.
10. Define *specific gravity*.
11. Identify the relationships between specific weight, specific gravity, and density, and solve problems using these relationships.

**1.3**
**DIFFERENCE BETWEEN**
**LIQUIDS AND GASES**

When a liquid is held in a container, it tends to take the shape of the container, covering the bottom and sides. The top surface, in contact with the atmosphere above it, maintains a uniform level. As the container is tipped, the liquid tends to pour out; the rate of pouring is dependent on a property called *viscosity*, which will be defined later.

When a gas is held in a closed container, it tends to expand and completely fill the container. If the container is opened, the gas tends to continue to expand and escape from the container.

In addition to these familiar differences between gases and liquids, another difference is important in the study of fluid mechanics:

- Liquids are only slightly compressible.
- Gases are readily compressible.

Compressibility refers to the change in volume of a substance when the pressure on it changes. These distinctions are sufficient for most purposes.

The continuing discussion of pressure, compressibility, and other properties of fluids requires an understanding of the units in which these quantities are measured as presented in the following sections.

**1.4**
**FORCE AND MASS**

An understanding of fluid properties requires a careful distinction between *mass* and *weight*. The following definitions apply.

> **Mass** *is the property of a body of fluid that is a measure of its inertia or resistance to a change in motion. It is also a measure of the quantity of fluid.*

We will use the symbol *m* for mass.

> **Weight** *is the amount that a body weighs, that is, the force with which a body is attracted toward the earth by gravitation.*

We will use the symbol *w* for weight.

Weight is related to mass and the acceleration due to gravity, *g*, by Newton's law of gravitation,

 WEIGHT-MASS RELATIONSHIP

$$w = mg \qquad\qquad (1\text{--}1)$$

In this book, we will use $g = 9.81$ m/s$^2$ in the SI system and $g = 32.2$ ft/s$^2$ in the U.S. Customary System. These are the standard values for *g* to three significant digits. To a greater degree of precision, the standard values of $g = 9.806\ 65$ m/s$^2$ or $g = 32.1740$ ft/s$^2$. For high precision work and at high elevations (such as aerospace operations) where the actual value of *g* is different from the standard, the local value should be used.

**1.5**
**THE INTERNATIONAL
SYSTEM OF UNITS (SI)**

In any technical work the units in which physical properties are measured must be stated. A system of units specifies the units of the basic quantities of length, time, force, and mass. The units of other terms are then derived from these.

Le Système International d'Unités, or the International System of Units (abbreviated SI), is used in this book. The units for the basic quantities are

$$length = meter^* \text{ (m)}$$
$$time = second \text{ (s)}$$
$$mass = kilogram \text{ (kg) or N}\cdot\text{s}^2/\text{m}$$
$$force = newton \text{ (N) or kg}\cdot\text{m/s}^2$$

An equivalent unit for force is the kg·m/s$^2$, as indicated above. This is derived from the relationship between force and mass from physics,

$$F = ma$$

where *a* is the acceleration expressed in units of m/s$^2$. Therefore, the derived unit for force is

$$F = ma = \text{kg}\cdot\text{m/s}^2 = \text{newton}$$

Thus a force of 1.0 N would give a mass of 1.0 kg an acceleration of 1.0 m/s$^2$. This means that either newtons (N) or kg·m/s$^2$ may be used as the units for force. In fact, some calculations in this book require that you be able to use both or to convert from one to the other.

---

*The American National Metric Council (ANMC) recommends the spellings "meter" and "liter" instead of "metre" and "litre." We will use the preferred notation of the ANMC throughout this book.

For example, we may say that a rock with a mass of 5.60 kg is suspended by a wire. Then, in order to determine what force is exerted on the wire, Newton's law of gravitation ($w = mg$) should be used:

$$w = mg = \text{mass} \times \text{acceleration of gravity}$$

But, under standard conditions at sea level, $g = 9.81 \text{ m/s}^2$. Then we have

$$w = 5.60 \text{ kg} \times 9.81 \text{ m/s}^2 = 54.9 \text{ kg·m/s}^2 = 54.9 \text{ N}$$

Thus, 5.60 kg of the rock weighs 54.9 N.

Similarly, besides using the kg as the standard unit mass, an equivalent unit is $\text{N·s}^2/\text{m}$. This can be derived again from $F = ma$.

$$m = \frac{F}{a} = \frac{\text{N}}{\text{m/s}^2} = \frac{\text{N·s}^2}{\text{m}}$$

Therefore, either kg or $\text{N·s}^2/\text{m}$ can be used for the unit of mass. For example, a metal housing for a valve weighs 8.25 N. What is its mass?

$$w = mg$$

$$m = \frac{w}{g} = \frac{8.25 \text{ N}}{9.81 \text{ m/s}^2} = \frac{0.841 \text{ N·s}^2}{\text{m}} = 0.841 \text{ kg}$$

**SI Unit prefixes**

Because the actual size of physical quantities in the study of fluid mechanics covers a wide range, prefixes are added to the basic quantities. Table 1.1 shows these prefixes. Standard usage in the SI system calls for only those prefixes varying in steps of $10^3$ as shown. Results of calculations should normally be adjusted so that the number is between 0.1 and 10 000 times some multiple of $10^3$.† Then the proper unit with a prefix can be specified. Some examples follow.

| Computed Result | Reported Result |
|---|---|
| 0.004 23 m | $4.23 \times 10^{-3}$ m, or 4.23 mm (millimeters) |
| 15 700 kg | $15.7 \times 10^3$ kg, or 15.7 Mg (megagrams) |
| 86 330 N | $86.33 \times 10^3$ N, or 86.33 kN (kilonewtons) |

†Because commas are used as decimal markers in many countries, we will not use commas to separate groups of digits. We will separate the digits into groups of three, counting both to the left and to the right from the decimal point, and use a space to separate the groups of three digits. We will not use a space if there are only four digits to the left or right of the decimal point unless required in tabular matter.

**TABLE 1.1**    SI unit prefixes

| Prefix | SI symbol | Factor |
|---|---|---|
| giga | G | $10^9 = 1\ 000\ 000\ 000$ |
| mega | M | $10^6 = 1\ 000\ 000$ |
| kilo | k | $10^3 = 1\ 000$ |
| milli | m | $10^{-3} = 0.001$ |
| micro | $\mu$ | $10^{-6} = 0.000\ 001$ |

**1.6
THE U.S. CUSTOMARY
SYSTEM**

Sometimes called the *English gravitational unit system* or the *pound-foot-second* system, the U.S. Customary System defines the basic quantities as follows:

$$length = foot\ (ft)$$
$$time = second\ (s)$$
$$force = pound\ (lb)$$
$$mass = slug\ or\ lb\text{-}s^2/ft$$

Probably the most difficult of these units to understand is the slug, since we are familiar with measuring in terms of pounds, seconds, and feet. It may help to note the relationship between force and mass from physics,

$$F = ma$$

where $a$ is acceleration expressed in units of $ft/s^2$. Therefore, the derived unit for mass is

$$m = \frac{F}{a} = \frac{lb}{ft/s^2} = \frac{lb\text{-}s^2}{ft} = slug$$

This means that you may use either slugs or $lb\text{-}s^2/ft$ for the unit of mass. In fact, some calculations in this book require that you be able to use both or to convert from one to the other.

### Mass Expressed as lbm (Pounds-Mass)

Some professionals, in the analysis of fluid systems, have used the unit lbm (pounds-mass) for the unit of mass instead of the unit of slugs. In this system, an object or a quantity of fluid having a weight of 1.0 lb would have a mass of 1.0 lbm. The pound-force is then sometimes designated lbf. It must be noted that the numerical equivalence of lbf and lbm applies *only* when the value of $g$ is equal to the standard value.

This system is avoided in this book because it is not a coherent system. When one tries to relate force and mass units using Newton's law, the following is obtained.

$$F = m \cdot a = lbm(ft/s^2) = lbm\text{-}ft/s^2$$

This is *not* the same as the lbf.

To overcome this difficulty, a conversion constant, commonly called $g_c$, is defined having both a numerical value and units. That is,

$$g_c = \frac{32.2\ lbm}{lbf/(ft/s^2)} = \frac{32.2\ lbm\text{-}ft/s^2}{lbf}$$

Then to convert from lbm to lbf, a modified form of Newton's law is used.

$$F = m(a/g_c)$$

Letting the acceleration $a = g$,

$$F = m(g/g_c)$$

For example, to determine the weight of material in lbf that has a mass of 100 lbm, and assuming that the local value of $g$ is equal to the standard value of $32.2\ ft/s^2$,

$$w = F = m\frac{g}{g_c} = 100\ lbm\ \frac{32.2\ ft/s^2}{32.2\ \dfrac{lbm\text{-}ft/s^2}{lbf}} = 100\ lbf$$

This shows that weight in lbf is numerically equal to mass in lbm *provided* $g = 32.2 \text{ ft/s}^2$.

But if the analysis were to be done for an object or fluid on the earth's moon where $g$ is approximately $\frac{1}{6}$ that on the earth, say, $5.4 \text{ ft/s}^2$,

$$w = F = m\frac{g}{g_c} = 100 \text{ lbm} \frac{5.4 \text{ ft/s}^2}{32.2 \dfrac{\text{lbm-ft/s}^2}{\text{lbf}}} = 16.8 \text{ lbf}$$

This is a dramatic difference.

In summary, because of the cumbersome nature of the relationship between lbm and lbf, we avoid the use of lbm in this book. Mass will be expressed in the unit of slugs when problems are in the U.S. Customary System of units.

**1.7**
**CONSISTENT UNITS IN AN EQUATION**

The analyses required in fluid mechanics involve the algebraic manipulation of several terms. The equations are often complex, and it is extremely important that the results be dimensionally correct. That is, they must have their proper units. Indeed, answers will have the wrong numerical value if the units in the equation are not consistent. Tables 1.2 and 1.3 summarize standard and other common units for the quantities used in fluid mechanics.

A simple straightforward procedure called *unit cancellation* will ensure proper units in any kind of calculation, not only in fluid mechanics but also in virtually all your technical work. The six steps of the procedure are listed below.

**UNIT CANCELLATION PROCEDURE**

**1.** Solve the equation algebraically for the desired term.
**2.** Decide on the proper units for the result.
**3.** Substitute known values, including units.
**4.** Cancel units that appear in both the numerator and denominator of any term.
**5.** Use conversion factors to eliminate unwanted units and obtain the proper units as decided in Step 2.
**6.** Perform the calculation.

**TABLE 1.2**   SI Units for Common Quantities Used in Fluid Mechanics

| Quantity | Basic Definition | Standard SI Units | Other Units Often Used |
|---|---|---|---|
| Length | — | meter (m) | millimeter (mm); kilometer (km) |
| Time | — | second (s) | hour (h); minute (min) |
| Mass | Quantity of a substance | kilogram (kg) | $N{\cdot}s^2/m$ |
| Force or weight | Push or pull on an object | newton (N) | $kg{\cdot}m/s^2$ |
| Pressure | Force/area | $N/m^2$ or pascal (Pa) | kilopascals (kPa); bar |
| Energy | Force times distance | $N{\cdot}m$ or Joule (J) | $kg{\cdot}m^2/s^2$ |
| Power | Energy/time | $N{\cdot}m/s$ or $J/s$ | watt (W); kW |
| Volume | $(Length)^3$ | $m^3$ | liter (L) |
| Area | $(Length)^2$ | $m^2$ | $mm^2$ |
| Volume flow rate | Volume/time | $m^3/s$ | L/s; L/min |
| Weight flow rate | Weight/time | N/s | kN/s; kN/min |
| Mass flow rate | Mass/time | kg/s | kg/h |
| Specific weight | Weight/volume | $N/m^3$ | $kg/m^2{\cdot}s^2$ |
| Density | Mass/volume | $kg/m^3$ | $N{\cdot}s^2/m^4$ |

**TABLE 1.3**   U.S. Customary Units for Common Quantities Used in Fluid Mechanics

| Quantity | Basic Definition | Standard U.S. Units | Other Units Often Used |
|---|---|---|---|
| Length | — | feet (ft) | inches (in), miles (mi) |
| Time | — | second (s) | hour (h); minute (min) |
| Mass | Quantity of a substance | slugs | $lb \cdot s^2/ft$ |
| Force or weight | Push or pull on an object | pound (lb) | kip (1000 lb) |
| Pressure | Force/area | $lb/ft^2$ or psf | $lb/in^2$ or psi; $kip/in^2$ or ksi |
| Energy | Force times distance | $lb \cdot ft$ | $lb \cdot in$ |
| Power | Energy/time | $lb \cdot ft/s$ | horsepower (hp) |
| Volume | $(Length)^3$ | $ft^3$ | gallon (gal) |
| Area | $(Length)^2$ | $ft^2$ | $in^2$ |
| Volume flow rate | Volume/time | $ft^3/s$ or cfs | gal/min (gpm); $ft^3/min$ (cfm) |
| Weight flow rate | Weight/time | lb/s | lb/min; lb/h |
| Mass flow rate | Mass/time | slugs/s | slugs/min; slugs/h |
| Specific weight | Weight/volume | $lb/ft^3$ | |
| Density | Mass/volume | $slugs/ft^3$ | |

This procedure, properly executed, will work for any equation. It is really very simple, but some practice may be required to use it. We are going to borrow some material from elementary physics with which you should be familiar to illustrate the method. However, the best way to learn how to do something is to do it. The following example problems are presented in a form called *programmed instruction*. You will be guided through the problems in a step-by-step fashion with your participation required at each step.

To proceed with the program you should cover all material below the heading "Programmed Example Problem," using a heavy piece of paper. You should have a blank piece of paper handy on which to perform the requested operations. Then successively uncover one panel at a time down to the heavy line that runs across the page. The first panel presents a problem and asks you to perform some operation or to answer a question. After doing what is asked, uncover the next panel, which will contain information that you can use to check your result. Then continue with the next panel, and so on through the program.

Remember, the purpose of this is to help you learn how to get correct answers using the unit cancellation method. You may want to refer to the table of conversion factors in Appendix K.

## PROGRAMMED EXAMPLE PROBLEM

☐ **EXAMPLE PROBLEM 1.1**   Imagine you are traveling in a car at a constant speed of 80 kilometers per hour (km/h). How many seconds (s) would it take to travel 1.5 km?

For the solution, use the equation

$$s = vt$$

where $s$ is the distance traveled, $v$ is the speed, and $t$ is the time. Using the unit cancellation procedure outlined above, what is the first thing to do?

The first step is to solve for the desired term. Since you were asked to find time, you should have written

$$t = \frac{s}{v}$$

Now perform Step 2 of the procedure described above.

---

Step 2 is to decide on the proper units for the result, in this case time. From the problem statement the proper unit is seconds. If no specification had been given for units, you could choose any acceptable time unit such as hours.

Proceed to Step 3.

---

The result should look something like this:

$$t = \frac{s}{v} = \frac{1.5 \text{ km}}{80 \text{ km/h}}$$

For the purpose of cancellation it is not convenient to have the units in the form of a compound fraction as we have above. To clear this to a simple fraction, write it in the form

$$t = \frac{\dfrac{1.5 \text{ km}}{1}}{\dfrac{80 \text{ km}}{\text{h}}}$$

This can be reduced to

$$t = \frac{1.5 \text{ km} \cdot \text{h}}{80 \text{ km}}$$

After some practice, equations may be written in this form directly. Now perform Step 4 of the procedure.

---

The result should now look like this:

$$t = \frac{1.5 \,\cancel{\text{km}} \cdot \text{h}}{80 \,\cancel{\text{km}}}$$

This illustrates that units can be cancelled just as numbers can if they appear in both the numerator and denominator of a term in an equation.

Now do Step 5.

---

The answer looks like this:

$$t = \frac{1.5 \,\cancel{\text{km}} \cdot \cancel{\text{h}}}{80 \,\cancel{\text{km}}} \times \frac{3600 \text{ s}}{1 \,\cancel{\text{h}}}$$

The equation in the preceding panel showed the result for time in hours after kilometer units were cancelled. Although hours is an acceptable time unit, our desired unit is seconds as determined in Step 2. Thus the conversion factor 3600 s/1 h is required.

How did we know to multiply by 3600 instead of dividing?

---

The units determine this. Our objective in using the conversion factor was to eliminate the hour unit and obtain the second unit. Since the unwanted hour unit was in the numerator of the original equation, the hour unit in the conversion factor must be in the denominator in order to cancel.

Now that we have the time unit of seconds we can proceed with Step 6.

The correct answer is $t = 67.5$ s.

■

**1.8**
**THE DEFINITION**
**OF PRESSURE**

⇨ PRESSURE

*Pressure* is defined as the amount of force exerted on a unit area of a substance. This can be stated by the equation

$$p = \frac{F}{A} \qquad (1-2)$$

Two important principles about pressure were described by Blaise Pascal, a seventeenth-century scientist.

- Pressure acts uniformly in all directions on a small volume of a fluid.
- In a fluid confined by solid boundaries, pressure acts perpendicular to the boundary.

These principles, sometimes called *Pascal's laws*, are illustrated in Figs. 1.2 and 1.3.

Using Eq. (1–2) and the second of Pascal's laws, we can compute the magnitude of the pressure in a fluid if we know the amount of force exerted on a given area.

□ **EXAMPLE PROBLEM 1.2**    Figure 1.4 shows a container of liquid with a movable piston supporting a load. Compute the magnitude of the pressure in the liquid under the piston if the total weight of the piston and the load is 500 N and the area of the piston is 2500 mm².

*Solution*    It is reasonable to assume that the entire surface of the fluid under the piston is sharing in the task of supporting the load. The second of Pascal's laws states that the fluid pressure acts perpendicular to the piston. Then, using Eq. (1–2),

$$p = \frac{F}{A} = \frac{500 \text{ N}}{2500 \text{ mm}^2} = 0.20 \text{ N/mm}^2$$

**FIGURE 1.2**   Pressure acting uniformly in all directions on a small volume of fluid.

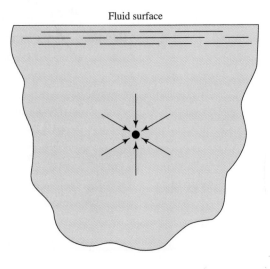
Fluid surface

**FIGURE 1.3**   Direction of fluid pressure on boundaries.

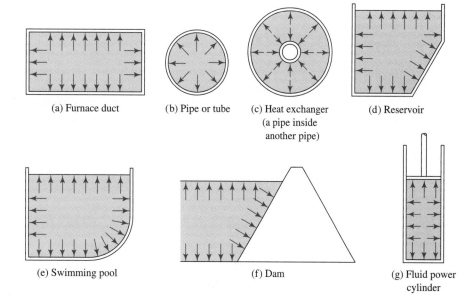

(a) Furnace duct          (b) Pipe or tube          (c) Heat exchanger          (d) Reservoir
                                                    (a pipe inside
                                                    another pipe)

(e) Swimming pool                    (f) Dam                    (g) Fluid power
                                                                 cylinder

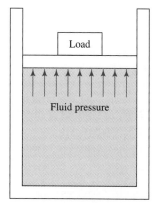

Load

Fluid pressure

**FIGURE 1.4**   Illustration of fluid pressure supporting a load.

The standard unit of pressure in the SI system is the N/m², called the *pascal* (Pa) in honor of Blaise Pascal. The conversion can be made by using the factor $10^3$ mm = 1 m.

$$p = \frac{0.20 \text{ N}}{\text{mm}^2} \times \frac{(10^3 \text{ mm})^2}{\text{m}^2} = 0.20 \times 10^6 \text{ N/m}^2 = 0.20 \text{ MPa}$$

Note that the pressure in N/mm² is numerically equal to pressure in MPa. It is not unusual to encounter pressure in the range of several megapascals (MPa) or several hundred kilopascals (kPa).

Pressure in the U.S. Customary System is illustrated in the following example problem.

☐ **EXAMPLE PROBLEM 1.3**   A load of 200 pounds (lb) is exerted on a piston confining an oil in a circular cylinder with an inside diameter of 2.50 inches (in). Compute the pressure in the oil at the piston. See Fig. 1.4.

*Solution*   Using Eq. (1–2), we must compute the area of the piston.

$$A = \pi D^2/4 = \pi (2.50 \text{ in})^2/4 = 4.91 \text{ in}^2$$

Then,

$$P = \frac{F}{A} = \frac{200 \text{ lb}}{4.91 \text{ in}^2} = 40.7 \text{ lb/in}^2$$

Although the standard unit for pressure in the U.S. Customary System is pounds per square foot (lb/ft²), it is not often used because it is inconvenient. Length measurements are more conveniently made in inches, and pounds per square inch (lb/in²), abbreviated psi, is used most often for pressure in this system. The pressure in the oil is 40.7 psi. This is a fairly low pressure; it is not unusual to encounter pressures of several hundred or several thousand psi.

■

The *bar* is another unit used by some people working in fluid mechanics and thermodynamics. The bar is defined as $10^5$ Pa or $10^5$ N/m². Another way of expressing the bar is $100 \times 10^3$ N/m², which is equivalent to 100 kPa. Since atmospheric pressure near sea level is very nearly the same, the bar has a convenient point of physical reference. This, plus the fact that pressures expressed in bars yield smaller numbers, makes this unit attractive to some practitioners. You must realize, however, that the bar is not a part of the coherent SI system and that you must carefully convert to N/m² (pascals) in problem solving.

### 1.9 COMPRESSIBILITY

*Compressibility* refers to the change in volume ($V$) of a substance that is subjected to a change in pressure on it. The usual quantity used to measure this phenomenon is the *bulk modulus of elasticity* or, simply, *bulk modulus, E*.

⇨ **BULK MODULUS**

$$E = \frac{-\Delta p}{(\Delta V)/V} \qquad (1-3)$$

Because the quantities $\Delta V$ and $V$ would have the same units, the denominator of Eq. (1–3) is dimensionless. Therefore, the units for $E$ are the same as those for the pressure.

As stated before, liquids are very slightly compressible, indicating that it would take a very large change in pressure to produce a small change in volume. Thus, the magnitudes of $E$ for liquids, as shown in Table 1.4, are very high. For this reason, liquids will be considered incompressible in this book, unless stated otherwise.

The term *bulk modulus* is not usually applied to gases, and the principles of thermodynamics must be applied to determine the change in volume of a gas with a change in pressure.

---

☐ **EXAMPLE PROBLEM 1.4**

Compute the change in pressure that must be applied to water to change its volume by 1.0 percent.

**Solution**

The 1.0-percent volume change indicates that $\Delta V/V = -0.01$. Then, the required change in pressure is

$$\Delta p = -E[(\Delta V)/V] = [-316\,000 \text{ psi}][-0.01] = 3160 \text{ psi}$$

■

---

**TABLE 1.4**   Values for bulk modulus for selected liquids at atmospheric pressure and 68°F (20°C)

|               | Bulk Modulus |           |
|---------------|--------------|-----------|
| Liquid        | (psi)        | (MPa)     |
| Ethyl alcohol | −130 000     | ⁻896      |
| Benzene       | 154 000      | 1 062     |
| Machine oil   | 189 000      | 1 303     |
| Water         | ⁻316 000     | ⁻2 179    |
| Glycerine     | 654 000      | 4 509     |
| Mercury       | 3 590 000    | 24 750    |

Because the study of fluid mechanics typically deals with a continuously flowing fluid or with a small amount of fluid at rest, it is most convenient to relate the mass and weight of the fluid to a given volume of the fluid. Thus, the properties of density and specific weight are defined as follows.

**Density** *is the amount of mass per unit volume of a substance.*

Therefore, using the greek letter $\rho$ (rho) for density,

⇨ DENSITY

$$\rho = m/V \tag{1-4}$$

where $V$ is the volume of the substance having a mass $m$. The units for density are kilograms per cubic meter in the SI system and slugs per cubic foot in the U.S. Customary System.

The American Society for Testing and Materials (ASTM) has published several standard test methods for measuring density that describe vessels having precisely known volumes, called *pycnometers*. The proper filling, handling, temperature control, and reading of these devices are prescribed. Two types are known as the *Bingham pycnometer* and the *Lipkin bicapillary pycnometer*. The standards also call for the precise determination of the mass of the fluids in the pycnometers to the nearest 0.1 mg using an analytical balance. See References 2, 4, and 5.

**Specific weight** *is the amount of weight per unit volume of a substance.*

Using the Greek letter $\gamma$ (gamma) for specific weight,

⇨ SPECIFIC WEIGHT

$$\gamma = w/V \tag{1-5}$$

where $V$ is the volume of a substance having the weight $w$. The units for specific weight are newtons per cubic meter ($N/m^3$) in the SI system and pounds per cubic foot ($lb/ft^3$) in the U.S. Customary System.

It is often convenient to indicate the specific weight or density of a fluid in terms of its relationship to the specific weight or density of a common fluid. When the term *specific gravity* is used in this book, the reference fluid is pure water at 4°C. At that temperature, water has its greatest density. Then, specific gravity can be defined in either of two ways:

**a.** *Specific gravity* is the ratio of the density of a substance to the density of water at 4°C.

**b.** *Specific gravity* is the ratio of the specific weight of a substance to the specific weight of water at 4°C.

These definitions for specific gravity can be shown mathematically as

⇨ SPECIFIC GRAVITY

$$sg = \frac{\gamma_s}{\gamma_w \text{ @ } 4°C} = \frac{\rho_s}{\rho_w \text{ @ } 4°C} \tag{1-6}$$

where the subscript $s$ refers to the substance whose specific gravity is being determined and the subscript $w$ refers to water. The properties of water at 4°C are constant, having the values shown below.

$$\gamma_w \text{ @ } 4°C = 9.81 \text{ kN/m}^3 \qquad \gamma_w \text{ @ } 4°C = 62.4 \text{ lb/ft}^3$$

or

$$\rho_w \text{ @ } 4°C = 1000 \text{ kg/m}^3 \qquad \rho_w \text{ @ } 4°C = 1.94 \text{ slugs/ft}^3$$

Therefore, the mathematical definition of specific gravity can be written

$$sg = \frac{\gamma_s}{9.81 \text{ kN/m}^3} = \frac{\rho_s}{1000 \text{ kg/m}^3} \text{ or } sg = \frac{\gamma_s}{62.4 \text{ lb/ft}^3} = \frac{\rho_s}{1.94 \text{ slugs/ft}^3} \tag{1-7}$$

This definition holds regardless of the temperature at which the specific gravity is being determined.

The properties of fluids do, however, vary with temperature. In general, the density (and therefore, the specific weight and specific gravity) decreases with increasing temperature. The properties of water at various temperatures are listed in Appendix A. The properties of other liquids at a few selected temperatures are listed in Appendix B and Appendix C.

You should seek other references, such as References 7 and 10, for data on specific gravity at specified temperatures if they are not reported in the appendix and if high precision is desired. One estimate that gives reasonable accuracy for petroleum oils, as presented more fully in References 7 and 9, is that the specific gravity of oils decreases approximately 0.036 for a 100°F (37.8°C) rise in temperature. This applies for nominal values of sg from 0.80 to 1.00 and for temperatures in the range from approximately 32°F to 400°F (0°C to 204°C).

Some industry sectors prefer modified definitions for specific gravity. Instead of using the properties of water at 4°C (39.2°F) as the basis, the petroleum industry and others use water at 60°F (15.6°C). This makes little difference for typical design and analysis. Although the density of water at 4°C is 1000.00 kg/m³, at 60°F it is 999.04 kg/m³. The difference is less than 0.1 percent. References 2, 3, 5, 6, 7, and 10 contain more extensive tables of the properties of water at temperatures from 0°C to 100°C (32°F to 212°F).

Specific gravity in the Baumé and API scales is discussed in Section 1.10.2. We will continue to use water at 4°C as the basis for specific gravity in this book.

The ASTM also refers to the property of specific gravity as *relative density*. See References 2–5.

### 1.10.1
### Relation Between Density and Specific Weight

$\gamma$–$\rho$ RELATION

Quite often the specific weight of a substance must be found when its density is known and vice versa. The conversion from one to the other can be made in the following equation:

$$\gamma = \rho g \qquad (1\text{--}8)$$

where $g$ is the acceleration due to gravity. This equation can be justified by referring to the definitions of density and specific gravity and by using the equation relating mass to weight, $w = mg$.

The definition of specific weight is

$$\gamma = \frac{w}{V}$$

By multiplying both the numerator and the denominator of this equation by $g$ we obtain

$$\gamma = \frac{wg}{Vg}$$

But $m = w/g$. Therefore, we have

$$\gamma = \frac{mg}{V}$$

Since $\rho = m/V$, we get

$$\gamma = \rho g$$

The following problems illustrate the definitions of the basic fluid properties presented above and the relationships between the various properties.

☐ **EXAMPLE PROBLEM 1.5**   Calculate the weight of a reservoir of oil if it has a mass of 825 kg.

*Solution*   Since $w = mg$,

$$w = 825 \text{ kg} \times 9.81 \text{ m/s}^2 = 8093 \text{ kg·m/s}^2$$

Substituting the newton for the unit kg·m/s², we have

$$w = 8093 \text{ N} = 8.093 \times 10^3 \text{ N} = 8.093 \text{ kN}$$

☐ **EXAMPLE PROBLEM 1.6**   If the reservoir from Example Problem 1.5 has a volume of 0.917 m³, compute the density, specific weight, and the specific gravity of the oil.

*Solution*   Density:

$$\rho = \frac{m}{V} = \frac{825 \text{ kg}}{0.917 \text{ m}^3} = 900 \text{ kg/m}^3$$

$1.08 \text{ g/m}^3$
$1080 \text{ kg/m}^3$

Specific weight:

$$\gamma = \frac{w}{V} = \frac{8.093 \text{ kN}}{0.917 \text{ m}^3} = 8.83 \text{ kN/m}^3$$

Specific gravity:

$$sg = \frac{\rho_o}{\rho_w @ 4°C} = \frac{900 \text{ kg/m}^3}{1000 \text{ kg/m}^3} = 0.90$$

☐ **EXAMPLE PROBLEM 1.7**   Glycerine at 20°C has a specific gravity of 1.263. Compute its density and specific weight.

*Solution*   Density:

$$\rho_g = (sg)_g(1000 \text{ kg/m}^3) = (1.263)(1000 \text{ kg/m}^3) = 1263 \text{ kg/m}^3$$

Specific weight:

$$\gamma_g = (sg)_g(9.81 \text{ kN/m}^3) = (1.263)(9.81 \text{ kN/m}^3) = 12.39 \text{ kN/m}^3$$

☐ **EXAMPLE PROBLEM 1.8**   A pint of water weighs 1.041 lb. Find its mass.

*Solution*   Since $w = mg$, the mass is

$$m = \frac{w}{g} = \frac{1.041 \text{ lb}}{32.2 \text{ ft/s}^2} = \frac{1.041 \text{ lb-s}^2}{32.2 \text{ ft}}$$
$$= 0.0323 \text{ lb-s}^2/\text{ft} = 0.0323 \text{ slugs}$$

Remember that the units of slugs and lb-s²/ft are the same.

☐ **EXAMPLE PROBLEM 1.9**   One gallon of mercury has a mass of 3.51 slugs. Find its weight.

*Solution*   $$w = mg = 3.51 \text{ slugs} \times 32.2 \text{ ft/s}^2 = 113 \text{ slug-ft/s}^2$$

This is correct, but the units may seem confusing since weight is normally expressed in pounds. The units of mass may be rewritten as lb-s$^2$/ft.

$$w = mg = 3.51 \ \frac{\text{lb-s}^2}{\text{ft}} \times \frac{32.2 \ \text{ft}}{\text{s}^2} = 113 \ \text{lb}$$

■

**1.10.2
Specific Gravity in Degrees
Baumé or Degrees API**

The reference temperature for measurements in the Baumé or API scale is 60°F rather than 4°C as defined before. To emphasize this difference, the API or Baumé specific gravity is often reported as

$$\text{Specific gravity} \ \frac{60°}{60°} \ \text{F}$$

This notation indicates that both the reference fluid (water) and the oil are at 60°F.

Specific gravities of crude oils vary widely depending on where they are found. Those from the western U.S. range from approximately 0.87 to 0.92. Eastern U.S. oil fields produce oil at about 0.82 specific gravity. Mexican crude oil is among the highest at 0.97. A few heavy asphaltic oils have sg > 1.0. (See Reference 6.)

Most oils are distilled before use to enhance their quality of burning. The resulting gasolines, kerosenes, and fuel oils have specific gravities ranging from about 0.67 to 0.98.

The equation used to compute specific gravity when the degrees Baumé are known is different for fluids lighter than water and fluids heavier than water. For liquids heavier than water,

$$\text{sg} = \frac{145}{145 - \text{deg Baumé}} \tag{1-9}$$

Or, to compute the degrees Baumé for a given specific gravity,

$$\text{deg Baumé} = 145 - \frac{145}{\text{sg}} \tag{1-10}$$

For liquids lighter than water,

$$\text{sg} = \frac{140}{130 + \text{deg Baumé}} \tag{1-11}$$

$$\text{deg Baumé} = \frac{140}{\text{sg}} - 130 \tag{1-12}$$

The American Petroleum Institute (API) has developed the API scale, slightly different from the Baumé scale, for liquids lighter than water. The formulas are

$$\text{sg} = \frac{141.5}{131.5 + \text{deg API}} \tag{1-13}$$

$$\text{deg API} = \frac{141.5}{\text{sg}} - 131.5 \tag{1-14}$$

Degrees API for oils may range from 10 to 80. Most fuel grades will fall in the range of API 20 to 70, corresponding to specific gravities from 0.93 to 0.70. Note that the heavier oils have the lower values of degrees API.

**FIGURE 1.5**  Hydrometer with
built-in thermometer (thermohydrom-
eter).

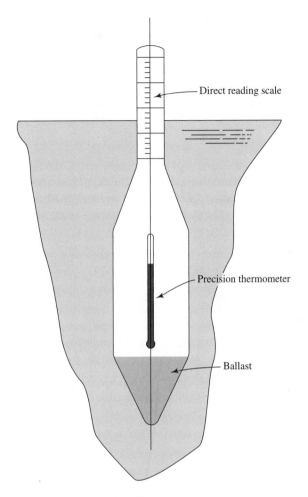

Direct reading scale

Precision thermometer

Ballast

ASTM Standards D 287 and D 1298 (References 1 and 3) describe standard test methods for determining API gravity using a *hydrometer*. Figure 1.5 shows a sketch of a typical hydrometer incorporating a weighted glass bulb with a smaller diameter stem at the top that is designed to float upright in the test liquid. Based on the principles of buoyancy (see Chapter 5), the hydrometer rests at a position that is dependent on the density of the liquid. The stem is marked with a calibrated scale from which the direct reading of density, specific gravity, or API gravity can be made. Because of the importance of temperature to an accurate measurement of density, some hydrometers, called *thermohydrometers*, have a built-in precision thermometer.

**1.11**
**SURFACE TENSION**

You may have experimented with the surface tension of water by trying to cause an object to be supported on the surface when you would otherwise predict it would sink. For example, it is fairly easy to place a small needle on a still water surface and it will be supported by the surface tension of the water. Note that it is not significantly supported by buoyancy. If the needle is submerged, it will readily sink to the bottom.

Then, if you place a very small amount of dishwashing detergent in the water when the needle is supported, it will almost immediately sink. The detergent lowers the surface tension dramatically.

Surface tension acts somewhat like a film at the interface between the liquid water surface and the air above it. The water molecules beneath the surface are attracted to each other and to those at the surface. Quantitatively, surface tension is measured as the work per unit area required to move lower molecules to the surface of the liquid. The resulting units are force per unit length, such as N/m.

Surface tension is also the reason that water droplets assume a nearly spherical shape. And the phenomenon of capillarity depends on the surface tension. The surface of a liquid in a small-diameter tube will assume a curved shape that depends on the surface tension of the liquid. Mercury will form virtually an extended bulbous shape. But the surface of water will settle into a depressed cavity with the liquid seeming to climb the walls of the tube by a small amount. Adhesion of the liquid to the walls of the tube contributes to this behavior.

The movement of liquids within small spaces depends on this capillary action. *Wicking* is the term often used to describe the rise of a fluid from a liquid surface into a woven material. The movement of liquids within soils is also affected by surface tension and the corresponding capillary action.

Table 1.5 gives the surface tension of water at atmospheric pressure at varying temperatures. The SI units used here are mN/m where 1 000 mN = 1.0 N. Similarly, U.S. Customary units are mlb/ft where 1 000 mlb = 1.0 lb force. Table 1.6 gives values for a variety of common liquids also at atmospheric pressure at selected temperatures.

**TABLE 1.5**   Surface Tension of Water

| Temperature (°F) | Surface Tension (mlb/ft) | Temperature (°C) | Surface Tension (mN/m) |
|---|---|---|---|
| 32 | 5.18 | 0 | 75.6 |
| 40 | 5.13 | 5 | 74.9 |
| 50 | 5.09 | 10 | 74.2 |
| 60 | 5.03 | 20 | 72.8 |
| 70 | 4.97 | 30 | 71.2 |
| 80 | 4.91 | 40 | 69.6 |
| 90 | 4.86 | 50 | 67.9 |
| 100 | 4.79 | 60 | 66.2 |
| 120 | 4.67 | 70 | 64.5 |
| 140 | 4.53 | 80 | 62.7 |
| 160 | 4.40 | 90 | 60.8 |
| 180 | 4.26 | 100 | 58.9 |
| 200 | 4.12 | | |
| 212 | 4.04 | | |

Adapted from data from CRC Handbook of Chemistry and Physics, CRC Press
Notes:
Values taken at atmospheric pressure
1.0 lb = 1 000 mlb;   1.0 N = 1 000 mN

**TABLE 1.6** **Surface Tension of Some Common Liquids**

| Liquid | Surface Tension at Stated Temperature | | | | | | | | | |
|---|---|---|---|---|---|---|---|---|---|---|
| | 10°C (mN/m) | 50°F (mlb/ft) | 25°C (mN/m) | 77°F (mlb/ft) | 50°C (mN/m) | 122°F (mlb/ft) | 75°C (mN/m) | 167°F (mlb/ft) | 100°C (mN/m) | 212°F (mlb/ft) |
| Water | 74.2 | 5.08 | 72.0 | 4.93 | 67.9 | 4.65 | 63.6 | 4.36 | 58.9 | 4.04 |
| Methanol | 23.2 | 1.59 | 22.1 | 1.51 | 20.1 | 1.38 | | | | |
| Ethanol | 23.2 | 1.59 | 22.0 | 1.51 | 19.9 | 1.36 | | | | |
| Ethylene Glycol | | | 48.0 | 3.29 | 45.8 | 3.14 | 43.5 | 2.98 | 41.3 | 2.83 |
| Acetone | 23.5 | 1.61 | 20.7 | 1.42 | | | | | | |
| Benzene | | | 28.2 | 1.93 | 25.0 | 1.71 | 21.8 | 1.49 | | |
| Mercury | 488 | 33.4 | 485 | 33.2 | 480 | 32.9 | 475 | 32.5 | 470 | 32.2 |

Adapted from data from CRC Handbook of Chemistry and Physics, CRC Press

Notes:
Values taken at atmospheric pressure
1.0 lb = 1 000 mlb   1.0 N = 1 000 mN

# REFERENCES

1. American Society for Testing and Materials (ASTM). 1997. *Standard D 287-92: Standard Test Method for API Gravity of Crude Petroleum and Petroleum Products (Hydrometer Method)*. West Conshohocken, PA: Author.

2. _____. 1997. *Standard D 1217-93: Standard Test Method for Density and Relative Density (Specific Gravity) of Liquids by Bingham Pycnometer*. West Conshohocken, PA: Author.

3. _____. 1997. *Standard D 1298-85 (1990 e1): Standard Test Method for Density, Relative Density (Specific Gravity), or API Gravity of Crude Petroleum and Liquid Petroleum Products by Hydrometer Method*. West Conshohocken, PA: Author.

4. _____. 1997. *Standard D 1480-93 (1977): Standard Test Method for Density and Relative Density (Specific Gravity) of Viscous Materials by Bingham Pycnometer*. West Conshohocken, PA: Author.

5. _____. 1997. *Standard D 1481-93: Standard Test Method for Density and Relative Density (Specific Gravity) of Viscous Materials by Lipkin Bicapillary Pycnometer*. West Conshohocken, PA: Author.

6. Avallone, Eugene A., and Theodore Baumeister III, eds. 1996. *Marks' Standard Handbook for Mechanical Engineers*. 10th ed. New York: McGraw-Hill.

7. Bolz, Ray E., and George L. Tuve, eds. 1973. *CRC Handbook of Tables for Applied Engineering Science*. 2nd ed. Boca Raton, Florida: CRC Press, Inc.

8. Cheremisinoff, N. P., ed. 1986. *Encyclopedia of Fluid Mechanics*. Vol. 1, *Flow Phenomena and Measurement*. Houston, Texas: Gulf Publishing Co.

9. Heald, C. C., ed. 1998. *Cameron Hydraulic Data*. 18th ed. Liberty Corner, New Jersey: Ingersoll-Dresser Pumps.

10. Lide, David R., ed. 1998. *CRC Handbook of Chemistry and Physics, 79th ed.* Boca Raton, Florida: CRC Press, Inc.

11. Miller, R. W. 1996: *Flow Measurement Engineering Handbook*. 3rd ed. New York: McGraw-Hill.

# PRACTICE PROBLEMS

## Conversion Factors

**1.1** Convert 1250 millimeters to meters.

**1.2** Convert 1600 square millimeters to square meters.

**1.3** Convert $3.65 \times 10^3$ cubic millimeters to cubic meters.

**1.4** Convert 2.05 square meters to square millimeters.

**1.5** Convert 0.391 cubic meters to cubic millimeters.

**1.6** Convert 55.0 gallons to cubic meters.

**1.7** An automobile is moving at 80 kilometers per hour. Calculate its speed in meters per second.

**1.8** Convert a length of 25.3 feet to meters.

**1.9** Convert a distance of 1.86 miles to meters.

**1.10** Convert a length of 8.65 inches to millimeters.

**1.11** Convert a distance of 2580 feet to meters.

**1.12** Convert a volume of 480 cubic feet to cubic meters.

**1.13** Convert a volume of 7390 cubic centimeters to cubic meters.

**1.14** Convert a volume of 6.35 liters to cubic meters.

**1.15** Convert 6.0 feet per second to meters per second.

**1.16** Convert 2500 cubic feet per minute to cubic meters per second.

(*Note:* In all Practice Problems sections in this book, the problems will use both SI and U.S. Customary System units. When SI units are used, "M" will follow the problem number and the problem will be printed in *italic*. When U.S. Customary System units are used, "E" will follow the problem number. When both systems of units are combined in a problem, "C" will follow the problem number.)

## Consistent Units in an Equation

A body moving with constant velocity obeys the relationship $s = vt$, where $s$ = distance, $v$ = velocity, and $t$ = time.

**1.17M** *A car travels 0.50 km in 10.6 s. Calculate its average speed in m/s.*

**1.18M** *In an attempt at a land speed record, a car travels 1.50 km in 5.2 s. Calculate its average speed in km/h.*

**1.19E** A car travels 1000 ft in 14 s. Calculate its average speed in mi/h.

**1.20E** In an attempt at a land speed record, a car travels a mile in 5.7 s. Calculate its average speed in mi/h.

A body starting from rest with constant acceleration moves according to the relationship $s = \frac{1}{2}at^2$, where $s$ = distance, $a$ = acceleration, and $t$ = time.

**1.21M** *If a body moves 3.2 km in 4.7 min while moving with constant acceleration, calculate the acceleration in m/s².*

**1.22M** *An object is dropped from a height of 13 m. Neglecting air resistance, how long would it take for the body to strike the ground? Use $a = g = 9.81\,m/s^2$.*

**1.23C** If a body moves 3.2 km in 4.7 min while moving with constant acceleration, calculate the acceleration in ft/s².

**1.24E** An object is dropped from a height of 53 in. Neglecting air resistance, how long would it take for the body to strike the ground? Use $a = g = 32.2$ ft/s².

The formula for kinetic energy is $KE = \frac{1}{2}mv^2$, where $m$ = mass and $v$ = velocity.

**1.25M** *Calculate the kinetic energy in N·m of a 15-kg mass if it has a velocity of 1.20 m/s.*

**1.26M** *Calculate the kinetic energy in N·m of a 3600-kg truck moving at 16 km/h.*

**1.27M** *Calculate the kinetic energy in N·m of a 75-kg box moving on a conveyor at 6.85 m/s.*

**1.28M** *Calculate the mass of a body in kg if it has a kinetic energy of 38.6 N·m when moving at 31.5 km/h.*

**1.29M** *Calculate the mass of a body in g if it has a kinetic energy of 94.6 mN·m when moving at 2.25 m/s.*

**1.30M** *Calculate the velocity in m/s of a 12-kg object if it has a kinetic energy of 15 N·m.*

**1.31M** *Calculate the velocity in m/s of a 175-g body if it has a kinetic energy of 212 mN·m.*

**1.32E** Calculate the kinetic energy in ft-lb of a 1-slug mass if it has a velocity of 4 ft/s.

**1.33E** Calculate the kinetic energy in ft-lb of an 8000-lb truck moving 10 mi/h.

**1.34E** Calculate the kinetic energy in ft-lb of a 150-lb box moving on a conveyor at 20 ft/s.

**1.35E** Calculate the mass of a body in slugs if it has a kinetic energy of 15 ft-lb when moving at 2.2 ft/s.

**1.36E** Calculate the weight of a body in lb if it has a kinetic energy of 38.6 ft-lb when moving at 19.5 mi/h.

**1.37E** Calculate the velocity in ft/s of a 30-lb object if it has a kinetic energy of 10 ft-lb.

**1.38E** Calculate the velocity in ft/s of a 6-oz body if it has a kinetic energy of 30 in-oz.

One measure of a baseball pitcher's performance is his earned run average, or ERA. It is the average number of earned runs he has allowed if all the innings he has pitched were converted to equivalent nine-inning games. Therefore, the units for ERA are runs per game.

**1.39** If a pitcher has allowed 39 runs during 141 innings, calculate his ERA.

**1.40** A pitcher has an ERA of 3.12 runs/game and has pitched 150 innings. How many earned runs has he allowed?

**1.41** A pitcher has an ERA of 2.79 runs/game and has allowed 40 earned runs. How many innings has he pitched?

**1.42** A pitcher has allowed 49 earned runs during 123 innings. Calculate his ERA.

## The Definition of Pressure

**1.43E** Compute the pressure produced in the oil in a closed cylinder by a piston exerting a force of 2500 lb on the enclosed oil. The piston has a diameter of 3.00 in.

**1.44E** A hydraulic cylinder must be able to exert a force of 8700 lb. The piston diameter is 1.50 in. Compute the required pressure in the oil.

**1.45M** *Compute the pressure produced in the oil in a closed cylinder by a piston exerting a force of 12.0 kN on the enclosed oil. The piston has a diameter of 75 mm.*

**1.46M** *A hydraulic cylinder must be able to exert a force of 38.8 kN. The piston diameter is 40 mm. Compute the required pressure in the oil.*

**1.47E** The hydraulic lift for an automobile service garage has a cylinder with a diameter of 8.0 in. What pressure must the oil have to be able to lift 6000 lb?

**1.48E** A coining press is used to produce commemorative coins with the likenesses of all the U.S. presidents. The coining process requires a force of 18 000 lb. The hydraulic cylinder has a diameter of 2.50 in. Compute the required oil pressure.

**1.49M** *The maximum pressure that can be developed for a certain fluid power cylinder is 20.5 MPa. Compute the force it can exert if its piston diameter is 50 mm.*

**1.50E** The maximum pressure that can be developed for a certain fluid power cylinder is 6000 psi. Compute the force it can exert if its piston diameter is 2.00 in.

**1.51E** The maximum pressure that can be developed for a certain fluid power cylinder is 5000 psi. Compute the required diameter for the piston if the cylinder must exert a force of 20 000 lb.

**1.52M** *The maximum pressure that can be developed for a certain fluid power cylinder is 15.0 MPa. Compute the required diameter for the piston if the cylinder must exert a force of 30 kN.*

**1.53E** A line of fluid power cylinders has a range of diameters in 1.00-in increments from 1.00 to 8.00 in. Compute the force that could be exerted by each cylinder with a fluid pressure of 500 psi. Draw a graph of the force versus cylinder diameter.

**1.54E** A line of fluid power cylinders has a range of diameters in 1.00-in increments from 1.00 to 8.00 in. Compute the pressure required by each cylinder if it must exert a force of 5000 lb. Draw a graph of the pressure versus cylinder diameter.

**1.55C** Determine your own weight in newtons. Then, compute the pressure in pascals that would be created in an oil if you stood on a piston in a cylinder with a 20-mm diameter. Convert the resulting pressure to psi.

**1.56C** For the pressure you computed in Problem 1.55, compute the force in newtons that could be exerted on a piston with 250-mm diameter. Then, convert the resulting force to pounds.

## Bulk Modulus

**1.57C** Compute the pressure change required to cause a decrease in the volume of ethyl alcohol by 1.00 percent. Express the result in both psi and MPa.

**1.58C** Compute the pressure change required to cause a decrease in the volume of mercury by 1.00 percent. Express the result in both psi and MPa.

**1.59C** Compute the pressure change required to cause a decrease in the volume of machine oil by 1.00 percent. Express the result in both psi and MPa.

**1.60E** For the conditions described in Problem 1.59, assume that the 1.00 percent volume change occurred in a cylinder with an inside diameter of 1.00 in and a length of 12.00 in. Compute the axial distance the piston would travel as the volume change occurs.

**1.61E** A certain hydraulic system operates at 3000 psi. Compute the percentage change in the volume of the oil in the system as the pressure is increased from zero to 3000 psi if the oil is similar to the machine oil listed in Table 1.4.

**1.62M** *A certain hydraulic system operates at 20.0 MPa. Compute the percentage change in the volume of the oil in the system if the oil is similar to the machine oil listed in Table 1.4.*

**1.63E** A measure of the stiffness of a linear actuator system is the amount of force required to cause a certain linear deflection. For an actuator that has an inside diameter of 0.50 in and a length of 42.0 in and that is filled with a machine oil, compute the stiffness in lb/in.

**1.64E** Repeat Problem 1.63 but change the length of the cylinder to 10.0 in. Compare the results.

**1.65E** Repeat Problem 1.63 but change the cylinder diameter to 2.00 in. Compare the results.

**1.66E** Using the result of Problems 1.63, 1.64, and 1.65 together, generate a statement about the general design approach to achieve a very stiff system.

## Force and Mass

**1.67M** *Calculate the mass of a can of oil if it weighs 610 N.*

**1.68M** *Calculate the weight of a tank of gasoline if it weighs 1.35 kN.*

**1.69M** *Calculate the weight of 1 m³ of kerosene if it has a mass of 825 kg.*

**1.70M** *Calculate the weight of a jar of castor oil if it has a mass of 450 g.*

**1.71E** Calculate the mass of 1 gal of oil if it weighs 7.8 lb.

**1.72E** Calculate the mass of 1 ft³ of gasoline if it weighs 42.0 lb.

**1.73E** Calculate the weight of 1 ft³ of kerosene if it has a mass of 1.58 slugs.

**1.74E** Calculate the weight of 1 gal of water if it has a mass of 0.258 slug.

**1.75C** Assume that a man weighs 160 lb (force).
**a.** Compute his mass in slugs.
**b.** Compute his weight in N.
**c.** Compute his mass in kg.

**1.76C** In the United States, hamburger and other meats are sold by the pound. Assuming that this is 1.00-lb force, compute the mass in slugs, the mass in kg, and the weight in N.

**1.77M** *The metric ton is 1000 kg (mass). Compute the force in Newtons required to lift it.*

**1.78C** Convert the force found in Problem 1.77M to lb.

**1.79C** Determine your own weight in lb and N and your own mass in slugs and kg.

## Density, Specific Weight, and Specific Gravity

**1.80M**  *The specific gravity of benzene is 0.876. Calculate its specific weight and its density in SI units.*

**1.81M**  *Air at 16°C and standard atmospheric pressure has a specific weight of 12.02 N/m³. Calculate its density.*

**1.82M**  *Carbon dioxide has a density of 1.964 kg/m³ at 0°C. Calculate its specific weight.*

**1.83M**  *A certain medium lubricating oil has a specific weight of 8.860 kN/m³ at 5°C and 8.483 kN/m³ at 50°C. Calculate its specific gravity at each temperature.*

**1.84M**  *At 100°C mercury has a specific weight of 130.4 kN/m³. What volume of the mercury would weigh 2.25 kN?*

**1.85M**  *A cylindrical can, 150 mm in diameter, is filled to a depth of 100 mm with a fuel oil. The oil has a mass of 1.56 kg. Calculate its density, specific weight, and specific gravity.*

**1.86M**  *Glycerine has a specific gravity of 1.258. How much would 0.50 m³ of glycerine weigh? What would be its mass?*

**1.87M**  *The fuel tank of an automobile holds 0.095 m³. If it is full of gasoline having a specific gravity of 0.68, calculate the weight of the gasoline.*

**1.88M**  *The density of muriatic acid is 1200 kg/m³. Calculate its specific weight and its specific gravity.*

**1.89M**  *Liquid ammonia has a specific gravity of 0.826. Calculate the volume of ammonia that would weigh 22.0 N.*

**1.90M**  *Vinegar has a density of 1080 kg/m³. Calculate its specific weight and its specific gravity.*

**1.91M**  *Methyl alcohol has a specific gravity of 0.789. Calculate its density and its specific weight.*

**1.92M**  *A cylindrical container is 150 mm in diameter and weighs 2.25 N when empty. When filled to a depth of 200 mm with a certain oil, it weighs 35.4 N. Calculate the specific gravity of the oil.*

**1.93M**  *A storage vessel for gasoline (sg = 0.68) is a vertical cylinder 10 m in diameter. If it is filled to a depth of 6.75 m, calculate the weight and mass of the gasoline.*

**1.94M**  *What volume of mercury (sg = 13.54) would weigh the same as 0.020 m³ of castor oil, which has a specific weight of 9.42 kN/m³?*

**1.95M**  *A rock has a specific gravity of 2.32 and a volume of 1.42 × 10⁻⁴ m³. How much does it weigh?*

**1.96E**  The specific gravity of benzene is 0.876. Calculate its specific weight and its density in U.S. Customary System units.

**1.97E**  Air at 59°F and standard atmospheric pressure has a specific weight of 0.0765 lb/ft³. Calculate its density.

**1.98E**  Carbon dioxide has a density of 0.003 81 slug/ft³ at 32°F. Calculate its specific weight.

**1.99E**  A certain medium lubricating oil has a specific weight of 56.4 lb/ft³ at 40°F and 54.0 lb/ft³ at 120°F. Calculate its specific gravity at each temperature.

**1.100E**  At 212°F mercury has a specific weight of 834 lb/ft³. What volume of the mercury would weigh 500 lb?

**1.101E**  One gal of a certain fuel oil weighs 7.50 lb. Calculate its specific weight, its density, and its specific gravity.

**1.102E**  Glycerine has a specific gravity of 1.258. How much would 50 gal of glycerine weigh?

**1.103E**  The fuel tank of an automobile holds 25.0 gal. If it is full of gasoline having a density of 1.32 slugs/ft³, calculate the weight of the gasoline.

**1.104C**  The density of muriatic acid is 1.20 g/cm³. Calculate its density in slugs/ft³, its specific weight in lb/ft³, and its specific gravity. (Note that specific gravity and density in g/cm³ are numerically equal.)

**1.105C**  Liquid ammonia has a specific gravity of 0.826. Calculate the volume in cm³ that would weigh 5.0 lb.

**1.106C**  Vinegar has a density of 1.08 g/cm³. Calculate its specific weight in lb/ft³.

**1.107C**  Alcohol has a specific gravity of 0.79. Calculate its density both in slugs/ft³ and g/cm³.

**1.108E**  A cylindrical container has a 6.0-in diameter and weighs 0.50 lb when empty. When filled to a depth of 8.0 in with a certain oil, it weighs 7.95 lb. Calculate the specific gravity of the oil.

**1.109E**  A storage vessel for gasoline (sg = 0.68) is a vertical cylinder 30 ft in diameter. If it is filled to a depth of 22 ft, calculate the number of gallons in the tank and the weight of the gasoline.

**1.110E**  How many gal of mercury (sg = 13.54) would weigh the same as 5 gal of castor oil, which has a specific weight of 59.69 lb/ft³?

**1.111E**  A rock has a specific gravity of 2.32 and a volume of 8.64 in³. How much does it weigh?

---

## COMPUTER PROGRAMMING ASSIGNMENTS

**1.** Write a program that computes the specific weight of water for a given temperature using the data from Appendix A. Such a program could be part of a more comprehensive program to be written later. The following options could be used:

**a.** Enter the table data for specific weight as a function of temperature into an array. Then, for a specified temperature, search the array for the corresponding specific weight. Interpolate temperatures between values given in the table.

**b.** Include data in both SI units and U.S. Customary System units.

**c.** Include density.

**d.** Include checks in the program to ensure that the specified temperature is within the range given in the tables (i.e., above freezing point and below boiling point).

**e.** Instead of using the table look-up approach, use a curve-fit technique to obtain equations of the properties of water versus temperature. Then compute the desired property value for any specified temperature.

**2.** Use a spreads\heet to display the values of specific weight and density of water from Appendix A. Then create curve fit equations for specific weight vs. temperature and density vs. temperature using the *Trendlines* feature of the spreadsheet chart. Add this equation to the spreadsheet to produce computed values of specific weight and density for any given temperature. Compute the percent difference between the table values and the computed values. Display graphs for specific weight vs. temperature and density vs. temperature on the spreadsheet showing the equations used.

1.92
1.101
1.104

# 2 Viscosity of Fluids

## 2.1 The Big Picture

### Discussion Map

- The ease with which a fluid pours is an indication of its viscosity.

- You are likely aware that a fluid such as a lubricating oil pours much more slowly when it is cold than when it is warm. Its viscosity is much higher when it is cold.

- You also have likely noticed that, at a given temperature, the oil pours more slowly than water. The oil has a much higher viscosity than water.

- We define *viscosity* as the property of a fluid that offers resistance to the relative motion of fluid molecules. The relative motion causes shearing stress to be developed in the fluid.

- In Chapter 8, we will use viscosity as an important parameter in determining the amount of energy lost from a fluid as it flows through a pipe, tube, or other shape of conduit.

- This chapter describes the nature of viscosity, defines both *dynamic viscosity* and *kinematic viscosity,* and describes a variety of methods for measuring the viscosity of fluids.

### Discover

- Obtain samples of at least three different fluids that appear to have distinctly different values of viscosity at room temperature. Consider water, some kind of lubricant, a detergent or cleaning fluid, or some liquid food product. The amount should be approximately 0.25 liter (250 mL, about 1.0 cup, 8 fluid ounces, 1/2 pint).
- Get some kind of volumetric measuring device such as a kitchen measuring cup that has graduations marked on it.
- Locate a disposable small container such as a metal food can or a plastic cup and make a small hole in its bottom.
- Cover the hole while approximately 0.25 liter of one of the fluids is poured into the container.
- Hold the container above the measuring device.
- Have a partner stand ready to measure time to the nearest second.
- Quickly uncover the hole and measure the amount of time required to collect in the measuring cup a specific volume of the fluid as it flows through the hole in the container. This method of testing fluids is similar to that used in the Saybolt viscometer as described in Section 2.6.5 of this chapter.
- Repeat these steps for each of the other fluids.
- The comparisons of the times required to collect the same volume of the fluids is a relative indication of its viscosity.
- Then place the three fluids in a refrigerator or in some other cold place for several hours.
- Repeat the measurement of the time required to collect the same volume of each fluid while it is at this cold temperature.
- Then bring the three fluids to a very warm temperature by carefully heating them on a stove or in a microwave oven. The fluids should not be allowed to boil and great care should be exercised to ensure that nobody gets burned. For consistency, it is advised to use a thermometer to measure the temperature of each fluid and try to make the temperature of each the same.

- Repeat the measurement of the time required to collect the same volume of each fluid while it is at this warm temperature.
- Compare the times to collect the cold and the warm fluids with those found at room temperature. The change in time is an indication of the degree to which the viscosity changes with temperature. This is discussed in Section 2.5 of this chapter. Also study the graphs in Appendix D that show data for the variation of viscosity with temperature for many kinds of fluids.

Have you noticed the viscosity ratings given for the kinds of oil that are used in automotive engines? They might be called *SAE 30*, *SAE 40*, *SAE 5W-40*, or similar designations. Perhaps a designation like *ISO viscosity grade 15* is used. Such designations come from standards of the Society of Automotive Engineers (SAE), the International Standards Organization (ISO), or the American Society for Testing and Materials (ASTM), as discussed in Section 2.7 and 2.8 of this chapter. The lubricating ability of an oil is highly dependent on its viscosity and, therefore, it must be carefully controlled throughout the range of operating conditions of the engine.

> Now, as you study the rest of this chapter, concentrate on the analytical definitions of viscosity, the units used to report the values of viscosity in standard U.S. Customary System units, standard SI units, and the several other unit systems that are commonly used. Also, become familiar with the many types of data given in the Appendix because they are used in problems in several parts of this book.

## 2.2 OBJECTIVES OF THIS CHAPTER

After completing this chapter, you should be able to:

1. Define *dynamic viscosity*.
2. Define *kinematic viscosity*.
3. Identify the units of viscosity in both the SI system and the U.S. Customary System.
4. Describe the difference between a *newtonian fluid* and a *nonnewtonian fluid*.
5. Describe the methods of viscosity measurement using the *rotating drum viscometer*, the *capillary tube viscometer*, the *falling ball viscometer*, and the *Saybolt Universal viscometer*.
6. Describe the variation of viscosity with temperature for both liquids and gases.
7. Define *viscosity index*.
8. Describe the viscosity of lubricants using the SAE viscosity grades and the ISO viscosity grades.

## 2.3 DYNAMIC VISCOSITY

As a fluid moves there is developed in it a shear stress, the magnitude of which depends on the viscosity of the fluid. *Shear stress*, denoted by the Greek letter $\tau$ (tau), can be defined as the force required to slide one unit area layer of a substance over another. Thus, $\tau$ is a force divided by an area and can be measured in the units of newtons per square meter or lb/ft$^2$. In a fluid such as water, oil, alcohol, or other common liquids we find that the magnitude of the shearing stress is directly proportional to the change of velocity between different positions in the fluid.

Figure 2.1 illustrates the concept of velocity change in a fluid by showing a thin layer of fluid between two surfaces, one of which is stationary while the other

**FIGURE 2.1**   Velocity gradient in a moving fluid.

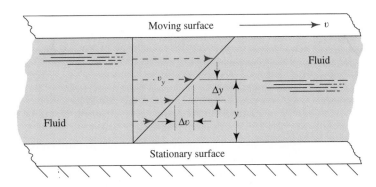

is moving. A fundamental condition that exists when a real fluid is in contact with a boundary surface is that the fluid has the same velocity as the boundary. In Fig. 2.1, then, the fluid in contact with the lower surface has a zero velocity and that in contact with the upper surface has the velocity $v$. If the distance between the two surfaces is small, then the rate of change of velocity with position $y$ is linear. That is, it varies in a straight-line manner. The *velocity gradient* is a measure of the velocity change and is defined as $\Delta v/\Delta y$. This is also called the *shear rate*.

The fact that the shear stress in the fluid is directly proportional to the velocity gradient can be stated mathematically as

$$\tau = \mu(\Delta v/\Delta y) \qquad (2–1)$$

where the constant of proportionality $\mu$ (the Greek letter mu) is called the *dynamic viscosity* of the fluid.

You can visualize the physical interpretation of Eq. (2–1) by stirring a fluid with a rod. The action of stirring causes a velocity gradient to be created in the fluid. A greater force is required to stir a cold oil with a high viscosity (a high value of $\mu$) than is required to stir water with a low viscosity. This is an indication of the higher shear stress in the cold oil.

The direct application of Eq. (2–1) is used in some types of viscosity measuring devices as will be explained later.

**2.3.1**
**Units for Dynamic Viscosity**

Many different unit systems are used to express viscosity. The systems used most frequently are described here for dynamic viscosity and in the next section for kinematic viscosity. Summary tables listing many conversion factors are included in Appendix K.

The definition of dynamic viscosity can be derived from Eq. (2–1) by solving for $\mu$.

⇨ DYNAMIC VISCOSITY

$$\mu = \frac{\tau}{\Delta v/\Delta y} = \tau\left(\frac{\Delta y}{\Delta v}\right) \qquad (2–2)$$

The units for $\mu$ can be derived by substituting the SI units only into Eq. (2–2) as follows.

$$\mu = \frac{N}{m^2} \times \frac{m}{m/s} = \frac{N{\cdot}s}{m^2}$$

Since Pa is another name for $N/m^2$, we can also express $\mu$ as

$$\mu = Pa{\cdot}s$$

Sometimes, when units for $\mu$ are being combined with other terms—especially density—it is convenient to express $\mu$ in terms of kg rather than N. Since $1\ \text{N} = 1\ \text{kg·m/s}^2$, $\mu$ can be expressed as

$$\mu = \text{N} \times \frac{\text{s}}{\text{m}^2} = \frac{\text{kg·m}}{\text{s}^2} \times \frac{\text{s}}{\text{m}^2} = \frac{\text{kg}}{\text{m·s}}$$

Thus, either N·s/m², Pa·s, or kg/m·s may be used for $\mu$ in the SI system.

Table 2.1 lists the dynamic viscosity units in the three most widely used systems. The dimensions of force multiplied by time divided by length squared are evident in each system. The units of poise and centipoise are listed here because much published data are given in these units. They are part of the obsolete metric system called cgs, derived from its base units of centimeter, gram, and second. Conversion factors are given in Appendix K.

**TABLE 2.1**

| Unit System | Dynamic Viscosity Units |
|---|---|
| International System (SI) | N·s/m², Pa·s, or kg/(m·s) |
| U.S. Customary System | lb·s/ft² or slug/(ft·s) |
| cgs system (obsolete) | poise = dyne·s/cm² = g/(cm·s) = 0.1 Pa·s |
|  | centipoise = poise/100 = 0.001   Pa·s = 1.0 mPa·s |

## 2.4
## KINEMATIC VISCOSITY

 KINEMATIC VISCOSITY

Many calculations in fluid mechanics involve the ratio of the dynamic viscosity to the density of the fluid. As a matter of convenience, the kinematic viscosity $\nu$ (the Greek letter nu) is defined as

$$\nu = \mu/\rho \qquad (2\text{–}3)$$

Since $\mu$ and $\rho$ are both properties of the fluid, $\nu$ is also a property.

### 2.4.1
### Units for Kinematic Viscosity

We can derive the SI units for kinematic viscosity by substituting the previously developed units for $\mu$ and $\rho$:

$$\nu = \frac{\mu}{\rho} = \mu\left(\frac{1}{\rho}\right)$$

$$\nu = \frac{\text{kg}}{\text{m·s}} \times \frac{\text{m}^3}{\text{kg}}$$

$$\nu = \text{m}^2/\text{s}$$

Table 2.2 lists the kinematic viscosity units in the three most widely used systems. The basic dimensions of length squared divided by time are evident in each system. The units of stoke and centistoke are listed because published data often employ these units. Appendix K lists conversion factors.

**TABLE 2.2**

| Unit System | Kinematic Viscosity Units |
|---|---|
| International System (SI) | m²/s |
| U.S. Customary System | ft²/s |
| cgs system (obsolete) | stoke = cm²/s = 1 × 10⁻⁴ m²/s |
|  | centistoke = stoke/100 = 1 × 10⁻⁶ m²/s = 1 mm²/s |

## 2.5
## NEWTONIAN FLUIDS
## AND NONNEWTONIAN
## FLUIDS

The study of the deformation and flow characteristics of substances is called *rheology*, which is the field from which we learn about viscosity of fluids. One important distinction you should understand is between a *newtonian fluid* and a *nonnewtonian fluid*. Any fluid that behaves in accordance with Eq. (2–1) is called a *newtonian fluid*. The viscosity $\mu$ is a function only of the condition of the fluid, particularly its temperature. The magnitude of the velocity gradient $\Delta v/\Delta y$ has no effect on the magnitude of $\mu$. Most common fluids such as water, oil, gasoline, alcohol, kerosene, benzene, and glycerine are classified as newtonian fluids.

Conversely, a fluid that does not behave in accordance with Eq. (2–1) is called a *nonnewtonian fluid*. The difference between the two is shown in Fig. 2.2. The viscosity of the nonnewtonian fluid is dependent on the velocity gradient in addition to the condition of the fluid.

**FIGURE 2.2**   Newtonian and nonnewtonian fluids.

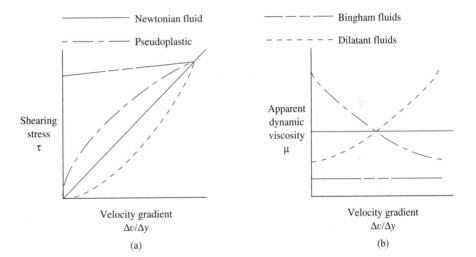

Note that in Fig. 2.2(a), the slope of the curve for shear stress versus the velocity gradient is a measure of the *apparent viscosity* of the fluid. The steeper the slope, the higher is the apparent viscosity. Because the newtonian fluids have a linear relationship between shear stress and velocity gradient, the slope is constant and, therefore, the viscosity is constant. The slope of the curves for nonnewtonian fluids varies. Figure 2.2(b) shows how viscosity changes with velocity gradient.

Two major classifications of nonnewtonian fluids are *time-independent* and *time-dependent*. As their name implies, time-independent fluids have a viscosity at any given shear stress that does not vary with time. The viscosity of time-dependent fluids, however, will change with time (see Reference [10]).

Three types of time-independent fluids can be defined:

- *Pseudoplastic*   The plot of shear stress versus velocity gradient lies above the straight, constant sloped line for newtonian fluids, as shown in Fig. 2.2. The curve begins steeply, indicating a high apparent viscosity. Then the slope decreases with increasing velocity gradient. Examples of such fluids are blood plasma, molten polyethylene, and water suspensions of clay.
- *Dilatant Fluids*   The plot of shear stress versus velocity gradient lies below the straight line for newtonian fluids. The curve begins with a low slope, indicating a low apparent viscosity. Then, the slope increases with increasing velocity gradient. Examples of dilatant fluids are corn starch in ethylene glycol, starch in water, and titanium dioxide.

■ *Bingham Fluids*    Sometimes called *plug-flow fluids*, Bingham fluids require the development of a significant level of shear stress before flow will begin, as illustrated in Fig. 2.2. Once flow starts, there is an essentially linear slope to the curve indicating a constant apparent viscosity. Examples of Bingham fluids are chocolate, catsup, mustard, mayonnaise, toothpaste, paint, asphalt, some greases, and water suspensions of fly ash or sewage sludge.

Time-dependent fluids are very difficult to analyze because apparent viscosity varies with time as well as with velocity gradient and temperature. Examples of time-dependent fluids are some crude oils at low temperatures, printer's ink, nylon, some jellies, flour dough, and several polymer solutions. Such fluids are called *thixotropic* fluids.

*Electrorheological fluids* are being developed that offer very unique properties that are controllable by the application of an electric current. Sometimes referred to as "ER fluids," they are suspensions of fine particles, such as starch, polymers, and ceramics, in a nonconducting oil, such as mineral oil or silicone oil. When no current is applied they behave like other liquids. But when a current is applied they turn to a gel and behave more like a solid. The change can occur in less than $1/1000$ of a second. Potential applications for such fluids are as a replacement for conventional valves, in clutches, in suspension systems for vehicles and machinery, and in automation actuators. (See Reference 11.)

**2.6**
**VARIATION OF**
**VISCOSITY WITH**
**TEMPERATURE**

You are probably familiar with some examples of the variation of fluid viscosity with temperature. Engine oil is generally quite difficult to pour when it is cold, indicating that it has a high viscosity. As the temperature of the oil is increased, its viscosity decreases noticeably.

All fluids exhibit this behavior to some extent. Appendix D gives two graphs of dynamic viscosity versus temperature for many common liquids. Notice that viscosity is plotted on a logarithmic scale because of the large range of numerical values. In order to check your ability to interpret these graphs, a few examples are listed in Table 2.3.

Gases behave differently from liquids in that the viscosity increases as the temperature increases. Also, the amount of change is generally smaller than that for liquids.

**TABLE 2.3**

| Fluid | Temperature (°C) | Dynamic Viscosity (N·s/m² or Pa·s) |
|---|---|---|
| Water | 20 | $1.0 \times 10^{-3}$ |
| Gasoline | 20 | $3.1 \times 10^{-4}$ |
| SAE 30 oil | 20 | $3.5 \times 10^{-1}$ |
| SAE 30 oil | 80 | $1.9 \times 10^{-2}$ |

**2.6.1**
**Viscosity Index**

A measure of how greatly the viscosity of a fluid changes with temperature is given by its viscosity index, sometimes referred to as VI. This is especially important for lubricating oils and hydraulic fluids used in equipment that must operate at wide extremes of temperature.

> *A fluid with a high viscosity index exhibits a small change in viscosity with temperature. A fluid with a low viscosity index exhibits a large change in viscosity with temperature.*

**FIGURE 2.3**  Typical viscosity index curves.

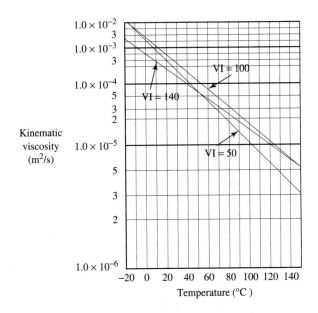

Typical curves for oils with VI values of 50, 100, and 140 are shown in Figure 2.3 on chart paper created specially for viscosity index that results in the curves being straight lines. Viscosity index is determined by measuring the viscosity of the sample fluid at 40°C and 100°C (104°F and 212°F) and comparing these values with those of certain reference fluids that were assigned VI values of 0 and 100. Standard ASTM D 2270 of the American Society for Testing and Materials (ASTM) gives the complete method. See Reference 3.

Lubricants and hydraulic fluids with a high VI should be used in construction equipment used outdoors where temperatures at critical areas vary over wide ranges. Even in a given day the oil could start at a very low temperature after soaking overnight in cold winter weather, then warm up as equipment is run, and finally reach quite high temperatures during times of heavy use. The low slope of the viscosity-temperature curve for High-VI fluids indicates that the rate of change of viscosity with temperature is significantly lower than for those with low values for VI.

The higher values of VI are obtained by blending selected oils with high paraffin content or by adding special polymers that increase VI while maintaining good lubricating properties, and good performance in pumps, valves, and actuators.

**2.7**
**VISCOSITY**
**MEASUREMENT**

Procedures and equipment for measuring viscosity are numerous. Some employ fundamental principles of fluid mechanics to indicate viscosity in its basic units. Others indicate only relative values for viscosity which can be used to compare different fluids. In this section we will describe several common methods used for viscosity measurement.

The American Society for Testing and Materials (ASTM) generates standards for viscosity measurement and reporting. Specific standards are cited in the sections that follow.

**2.7.1**
**Rotating Drum Viscometer**

The apparatus shown in Fig. 2.4(a) measures viscosity by the definition of dynamic viscosity given in Eq. (2–2):

$$\mu = \tau/(\Delta v/\Delta y) \qquad (2-2)$$

The outer cup is held stationary while the motor in the meter drives the rotating drum. The space $\Delta y$ between the rotating drum and the cup is small. The part of

32    Chapter 2   Viscosity of Fluids

**FIGURE 2.4**   Rotating drum vis-
cometer. (Source of photo: Extech
Instruments Corporation, Waltham,
MA)

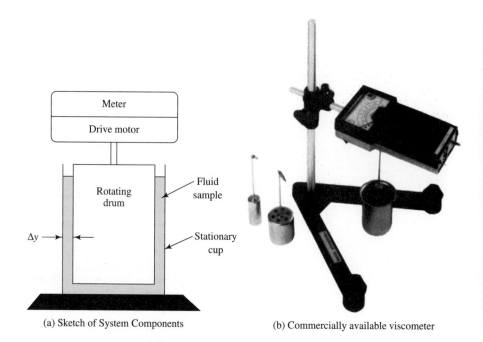

(a) Sketch of System Components        (b) Commercially available viscometer

the fluid in contact with the outer cup is stationary while the fluid in contact with
the surface of the inner drum is moving with a velocity equal to the surface speed
of the drum. Therefore, a known velocity gradient, $\Delta v/\Delta y$, is set up in the fluid.
The fluid viscosity causes shearing stress $\tau$ in the fluid that exerts a drag torque on
the rotating drum. The meter senses the drag and indicates viscosity directly on the
analog display. Special consideration is given to the fluid in contact with the bottom of the drum because its velocity varies from zero at the center to the higher
value at the outer diameter. Different models for the commercially available tester,
shown in part (b) of the figure, and different rotors for each tester allow measurement of a wide range of viscosity levels from 2.0 cP (mPa·s) up to 400 000 cP
(400 000 mPa·s or 400 Pa·s). This kind of tester can be used for a variety of fluids
such as paint, ink, food, petroleum products, cosmetics, and adhesives. The tester
is battery operated and can be either mounted on a stand as shown or hand-held for
in-plant operation.

A variant of the rotating drum viscometer is used in ASTM Standard D 5293:
*Standard Test Method for Apparent Viscosity of Engine Oils Between −5 and −30°C
Using the Cold-Cranking Simulator.* In this apparatus a universal motor drives a rotor which is closely fitted inside a stator. The rotor speed is related to the viscosity
of the test oil that fills the space between the stator and the rotor because of the viscous drag produced by the oil. Speed measurement is correlated to viscosity in centipoise (cP or mPa·s) by reference to a calibration chart prepared by running a set
of at least five standard calibration oils of known viscosity on the particular apparatus being used. The resulting data are used by engine designers and users to ensure the proper operation of the engine at cold temperatures.

The Society of Automotive Engineers (SAE) specifies that a test to determine
the maximum *borderline pumping temperature* (BPT) of engine oils be run. See
Section 2.8 in this chapter. The test method and apparatus is described by ASTM
Standard D 3829: *Standard Test Method for Predicting the Borderline Pumping
Temperature of Engine Oil.* The apparatus incorporates a rotary viscometer having

a calibrated rotor-stator assembly. The time required for the rotor to make one revolution is measured with the test oil in the space between the rotor and stator and with the system at a known temperature. After a series of tests, the temperature at which the test oil has an apparent viscosity of 30 Pa·s (30 000 cP) is reported as the BPT. Another part of the standard method calls for the determination of a *critical yield stress*. Refer to the ASTM standard for this procedure. (See Reference 7.)

**2.7.2**
**Capillary Tube Viscometer**

Figure 2.5 shows two reservoirs connected by a long, small-diameter tube called a *capillary tube*. As the fluid flows through the tube with a constant velocity, some energy is lost from the system, causing a pressure drop that can be measured by using manometers. The magnitude of the pressure drop is related to the fluid viscosity by the following equation, which is developed in a later chapter of this book:

$$\mu = \frac{(p_1 - p_2)D^2}{32vL} \tag{2–4}$$

In Eq. (2–4), $D$ is the inside diameter of the tube, $v$ is the fluid velocity, and $L$ is the length of the tube between points 1 and 2 where the pressure is measured.

**FIGURE 2.5** Capillary tube viscometer.

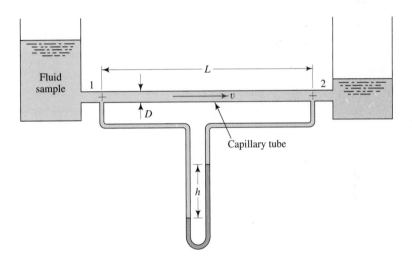

**2.7.3**
**Standard Calibrated Glass**
**Capillary Viscometers**

ASTM Standards D 445 and D 446 (References 1 and 2) describe the use of standard glass capillary viscometers to measure the kinematic viscosity of transparent and opaque liquids. Figure 2.6 and 2.7 show two of the 17 types of viscometers discussed in the standard. Figure 2.8 shows a commercially available bath for holding the tubes and maintaining test temperatures within 0.01°C (0.02°F) throughout the bath.

In preparation for the viscosity test, the viscometer tube is charged with a specified quantity of test fluid.

After stabilizing the test temperature, suction is used to draw fluid through the bulb and slightly above the upper timing mark. The suction is removed and the fluid is allowed to flow by gravity. The working section of the tube is the capillary below the lower timing mark. The time required for the leading edge of the meniscus to pass from the upper timing mark to the lower timing mark is recorded. The kinematic viscosity is computed by multiplying the flow time by the calibration

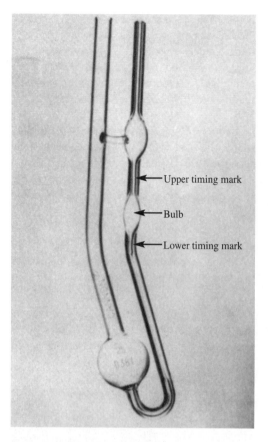

**FIGURE 2.6**   Cannon-Fenske routine viscometer. (Source: Fisher Scientific, Pittsburgh, PA)

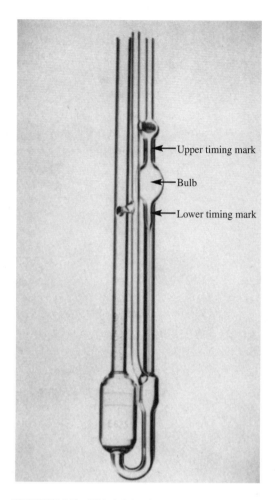

**FIGURE 2.7**   Ubbelohde viscometer. (Source: Fisher Scientific, Pittsburgh, PA)

constant of the viscometer supplied by the vendor. The viscosity unit used in these tests is the centistoke (cSt), which is equivalent to $mm^2/s$. This value must be multiplied by $10^{-6}$ to obtain the SI standard unit of $m^2/s$ that is used for calculations in this book.

**2.7.4**
**Falling Ball Viscometer**

As a body falls in a fluid under the influence of gravity only, it will accelerate until the downward force (its weight) is just balanced by the buoyant force and the viscous drag force acting upward. Its velocity at that time is called the *terminal velocity*. The falling ball viscometer sketched in Fig. 2.9 uses this principle by causing a spherical ball to fall freely through the fluid and measuring the time required for the ball to drop a known distance. Thus, the velocity can be calculated. Figure 2.10 shows a free body diagram of the ball where $w$ is the weight of the ball, $F_b$ is the buoyant force, and $F_d$ is the viscous drag force on the ball. When the ball has reached its terminal velocity, it is in equilibrium. Therefore, we have

$$w - F_b - F_d = 0. \qquad (2-5)$$

**FIGURE 2.8** Kinematic viscosity bath for holding standard calibrated glass capillary viscometers. (Source: Precision Scientific Petroleum Instruments Company, Bellwood, IL)

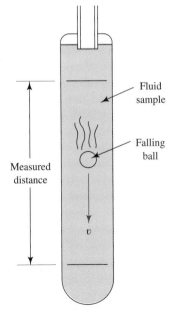

Fluid sample

Falling ball

Measured distance

$v$

**FIGURE 2.9** Falling ball viscometer.

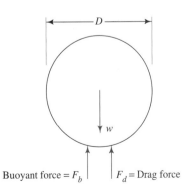

$D$

$w$

Buoyant force = $F_b$    $F_d$ = Drag force

**FIGURE 2.10** Free body diagram of ball in falling ball viscometer.

If $\gamma_s$ is the specific weight of the sphere, $\gamma_f$ is the specific weight of the fluid, $V$ is the volume of the sphere, and $D$ is the diameter of the sphere, we have

$$w = \gamma_s V = \gamma_s \pi D^3/6 \tag{2–6}$$

$$F_b = \gamma_f V = \gamma_f \pi D^3/6 \tag{2–7}$$

For very viscous fluids and a small velocity, the drag force on the sphere is

$$F_d = 3\pi\mu vD \tag{2–8}$$

(This will be discussed in Chapter 17.) Equation (2–5) then becomes

$$\mu = \frac{(\gamma_s - y_f)D^2}{18v} \tag{2–9}$$

## 2.7.5 Saybolt Universal Viscometer

The ease with which a fluid flows through a small-diameter orifice is an indication of its viscosity. This is the principle on which the Saybolt viscometer is based. The fluid sample is placed in an apparatus similar to that sketched in Fig. 2.11(a). After flow is established, the time required to collect 60 mL of the fluid is measured. The resulting time is reported as the viscosity of the fluid in Saybolt Seconds Universal (SSU or sometimes SUS). Since the measurement is not based on the basic definition of viscosity, the results are only relative. However, they do serve to compare the viscosities of different fluids. The advantage of this procedure is that it is simple and requires relatively unsophisticated equipment. Approximate conversion can be made from SSU to kinematic viscosity as shown in Appendix K. Figure 2.11(b) and (c) show a commercially available Saybolt viscometer and the 60 mL flask used to collect the sample. The use of the Saybolt viscometer was formerly covered by Standard ASTM D 88. However, that standard is no longer supported by ASTM. Preference is now given to the use of the glass capillary viscometers described in ASTM D 445 and D 446 and discussed in Section 2.7.3 of this chapter.

## 2.8 SAE VISCOSITY GRADES

The Society of Automotive Engineers (SAE) has developed a rating system for engine oils (Table 2.4) and automotive gear lubricants (Table 2.5) which indicates the viscosity of the oils at specified temperatures. Oils with a suffix W are based on maximum absolute viscosity at specified cold temperatures under conditions that simulate both the cranking of an engine and the pumping of the oil by the oil pump. They must also exhibit a kinematic viscosity above a specified minimum at 100°C. Those without the suffix W are rated at high temperatures by two different methods. The kinematic viscosity under low-shear rate conditions at 100°C must be in the indicated range from Table 2.4. The dynamic viscosity under high-shear rate conditions at 150°F must be greater than the minimum listed in the last column of Table 2.4. This rating simulates the conditions in journal bearings and for sliding surfaces. Note the two different high-shear ratings for SAE 40 grade. The first is typical of multiviscosity-grade oils used in light-duty engines. The second is typical of single-grade SAE 40 and multiviscosity-grade oils used in heavy-duty engines. Multiviscosity-grade oils, such as SAE 10W-40, must meet the standards at both the low- and high-temperature conditions.

The specification of maximum low-temperature viscosity values for oils is related to the ability of the oil to flow to the surfaces needing lubrication at the engine speeds encountered during starting at cold temperatures. The pumping viscosity indicates the ability of the oil to flow into the oil pump inlet of an engine. The high temperature viscosity range specifications relate to the ability of the oil to pro-

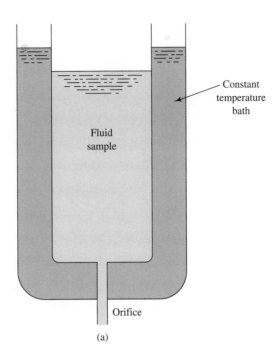

Constant
temperature
bath

Fluid
sample

Orifice

(a)

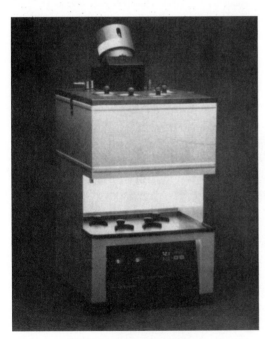

(b) Universal Saybolt viscometer

(c) 60 mL flask for collecting Saybolt sample

**FIGURE 2.11** Saybolt viscometer. [Sources of photos: (b) Precision Scientific Petroleum
Instruments Co., Bellwood, IL; (c) Corning, Inc., Corning, NY]

**TABLE 2.4**   SAE viscosity grades for engine oils

| SAE Viscosity Grade | Low Temperature—Dynamic Viscosity | | High-Temperature Kinematic Viscosity at 100°C (cSt)$^+$ | | High Temperature High-Shear Rate Dynamic Viscosity$^\Diamond$ at 150°C (cP) Min. |
|---|---|---|---|---|---|
| | Cranking Condition* (cP) Max. at (°C) | Pumping Condition$^\#$ (cP) Max. at (°C) | Min. | Max. | |
| 0W | 3250 at −30 | 60 000 at −40 | 3.8 | — | — |
| 5W | 3500 at −25 | 60 000 at −35 | 3.8 | — | — |
| 10W | 3500 at −20 | 60 000 at −30 | 4.1 | — | — |
| 15W | 3500 at −15 | 60 000 at −25 | 5.6 | — | — |
| 20W | 4500 at −10 | 60 000 at −20 | 5.6 | — | — |
| 25W | 6000 at −5 | 60 000 at −15 | 9.3 | — | — |
| 20 | — | — | 5.6 | <9.3 | 2.6 |
| 30 | — | — | 9.3 | <12.5 | 2.9 |
| 40 | — | — | 12.5 | <16.3 | 2.9$^\perp$ |
| 40 | — | — | 12.5 | <16.3 | 3.7$^\lceil$ |
| 50 | — | — | 16.3 | <21.9 | 3.7 |
| 60 | — | — | 21.9 | <26.1 | 3.7 |

**Source:**   Reprinted with permission from SAE J300. Society of Automotive Engineers, Inc. (See Reference 14.)

\*   Using modified ASTM Standard D 5293

$^\#$   Using ASTM D 4684

$^+$   Using ASTM D 445

$^\Diamond$   Using ASTM D 4683, D 4741, or D 5481

$^\perp$   When used in these multiviscosity grades: 0W-40, 5W-40, 10W-40

$^\lceil$   When used single grade SAE 40 and in these multiviscosity grades: 15W-40, 20W-40, 25W-40

**TABLE 2.5**   SAE viscosity grades for automotive gear lubricants

| SAE Viscosity Grade | Maximum Temperature for Dynamic Viscosity of 150 000 cP* (°C) | Kinematic Viscosity at 100°C (cSt)$^\#$ | |
|---|---|---|---|
| | | Min. | Max. |
| 70W | −55 | 4.1 | — |
| 75W | −40 | 4.1 | — |
| 80W | −26 | 7.0 | — |
| 85W | −12 | 11.0 | — |
| 80 | — | 7.0 | <11.0 |
| 85 | — | 11.0 | <13.5 |
| 90 | — | 13.5 | <24.0 |
| 140 | — | 24.0 | <41.0 |
| 250 | — | 41.0 | |

**Source:** Reprinted with permission from SAE J306. Society of Automotive Engineers, Inc. (See Reference 15.)

\*   Using ASTM D 2983

$^\#$   Using ASTM D 445

vide a satisfactory oil film to carry expected loads while not having an excessively high viscosity that would increase friction and energy losses generated by moving parts.

The following standards apply to the SAE classifications and the methods of testing:

| | |
|---|---|
| SAE J300 | *Engine Oil Viscosity Classification* |
| SAE J306 | *Automotive Gear Lubricant Viscosity Classification* |
| ASTM D 445 | *Standard Test Method for Kinematic Viscosity of Transparent and Opaque Liquids* |
| ASTM D 446 | *Standard Specifications for Glass Capillary Kinematic Viscometers* |
| ASTM D 5293 | *Standard Test Method for Apparent Viscosity of Engine Oils Between −5 and −30°C Using the Cold-Cranking Simulator* |
| ASTM D 2983 | *Standard Test Method for Low-Temperature Viscosity of Automotive Fluid Lubricants Measured by the Brookfield Viscometer* |
| ASTM D 3829 | *Standard Test Method for Predicting the Borderline Pumping Temperature of Engine Oil* |
| ASTM D 4683 | *Standard Test Method for Measuring Viscosity at High Temperature and High-Shear Rate by Tapered-Plug Viscometer* |
| ASTM D 4684 | *Standard Test Method for Determination of Yield Stress and Apparent Viscosity of Engine Oils at Low Temperature* |

Consult the latest revision of these standards.

See also Appendix C for typical properties of petroleum lubricating oils used in hydraulic systems and machine tool applications.

Note that oils designed to operate at wide ranges of temperatures have special additives to increase the viscosity index. An example is a multiviscosity engine oil that must meet stringent low-temperature viscosity limits while maintaining a sufficiently high viscosity at higher engine operating temperatures for effective lubrication. Also, automotive hydraulic system oils that must operate with similar performance in cold as well as warm climates and machine tool hydraulic system oils that must operate outdoors as well as indoors must have high viscosity indexes. See Appendix C for representative values.

Achieving a high viscosity index in an oil often calls for the blending of polymeric materials with the petroleum. The resulting blend may exhibit nonnewtonian characteristics, particularly at the lower temperatures.

**2.9
ISO VISCOSITY GRADES**

Lubricants used in industrial applications must be available in a wide range of viscosities to meet the needs of production machinery, bearings, gear drives, electrical machines, fans and blowers, fluid power systems, mobile equipment, and many other devices. The designers of such systems must ensure that the lubricant can withstand the temperatures to be experienced while providing sufficient load-carrying ability. The result is a need for a wide range of viscosities.

To meet these requirements and still have an economical and manageable number of options, ASTM Standard D 2422 *Standard Classification of Industrial Fluid*

*Lubricants by Viscosity System* defines a set of 20 ISO viscosity grades. The standard designation includes the prefix ISO VG followed by a number representing the nominal kinematic viscosity in cSt (mm$^2$/s) for a temperature of 40°C. Table 2.6 gives the data. The maximum and minimum values are $+/-$ 10 percent from the nominal. Although the standard is voluntary, the intent is to encourage producers and users of lubricants to agree on the specification of viscosities from the list. This system is gaining favor throughout the world markets.

TABLE 2.6    ISO viscosity grades

| Grade ISO VG | Kinematic Viscosity at 40°C (cSt) or (mm$^2$/s) | | |
|---|---|---|---|
| | Nominal | Minimum | Maximum |
| 2 | 2.2 | 1.98 | 2.40 |
| 3 | 3.2 | 2.88 | 3.52 |
| 5 | 4.6 | 4.14 | 5.06 |
| 7 | 6.8 | 6.12 | 7.48 |
| 10 | 10 | 9.00 | 11.0 |
| 15 | 15 | 13.5 | 16.5 |
| 22 | 22 | 19.8 | 24.2 |
| 32 | 32 | 28.8 | 35.2 |
| 46 | 46 | 41.4 | 50.6 |
| 68 | 68 | 61.2 | 74.8 |
| 100 | 100 | 90.0 | 110 |
| 150 | 150 | 135 | 165 |
| 220 | 220 | 198 | 242 |
| 320 | 320 | 288 | 352 |
| 460 | 460 | 414 | 506 |
| 680 | 680 | 612 | 748 |
| 1000 | 1000 | 900 | 1100 |
| 1500 | 1500 | 1350 | 1650 |
| 2200 | 2200 | 1980 | 2420 |
| 3200 | 3200 | 2880 | 3520 |

**Source:**    American Society for Testing and Materials. ASTM Standard D 2422. West Conshohocken, PA. (See Reference 4.) Copyright ASTM. Reprinted with permission.

## 2.10 HYDRAULIC FLUIDS FOR FLUID POWER SYSTEMS

Fluid power systems use fluids under pressure to actuate linear or rotary devices used in construction equipment, industrial automation systems, agricultural equipment, automotive braking systems, and many others. Fluid power includes both air-type systems, commonly called *pneumatics*, and liquid-type, usually referred to as hydraulic systems. This section will deal with the liquid-type systems.

There are several types of hydraulic fluids in common use, such as,

- Petroleum oils
- Water-glycol fluids
- High water-based fluids (HWBF)
- Silicone fluids
- Synthetic oils

The primary characteristics of such fluids for operation in fluid power systems are

- Adequate viscosity for the purpose
- High lubricating capability, sometimes called lubricity
- Cleanliness
- Chemical stability at operating temperatures
- Non-corrosiveness with the materials used in fluid power systems
- Inability to support bacteria growth
- Ecologically acceptable
- High bulk modulus (low compressibility)

You should examine carefully the environment in which the fluid power system is to be used and select a fluid that is optimal for the application. Trade-offs will typically be required so that the combination of properties is acceptable. Suppliers of components, particularly pumps and valves, should be consulted for appropriate fluids to use with their products.

Viscosity is one of the most important properties as it relates to lubricity and the ability of the fluid to be pumped and to flow through the tubing, piping, actuators, valves, and other control devices found in fluid power systems.

Common industrial fluid power systems require fluids with viscosities in the range of ISO grades 32, 46, or 68. See Table 2.6 for the kinematic viscosity ranges for such fluids. In general, the ISO grade number is the nominal kinematic viscosity in the unit of centistokes (cSt or mm$^2$/s).

Special care is needed when extremes of temperature are encountered. Consider the case of the fluid power system on a piece of construction equipment that is kept outdoors throughout the year. In winter, the temperature many range to $-20°F$ ($-29°C$). When starting the system at that temperature you must consider the ability of the fluid to flow into the intake ports of the pumps, through the piping systems, and through the control valves. The fluid viscosity may be greater than 800 cSt ($8.0 \times 10^{-4}$ m$^2$/s). Then, when the system has warmed to approximately 150°F (66°C), the fluid viscosity may be as low as 15 cSt ($1.5 \times 10^{-5}$ m$^2$/s). The performance of the pumps and valves is likely to be remarkably different under this range of conditions. Also, as you will learn in Chapter 8, the very nature of the flow may change as the viscosities change. At the cold temperatures the fluid flow will likely be laminar, while at the higher temperatures with the decreased viscosities the flow may be turbulent. Hydraulic fluids for operation at these ranges of temperatures should have a high viscosity index, as described earlier in this chapter.

*Petroleum oils* may be very similar to the automotive engine oils discussed earlier in this chapter. SAE 10W or SAE 20-20W are appropriate. However, there should be a number of additives required to inhibit the growth of bacteria, to ensure compatibility with seals and other parts of fluid power components, to improve its anti-wear performance in pumps, and to improve the viscosity index. Suppliers of hydraulic fluids should be consulted for recommendations of specific formulations. Some of the additives used to improve viscosity are polymeric materials, and they may change the flow characteristics dramatically under certain high-pressure conditions that may occur within valves and pumps. The oils may behave as nonnewtonian fluids.

*Silicone fluids* are desirable when high temperatures are to be encountered, as in work near furnaces, hot processes, and some vehicle braking systems. These fluids exhibit very high thermal stability. Compatibility with the pumps and valves of the system must be checked.

*High water-based fluids* (HWBF) are desirable where fire resistance is needed. Water-in-oil emulsions contain approximately 40% oil blended in water with a significant variety and quantity of additives to tailor the fluid properties to the application. A different class of fluids, called oil-in-water emulsions, contains 90–95% water with the balance being oil and additives. Such emulsions typically appear to be milky-white because the oil is dispersed in the form of very small droplets.

*Water-glycol fluids* are also fire resistant, containing approximately 35 to 50% water, with the balance being any of several glycols along with additives suitable for the environment in which the system is to be operated.

---

## REFERENCES

1. American Society for Testing and Materials (ASTM). 1997 *ASTM D 445-97. Standard Test Method for Kinematic Viscosity of Transparent and Opaque Liquids.* West Conshohocken, PA: Author.

2. ———. 1997. *ASTM D 446-93: Standard Specifications for Glass Capillary Kinematic Viscometers.* West Conshohocken, PA: Author.

3. ———. 1997. *ASTM D 2270-95(1998): Standard Practice for Calculating Viscosity Index from Kinematic Viscosity at 40 and 100°C.* West Conshohocken, PA: Author.

4. ———. 1997. *ASTM D 2422-97: Standard Classification of Industrial Lubricants by Viscosity System.* West Conshohocken, PA: Author.

5. ———. 1997. *ASTM D 5293-98: Standard Test Method for Apparent Viscosity of Engine Oils Between −5 and −30°C Using the Cold-Cranking Simulator.* West Conshohocken, PA: Author.

6. ———. 1997. *ASTM D 2983-87: Standard Test Method for Low-Temperature Viscosity of Automotive Fluid Lubricants Measured by Brookfield Viscometer.* West Conshohocken, PA: Author.

7. ———. 1997. *ASTM D 3829-93e1: Standard Test Method for Predicting the Borderline Pumping Temperature of Engine Oil.* West Conshohocken, PA: Author.

8. ———. 1997. *ASTM D 4683: Standard Test Method for Measuring Viscosity at High Temperature and High-Shear Rate by Tapered-Plug Viscometer.* West Conshohocken, PA: Author.

9. ———. 1997. *ASTM D 4684-98: Standard Test Method for Determination of Yield Stress and Apparent Viscosity of Engine Oils at Low Temperature.* West Conshohocken, PA: Author.

10. Avallone, Eugene A., and Theodore Baumeister, eds. 1996. *Marks' Standard Handbook for Mechanical Engineers.* 10th ed. New York: McGraw-Hill.

11. Cheremisinoff, N. P., ed. 1986. *Encyclopedia of Fluid Mechanics.* Vol. 1, *Flow Phenomena and Measurement.* Houston, Texas: Gulf Publishing Co.

12. Goldstein, Gina. 1990 (October). "Electrorheological Fluids," *Mechanical Engineering Magazine* 112(10): 48–52.

13. Miller, R. W. 1996. *Flow Measurement Engineering Handbook.* 3rd ed. New York: McGraw-Hill.

14. Society of Automotive Engineers (SAE). 1997. *SAE Standard J300: Engine Oil Viscosity Classification.* Warrendale, PA: Author.

15. ———. 1991. *SAE Standard J306: Automotive Gear Lubricant Viscosity Classification.* Warrendale, PA: Author.

---

## PRACTICE PROBLEMS

**2.1** Define *shear stress* as it applies to a moving fluid.

**2.2** Define *velocity gradient*.

**2.3** State the mathematical definition for *dynamic viscosity*.

**2.4** Which would have the greater dynamic viscosity, a cold lubricating oil or fresh water? Why?

**2.5** State the standard units for dynamic viscosity in the SI system.

**2.6** State the standard units for dynamic viscosity in the U.S. Customary System.

**2.7** State the equivalent units for *poise* in terms of the basic quantities in the cgs system.

**2.8** Why are the units of poise and centipoise considered obsolete?

**2.9** State the mathematical definition for *kinematic viscosity*.

**2.10** State the standard units for kinematic viscosity in the SI system.

**2.11** State the standard units for kinematic viscosity in the U.S. Customary System.

**2.12** State the equivalent units for *stoke* in terms of the basic quantities in the cgs system.

**2.13** Why are the units of stoke and centistoke considered obsolete?

**2.14** Define a *newtonian fluid*.

**2.15** Define a *nonnewtonian fluid*.

**2.16** Give five examples of newtonian fluids.

**2.17** Give four examples of the types of fluids that are non-newtonian.

Appendix D gives dynamic viscosity for a variety of fluids as a function of temperature. Using Appendix D, give the value of the viscosity for the following fluids:

**2.18M** *Water at 40°C.*

**2.19M** *Water at 5°C.*

**2.20M** *Air at 40°C.*

**2.21M** *Hydrogen at 40°C.*

**2.22M** *Glycerine at 40°C.*

**2.23M** *Glycerine at 20°C.*

**2.24E** Water at 40°F.

**2.25E** Water at 150°F.

**2.26E** Air at 40°F.

**2.27E** Hydrogen at 40°F.

**2.28E** Glycerine at 60°F.

**2.29E** Glycerine at 110°F.

**2.30E** Mercury at 60°F.

**2.31E** Mercury at 210°F.

**2.32E** SAE 10 oil at 60°F.

**2.33E** SAE 10 oil at 210°F.

**2.34E** SAE 30 oil at 60°F.

**2.35E** SAE 30 oil at 210°F.

**2.36** Define *viscosity index* (VI).

**2.37** If you want to choose a fluid that exhibits a small change in viscosity as the temperature changes, would you choose one with a high VI or a low VI?

**2.38** Which type of viscosity measurement method uses the basic definition of dynamic viscosity for direct computation?

**2.39** In the *rotating drum viscometer*, describe how the velocity gradient is created in the fluid to be measured.

**2.40** In the *rotating drum viscometer*, describe how the magnitude of the shear stress is measured.

**2.41** What measurements must be taken to determine dynamic viscosity when using a *capillary tube viscometer*?

**2.42** Define the term *terminal velocity* as it applies to a falling ball viscometer.

**2.43** What measurements must be taken to determine dynamic viscosity when using the *falling ball viscometer*?

**2.44** Describe the basic features of the *Saybolt Universal viscometer*.

**2.45** Are the results of the Saybolt viscometer tests considered to be direct measurements of viscosity?

**2.46** Does the Saybolt viscometer produce data related to a fluid's dynamic viscosity or kinematic viscosity?

**2.47** On which type of viscometer is the SAE numbering system for viscosity at 100°C based?

**2.48** Describe the difference between an SAE 20 oil and an SAE 20W oil.

**2.49** What grades of SAE oil are suitable for lubricating the crankcases of engines?

**2.50** What grades of SAE oil are suitable for lubricating gear-type transmissions?

**2.51** If you were asked to check the viscosity of an oil that is described as SAE 40, at what temperatures would you make the measurements?

**2.52** If you were asked to check the viscosity of an oil that is described as SAE 10W, at what temperatures would you make the measurements?

**2.53** How would you determine the viscosity of an oil labeled SAE 5W-40 for comparison with SAE standards?

**2.54C** The viscosity of a lubricating oil is given as 500 SSU. Calculate the viscosity in $m^2/s$ and $ft^2/s$.

**2.55M** *Using the data from Table 2.4, report the standard values for viscosity in SI units for SAE 10W-30 oil (sg = 0.88) at both the low and high temperature points.*

**2.56C** Convert a dynamic viscosity measurement of 4500 centipoises to Pa·s and lb·s/ft².

**2.57C** Convert a kinematic viscosity measurement of 5.6 centistokes to $m^2/s$ and $ft^2/s$.

**2.58C** The viscosity of an oil is given at 80 SSU. Calculate the viscosity in $m^2/s$.

**2.59C** Convert a viscosity measurement of $6.5 \times 10^{-3}$ Pa·s to the units of lb·s/ft².

**2.60C** An oil container indicates that it has a viscosity of 0.12 poise at 60°C. Which oil in Appendix D has a similar viscosity?

**2.61M** *In a falling ball viscometer, a steel ball 1.6 mm in diameter is allowed to fall freely in a heavy fuel oil having a specific gravity of 0.94. Steel weighs 77 kN/m³. If the ball is observed to fall 250 mm in 10.4 s, calculate the viscosity of the oil.*

**2.62M** *A capillary tube viscometer similar to that shown in Fig. 2.5 is being used to measure the viscosity of an oil having a specific gravity of 0.90. The following data apply:*

> *Tube inside diameter = 2.5 mm = D*
>
> *Length between manometer taps = 300 mm = L*
>
> *Manometer fluid is mercury*
>
> *Manometer deflection = 177 mm = h*
>
> *Velocity of flow = 1.58 m/s = v*

*Determine the viscosity of the oil.*

**2.63E** In a falling ball viscometer, a steel ball with a diameter of 0.063 in is allowed to fall freely in a heavy fuel oil having a specific gravity of 0.94. Steel weighs 0.283 lb/in$^3$. If the ball is observed to fall 10.0 inches in 10.4 s, calculate the dynamic viscosity of the oil in lb·s$^2$/ft.

**2.64E** A capillary type viscometer similar to that shown in Fig. 2.5 is being used to measure the viscosity of an oil having a specific gravity of 0.90. The following data apply:

Tube inside diameter = 0.100 in = $D$

Length between manometer taps = 12.0 in = $L$

Manometer fluid is mercury

Manometer deflection = 7.00 in = $h$

Velocity of flow = 4.82 ft/s = $v$

Determine the dynamic viscosity of the oil in lb·s$^2$/ft.

---

## COMPUTER PROGRAMMING ASSIGNMENTS

1. Write a program to convert viscosity units from any given system to another system using the conversion factors and techniques from Appendix K. Note the special conditions for conversion of SSU data to kinematic viscosity in m$^2$/s when SSU < 100.

2. Write a program to determine the viscosity of water at a given temperature using data from Appendix A. This program could be joined with the one you wrote in Chapter 1, which used other properties of water. Use the same options described in Chapter 1.

3. Use a spreadsheet to display the values of kinematic viscosity and dynamic viscosity of water from Appendix A. Then create curve-fit equations for both types of viscosity vs. temperature using the *Trendlines* feature of the spreadsheet chart. Display graphs for both viscosities vs. temperature on the spreadsheet showing the equations used.

# 3 Pressure Measurement

**3.1
The
Big
Picture**

**Discussion Map**

▨ Review the definition of pressure from Chapter 1:

$$p = F/A \qquad (3\text{–}1)$$

Pressure equals force per unit area.

▨ Standard units for pressure are:

In the SI metric system: pascal (Pa) or $N/m^2$

In the U.S. Customary System: $lb/ft^2$

But see Tables 1.2 and 1.3 in Chapter 1 for other commonly used units. We will use the psi ($lb/in^2$) and kPa (kilopascals) most often in this book because they represent convenient values of typical pressures found in industrial work and product design. Other units will be defined in this chapter.

▨ Discuss pressure measurement with friends and colleagues. List as many situations as you can in which you have seen the need for pressure measurement. What was the purpose of measuring the pressure? For what kind of fluid was the pressure being measured? What kind of equipment was generating the pressure? Describe the kind of pressure measurement device that was being used. Can you recall the general order of magnitude of the pressure being measured?

*Discover*

*Here are a few examples to get you started.*

■ Have you measured the air pressure in tires for bicycles or automobiles?
■ Have you observed the pressure reading on some kind of steam or hot-water boiler?
■ Have you measured the pressure in a water supply system?
■ Have you seen pressure gages mounted on hydraulic or pneumatic fluid power systems?

In this chapter you will learn about absolute pressure (measured relative to a perfect vacuum) and gage pressure (measured relative to the local atmospheric pressure). You will learn how to compute the change of pressure with changes in elevation in a static fluid. And you will see photographs and descriptions of several kinds of commercially available pressure measuring devices.

**3.2**
**OBJECTIVES**

After completing this chapter, you should be able to:

1. Define the relationship between *absolute pressure, gage pressure,* and *atmospheric pressure.*
2. Describe the degree of variation of atmospheric pressure near the earth's surface.
3. Describe the properties of air at standard atmospheric pressure.
4. Describe the properties of the atmosphere at elevations from sea level to 30 000 m.
5. Define the relationship between a change in elevation and the change in pressure in a fluid.
6. Describe how a *manometer* works and how it is used to measure pressure.
7. Describe a U-tube manometer, a differential manometer, a well-type manometer, and an inclined well-type manometer.
8. Describe a *barometer* and how it indicates the value of the local atmospheric pressure.
9. Describe various types of pressure gages and pressure transducers.

**3.3**
**ABSOLUTE AND**
**GAGE PRESSURE**

When making calculations involving pressure in a fluid, you must make the measurements relative to some reference pressure. Normally the reference pressure is that of the atmosphere, and the resulting measured pressure is called *gage pressure.* Pressure measured relative to a perfect vacuum is called *absolute pressure.* It is extremely important for you to know the difference between these two ways of measuring pressure and to be able to convert from one to the other.

A simple equation relates the two pressure measuring systems:

▷  ABSOLUTE AND GAGE PRESSURE

$$p_{abs} = p_{gage} + p_{atm} \qquad (3-2)$$

where
$$p_{abs} = \text{absolute pressure}$$
$$p_{gage} = \text{gage pressure}$$
$$p_{atm} = \text{atmospheric pressure}$$

Figure 3.1 shows an interpretation of this equation graphically. A few basic concepts may help you to understand the equation.

1. A perfect vacuum is the lowest possible pressure. Therefore, an absolute pressure will always be positive.
2. A gage pressure above atmospheric pressure is positive.
3. A gage pressure below atmospheric pressure is negative, sometimes called *vacuum.*
4. Gage pressure will be indicated in the units of Pa(gage) or psig.
5. Absolute pressure will be indicated in the units of Pa(abs) or psia.
6. The actual magnitude of the atmospheric pressure varies with location and with climatic conditions. The barometric pressure as broadcast in weather reports is an indication of the continually varying atmospheric pressure.
7. The range of normal variation of atmospheric pressure near the earth's surface is approximately from 95 kPa(abs) to 105 kPa(abs) or from 13.8 psia to 15.3 psia. At sea level, the standard atmospheric pressure is 101.3 kPa(abs) or 14.69 psia. Unless the prevailing atmospheric pressure is given, we will assume it to be 101 kPa(abs) or 14.7 psia in this book.

**FIGURE 3.1**  Comparison between absolute and gage pressure.

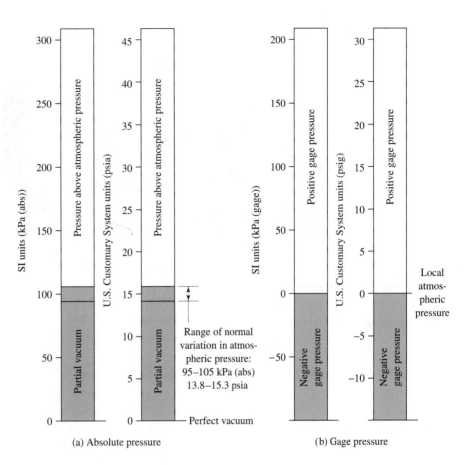

(a) Absolute pressure          (b) Gage pressure

---

☐ **EXAMPLE PROBLEM 3.1**      Express a pressure of 155 kPa(gage) as an absolute pressure. The local atmospheric pressure is 98 kPa(abs).

*Solution*
$$p_{abs} = p_{gage} + p_{atm}$$
$$p_{abs} = 155 \text{ kPa(gage)} + 98 \text{ kPa(abs)} = 253 \text{ kPa(abs)}$$

Notice that the units in this calculation are kilopascals (kPa) for each term and are consistent. The indication of gage or absolute is for convenience and clarity.

---

☐ **EXAMPLE PROBLEM 3.2**      Express a pressure of 225 kPa(abs) as a gage pressure. The local atmospheric pressure is 101 kPa(abs).

*Solution*
$$p_{abs} = p_{gage} + p_{atm}$$

Solving algebraically for $p_{gage}$ gives

$$p_{gage} = p_{abs} - p_{atm}$$
$$p_{gage} = 225 \text{ kPa(abs)} - 101 \text{ kPa(abs)} = 124 \text{ kPa(gage)}$$

---

☐ **EXAMPLE PROBLEM 3.3**

Express a pressure of 10.9 psia as a gage pressure. The local atmospheric pressure is 15.0 psia.

*Solution*

$$p_{abs} = p_{gage} + p_{atm}$$
$$p_{gage} = p_{abs} - p_{atm}$$
$$p_{gage} = 10.9 \text{ psia} - 15.0 \text{ psia} = -4.1 \text{ psig}$$

Notice that the result is negative. This can also be read "4.1 psi below atmospheric pressure" or "4.1 psi vacuum."

---

☐ **EXAMPLE PROBLEM 3.4**

Express a pressure of −6.2 psig as an absolute pressure.

*Solution*

$$p_{abs} = p_{gage} + p_{atm}$$

Since no value was given for the atmospheric pressure, we will use $p_{atm} = 14.7$ psia:

$$p_{abs} = -6.2 \text{ psig} + 14.7 \text{ psia} = 8.5 \text{ psia}$$

■

---

**3.4
RELATIONSHIP
BETWEEN PRESSURE
AND ELEVATION**

You are probably familiar with the fact that as you go deeper in a fluid, such as in a swimming pool, the pressure increases. There are many situations in which it is important to know just how the pressure varies with a change in depth or elevation.

In this book the term *elevation* means the vertical distance from some reference level to a point of interest and is called *z*. A *change* in elevation between two points is called *h*. Elevation will always be measured positively in the upward direction. In other words, a higher point has a larger elevation than a lower point.

The reference level can be taken at any level as illustrated in Fig. 3.2, which shows a submarine under water. In part (a) of the figure the sea bottom is taken as reference, while in (b) the position of the submarine is the reference level. Since fluid mechanics calculations usually consider differences in elevation, it is advisable to choose the lowest point of interest in a problem as the reference level to eliminate the use of negative values for *z*. This will be especially important in later work.

**FIGURE 3.2**   Illustration of reference level for elevation.

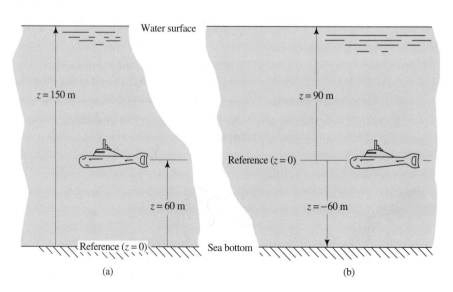

(a)                                    (b)

The change in pressure in a homogeneous liquid at rest due to a change in elevation can be calculated from

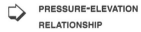

**PRESSURE-ELEVATION**
**RELATIONSHIP**

$$\Delta p = \gamma h \qquad (3-3)$$

where
$\Delta p$ = change in pressure
$\gamma$ = specific weight of liquid
$h$ = change in elevation

Some general conclusions from Eq. (3–3) will help you to apply it properly:

1. The equation is valid only for a homogeneous liquid at rest.
2. Points on the same horizontal level have the same pressure.
3. The change in pressure is directly proportional to the specific weight of the liquid.
4. Pressure varies linearly with the change in elevation or depth.
5. A decrease in elevation causes an increase in pressure. (This is what happens when you go deeper in a swimming pool.)
6. An increase in elevation causes a decrease in pressure.

Equation (3–3) does not apply to gases because the specific weight of a gas changes with a change in pressure. However, it requires a large change in elevation to produce a significant change in pressure in a gas. For example, an increase in elevation of 300 m (about 1000 ft) in the atmosphere causes a decrease in pressure of only 3.4 kPa (about 0.5 psi). *In this book we assume that the pressure in a gas is uniform unless otherwise specified.*

---

☐ **EXAMPLE PROBLEM 3.5**    Calculate the change in water pressure from the surface to a depth of 5 m.

*Solution*    Using Eq. (3–3), $\Delta p = \gamma h$, let $\gamma = 9.81$ kN/m$^3$ for water and $h = 5$ m. Then we have

$$\Delta p = (9.81 \text{ kN/m}^3)(5.0 \text{ m}) = 49.05 \text{ kN/m}^2 = 49.05 \text{ kPa}$$

If the surface of the water is exposed to the atmosphere, the pressure there is 0 Pa(gage). Descending in the water (decreasing elevation) produces an increase in pressure. Therefore, at 5 m the pressure is 49.05 kPa(gage).

---

☐ **EXAMPLE PROBLEM 3.6**    Calculate the change in water pressure from the surface to a depth of 15 ft.

*Solution*    Using Eq. (3–3), $\Delta p = \gamma h$, let $\gamma = 62.4$ lb/ft$^3$ for water and $h = 15$ ft. Then we have

$$\Delta p = \frac{62.4 \text{ lb}}{\text{ft}^3} \times 15 \text{ ft} \times \frac{1 \text{ ft}^2}{144 \text{ in}^2} = 6.5 \frac{\text{lb}}{\text{in}^2}$$

If the surface of the water is exposed to the atmosphere, the pressure there is 0 psig. Descending in the water (decreasing elevation) produces an increase in pressure. Therefore, at 15 ft the pressure is 6.5 psig.

---

☐ **EXAMPLE PROBLEM 3.7**    Figure 3.3 shows a tank of oil with one side open to the atmosphere and the other side sealed with air above the oil. The oil has a specific gravity of 0.90. Calculate the gage pressure at points A, B, C, D, E, and F and the air pressure in the right side of the tank.

***Solution***   At *point A* the oil is exposed to the atmosphere, and therefore

$$p_A = 0 \text{ Pa(gage)}$$

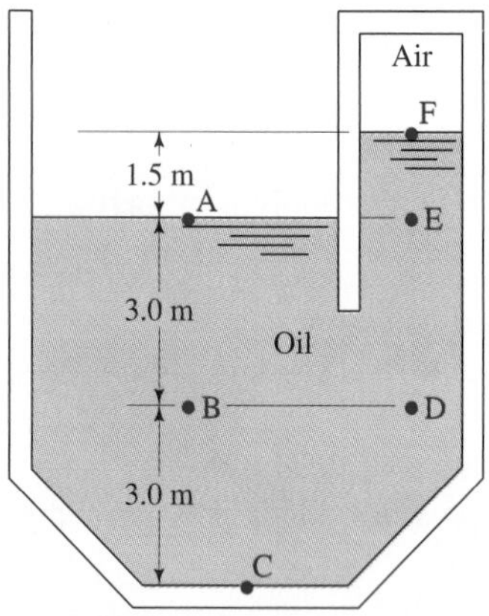

**FIGURE 3.3**   Tank for Example Problem 3.7.

*Point B:*   The change in elevation between point A and point B is 3.0 m, with B lower than A. To use Eq. (3–3) we need the specific weight of the oil:

$$\gamma_{oil} = (sg)_{oil}(9.81 \text{ kN/m}^3) = (0.90)(9.81 \text{ kN/m}^3) = 8.83 \text{ kN/m}^3$$

Then, we have

$$\Delta p_{A-B} = \gamma h = (8.83 \text{ kN/m}^3)(3.0 \text{ m}) = 26.5 \text{ kN/m}^2 = 26.5 \text{ kPa}$$

Now, the pressure at B is

$$p_B = p_A + \Delta p_{A-B} = 0 \text{ Pa(gage)} + 26.5 \text{ kPa} = 26.5 \text{ kPa(gage)}$$

*Point C:*   The change in elevation from point A to point C is 6.0 m, with C lower than A. Then, the pressure at point C is

$$\Delta p_{A-C} = \gamma h = (8.83 \text{ kN/m}^3)(6.0 \text{ m}) = 53.0 \text{ kN/m}^2 = 53.0 \text{ kPa}$$
$$p_C = p_A + \Delta p_{A-C} = 0 \text{ Pa(gage)} + 53.0 \text{ kPa} = 53.0 \text{ kPa(gage)}$$

*Point D:*   Since point D is at the same level as point B, the pressure is the same. That is, we have

$$p_D = p_B = 26.5 \text{ kPa(gage)}$$

*Point E:*   Since point E is at the same level as point A, the pressure is the same. That is, we have

$$p_E = p_A = 0 \text{ Pa(gage)}$$

*Point F:*   The change in elevation between point A and point F is 1.5 m, with F higher than A. Then, the pressure at F is

$$\Delta p_{A-F} = -\gamma h = (-8.83 \text{ kN/m}^3)(1.5 \text{ m}) = -13.2 \text{ kN/m}^2 = -13.2 \text{ kPa}$$
$$p_F = p_A + \Delta p_{A-F} = 0 \text{ Pa(gage)} + (-13.2 \text{ kPa}) = -13.2 \text{ kPa}$$

*Air pressure:*   Since the air in the right side of the tank is exposed to the surface of the oil where $p_F = -13.2$ kPa, the air pressure is also −13.2 kPa, or 13.2 kPa below atmospheric pressure.

■

**3.4.1
Summary of Observations
from Example Problem 3.7**

The results from Problem 3.7 illustrate the general conclusions listed below Eq. (3–3) on page 49.

a. The pressure increases as the depth in the fluid increases. This result can be seen from $p_C > p_B > p_A$.
b. Pressure varies linearly with a change in elevation; that is, $p_C$ is two times greater than $p_B$, and C is at twice the depth of B.
c. Pressure on the same horizontal level is the same. Note that $p_E = p_A$ and $p_D = p_B$.
d. The decrease in pressure from E to F occurs because point F is at a higher elevation than point E. Note that $p_F$ is negative; that is, it is below the atmospheric pressure that exists at A and E.

**3.5**
**DEVELOPMENT OF THE**
**PRESSURE-ELEVATION**
**RELATION**

The relationship between a change in elevation in a liquid, $h$, and a change in pressure, $\Delta p$, is

$$\Delta p = \gamma h \tag{3–3}$$

where $\gamma$ is the specific weight of the liquid. This section presents the basis for this equation.

Figure 3.4 shows a body of static fluid with a specific weight $\gamma$. Consider a small volume of the fluid somewhere below the surface. In Fig. 3.4, the small volume is illustrated as a cylinder, but the actual shape is arbitrary.

**FIGURE 3.4**   Small volume of fluid within a body of static fluid.

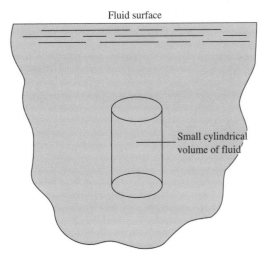

Because the entire body of fluid is stationary and in equilibrium, the small cylinder of the fluid is also in equilibrium. From physics, we know that for a body in static equilibrium, the sum of forces acting on it in all directions must be zero.

First consider the forces acting in the horizontal direction. A thin ring around the cylinder is shown at some arbitrary elevation in Fig. 3.5. The vectors acting on

**FIGURE 3.5**   Pressure forces acting in a horizontal plane on a thin ring.

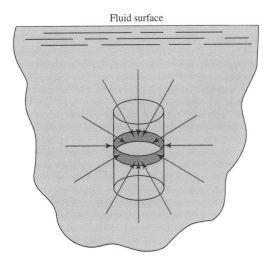

the ring represent the forces exerted on it by the fluid pressure. Recall from previous work that the pressure at any horizonal level in a static fluid is the same. Also recall that the pressure at a boundary, and therefore the force due to the pressure, acts perpendicular to the boundary. Therefore, the forces are completely balanced around the sides of the cylinder.

Now consider Fig. 3.6. The forces acting on the cylinder in the vertical direction are shown. The following concepts are illustrated in the figure:

1. The fluid pressure at the level of the bottom of the cylinder is called $p_1$.
2. The fluid pressure at the level of the top of the cylinder is called $p_2$.
3. The elevation difference between the top and the bottom of the cylinder is called $dz$, where $dz = z_2 - z_1$.
4. The pressure change that occurs in the fluid between the level of the bottom and the top of the cylinder is called $dp$. Therefore, $p_2 = p_1 + dp$.
5. The area of the top and bottom of the cylinder is called $A$.
6. The volume of the cylinder is the product of the area, $A$, and the height of the cylinder, $dz$. That is, $V = A(dz)$.
7. The weight of the fluid within the cylinder is the product of the specific weight of the fluid, $\gamma$, and the volume of the cylinder. That is, $w = \gamma V = \gamma A(dz)$. The weight is a force acting on the cylinder in the downward direction through the centroid of the cylindrical volume.
8. The force acting on the bottom of the cylinder due to the fluid pressure $p_1$ is the product of the pressure and the area, $A$. That is, $F_1 = p_1 A$. This force acts vertically upward, perpendicular to the bottom of the cylinder.
9. The force acting on the top of the cylinder due to the fluid pressure $p_2$ is the product of the pressure and the area, $A$. That is, $F_2 = p_2 A$. This force acts vertically downward, perpendicular to the top of the cylinder. Because $p_2 = p_1 + dp$, another expression for the force $F_2$ is

$$F_2 = (p_1 + dp)A \qquad\qquad (3\text{--}4)$$

**FIGURE 3.6**   Forces acting in the vertical direction.

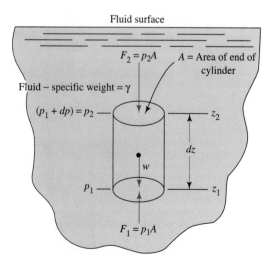

Now we can apply the principle of static equilibrium, which states that the sum of the forces in the vertical direction must be zero. Using upward forces as positive, we get

$$\sum F_v = 0 = F_1 - F_2 - w \tag{3–5}$$

Substituting from Steps 7, 8, and 9 gives

$$p_1 A - (p_1 + dp)A - \gamma(dz)A = 0 \tag{3–6}$$

Notice that the area, $A$, appears in all terms on the left side of Eq. (3–6). It can be eliminated by dividing all terms by $A$. The resulting equation is

$$p_1 - p_1 - dp - \gamma(dz) = 0 \tag{3–7}$$

Now the $p_1$ term can be cancelled out. Solving for $dp$ gives

$$dp = -\gamma(dz) \tag{3–8}$$

Equation (3–8) is the controlling relationship between a change in elevation and a change in pressure. The use of Eq. (3–8), however, depends on the kind of fluid. Remember that the equation was developed for a very small element of the fluid. The process of integration extends Eq. (3–8) to large changes in elevation, as indicated below:

$$\int_{p_1}^{p_2} dp = \int_{z_1}^{z_2} -\gamma(dz) \tag{3–9}$$

To complete the analysis, we must define how the specific weight of the fluid varies with a change in pressure. Eq. (3–9) is developed differently for liquids and for gases.

**3.5.1 Liquids**    A liquid is considered to be incompressible. Thus, its specific weight, $\gamma$, is a constant. This allows $\gamma$ to be taken outside the integral sign in Eq. (3–9). Then,

$$\int_{p_1}^{p_2} dp = -\gamma \int_{z_1}^{z_2} (dz) \tag{3–10}$$

Completing the integration process and applying the limits gives

$$p_2 - p_1 = -\gamma(z_2 - z_1) \tag{3–11}$$

For convenience, we define $\Delta p = p_2 - p_1$ and $h = z_1 - z_2$. Equation (3–11), then, becomes

$$\Delta p = \gamma h$$

which is identical to Eq. (3–3). The signs for $\Delta p$ and $h$ can be assigned at the time of use of the formula by recalling that pressure increases as depth in the fluid increases and vice versa.

**3.5.2 Gases**    Because a gas is compressible, its specific weight changes as pressure changes. To complete the integration process called for in Eq. (3–9), you must know the relationship between the change in pressure and the change in specific weight. The relationship is different for different gases, but a complete discussion of those relationships is beyond the scope of this text and requires the study of thermodynamics.

Appendix E describes the properties of air in the *standard atmosphere* as defined by the U.S. National Oceanic and Atmospheric Administration (NOAA).

**3.6**
**PASCAL'S PARADOX**

In the development of the relationship $\Delta p = \gamma h$, the size of the small volume of fluid does not affect the result. The change in pressure depends only on the change in elevation and the type of fluid, not on the size of the fluid container. Therefore, all the containers shown in Fig. 3.7 would have the same pressure at the bottom, even though they contain vastly different amounts of fluid. This observation is called *Pascal's paradox*.

**FIGURE 3.7**   Illustration of Pascal's paradox.

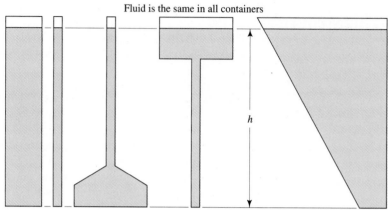

Fluid is the same in all containers

$h$

Pressure is the same at the bottom of all containers

This phenomenon is useful when a consistently high pressure must be produced on a system of interconnected pipes and tanks. Water systems for cities often include water towers located on high hills, as shown in Fig. 3.8. Besides providing a reserve supply of water, the primary purpose of such tanks is to maintain a sufficiently high pressure in the water system for satisfactory delivery of the water to residential, commercial, and industrial users.

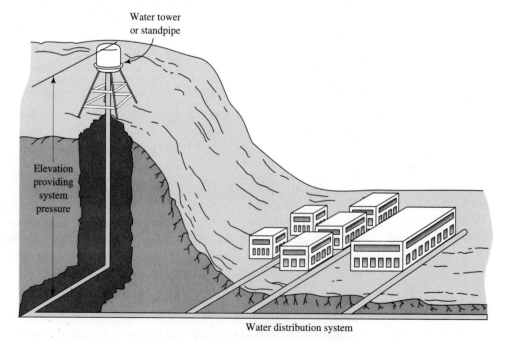

Water tower or standpipe

Elevation providing system pressure

Water distribution system

**FIGURE 3.8**   Use of a water tower or standpipe to maintain water system pressure.

In industrial or laboratory applications, a standpipe containing a static liquid can be used to create a stable pressure on a particular process or system. It is positioned at a high elevation relative to the system and is connected to the system by pipes. Raising or lowering the level of the fluid in the standpipe changes the pressure in the system. Standpipes are sometimes placed on the roofs of buildings to maintain water pressure in local firefighting systems.

**3.7
MANOMETERS**

This and following sections describe several types of pressure measurement devices. The first is the *manometer*, which uses the relationship between a change in pressure and a change in elevation in a static fluid, $\Delta p = \gamma h$ (see Sections 3.4 and 3.5). Photographs of commercially available manometers are shown in Figs. 3.9, 3.12, and 3.13.

The simplest kind of manometer is the U-tube (Fig. 3.9). One end of the U-tube is connected to the pressure that is to be measured, while the other end is left open to the atmosphere. The tube contains a liquid called the *gage fluid*, which does not mix with the fluid whose pressure is to be measured. Typical gage fluids are water, mercury, and colored light oils.

**FIGURE 3.9**   U-tube manometer.
(Source of photo: Dwyer
Instruments, Inc. Michigan City, IN)

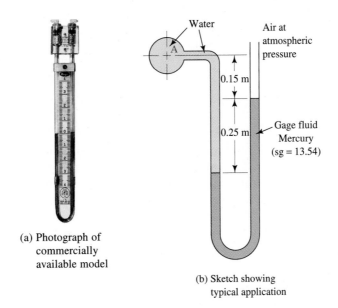

(a) Photograph of
    commercially
    available model

(b) Sketch showing
    typical application

Under the action of the pressure to be measured, the gage fluid is displaced from its normal position. Since the fluids in the manometer are at rest, the equation $\Delta p = \gamma h$ can be used to write expressions for the changes in pressure that occur throughout the manometer. These expressions can then be combined and solved algebraically for the desired pressure. Because manometers are used in many real situations such as those described in this book, you should learn the following step-by-step procedure:

**PROCEDURE FOR WRITING THE EQUATION FOR A MANOMETER**

**1.** Start from a convenient point and write this pressure in symbol form (e.g., $p_A$ refers to the pressure at point A). Usually this point will be at one of the end-points of the manometer system. If the manometer is of the open type it is convenient to start at the open end.

2. Using $\Delta p = \gamma h$, write expressions for the changes in pressure that occur from the starting point to the point at the opposite end of the manometer system, being careful to include the correct algebraic sign for each term.
3. Equate the expression from Step 2 to the pressure at the final point.
4. Substitute known values and solve for the desired pressure.

Working several practice problems will help you to apply this procedure correctly. The following problems are written in the programmed instruction format. To work through the program, cover the material below the heading "Programmed Example Problems" and then uncover one panel at a time.

## PROGRAMMED EXAMPLE PROBLEMS

☐ **EXAMPLE PROBLEM 3.8**    Using Fig. 3.9, calculate the pressure at point A. Perform Step 1 of the procedure before going to the next panel.

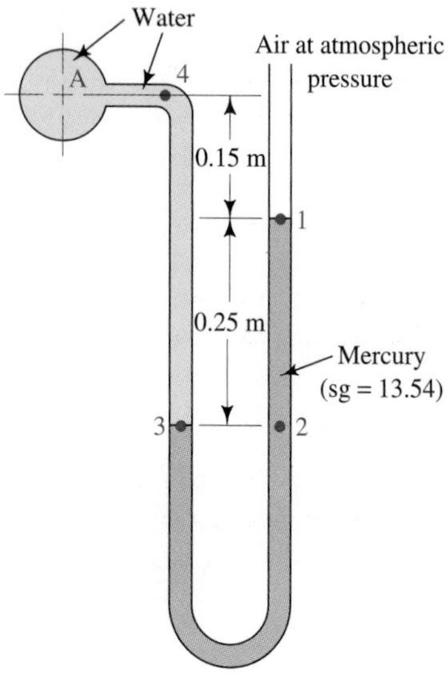

**FIGURE 3.10**   U-tube manometer.

Figure 3.10 is identical to Fig. 3.9(b) except that certain key points have been numbered for use in the problem solution.

The only point for which the pressure is known is the surface of the mercury in the right leg of the manometer, point 1. Now, how can an expression be written for the pressure that exists within the mercury, 0.25 m below this surface at point 2?

The expression is

$$p_1 + \gamma_m(0.25 \text{ m})$$

The term $\gamma_m(0.25 \text{ m})$ is the change in pressure between points 1 and 2 due to a change in elevation, where $\gamma_m$ is the specific weight of mercury, the gage fluid. This pressure change is added to $p_1$ because there is an increase in pressure as we descend in a fluid.

So far we have an expression for the pressure at point 2 in the right leg of the manometer. Now write the expression for the pressure at point 3 in the left leg.

This is the expression:

$$p_1 + \gamma_m(0.25 \text{ m})$$

Because points 2 and 3 are on the same level in the same fluid at rest, their pressures are equal.

Continue and write the expression for the pressure at point 4.

$$p_1 + \gamma_m(0.25 \text{ m}) - \gamma_w(0.40 \text{ m})$$

where $\gamma_w$ is the specific weight of water. Remember, there is a decrease in pressure between points 3 and 4, so this last term must be subtracted from our previous expression.

What must you do to get an expression for the pressure at point A?

Nothing. Because points A and 4 are on the same level, their pressures are equal. Now perform Step 3 of the procedure.

You should now have

$$p_1 + \gamma_m(0.25 \text{ m}) - \gamma_w(0.40 \text{ m}) = p_A$$

or

$$p_A = p_1 + \gamma_m(0.25 \text{ m}) - \gamma_w(0.40 \text{ m})$$

Be sure to write the complete equation for the pressure at point A. Now do Step 4.

---

Several calculations are required here:

$$p_1 = p_{\text{atm}} = 0 \text{ Pa(gage)}$$
$$\gamma_m = (\text{sg})_m(9.81 \text{ kN/m}^3) = (13.54)(9.81 \text{ kN/m}^3) = 132.8 \text{ kN/m}^3$$
$$\gamma_w = 9.81 \text{ kN/m}^3$$

Then, we have

$$p_A = p_1 + \gamma_m(0.25 \text{ m}) - \gamma_w(0.40 \text{ m})$$
$$= 0 \text{ Pa(gage)} + (132.8 \text{ kN/m}^3)(0.25 \text{ m}) - (9.81 \text{ kN/m}^3)(0.40 \text{ m})$$
$$= 0 \text{ Pa(gage)} + 33.20 \text{ kN/m}^2 - 3.92 \text{ kN/m}^2$$
$$p_A = 29.28 \text{ kN/m}^2 = 29.28 \text{ kPa(gage)}$$

Remember to include the units in your calculations. Review this problem to be sure you understand every step before going to the next panel for another problem.

---

☐ **EXAMPLE PROBLEM 3.9**

Calculate the difference in pressure between points A and B in Fig. 3.11 and express it as $p_B - p_A$.

This type of manometer is called a *differential manometer* because it indicates the difference between the pressure at two points but not the actual value of either one. Do Step 1 of the procedure to write the equation for the manometer.

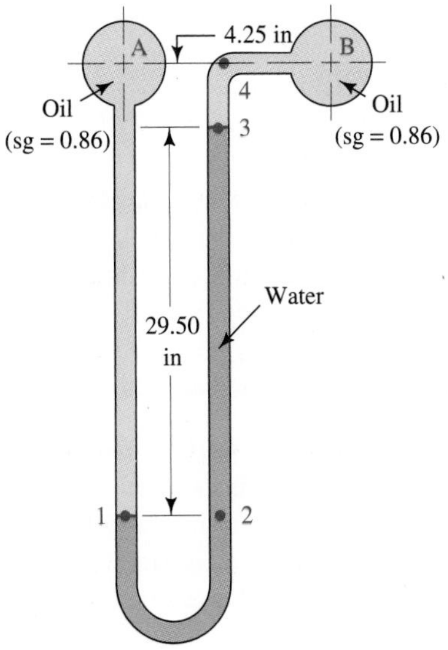

---

We could start either at point A or point B. Let's start at A and call the pressure there $p_A$. Now write the expression for the pressure at point 1 in the left leg of the manometer.

---

You should have
$$p_A + \gamma_o(33.75 \text{ in})$$

where $\gamma_o$ is the specific weight of the oil.

What is the pressure at point 2?

---

It is the same as the pressure at point 1 because the two points are on the same level. Go on to point 3 in the manometer.

---

The expression should now look like this:
$$p_A + \gamma_o(33.75 \text{ in}) - \gamma_w(29.5 \text{ in})$$

Now write the expression for the pressure at point 4.

---

This is the desired expression:
$$p_A + \gamma_o(33.75 \text{ in}) - \gamma_w(29.5 \text{ in}) - \gamma_o(4.25 \text{ in})$$

**FIGURE 3.11** Differential manometer.

This is also the expression for the pressure at B since points 4 and B are on the same level. Now do Steps 3 and 4 of the procedure.

---

Our final expression should be the complete manometer equation,

$$p_A + \gamma_o(33.75 \text{ in}) - \gamma_w(29.5 \text{ in}) - \gamma_o(4.25 \text{ in}) = p_B$$

or, solving for the differential pressure $p_B - p_A$,

$$p_B - p_A = \gamma_o(33.75 \text{ in}) - \gamma_w(29.5 \text{ in}) - y_o(4.25 \text{ in})$$

The known values are

$$\gamma_o = (sg)_o(62.4 \text{ lb/ft}^3) = (0.86)(62.4 \text{ lb/ft}^3) = 53.7 \text{ lb/ft}^3$$
$$\gamma_w = 62.4 \text{ lb/ft}^3$$

In this case it may help to simplify the expression before substituting known values. Since two terms are multiplied by $\gamma_o$ they can be combined as follows:

$$p_B - p_A = \gamma_o(29.5 \text{ in}) - \gamma_w(29.5 \text{ in})$$

Factoring out the common term gives

$$p_B - p_A = (29.5 \text{ in})(\gamma_o - \gamma_w)$$

This looks simpler than the original equation. The difference between $p_B$ and $p_A$ is a function of the *difference* between the specific weights of the two fluids.

The pressure at B, then, is

$$p_B - p_A = (29.5 \text{ in})(53.7 - 62.4)\frac{\text{lb}}{\text{ft}^3} \times \frac{1 \text{ ft}^3}{1728 \text{ in}^3}$$
$$= \frac{(29.5)(-8.7)\text{lb/in}^2}{1728}$$
$$p_B - p_A = -0.15 \text{ lb/in}^2$$

The negative sign indicates that the magnitude of $p_A > p_B$. Notice that using a gage fluid with a specific weight very close to that of the fluid in the system makes the manometer very sensitive. A large displacement of the column of gage fluid is caused by a small differential pressure and this allows a very accurate reading.

■

---

Figure 3.12 shows another type of manometer called the *well-type*. When a pressure is applied to a well-type manometer, the fluid level in the well drops a small amount, while the level in the right leg rises a larger amount in proportion to the ratio of the areas of the well and the tube. A scale is placed alongside the tube so that the deflection can be read directly. The scale is calibrated to account for the small drop in the well level.

The *inclined well-type manometer,* shown in Fig. 3.13, has the same features as the well-type but offers a greater sensitivity by placing the scale along the inclined tube. The scale length is increased as a function of the angle of inclination of the tube, $\theta$. For example, if the angle $\theta$ in Fig. 3.13(b) is 15°, the ratio of scale length $L$ to manometer deflection $h$ is

$$\frac{h}{L} = \sin \theta$$

**FIGURE 3.12** Well-type manometer. (Source of photo: Dwyer Instruments, Inc. Michigan City, IN)

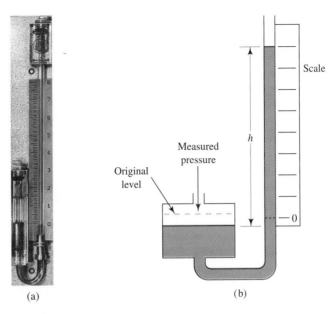

(a)

(b)

**FIGURE 3.13** Inclined well-type manometer. (Source of photo: Dwyer Instruments, Inc. Michigan City, IN)

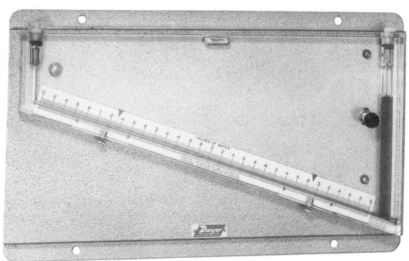

(a)

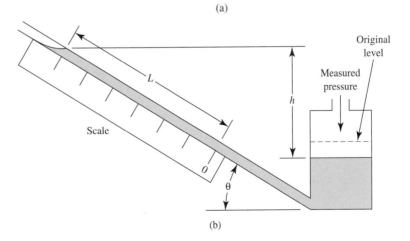

(b)

or

$$\frac{L}{h} = \frac{1}{\sin \theta} = \frac{1}{\sin 15°} = \frac{1}{0.259} = 3.86$$

The scale would be calibrated so that the deflection could be read directly.

### 3.8 BAROMETERS

A device for measuring the atmospheric pressure is called a *barometer*. A simple type is shown in Fig. 3.14. It consists of a long tube closed at one end which is initially filled completely with mercury. The open end is then submerged under the surface of a container of mercury and allowed to come to equilibrium, as shown in Fig. 3.14. A void is produced at the top of the tube which is very nearly a perfect vacuum, containing mercury vapor at a pressure of only 0.17 Pa at 20°C. By starting at this point and writing an equation similar to those for manometers, we get

$$0 + \gamma_m h = p_{atm}$$

or

$$p_{atm} = \gamma_m h \qquad (3-12)$$

Since the specific weight of mercury is approximately constant, a change in atmospheric pressure will cause a change in the height of the mercury column. This height is often reported as the barometric pressure. To obtain true atmospheric pressure it is necessary to multiply $h$ by $\gamma_m$.

Precision measurement of the atmospheric pressure with a mercury manometer requires that the specific weight of the mercury be adjusted for changes in temperature. But in this book, we will use the values given in Appendix B.

In SI units:

$$\gamma = 132.8 \text{ kN/m}^3$$

In U.S. Customary System units:

$$\gamma = 844.9 \text{ lb/ft}^3$$

The atmospheric pressure varies from time to time, as you hear on weather reports. The atmospheric pressure also varies with altitude. A decrease of approximately 1.0 in of mercury occurs per 1000 ft of increase in altitude. In SI units, the decrease is approximately 85 mm of mercury per 1000 m. See also Appendix E for variations in atmospheric pressure with altitude.

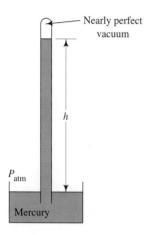

Nearly perfect vacuum

$h$

$P_{atm}$

Mercury

**FIGURE 3.14**    Barometer.

---

☐ **EXAMPLE PROBLEM 3.10**

A news broadcaster reports that the barometric pressure is 772 mm of mercury. Calculate the atmospheric pressure in kPa(abs).

*Solution*

In Eq. (3–12),

$$p_{atm} = \gamma_m h$$
$$\gamma_m = 132.8 \text{ kN/m}^3$$
$$h = 0.772 \text{ m}$$

Then we have

$$p_{atm} = (132.8 \text{ kN/m}^3)(0.772 \text{ m}) = 102.5 \text{ kN/m}^2 = 102.5 \text{ kPa(abs)}$$

---

☐ **EXAMPLE PROBLEM 3.11**    The standard atmospheric pressure is 101.3 kPa. Calculate the height of a mercury column equivalent to this pressure.

*Solution*    In Eq. (3–12),

$$p_{atm} = \gamma_m h$$
$$h = \frac{p_{atm}}{\gamma_m} = \frac{101.3 \times 10^3 \text{ N}}{\text{m}^2} \times \frac{\text{m}^3}{132.8 \times 10^3 \text{ N}} = 0.763 \text{ m} = 763 \text{ mm}$$

☐ **EXAMPLE PROBLEM 3.12**    A news broadcaster reports that the barometric pressure is 30.40 in of mercury. Calculate the pressure in psia.

*Solution*    In Eq. (3–12),

$$p_{atm} = \gamma_m h$$
$$\gamma_m = 844.9 \text{ lb/ft}^3$$
$$h = 30.40 \text{ in}$$

Then we have

$$p_{atm} = \frac{844.9 \text{ lb}}{\text{ft}^3} \times 30.40 \text{ in} \times \frac{1 \text{ ft}^3}{1728 \text{ in}^3} = 14.9 \text{ lb/in}^2$$
$$p_{atm} = 14.9 \text{ psia}$$

☐ **EXAMPLE PROBLEM 3.13**    The standard atmospheric pressure is 14.7 psia. Calculate the height of a mercury column equivalent to this pressure.

*Solution*    In Eq. (3–12),

$$p_{atm} = \gamma_m h$$
$$h = \frac{p_{atm}}{\gamma_m} = \frac{14.7 \text{ lb}}{\text{in}^2} \times \frac{\text{ft}^3}{844.9 \text{ lb}} \times \frac{1728 \text{ in}^3}{\text{ft}^3} = 30.06 \text{ in}$$

**3.9 PRESSURE GAGES AND TRANSDUCERS**

A widely used pressure measuring device is the *Bourdon tube pressure gage** (Fig. 3.15). The pressure to be measured is applied to the inside of a flattened tube which is normally shaped as a segment of a circle or a spiral. The increased pressure inside the tube causes it to be straightened somewhat. The movement of the end of the tube is transmitted through a linkage which causes a pointer to rotate.

The scale of the gage normally reads zero when the gage is open to atmospheric pressure and is calibrated in pascals (Pa) or other units of pressure above zero. Therefore, this type of gage reads *gage pressure* directly. Some gages are capable of reading pressures below atmospheric.

Figure 3.16 shows a pressure gage using an actuation means called *magnehelic*®.† The pointer is attached to a helix made from a material having a high magnetic permeability that is supported in sapphire bearings. A leaf spring is driven up and down by the motion of a flexible diaphragm, not shown in the figure. At the

* Note that two spellings, *gage* and *gauge*, are often used interchangeably.

† Magnehelic® is a registered trade name of Dwyer Instruments, Inc., Michigan City, Indiana.

**FIGURE 3.15** Bourdon tube pressure gage. (Source of photo: Ametek/U.S. Gauge, Sellersville, PA)

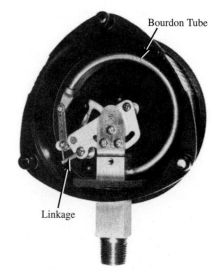

(a) Front view        (b) Rear view with back of case removed

**FIGURE 3.16** Magnehelic® pressure gage. (Source: Dwyer Instruments, Inc., Michigan City, IN)

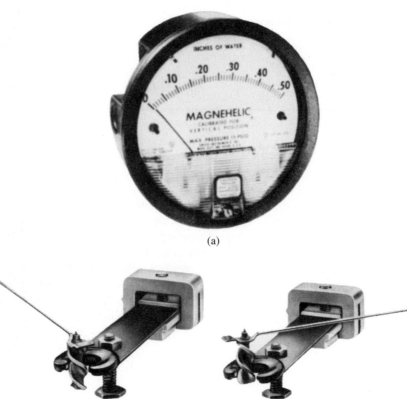

(a)

(b)          (c)

end of the spring, the C-shaped element contains a strong magnet placed in close proximity to the outer surface of the helix. As the leaf spring moves up and down, the helix rotates to follow the magnet, moving the pointer. Note that there is no physical contact between the magnet and the helix. Calibration of the gage is accomplished by adjusting the length of the spring at its clamped end.

**3.10**
**PRESSURE**
**TRANSDUCERS**

A *transducer* is an instrument that measures some physical quantity and generates an electrical signal that has a predictable relationship to the measured quantity. The level of the electrical signal then can be recorded, displayed on a meter, or stored in a computer memory for later display or analysis. This section describes several types of pressure transducers.

**3.10.1**
**Strain Gage Pressure**
**Transducer**

Figure 3.17 shows a *strain gage pressure transducer*. The pressure to be measured is introduced through the pressure port and acts on a diaphragm to which foil strain gages are bonded. As the strain gages sense the deformation of the diaphragm, their resistance changes. By passing an electrical current through the gages and connecting them into a network, called a *Wheatstone bridge*, a change in electrical voltage is produced. The readout device is typically a digital voltmeter, calibrated in pressure units.

**FIGURE 3.17** Strain gage pressure transducer and indicator. (Source of photos: Sensotec, Inc. Columbus, OH)

(a) Strain gage pressure transducer

(b) Digital electronic amplifier/indicator

### 3.10.2
### LVDT-Type Pressure
### Transducer

A *linear variable differential transformer* (LVDT) is composed of a cylindrical electric coil with a movable rod-shaped core. As the core moves along the axis of the coil, a voltage change occurs in relation to the position of the core. This type of transducer is applied to pressure measurement by attaching the core rod to a flexible diaphragm (Fig. 3.18). For gage pressure measurements, one side of the diaphragm is exposed to atmospheric pressure, while the other side is exposed to the pressure to be measured. Changes in pressure cause the diaphragm to move, thus moving the core of the LVDT. The resulting voltage change is recorded or indicated on a meter calibrated in pressure units. Differential pressure measurements can be made by introducing the two pressures to the opposite sides of the diaphragm.

**FIGURE 3.18**   LVDT-type pressure transducer. (Source of photo: Schaevitz Engineering. Pennsauken, NJ)

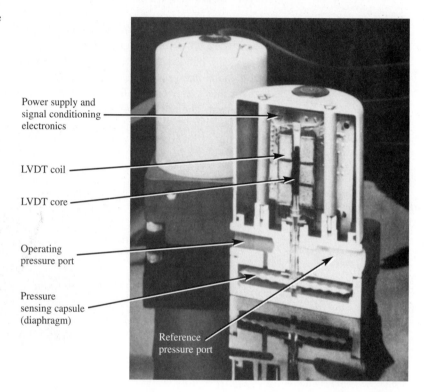

Power supply and
signal conditioning
electronics

LVDT coil

LVDT core

Operating
pressure port

Pressure
sensing capsule
(diaphragm)

Reference
pressure port

### 3.10.3
### Piezoelectric Pressure
### Transducers

Certain crystals, such as quartz and barium titanate, exhibit the *piezoelectric effect*, in which the electrical charge across the crystal varies with stress in the crystal by causing the pressure to exert a force, either directly or indirectly, on the crystal. A voltage change, which is related to the pressure change, is then produced.

Figure 3.19 shows a commercially available pressure gage that incorporates a piezoelectric pressure transducer. One can indicate pressure or vacuum in any of 18 different display units by simply pressing the *units* button. The gage also incorporates a calibrator signal that indicates a DC milliamp current reading for field calibration of remote transmitters. The *zero* key allows the setting of the reference pressure in the field.

### 3.10.4
### Quartz Resonator Pressure
### Transducers

A quartz crystal resonates at a frequency that is dependent on the stress in the crystal. Under increasing tension, the resonant frequency increases. Conversely, the resonant frequency decreases with compression. The changes in frequency can be meas-

**FIGURE 3.19**   Digital pressure gage. (Source: Rochester Instrument Systems, Rochester, NY)

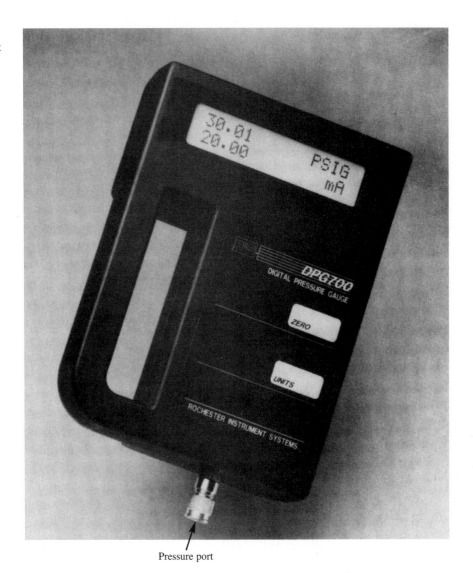

Pressure port

ured very precisely by digital electronic systems. Pressure transducers can use this phenomenon by linking bellows, diaphragms, or Bourdon tubes to quartz crystals. Such devices can provide pressure measurement accuracy of 0.01 percent or better.

**3.10.5**
**Solid State Pressure Sensors**

Solid state technology permits very small pressure sensors to be made from silicon. Thin film silicon resistors can be used instead of foil strain gages for a Wheatstone bridge-type system. Another type uses two parallel plates whose surfaces are composed of an etched pattern of silicon. Pressure applied to one plate causes it to deflect, changing the air gap between the plates. The resulting change in capacitance can be detected with an oscillator circuit.

**3.11**
**PRESSURE EXPRESSED**
**AS THE HEIGHT OF A**
**COLUMN OF LIQUID**

When measuring pressures in some fluid flow systems, such as air flow in heating ducts, the actual magnitude of the pressure reading is often small. Manometers are sometimes used to measure these pressures and their readings are given in units such as *"inches of water"* rather than the conventional units of psi or Pa.

3.14
3.35
3.44
3.49
3.52
3.56
3.62
3.63
3.67
3.68

To convert from such units to those needed in calculations, the pressure-elevation relationship must be used. For example, a pressure of 1.0 in of water (1.0 in $H_2O$) expressed in psi units is,

$$p = \gamma h$$

$$p = \frac{62.4 \text{ lb}}{\text{ft}^3} (1.0 \text{ in } H_2O) \frac{1 \text{ ft}^3}{1728 \text{ in}^3} = 0.0361 \text{ lb/in}^2 = 0.0361 \text{ psi}$$

Then we can use this as a conversion factor,

$$1.0 \text{ in of water} = 0.0361 \text{ psi}$$

Converting this to Pa gives

$$1.0 \text{ in of water} = 249.0 \text{ Pa}$$

Similarly, somewhat higher pressures are measured with a mercury manometer. Using $\gamma = 132.8$ kN/m$^3$ or $\gamma = 844.9$ lb/ft$^3$, we can develop the conversion factors,

$$1.0 \text{ in of mercury} = 0.489 \text{ psi}$$
$$1.0 \text{ mm of mercury} = 0.01926 \text{ psi}$$
$$1.0 \text{ mm of mercury} = 132.8 \text{ Pa}$$

You must remember that temperature of the gage fluid can affect its specific weight and therefore the accuracy of these factors.

---

## REFERENCES

1. Avallone, Eugene A., and Theodore Baumeister III, eds. 1996. *Marks' Standard Handbook for Mechanical Engineers.* 10th ed. New York: McGraw-Hill.
2. Busse, Donald W. 1987 (March). Quartz Transducers for Precision Under Pressure. *Mechanical Engineering Magazine* 109(5):52–56.
3. Walters, Sam. 1987 (March). Inside Pressure Measurement. *Mechanical Engineering Magazine* 109(5):41–47.
4. Worden, Roy D. 1987 (March). Designing a Fused-Quartz Pressure Transducer. *Mechanical Engineering Magazine* 109(5):48–51.

---

## PRACTICE PROBLEMS

### Absolute and Gage Pressure

**3.1** Write the expression for computing the pressure in a fluid.

**3.2** Define *absolute pressure.*

**3.3** Define *gage pressure.*

**3.4** Define *atmospheric pressure.*

**3.5** Write the expression relating gage pressure, absolute pressure, and atmospheric pressure.

State whether statements 3.6 through 3.10 are (or can be) true or false. For those that are false, tell why.

**3.6** The value for the absolute pressure will always be greater than that for the gage pressure.

**3.7E** As long as you stay on the surface of the earth, the atmospheric pressure will be 14.7 psia.

**3.8M** *The pressure in a certain tank is −55.8 Pa(abs).*

**3.9E** The pressure in a certain tank is −4.65 psig.

**3.10M** *The pressure in a certain tank is −150 kPa(gage).*

**3.11E** If you were to ride in an open-cockpit airplane to an elevation of 4000 ft above sea level, what would the atmospheric pressure be if it conforms to the standard atmosphere?

**3.12E** The peak of a certain mountain is 13 500 ft above sea level. What is the approximate atmospheric pressure?

**3.13** Expressed as a gage pressure, what is the pressure at the surface of a glass of milk?

Problems 3.14 to 3.33 require that you convert the given pressure from gage to absolute pressure or from absolute to gage pressure as indicated. The value of the atmospheric pressure is given.

| Problem | Given Pressure | | $p_{atm}$ | Express Result As: |
|---------|------|-----|-----------|--------------------|
| 3.14M | 583 | kPa(abs) | 103 kPa(abs) | Gage pressure |
| 3.15M | 157 | kPa(abs) | 101 kPa(abs) | Gage pressure |
| 3.16M | 30 | kPa(abs) | 100 kPa(abs) | Gage pressure |
| 3.17M | 74 | kPa(abs) | 97 kPa(abs) | Gage pressure |
| 3.18M | 101 | kPa(abs) | 104 kPa(abs) | Gage pressure |
| 3.19M | 284 | kPa(gage) | 100 kPa(abs) | Absolute pressure |
| 3.20M | 128 | kPa(gage) | 98.0 kPa(abs) | Absolute pressure |
| 3.21M | 4.1 | kPa(gage) | 101.3 kPa(abs) | Absolute pressure |
| 3.22M | −29.6 | kPa(gage) | 101.3 kPa(abs) | Absolute pressure |
| 3.23M | −86.0 | kPa(gage) | 99.0 kPa(abs) | Absolute pressure |
| 3.24E | 84.5 psia | | 14.9 psia | Gage pressure |
| 3.25E | 22.8 psia | | 14.7 psia | Gage pressure |
| 3.26E | 4.3 psia | | 14.6 psia | Gage pressure |
| 3.27E | 10.8 psia | | 14.0 psia | Gage pressure |
| 3.28E | 14.7 psia | | 15.1 psia | Gage pressure |
| 3.29E | 41.2 psig | | 14.5 psia | Absolute pressure |
| 3.30E | 18.5 psig | | 14.2 psia | Absolute pressure |
| 3.31E | 0.6 psig | | 14.7 psia | Absolute pressure |
| 3.32E | −4.3 psig | | 14.7 psia | Absolute pressure |
| 3.33E | −12.5 psig | | 14.4 psia | Absolute pressure |

## Pressure-Elevation Relationship

**3.34M** *If milk has a specific gravity of 1.08, what is the pressure at the bottom of a milk can 550 mm deep?*

**3.35E** The pressure in an unknown fluid at a depth of 4.0 ft is measured to be 1.820 psig. Compute the specific gravity of the fluid.

**3.36M** *The pressure at the bottom of a tank of propyl alcohol at 25°C must be maintained at 52.75 kPa(gage). What depth of alcohol should be maintained?*

**3.37E** When you dive to a depth of 12.50 ft in seawater, what is the pressure?

**3.38E** A water storage tank is on the roof of a factory building, and the surface of the water is 50.0 ft above the floor of the factory. If a pipe connects the storage tank to the floor level and the pipe is full of static water, what is the pressure in the pipe at floor level?

**3.39M** *An open tank contains ethylene glycol at 25°C. Compute the pressure at a depth of 3.0 m.*

**3.40M** *For the tank of ethylene glycol described in Problem 3.39, compute the pressure at a depth of 12.0 m.*

**3.41E** Figure 3.20 shows a diagram of the hydraulic system for a vehicle lift. An air compressor maintains pressure above

**FIGURE 3.20** Vehicle lift for Problem 3.41.

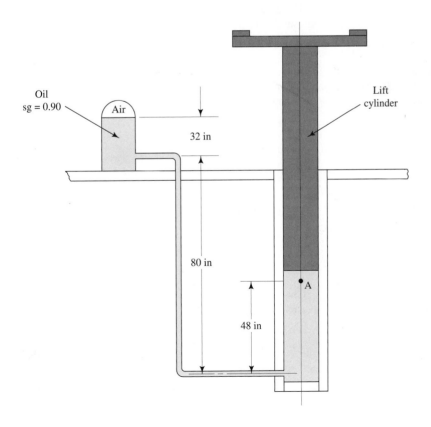

**FIGURE 3.21**   Washing machine
for Problem 3.42.

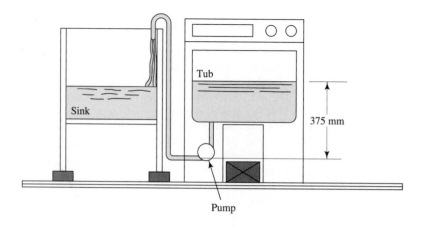

the oil in the reservoir. What must the air pressure be if
the pressure in point A must be at least 180 psig?

**3.42E**  Figure 3.21 shows a clothes washing machine. The pump
draws fluid from the tub and delivers it to the disposal
sink. Compute the pressure at the inlet to the pump when
the water is static (no flow). The soapy water solution has
a specific gravity of 1.15.

**3.43M**  *An airplane is flying at 10.6 km altitude. In its nonpres-
surized cargo bay is a container of mercury 325 mm deep.
The container is vented to the local atmosphere. What is
the absolute pressure at the surface of the mercury and
at the bottom of the container? Assume the conditions of
the standard atmosphere prevail for pressure. Use
sg = 13.54 for the mercury.*

**3.44E**  For the tank shown in Fig. 3.22, determine the reading
of the bottom pressure gage in psig if the top of the tank
is vented to the atmosphere and the depth of the oil, $h$, is
28.50 ft.

**3.45E**  For the tank shown in Fig. 3.22, determine the reading
of the bottom pressure gage in psig if the top of the tank

is sealed, the top gage reads 50.0 psig, and the depth of
the oil, $h$, is 28.50 ft.

**3.46E**  For the tank shown in Fig. 3.22, determine the reading
of the bottom pressure gage in psig if the top of the tank
is sealed, the top gage reads −10.8 psig, and the depth
of the oil, $h$, is 6.25 ft.

**3.47E**  For the tank shown in Fig. 3.22, determine the depth of
the oil, $h$, if the reading of the bottom pressure gage is
35.5 psig, the top of the tank is sealed, and the top gage
reads 30.0 psig.

**3.48M**  *For the tank in Fig. 3.23, compute the depth of the oil if
the depth of the water is 2.80 m and the gage at the bottom of the tank reads 52.3 kPa(gage).*

**3.49M**  *For the tank in Fig. 3.23, compute the depth of the water if the depth of the oil is 6.90 m and the gage at the
bottom of the tank reads 125.3 kPa(gage).*

**3.50M**  *Figure 3.23 represents an oil storage drum which is open
to the atmosphere at the top. Some water was acciden-
tally pumped into the tank and settled to the bottom as
shown in the figure. Calculate the depth of the water $h_2$*

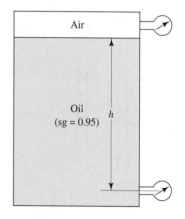

**FIGURE 3.22**   Problems 3.44 to 3.47.

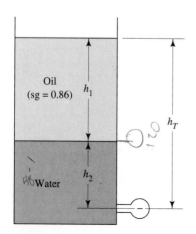

**FIGURE 3.23**   Problems 3.48 to 3.50.

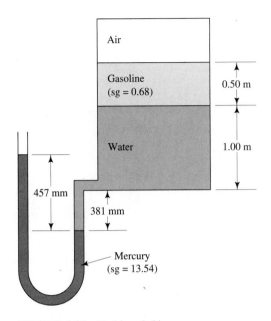

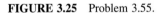

FIGURE 3.24    Problem 3.54.

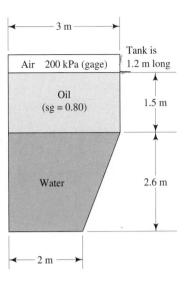

FIGURE 3.26    Problem 3.56.

*if the pressure gage at the bottom reads 158 kPa(gage). The total depth $h_T = 18.0$ m.*

**3.51M** *A storage tank for sulfuric acid is 1.5 m in diameter and 4.0 m high. If the acid has a specific gravity of 1.80, calculate the pressure at the bottom of the tank. The tank is open to the atmosphere at the top.*

**3.52E** *A storage drum for crude oil (sg = 0.89) is 32 ft deep and open at the top. Calculate the pressure at the bottom.*

**3.53M** *The greatest known depth in the ocean is approximately 11.0 km. Assuming that the specific weight of the water is constant at 10.0 kN/m³, calculate the pressure at this depth.*

**3.54M** *Figure 3.24 shows a closed tank that contains gasoline floating on water. Calculate the air pressure above the gasoline.*

**3.55M** *Figure 3.25 shows a closed container holding water and oil. Air at 34 kPa below atmospheric pressure is above the oil. Calculate the pressure at the bottom of the container in kPa(gage).*

**3.56M** *Determine the pressure at the bottom of the tank in Fig. 3.26.*

FIGURE 3.25    Problem 3.55.

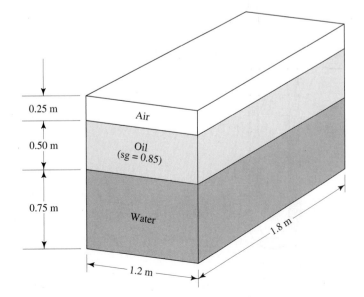

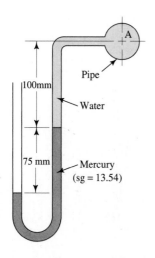

**FIGURE 3.27**   Problem 3.62.

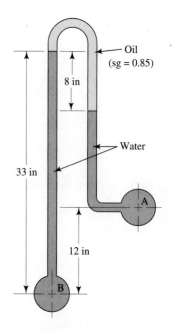

**FIGURE 3.29**   Problem 3.64.

## Manometers

**3.57** Describe a simple U-tube manometer.

**3.58** Describe a differential U-tube manometer.

**3.59** Describe a well-type manometer.

**3.60** Describe an inclined well-type manometer.

**3.61** Describe a compound manometer.

**3.62M** *Water is in the pipe shown in Fig. 3.27. Calculate the pressure at point A in kPa(gage).*

**3.63E** For the differential manometer shown in Fig. 3.28, calculate the pressure difference between points A and B. The specific gravity of the oil is 0.85.

**3.64E** For the manometer shown in Fig. 3.29, calculate $(p_A - p_B)$.

**3.65M** *For the manometer shown in Fig. 3.30, calculate* $(p_A - p_B)$.

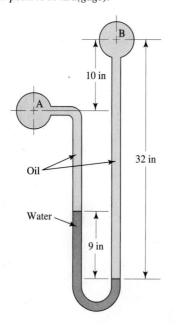

**FIGURE 3.28**   Problem 3.63.

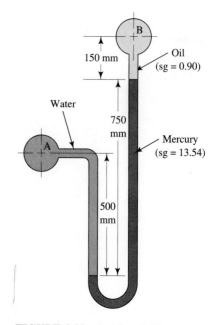

**FIGURE 3.30**   Problem 3.65.

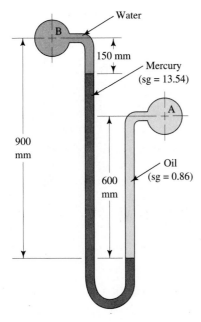

**FIGURE 3.31**    Problem 3.66.

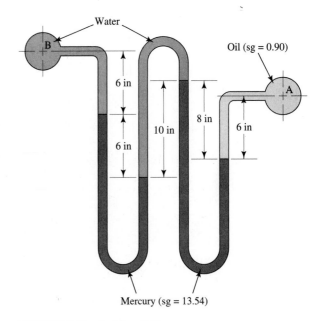

**FIGURE 3.33**    Problem 3.68.

**3.66M** *For the manometer shown in Fig. 3.31, calculate* $(p_A - p_B)$.

**3.67M** *For the compound manometer shown in Fig. 3.32, calculate the pressure at point A.*

**3.68E** For the compound differential manometer in Fig. 3.33, calculate $(p_A - p_B)$.

**3.69E** Figure 3.34 shows a manometer being used to indicate the difference in pressure between two points in a pipe. Calculate $(p_A - p_B)$.

**3.70E** For the well-type manometer in Fig. 3.35, calculate $p_A$.

**3.71M** *Figure 3.36 shows an inclined well-type manometer in which the distance L indicates the movement of the gage*

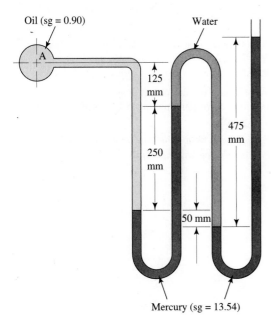

**FIGURE 3.32**    Problem 3.67.

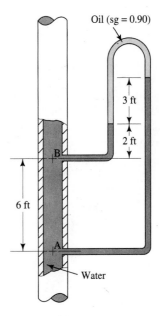

**FIGURE 3.34**    Problem 3.69.

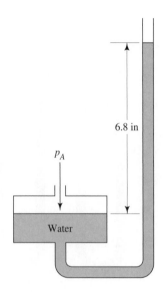

**FIGURE 3.35**  Problem 3.70.

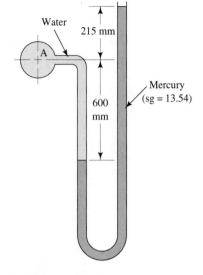

**FIGURE 3.37**  Problem 3.72.

*fluid level as the pressure $p_A$ is applied above the well. The gage fluid has a specific gravity of 0.87 and $L = 115$ mm. Neglecting the drop in fluid level in the well, calculate $p_A$.*

**3.72M** **a.** *Determine the gage pressure at point A in Fig. 3.37.*

      **b.** *If the barometric pressure is 737 mm of mercury, express the pressure at point A in kPa(abs).*

## Barometers

**3.73** What is the function of a barometer?

**3.74** Describe the construction of a barometer.

**3.75** Why is mercury a convenient fluid to use in a barometer?

**3.76** If water were to be used instead of mercury in a barometer, how high would the water column be?

**3.77E** What is the barometric pressure reading in inches of mercury corresponding to 14.7 psia?

**3.78M** *What is the barometric pressure reading in millimeters of mercury corresponding to 101.3 kPa(abs)?*

**3.79** Why must a barometric pressure reading be corrected for temperature?

**3.80E** By how much would the barometric pressure reading decrease from its sea-level value at an elevation of 1250 ft?

**3.81C** Denver, Colorado, is called the "Mile-High City" because it is located at an elevation of approximately 5200 ft. Assuming that the sea-level pressure is 101.3 kPa(abs), what would be the approximate atmospheric pressure in Denver?

**3.82E** The barometric pressure is reported to be 28.6 in of mercury. Calculate the atmospheric pressure in psia.

**3.83E** A barometer indicates the atmospheric pressure to be 30.65 in of mercury. Calculate the atmospheric pressure in psia.

**3.84E** What would be the reading of a barometer in inches of mercury corresponding to an atmospheric pressure of 14.2 psia?

**3.85M** *A barometer reads 745 mm of mercury. Calculate the barometric reading in kPa(abs).*

**FIGURE 3.36**  Problem 3.71.

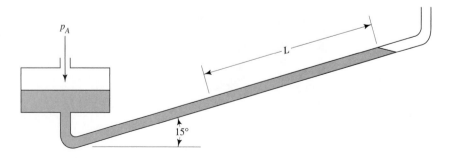

## Pressure Gages and Transducers

**3.86** Describe a Bourdon tube pressure gage.

**3.87** Describe a strain gage-type pressure transducer.

**3.88** Describe a quartz crystal pressure transducer that uses the piezoelectric effect.

**3.89** Describe a quartz crystal pressure transducer that uses the resonant frequency effect.

# 4 Forces Due to Static Fluids

## 4.1 The Big Picture

### Discussion Map

- We have defined pressure to be force divided by the area on which it acts. We can restate this principle to say that when a fluid is adjacent to a fixed surface, it exerts a force on the surface because of the pressure in the fluid.

- In some cases, the pressure is uniform over the surface of interest. In such cases, the following simple relationship applies:

  *Force = Pressure times Area*
  $$F = p \cdot A$$

- But there are many practical situations in which the pressure varies over the surface of interest. Other methods must be used to account for this variation in pressure before we can compute the magnitude of the resultant force on the surface.

### Discover

*Identify several examples of where the force exerted by a fluid on the surfaces that contain it may be of importance.*

*Discuss this with colleagues and try to describe how the force acts, how the pressure varies, if, indeed, it does vary at all, and the magnitude of the force. How is the design of the containing system affected by the forces created by the fluid pressure? What would be the consequence if the forces exceeded the ability of the system to withstand them? How would it fail?*

*Look at Figure 4.1 to see some examples that might spark your own thinking about cases you are familiar with.*

*Even such simple actions as swimming or bathing should give you a place to start. But think also of processes involved with cooking, your car, manufacturing systems and automation actuators, air and space vehicles, ships and submarines, pumps and compressors, dams and retaining walls, tanks storing liquids or gases, and many others.*

You should review the parts of Chapter 3 that deal with the variation of pressure with elevation in a static fluid. Take particular note of Equation 3–3 and the general conclusions related to it discussed in Section 3.3. These conclusions will help you visualize the variation of pressure on different types of areas along with the corresponding forces produced by those pressures.

Recall that the pressure everywhere on a horizontal plane in a static fluid is the same. This leads to the conclusion that the force on a horizontal surface due to that pressure can be computed from the simple product of pressure times the area of interest.

Also recall that the change in pressure with elevation in a static fluid is proportional to the specific weight of the fluid. Whenever we consider the force due to the pressure in a vessel containing a gas such as air, we can assume with reasonable accuracy, that the pressure is equal throughout the vessel. Therefore, for this special case, the force due to that pressure can be computed from the simple product of pressure times the area of interest.

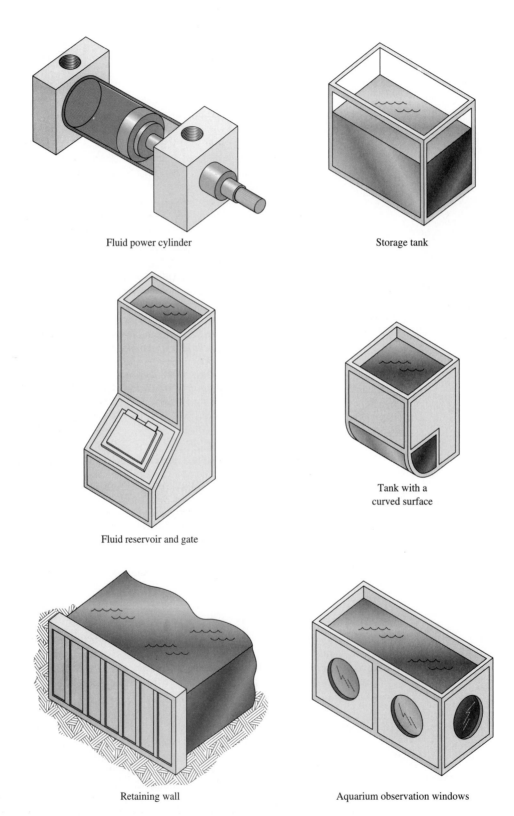

Fluid power cylinder

Storage tank

Fluid reservoir and gate

Tank with a
curved surface

Retaining wall

Aquarium observation windows

**FIGURE 4.1**  Examples of cases where forces on submerged areas must be computed.

You also learned about absolute and gage pressures in Chapter 3. Gage pressure is measured relative to the local atmospheric pressure. Then, if the pressure of interest is equal to the atmospheric pressure, we say the pressure is zero (gage) pressure. Often we are concerned with the force due to a pressure in the interior of a vessel as it acts on a window, a hatch, or some structural element of the vessel. Here we need to be concerned only with the gage pressure of the fluid because the atmospheric pressure acts on the outside of the surface of interest.

Sometimes we will consider areas that are rectangular in shape and are exposed to a liquid all the way from its free surface to some lower depth. Based on the pressure-elevation relationship we can say that the pressure varies linearly from zero at the top of the area to a maximum at the bottom. This allows a fairly simple computation of the resultant force on the area, as will be shown in the section dealing with *rectangular walls*.

At other times, the area of interest is completely submerged beneath the surface of the fluid. Consider an observation window in the side of an aquarium as an example. Here there is already a significant pressure on the top of the window and the pressure increases as the depth increases to the bottom of the window. Special techniques must be used to compute both the magnitude and location of the resultant force on the window.

The concept of *piezometric head* is also introduced in this chapter. This is the case when a vessel containing a liquid is closed at its top and the pressure above the fluid is either above or below atmospheric pressure. This causes the effective pressure on an area of interest in the vessel to be different from the case where the free surface of the fluid is exposed to the atmospheric. You will learn how to handle piezometric head in this chapter.

> This chapter will help you discover the principles governing the generation of forces due to fluids acting on plane (flat) and curved surfaces. Some of the solution procedures will be for special cases while others are more generally applicable.

**4.2**
**OBJECTIVES**

After completing this chapter, you should be able to:

1. Compute the force exerted on a plane area by a pressurized gas.
2. Compute the force exerted by any static fluid acting on a horizontal plane area.
3. Compute the resultant force exerted on a rectangular wall by a static liquid.
4. Define the term *center of pressure*.
5. Compute the resultant force exerted on any submerged plane area by a static liquid.
6. Show the vector representing the resultant force on any submerged plane area in its proper location and direction.
7. Visualize the distribution of force on a submerged curved surface.
8. Compute the total resultant force on the curved surface.
9. Compute the direction in which the resultant force acts and show its line of action on a sketch of the surface.
10. Include the effect of a pressure head over the liquid on the force on a plane or curved surface.

**4.3**
**GASES UNDER PRESSURE**

Figure 4.2 shows a pneumatic cylinder of the type used in automated machinery. The air pressure acts on the piston face, producing a force that causes the linear movement of the rod. The pressure also acts on the end of the cylinder, tending to pull it apart. This is the reason for the four tie rods between the end caps of the cylinder. The distribution of pressure within a gas is very nearly uniform. Therefore, we can calculate the force on the piston and the cylinder ends directly from $F = pA$.

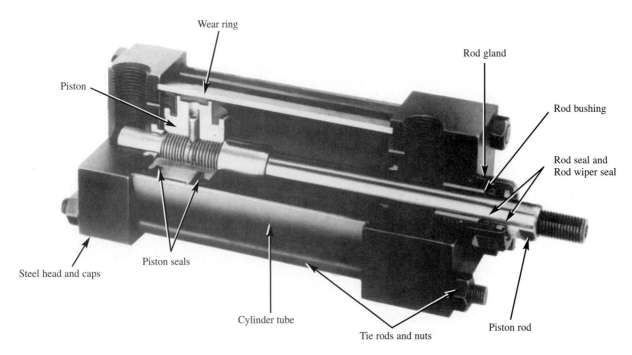

**FIGURE 4.2**   Fluid power cylinder. (Source of photo: Mosier Industries, Brookville, OH)

□ **EXAMPLE PROBLEM 4.1**   If the cylinder in Figure 4.2 has an internal diameter of 2 in and operates at a pressure of 300 psig, calculate the force on the ends of the cylinder.

**Solution**

$$F = pA$$

$$A = \frac{\pi D^2}{4} = \frac{\pi(2 \text{ in})^2}{4} = 3.14 \text{ in}^2$$

$$F = \frac{300 \text{ lb}}{\text{in}^2} \times 3.14 \text{ in}^2 = 942 \text{ lb}$$

Notice that gage pressure was used in the calculation of force instead of absolute pressure. The additional force due to atmospheric pressure acts on both sides of the area and is thus balanced. If the pressure on the outside surface is not atmospheric, then all external forces must be considered to determine a net force on the area. ■

**4.4**
**HORIZONTAL FLAT**
**SURFACES UNDER**
**LIQUIDS**

Figure 4.3 shows a cylindrical drum containing oil and water. The pressure in the water at the bottom of the drum is uniform across the entire area since it is a horizontal plane in a fluid at rest. Again, we can simply use $F = pA$ to calculate the force on the bottom.

□ **EXAMPLE PROBLEM 4.2**   If the drum in Fig. 4.3 is open to the atmosphere at the top, calculate the force on the bottom.

**Solution**  To use $F = pA$ we must first calculate the pressure at the bottom of the drum $p_B$ and the area of the bottom:

$$p_B = p_{atm} + \gamma_o(2.4 \text{ m}) + \gamma_w(1.5 \text{ m})$$

$$\gamma_o = (sg)_o(9.81 \text{ kN/m}^3) = (0.90)(9.81 \text{ kN/m}^3) = 8.83 \text{ kN/m}^3$$

$$p_B = 0 \text{ Pa(gage)} + (8.83 \text{ kN/m}^3)(2.4 \text{ m}) + (9.81 \text{ kN/m}^3)(1.5 \text{ m})$$

$$= (0 + 21.2 + 14.7) \text{ kPa} = 35.9 \text{ kPa(gage)}$$

$$A = \pi D^2/4 = \pi(3.0 \text{ m})^2/4 = 7.07 \text{ m}^2$$

$$F = p_B A = (35.9 \text{ kN/m}^2)(7.07 \text{ m}^2) = 253.8 \text{ kN}$$

☐ **EXAMPLE PROBLEM 4.3**   Would there be any difference between the force on the bottom of the drum in Fig. 4.3 and that on the bottom of the cone-shaped container in Fig. 4.4?

**Solution**  The force would be the same because the pressure at the bottom is dependent only upon the depth and specific weight of the fluid in the container. The total weight of fluid is not the controlling factor. Recall Pascal's paradox in Section 3.5.

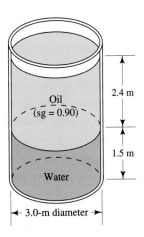

**FIGURE 4.3**   Cylindrical drum for Example Problem 4.2.

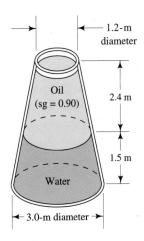

**FIGURE 4.4**   Cone-shaped container for Example Problem 4.3.

**4.5**
**RECTANGULAR WALLS**

The retaining walls shown in Fig. 4.5 are typical examples of rectangular walls exposed to a pressure varying from zero on the surface of the fluid to a maximum at the bottom of the wall. The force due to the fluid pressure tends to overturn the wall or break it at the place where it is fixed to the bottom.

The actual force is distributed over the entire wall, but for the purpose of analysis it is desirable to determine the resultant force and the place where it acts, called the *center of pressure*. That is, if the entire force were concentrated at a single point, where would that point be and what would the magnitude of the force be?

Figure 4.6 shows the pressure distribution on the vertical retaining wall. As indicated by the equation $\Delta p = \gamma h$, the pressure varies linearly (in a straight-line manner) with depth in the fluid. The lengths of the dashed arrows represent the

**FIGURE 4.5**   Rectangular walls.

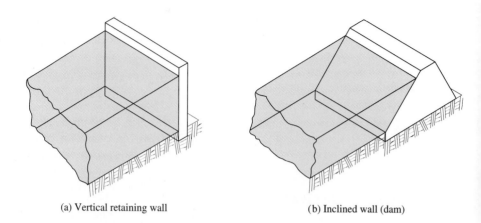

(a) Vertical retaining wall                    (b) Inclined wall (dam)

magnitude of the fluid pressure at various points on the wall. Because of this linear variation in pressure, the total resultant force can be calculated by the equation

$$F_R = p_{avg} \times A \qquad (4\text{–}1)$$

where $p_{avg}$ is the average pressure and $A$ is the total area of the wall. But the average pressure is that at the middle of the wall and can be calculated by the equation

$$p_{avg} = \gamma(d/2) \qquad (4\text{–}2)$$

where $d$ is the total depth of the fluid.

Therefore, we have

⇨ RESULTANT FORCE ON A
RECTANGULAR WALL

$$F_R = \gamma(d/2)A \qquad (4\text{–}3)$$

The pressure distribution shown in Fig. 4.6 indicates that a greater portion of the force acts on the lower part of the wall than on the upper part. The center of pressure is at the centroid of the pressure distribution triangle, one-third of the distance from the bottom of the wall. The resultant force $F_R$ acts perpendicular to the wall at this point.

The procedure for calculating the magnitude of the resultant force due to fluid pressure and the location of the center of pressure on a rectangular wall such as those shown in Fig. 4.5 is listed below. The procedure applies whether the wall is vertical or inclined.

**FIGURE 4.6**   Vertical rectangular wall.

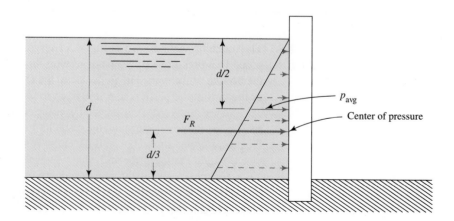

**PROCEDURE FOR COMPUTING THE FORCE ON A RECTANGULAR WALL**

**1.** Calculate the magnitude of the resultant force $F_R$ from

$$F_R = \gamma(d/2)A$$

where
$\gamma$ = specific weight of the fluid
$d$ = total depth of the fluid
$A$ = total area of the wall

**2.** Locate the center of pressure at a vertical distance of $d/3$ from the bottom of the wall.

**3.** Show the resultant force acting at the center of pressure perpendicular to the wall.

□ **EXAMPLE PROBLEM 4.4**   In Fig. 4.6, the fluid is gasoline (sg = 0.68) and the total depth is 12 ft. The wall is 40 ft long. Calculate the magnitude of the resultant force on the wall and the location of the center of pressure.

*Solution*   Step 1.

$$F_R = \gamma(d/2)A$$
$$\gamma = (0.68)(62.4 \text{ lb/ft}^3) = 42.4 \text{ lb/ft}^3$$
$$A = (12 \text{ ft})(40 \text{ ft}) = 480 \text{ ft}^2$$
$$F_R = \frac{42.4 \text{ lb}}{\text{ft}^3} \times \frac{12 \text{ ft}}{2} \times 480 \text{ ft}^2 = 122,000 \text{ lb}$$

Step 2.   The center of pressure is at a distance of

$$d/3 = 12 \text{ ft}/3 = 4 \text{ ft}$$

from the bottom of the wall.

Step 3.   The force $F_R$ acts perpendicular to the wall at the center of pressure as shown in Fig. 4.6.

□ **EXAMPLE PROBLEM 4.5**   Figure 4.7 shows a dam, 30.5 m long, which retains 8 m of fresh water and is inclined at an angle, $\theta$, of 60°. Calculate the magnitude of the resultant force on the dam and the location of the center of pressure.

**FIGURE 4.7**   Inclined rectangular wall.

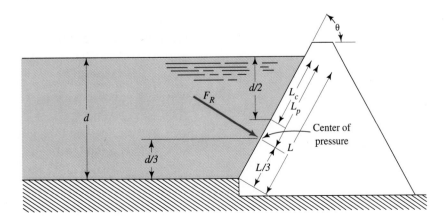

***Solution***   Step 1.

$$F_R = \gamma(d/2)A$$

To calculate the area of the dam we need the length of its face, called $L$ in Fig. 4.7:

$$\sin\theta = d/L$$

$$L = d/\sin\theta = 8 \text{ m}/\sin 60° = 9.24 \text{ m}$$

Then, the area of the dam is

$$A = (9.24 \text{ m})(30.5 \text{ m}) = 281.8 \text{ m}^2$$

Now we can calculate the resultant force:

$$F_R = \gamma(d/2)A = \frac{9.81 \text{ kN}}{\text{m}^3} \times \frac{8 \text{ m}}{2} \times 281.8 \text{ m}^2$$

$$= 11\,060 \text{ kN} = 11.06 \text{ MN}$$

Step 2.   The center of pressure is at a vertical distance of

$$d/3 = 8 \text{ m}/3 = 2.67 \text{ m}$$

from the bottom of the dam. Or, measured from the bottom of the dam along the face of the dam, the center of pressure is at:

$$L/3 = 9.24 \text{ m}/3 = 3.08 \text{ m}$$

Measured along the face of the dam we define:

$L_p$ = distance from the free surface of the fluid to the center of pressure

$L_p = L - L/3$

$L_p = 9.24 \text{ m} - 3.08 \text{ m} = 6.16 \text{ m}$

We show $F_R$ acting at the center of pressure perpendicular to the wall.

∎

**4.6
SUBMERGED PLANE
AREAS—GENERAL**

The procedure we will discuss in this section applies to problems dealing with plane areas, either vertical or inclined, that are completely submerged in the fluid. As in previous problems, the procedure will enable us to calculate the magnitude of the resultant force on an area and the location of the center of pressure where we can assume the resultant force to act.

Figure 4.8 shows a tank that has a rectangular window in an inclined wall. The standard dimensions and symbols used in the procedure described later are shown in the figure and defined below:

$F_R$       Resultant force on the area due to the fluid pressure

—       The *center of pressure* of the area is the point at which the resultant force can be considered to act

—       The *centroid* of the area is the point at which the area would be balanced if suspended from that point. It is equivalent to the center of gravity of a solid body.

$\theta$       Angle of inclination of the area

$d_c$       Depth of fluid from the free surface to the centroid of the area

$L_c$       Distance from the level of the free surface of the fluid to the centroid of the area, measured along the angle of inclination of the area

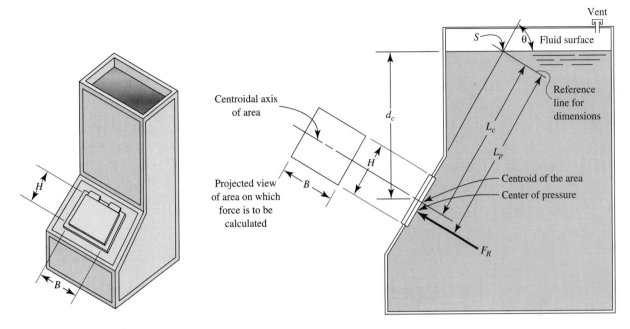

**FIGURE 4.8**   Force on submerged plane area.

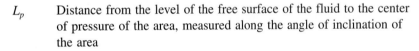

**FIGURE 4.9**   Properties of a rectangle.

$L_p$     Distance from the level of the free surface of the fluid to the center of pressure of the area, measured along the angle of inclination of the area

$B, H$    Dimensions of the area

Figure 4.9 shows the location of the centroid of a rectangle. Other shapes are described in Appendix L.

The following procedure will help you to calculate the magnitude of the resultant force on a submerged plane area due to fluid pressure and the location of the center of pressure.

**PROCEDURE FOR COMPUTING THE FORCE ON A SUBMERGED PLANE AREA**

1. Identify the point where the angle of inclination of the area of interest intersects the level of the free surface of the fluid. This may require the extension of the angled surface. Call this point $S$.
2. Locate the centroid of the area from its geometry.
3. Determine $d_c$ as the *vertical* distance from the level of the free surface down to the centroid of the area.
4. Determine $L_c$ as the *inclined* distance from the level of the free surface down to the centroid of the area. This is the distance from $S$ to the centroid. Note that $d_c$ and $L_c$ are related by:

$$d_c = L_c \sin \theta$$

5. Calculate the total area $A$ on which the force is to be determined.
6. Calculate the resultant force from

⇨ **RESULTANT FORCE ON A**
**SUBMERGED PLANE AREA**

$$F_R = \gamma d_c A \qquad\qquad (4\text{--}4)$$

where $\gamma$ is the specific weight of the fluid. This equation states that the resultant force is the product of the pressure at the centroid of the area and the total area.

7. Calculate $I_c$, the moment of inertia of the area about its centroidal axis.
8. Calculate the location of the center of pressure from

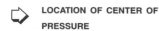

LOCATION OF CENTER OF
PRESSURE

$$L_p = L_c + \frac{I_c}{L_c A} \qquad (4\text{--}5)$$

Notice that the center of pressure is always below the centroid of an area which is inclined with the horizontal. In some cases it may be of interest to calculate only the difference between $L_p$ and $L_c$ from

$$L_p - L_c = \frac{I_c}{L_c A} \qquad (4\text{--}6)$$

9. Sketch the resultant force $F_R$ acting at the center of pressure, perpendicular to the area.
10. Show the dimension $L_p$ on the sketch in a manner similar to that used in Fig. 4.8.
11. Draw the dimension lines for $L_c$ and $L_p$ from a reference line drawn through point $S$ and perpendicular to the angle of inclination of the area.

We will now use the programmed instruction approach to illustrate the application of this procedure.

## PROGRAMMED EXAMPLE PROBLEM

□ **EXAMPLE PROBLEM 4.6**   The tank shown in Figure 4.8 contains a lubricating oil with a specific gravity of 0.91. The rectangular gate with the dimensions $B = 4$ ft and $H = 2$ ft is placed in the inclined wall of the tank ($\theta = 60°$). The centroid of the gate is at a depth of 5 ft from the surface of the oil. Calculate (a) the magnitude of the resultant force $F_R$ on the gate and (b) the location of the center of pressure.

Using the procedure described above, perform Steps 1 and 2 before going to the next panel.

---

Point $S$ is shown in Fig. 4.8.

The area of interest is the rectangular gate sketched as Fig. 4.10. The centroid is at the intersection of the axes of symmetry of the rectangle.

**FIGURE 4.10**   Rectangular gate for Example Problem 4.6.

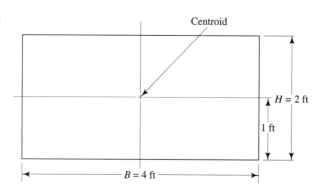

Now for Step 3, what is the distance $d_c$?

---

From the problem statement we know that $d_c = 5$ ft, the vertical depth from the free surface of the oil to the centroid of the gate.

Now calculate $L_c$. See Step 4.

---

The terms $L_c$ and $d_c$ are related in this case by

$$d_c = L_c \sin \theta$$

Therefore, we have

$$L_c = d_c/\sin \theta = 5 \text{ ft}/\sin 60° = 5.77 \text{ ft}$$

Both $d_c$ and $L_c$ will be needed for later calculations.

Go on to Step 5.

---

Since the area of the rectangle is $BH$,

$$A = BH = (4 \text{ ft})(2 \text{ ft}) = 8 \text{ ft}^2$$

Now do Step 6.

---

In the equation $F_R = \gamma d_c A$ we need the specific weight of the oil:

$$\gamma_o = (sg)_o(62.4 \text{ lb/ft}^3) = (0.91)(62.4 \text{ lb/ft}^3)$$
$$= 56.8 \text{ lb/ft}^3$$

Then we have

$$F_R = \gamma_o d_c A = \frac{56.8 \text{ lb}}{\text{ft}^3} \times 5 \text{ ft} \times 8 \text{ ft}^2 = 2270 \text{ lb}$$

The next steps concern the location of the center of pressure. Go on to Step 7 now.

---

From Fig. 4.9 we find that for a rectangle,

$$I_c = BH^3/12 = (4 \text{ ft})(2 \text{ ft})^3/12 = 2.67 \text{ ft}^4$$

Now we have all the data necessary to do Step 8.

---

Since $I_c = 2.67 \text{ ft}^4$, $L_c = 5.77$ ft, and $A = 8 \text{ ft}^2$,

$$L_p = L_c + \frac{I_c}{L_c A} = 5.77 \text{ ft} + \frac{2.67 \text{ ft}^4}{(5.77 \text{ ft})(8 \text{ ft}^2)}$$
$$L_p = 5.77 \text{ ft} + 0.058 \text{ ft} = 5.828 \text{ ft}$$

The result is $L_p = 5.828$ ft.

This means that the center of pressure is 0.058 ft (or 0.70 in) below the centroid of the gate.

Steps 9, 10, and 11 are already completed in Fig. 4.8. Be sure you understand how the dimension $L_p$ is drawn from the reference line.

■

**4.7**
**DEVELOPMENT OF THE GENERAL PROCEDURE FOR FORCES ON SUBMERGED PLANE AREAS**

Section 4.6 showed the use of the principles for computing the resultant force on a submerged plane area and for finding the location of the center of pressure. Equation (4–4) gives the resultant force, and Eq. (4–6) gives the distance between the centroid of the area of interest and the center of pressure. Figure 4.8 illustrates the various terms. This section shows the development of those relationships.

**4.7.1**
**Resultant Force**

The *resultant force* is defined as the summation of the forces on small elements of interest. Figure 4.11 illustrates the concept using the same rectangular window used in Fig. 4.8. Actually, the shape of the area is arbitrary. On any small area $dA$, there exists a force $dF$ acting perpendicular to the area owing to the fluid pressure, $p$. But the magnitude of the pressure at a depth $h$ in a static liquid of specific weight $\gamma$ is $p = \gamma h$. Then, the force is

$$dF = p(dA) = \gamma h(dA) \tag{4–7}$$

**FIGURE 4.11** Development of the general procedure for forces on submerged plane areas.

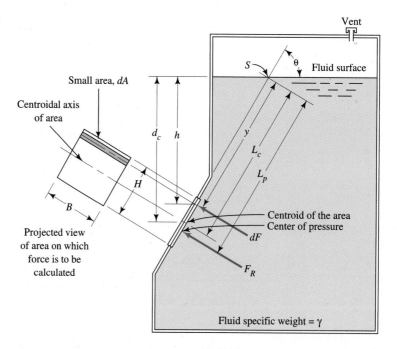

Because the area is inclined at an angle $\theta$ it is convenient to work in the plane of the area, using $y$ to denote the position on the area at any depth $h$. Note that

$$h = y \sin \theta \tag{4–8}$$

where $y$ is measured from the level of the free surface of the fluid along the angle of inclination of the area. Then,

$$dF = \gamma(y \sin \theta)(dA) \tag{4–9}$$

The summation of forces over the entire area is accomplished by the mathematical process of integration.

$$F_R = \int_A dF = \int_A \gamma(y \sin \theta)(dA) = \gamma \sin \theta \int_A y(dA)$$

From mechanics we learn that $\int y(dA)$ is equal to the product of the total area times the distance to the centroid of the area from the reference axis. That is,

$$\int_A y(dA) = L_c A$$

Then, the resultant force $F_R$ is

$$F_R = \gamma \sin \theta (L_c A) \tag{4-10}$$

Now we can substitute $d_c = L_c \sin \theta$, giving

$$F_R = \gamma d_c A \tag{4-11}$$

This is the same form as Eq. (4–4). Because each of the small forces, $dF$, acted perpendicular to the area, the resultant force also acts perpendicular to the area.

**4.7.2**
**Center of Pressure**

The *center of pressure* is that point on an area where the resultant force can be assumed to act to have the same effect as the distributed force over the entire area due to fluid pressure. We can express this effect in terms of the moment of a force with respect to an axis through $S$ perpendicular to the page.

See Fig. 4.11. The moment of each small force, $dF$, with respect to this axis is

$$dM = dF \cdot y$$

But, $dF = \gamma(y \sin \theta)(dA)$. Then,

$$dM = y[\gamma(y \sin \theta)(dA)] = \gamma \sin \theta(y^2 dA)$$

The moment of all the forces on the entire area is found by integrating over the area. Now, if we assume that the resultant force $F_R$ acts at the center of pressure, its moment with respect to the axis through $S$ is $F_R L_p$. Then,

$$F_R L_p = \int \gamma \sin \theta(y^2 dA) = \gamma \sin \theta \int (y^2 dA)$$

Again, from mechanics, we learn that $\int (y^2 dA)$ is defined as the moment of inertia $I$ of the entire area with respect to the axis from which $y$ is measured. Then,

$$F_R L_p = \gamma \sin \theta(I)$$

Solving for $L_p$ gives

$$L_p = \frac{\gamma \sin \theta(I)}{F_R}$$

Substituting for $F_R$ from Eq. (4–10) gives

$$L_p = \frac{\gamma \sin \theta(I)}{\gamma \sin \theta(L_c A)} = \frac{I}{L_c A} \tag{4-12}$$

A more convenient expression can be developed by using the transfer theorem for moment of inertia from mechanics. That is,

$$I = I_c + A L_c^2$$

where $I_c$ is the moment of inertia of the area of interest with respect to its own centroidal axis, and $L_c$ is the distance from the reference axis to the centroid. Equation (4–12) then becomes

$$L_p = \frac{I}{L_c A} = \frac{I_c + A L_c^2}{L_c A} = \frac{I_c}{L_c A} + L_c \tag{4-13}$$

Rearranging gives the same form as Eq. (4–6):

$$L_p - L_c = \frac{I_c}{L_c A}$$

**4.8**
**PIEZOMETRIC HEAD**

In all the problems demonstrated so far, the free surface of the fluid was exposed to the ambient pressure where $p = 0$ (gage). Therefore, our calculations for pressure within the fluid were also gage pressures. It was appropriate to use gage pressures for computing the magnitude of the net force on the areas of interest because the ambient pressure also acts outside the area.

A change is required in our procedure if the pressure above the free surface of the fluid is different from the ambient pressure outside the area. A convenient method uses the concept of *piezometric head*, in which the actual pressure above the fluid, $p_a$, is converted into an equivalent depth of the fluid, $d_a$, that would create the same pressure (Fig. 4.12):

⇨  PIEZOMETRIC HEAD

$$d_a = p_a/\gamma \qquad (4\text{–}14)$$

This depth is added to any depth $d$ below the free surface to obtain an *equivalent depth*, $d_e$. That is,

$$d_e = d + d_a \qquad (4\text{–}15)$$

Then, $d_e$ can be used in any calculation requiring a depth to compute pressure. For example, in Fig. 4.12, the equivalent depth to the centroid is

$$d_{ce} = d_c + d_a$$

**FIGURE 4.12**   Illustration of piezometric head for Example Problem 4.7.

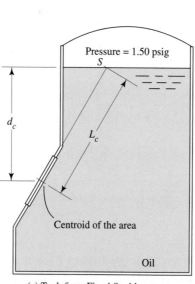

(a) Tank from Fig. 4.8 with pressure above the oil

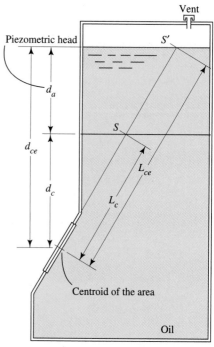

(b) Tank showing piezometric head equivalent to pressure above the oil

□ **EXAMPLE PROBLEM 4.7**  Repeat Example Problem 4.6, except consider that the tank shown in Fig. 4.8 is sealed at its top and that there is a pressure of 1.50 psig above the oil.

*Solution*  Several calculations in the solution to Example Problem 4.6 used the depth to the centroid, $d_c$, given to be 5.0 ft below the surface of the oil. With the pressure above the oil, we must add the piezometric head $d_a$ from Eq. (4–14). Using $\gamma = 56.8$ lb/ft$^3$, we get

$$d_a = \frac{p_a}{\gamma} = \frac{1.5 \text{ lb}}{\text{in}^2} \frac{144 \text{ in}^2}{\text{ft}^2} \frac{\text{ft}^3}{56.8 \text{ lb}} = 3.80 \text{ ft}$$

Then, the equivalent depth to the centroid is

$$d_{ce} = d_c + d_a = 5.00 \text{ ft} + 3.80 \text{ ft} = 8.80 \text{ ft}$$

The resultant force is then

$$F_R = \gamma d_{ce} A = (56.8 \text{ lb/ft}^3)(8.80 \text{ ft})(8.0 \text{ ft}^2) = 4000 \text{ lb}$$

Compare this with the value of 2270 lb found before for the open tank.
    The center of pressure also changes because the distance, $L_c$, changes to $L_{ce}$, as shown below.

$$L_{ce} = d_{ce}/\sin \theta = 8.80 \text{ ft}/\sin 60° = 10.16 \text{ ft}$$

$$L_{pe} - L_{ce} = \frac{I_c}{L_{ce}A} = \frac{2.67 \text{ ft}^4}{(10.16 \text{ ft})(8 \text{ ft}^2)} = 0.033 \text{ ft}$$

The corresponding distance from Example Problem 4.6 was 0.058 ft.

■

**4.9
DISTRIBUTION OF
FORCE ON A
SUBMERGED CURVED
SURFACE**

Figure 4.13 shows a tank holding a liquid with its top surface open to the atmosphere. Part of the left wall is vertical, and the lower portion is a segment of a cylinder. Here we are interested in the force acting on the curved surface due to the fluid pressure.
    One way to visualize the total force system involved is to isolate the volume of fluid directly above the surface of interest as a free body and show all the forces acting on it, as shown in Fig. 4.14. Our goal here is to determine the horizontal force $F_H$ and the vertical force $F_V$ exerted on the fluid by the curved surface and their resultant force $F_R$. The line of action of the resultant force acts through the center of curvature of the curved surface. This is because each of the individual force vectors due to the fluid pressure acts perpendicular to the boundary, which is then along the radius of curvature. Figure 4.14 shows the resulting force vectors.

**4.9.1
Horizontal Component**

The vertical solid wall at the left exerts horizontal forces on the fluid in contact with it in reaction to the forces due to the fluid pressure. This part of the system behaves in the same manner as the vertical walls studied earlier. The resultant force $F_1$ acting at a distance, $d/3$, from the bottom of the wall.
    The force $F_{2a}$ on the right side of the upper part to a depth of $d$ is equal to $F_1$ in magnitude and acts in the opposite direction. Then, they have no effect on the curved surface.
    By summing forces in the horizontal direction, you can see that $F_H$ must be equal to $F_{2b}$ acting on the lower part of the right side. The area on which $F_{2b}$ acts is the *projection* of the curved surface onto a vertical plane.

**FIGURE 4.13** Tank with a curved surface containing a static fluid.

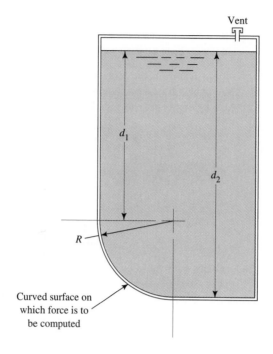

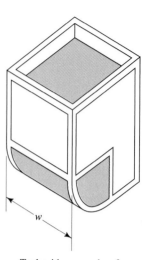

Tank with a curved surface

The magnitude of $F_{2b}$ and its location can be found using the procedures developed for plane surfaces. That is,

$$F_{2b} = \gamma d_c A \qquad (4\text{--}16)$$

where $d_c$ is the depth to the centroid of the projected area. For the type of surface shown in Fig. 4.14, the projected area is a rectangle. Calling the height of the rectangle $s$, you can see that $d_c = d + s/2$. Also, the area is $sw$ where $w$ is the width of the curved surface. Then,

$$F_{2b} = F_H = \gamma sw(d + s/2) \qquad (4\text{--}17)$$

The location of $F_{2b}$ is the center of pressure of the projected area. Again, using the principles developed earlier, we get

$$d_p - d_c = I_c/(d_c A)$$

But, for the rectangular projected area,

$$I_c = ws^3/12$$
$$A = sw$$

Then,

$$d_p - d_c = \frac{ws^3}{12(d_c)(sw)} = \frac{s^2}{12\,d_c} \qquad (4\text{--}18)$$

**4.9.2**
**Vertical Component**

The vertical component of the force exerted by the curved surface on the fluid can be found by summing forces in the vertical direction. Only the weight of the fluid acts downward, and only the vertical component, $F_V$, acts upward. Then, the weight and $F_V$ must be equal to each other in magnitude. The weight of the fluid is simply

**FIGURE 4.14** Free body diagram of volume of fluid above the curved surface.

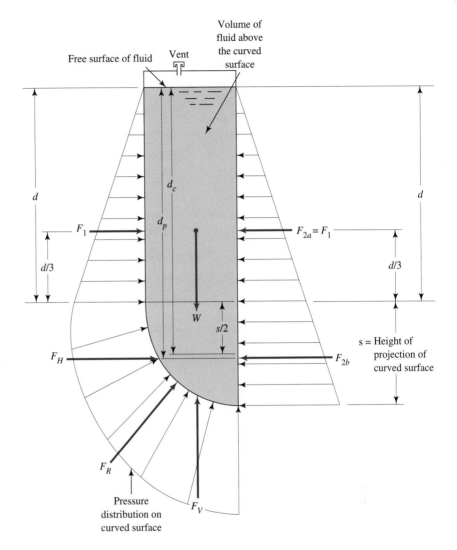

the product of its specific weight times the volume of the isolated body of fluid. The volume is the product of the cross-sectional area of the volume shown in Fig. 4.14 and the length of interest, $w$. That is,

$$F_V = \gamma(\text{volume}) = \gamma A w \tag{4–19}$$

**4.9.3**
**Resultant Force**

The total resultant force $F_R$ is

$$F_R = \sqrt{F_H^2 + F_V^2} \tag{4–20}$$

The resultant force acts at an angle, $\phi$, relative to the horizontal, found from

$$\phi = \tan^{-1}(F_V/F_H) \tag{4–21}$$

**4.9.4**
**Summary of the Procedure for Computing the Force on a Submerged Curved Surface**

Given a curved surface submerged beneath a static liquid similar to the configuration shown in Fig. 4.13, the following procedure can be used to compute the magnitude, direction, and location of the resultant force on the surface.

**1.** Isolate the volume of fluid above the surface.

2. Compute the weight of the isolated volume.
3. The magnitude of the vertical component of the resultant force is equal to the weight of the isolated volume. It acts in line with the centroid of the isolated volume.
4. Draw a projection of the curved surface onto a vertical plane and determine its height, called $s$.
5. Compute the depth to the centroid of the projected area from

$$d_c = d + s/2$$

where $d$ is the depth of the top of the projected area.
6. Compute the magnitude of the horizontal component of the resultant force from

$$F_H = \gamma sw(d + s/2) = \gamma swd_c$$

7. Compute the depth to the line of action of the horizontal component from

$$d_p = d_c + s^2/(12d_c)$$

8. Compute the resultant force from

$$F_R = \sqrt{F_V^2 + F_H^2}$$

9. Compute the angle of inclination of the resultant force relative to the horizontal from

$$\phi = \tan^{-1}(F_V/F_H)$$

10. Show the resultant force acting on the curved surface in such a direction that its line of action passes through the center of curvature of the surface.

---

☐ **EXAMPLE PROBLEM 4.8**    For the tank shown in Fig. 4.13, the following dimensions apply:

$$d_1 = 3.00 \text{ m}$$
$$d_2 = 4.50 \text{ m}$$
$$w = 2.50 \text{ m}$$
$$\gamma = 9.81 \text{ kN/m}^3 \quad \text{(water)}$$

Compute the horizontal and vertical components of the resultant force on the curved surface and the resultant force itself. Show these force vectors on a sketch.

**Solution**    Using the steps outlined above:

1. The volume above the curved surface is shown in Fig. 4.15.
2. The weight of the isolated volume is the product of the specific weight of the water times the volume. The volume is the product of the area times the length $w$.

$$\text{Area} = A_1 + A_2 = d_1 \cdot R + \tfrac{1}{4}(\pi R^2)$$
$$\text{Area} = (3.00 \text{ m})(1.50 \text{ m}) + \tfrac{1}{4}[\pi(1.50 \text{ m})^2] = 4.50 \text{ m}^2 + 1.767 \text{ m}^2$$
$$\text{Area} = 6.267 \text{ m}^2$$
$$\text{Volume} = \text{area} \cdot w = (6.267 \text{ m}^2)(2.50 \text{ m}) = 15.67 \text{ m}^3$$
$$\text{Weight} = \gamma V = (9.81 \text{ kN/m}^3)(15.67 \text{ m}^3) = 153.7 \text{ kN}$$

3. Then, $F_V = 153.7$ kN, acting upward through the centroid of the volume. The location of the centroid is found using the composite area technique. Refer to Fig. 4.15 for the

**FIGURE 4.15** Isolated volume above the curved surface for Example Problem 4.8.

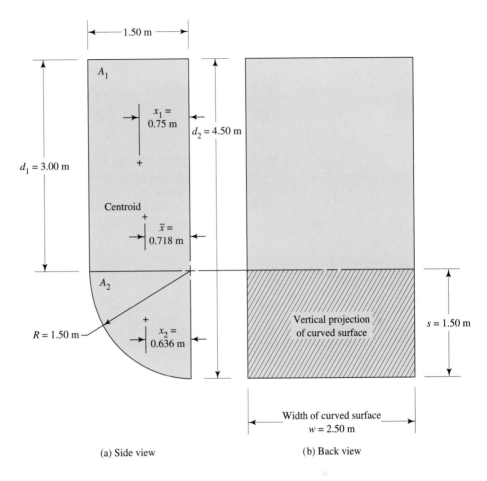

(a) Side view

(b) Back view

data. Each value should be obvious except $x_2$, the location of the centroid of the quadrant. From Appendix L,

$$x_2 = 0.424R = 0.424(1.50 \text{ m}) = 0.636 \text{ m}$$

Then, the location of the centroid for the composite area is

$$\bar{x} = \frac{A_1 x_1 + A_2 x_2}{A_1 + A_2} = \frac{(4.50)(0.75) + (1.767)(0.636)}{4.50 + 1.767} = 0.718 \text{ m}$$

4. The vertical projection of the curved surface is shown in Fig. 4.15. The height, $s$, equals 1.50 m.
5. The depth to the centroid of the projected area is

$$d_c = d_1 + s/2 = 3.00 \text{ m} + (1.50 \text{ m})/2 = 3.75 \text{ m}$$

6. The magnitude of the horizontal force is

$$F_H = \gamma sw(d_1 + s/2) = \gamma swd_c$$
$$F_H = (9.81 \text{ kN/m}^3)(1.50 \text{ m})(2.50 \text{ m})(3.75 \text{ m}) = 138.0 \text{ kN}$$

7. The depth to the line of action of the horizontal component is found from

$$d_p = d_c + s^2/(12d_c)$$
$$d_p = 3.75 \text{ m} + (1.50)^2/[(12)(3.75)] = 3.80 \text{ m}$$

**8.** The resultant force is computed from

$$F_R = \sqrt{F_V^2 + F_H^2}$$
$$F_R = \sqrt{(153.7 \text{ kN})^2 + (138.0 \text{ kN})^2} = 206.5 \text{ kN}$$

**9.** The angle of inclination of the resultant force relative to the horizontal is computed from

$$\phi = \tan^{-1}(F_V/F_H)$$
$$\phi = \tan^{-1}(153.7/138.0) = 48.1°$$

**10.** The horizontal component, the vertical component, and the resultant force are shown in Fig. 4.16. Note that the line of action of $F_R$ is through the center of curvature of the surface. Also, the vertical component is acting through the centroid of the volume of liquid above the surface. The horizontal component is acting through the center of pressure of the projected area at a depth $d_p$ from the level of the free surface of the fluid.

**FIGURE 4.16**   Results for Example Problem 4.8.

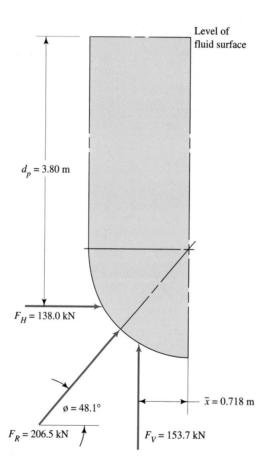

**4.10**
**EFFECT OF A PRESSURE**
**ABOVE THE FLUID**
**SURFACE**

In the preceding discussion of force on a submerged curved surface, the magnitude of the force was directly dependent on the depth of the static fluid above the surface of interest. If an additional pressure exists above the fluid or if the fluid itself is pressurized, the effect is to add to the actual depth a depth of fluid $d_a$ equivalent to $p/\gamma$. This is the same procedure, called *piezometric head*, used in Section 4.8. The new equivalent depth is used to compute both the vertical and horizontal forces.

**4.11**
**FORCES ON A CURVED**
**SURFACE WITH FLUID**
**BELOW IT**

To this point, problems have considered curved surfaces supporting a fluid above. An important concept presented for such problems was that the vertical force on the curved surface was equal to the weight of the fluid above the surface.

Now, consider the type of curved surface shown in Fig. 4.17, in which the fluid is restrained below the surface. Fluid pressure on such a surface causes forces that tend to push it upward and to the right. The surface and its connections then would have to exert reaction forces downward and to the left on the contained fluid.

The pressure in the fluid at any point is dependent on the depth of fluid to that point from the level of the free surface. This situation is equivalent to having the curved surface supporting a volume of liquid *above* it, except for the direction of the force vectors. Figure 4.18 shows that we can visualize an imaginary volume of fluid extending from the surface of interest to the level of the free surface or to the piezometric line if the fluid is under an additional pressure. Then, as before, the horizontal component of the force exerted by the curved surface on the fluid is the force on the projection of the curved surface on a vertical plane. The vertical component is equal to the weight of the imaginary volume of fluid above the surface.

**FIGURE 4.17**   Curved surface restraining a liquid below it.

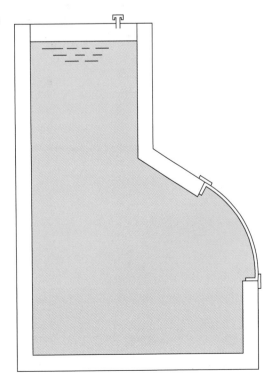

**FIGURE 4.18** Forces exerted by curved surface on the fluid.

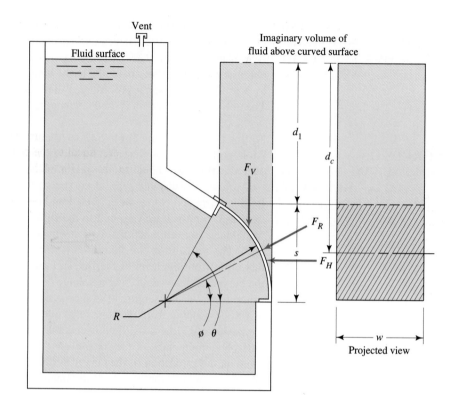

**FIGURE 4.19** Semicylindrical gate.

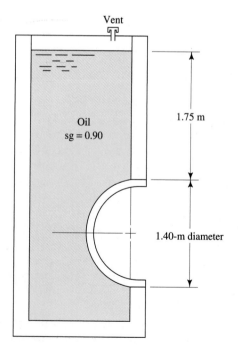

**4.12
FORCES ON CURVED
SURFACES WITH FLUID
ABOVE AND BELOW**

Figure 4.19 shows a semicylindrical gate projecting into a tank containing an oil. The force due to fluid pressure would have a horizontal component acting to the right on the gate. This force acts on the projection of the surface on a vertical plane and is computed in the same manner as used in Section 4.8.

In the vertical direction, the force on the top of the gate would act downward and would equal the weight of the oil above the gate. However, there is also a force acting upward on the bottom surface of the gate equal to the total weight of the fluid, both real and imaginary, above that surface. The net vertical force is the difference between the two forces, equal to the weight of the semicylindrical volume of fluid displaced by the gate itself (Fig. 4.20).

**FIGURE 4.20** Volumes used to compute net vertical force on the gate.

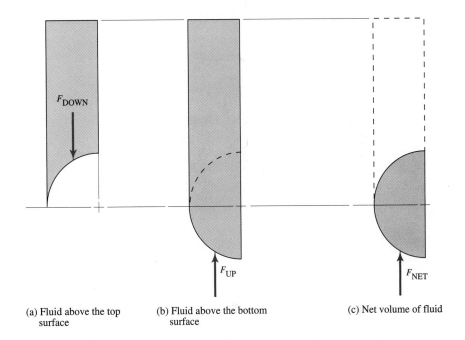

(a) Fluid above the top surface    (b) Fluid above the bottom surface    (c) Net volume of fluid

---

# PRACTICE PROBLEMS

## Forces Due to Gas Pressure

**4.1E** Figure 4.21 shows a vacuum tank which has a flat circular observation window in one end. If the pressure in the tank is 0.12 psia when the barometer reads 30.5 in of mercury, calculate the total force on the window.

**4.2E** The flat left end of the tank shown in Fig. 4.21 is secured with a bolted flange. If the inside diameter of the tank is 30 in and the internal pressure is raised to +14.4 psig, calculate the total force that must be resisted by the bolts in the flange.

**4.3E** An exhaust system for a room creates a partial vacuum in the room of 1.20 in of water relative to the atmospheric pressure outside the room. Compute the net force exerted on a 36-by-80-in door to this room.

**4.4E** A piece of 14-in Schedule 40 pipe is used as a pressure vessel by capping its ends. Compute the force on the caps

if the pressure in the pipe is raised to 325 psig. See Appendix F for the dimensions of the pipe.

**4.5M** *A pressure relief valve is designed so that the gas pressure in the tank acts on a piston with a diameter of 30 mm. How much spring force must be applied to the outside of the piston to hold the valve closed under a pressure of 3.50 MPa?*

**4.6M** *A gas-powered cannon shoots projectiles by introducing nitrogen gas at 20.5 MPa into a cylinder having an inside diameter of 50 mm. Compute the force exerted on the projectile.*

**4.7M** *The egress hatch of a manned spacecraft is designed so that the internal pressure in the cabin applies a force to help maintain the seal. If the internal pressure is 34.4 kPa(abs) and the external pressure is a perfect vacuum, calculate the force on a square hatch 800 mm on a side.*

**FIGURE 4.21**   Tank for Problems 4.1 and 4.2.

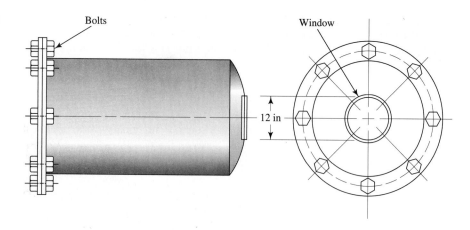

## Forces on Horizontal Flat Surfaces Under Liquids

**4.8E** A tank containing liquid ammonia at 77°F has a flat horizontal bottom. A rectangular door, 24 in by 18 in, is installed in the bottom to provide access for cleaning. Compute the force on the door if the depth of ammonia is 12.3 ft.

**4.9E** The bottom of a laboratory vat has a hole in it to allow the liquid mercury to pour out. The hole is sealed by a rubber stopper pushed in the hole and held by friction.

What force tends to push the 0.75-in-diameter stopper out of the hole if the depth of the mercury is 28.0 in?

**4.10M** *A simple shower for remote locations is designed with a cylindrical tank 500 mm in diameter and 1.800 m high as shown in Fig. 4.22. The water flows through a flapper valve in the bottom through a 75-mm-diameter opening. The flapper must be pushed upward in order to open the valve. How much force is required to open the valve?*

$$P = (9.81\,\text{kN}/\text{m}^3)(1.80\,\text{m})$$
$$P = 17.61\,\text{kN}/\text{m}^2$$
$$P = 17.61\,\text{KPa}$$

$$\frac{\pi D^2}{4} = A$$

$$\frac{\pi 95\,\text{mm}^2}{4} = A$$

$$74.16 = A\,\text{mm}^2$$
$$.75\,\text{m}^2 = A$$

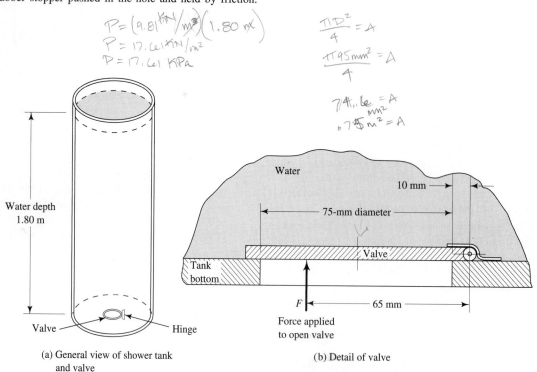

(a) General view of shower tank and valve

(b) Detail of valve

**FIGURE 4.22**   Shower tank and valve for Problem 4.10.

**4.11M** *Calculate the total force on the bottom of the closed tank shown in Fig. 4.23 if the air pressure is 52 kPa(gage).*

**FIGURE 4.23**   Problem 4.11.

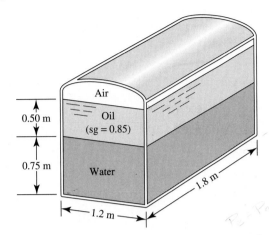

Air

Oil
(sg = 0.85)

0.50 m

0.75 m

Water

1.8 m

1.2 m

**4.12M** *If the length of the tank in Fig. 4.24 is 1.2 m, calculate the total force on the bottom of the tank.*

**FIGURE 4.24**   Problem 4.12.

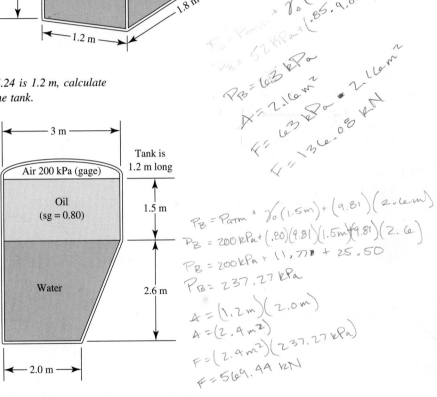

3 m

Tank is
1.2 m long

Air 200 kPa (gage)

Oil
(sg = 0.80)

1.5 m

Water

2.6 m

2.0 m

$P_B = P_{atm} + \gamma_o(.50\,m) + \gamma_w(.75\,m)$
$P_B = 52\,kPa + (.85 \cdot 9.81\frac{kN}{m^3})(.50\,m) + (9.81\frac{kN}{m^3})(.75\,m)$
$P_B = 63\,kPa$
$A = 2.16\,m^2$
$F = 63\,kPa \cdot 2.16\,m^2$
$F = 136.08\,kN$

$P_B = P_{atm} + \gamma_o(1.5\,m) + (9.81)(2.6\,m)$
$P_B = 200\,kPa + (.80)(9.81)(1.5\,m)(9.81)(2.6)$
$P_B = 200\,kPa + 11.77 + 25.50$
$P_B = 237.27\,kPa$
$A = (1.2\,m)(2.0\,m)$
$A = (2.4\,m^2)$
$F = (2.4\,m^2)(237.27\,kPa)$
$F = 569.44\,kN$

**4.13M** *An observation port in a small submarine is located in a horizontal surface of the sub. The shape of the port is shown in Fig. 4.25. Compute the total force acting on the port when the pressure inside the sub is 100 kPa(abs) and the sub is operating at a depth of 175 m in sea water.*

**FIGURE 4.25**   Port for Problem 4.13.

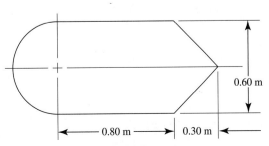

0.60 m

0.80 m

0.30 m

## Forces on Rectangular Walls

**4.14E** A rectangular gate is installed in a vertical wall of a reservoir, as shown in Fig. 4.26. Compute the magnitude of the resultant force on the gate and the location of the center of pressure. Also compute the force on each of the two latches shown.

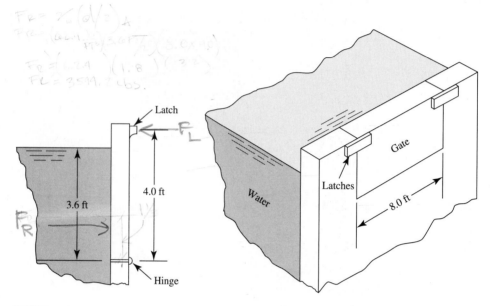

**FIGURE 4.26**   Gate in a reservoir wall for Problem 4.14.

**4.15E** A vat has a sloped side, as shown in Fig. 4.27. Compute the resultant force on this side if the vat contains 15.5 ft of glycerine. Also compute the location of the center of pressure, and show it on a sketch with the resultant force.

**FIGURE 4.27**   Vat for Problem 4.15.

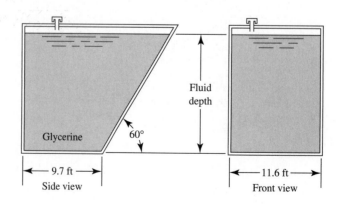

**4.16E** The wall shown in Fig. 4.28 is 20 ft long. (a) Calculate the total force on the wall due to water pressure and locate the center of pressure; and (b) calculate the moment due to this force at the base of the wall.

**FIGURE 4.28** Problem 4.16.

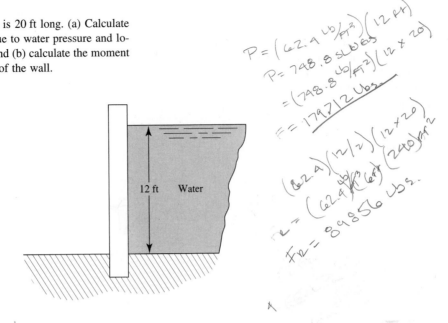

$$P = \left(62.9 \,\text{lb}/_{\text{FT}^3}\right)\left(12 \,\text{ft}\right)$$
$$P = 748.8 \,\text{slbers}$$
$$= \left(748.8 \,\text{lb}/_{\text{FT}^2}\right)\left(12 \times 20\right)$$
$$F = 179712 \,\text{lbs}$$

$$F_R = \left(62.4\right)\left(12/2\right)\left(12 \times 20\right)$$
$$= \left(62.4\right)\left(6 \,\text{ft}\right)\left(240\right)\text{ft}^2$$
$$F_R = 89856 \,\text{lbs}.$$

12 ft    Water

**4.17M** *If the wall in Fig. 4.29 is 4 m long, calculate the total force on the wall due to the oil pressure. Also determine the location of the center of pressure and show the resultant force on the wall.*

**FIGURE 4.29** Problem 4.17.

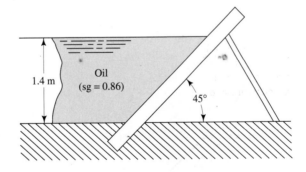

1.4 m    Oil (sg = 0.86)    45°

## Forces on Submerged Plane Areas

For each of the cases shown in Figs. 4.30 through 4.41, compute the magnitude of the resultant force on the indicated area and the location of the center of pressure. Show the resultant force on the area and clearly dimension its location.

**4.18E** Refer to Fig. 4.30.

**FIGURE 4.30**   Problem 4.18.

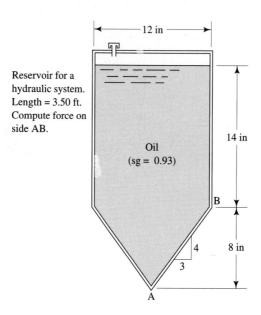

Reservoir for a hydraulic system. Length = 3.50 ft. Compute force on side AB.

**4.19M** *Refer to Fig. 4.31.*

**FIGURE 4.31**   Problems 4.19 and 4.43.

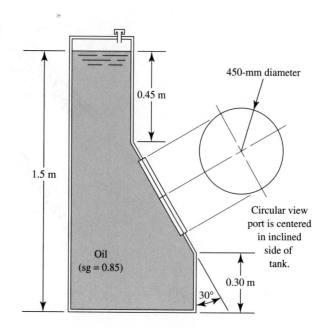

**4.20M** *Refer to Fig. 4.32.*

**FIGURE 4.32** Problems 4.20, 4.36, 4.37, and 4.44.

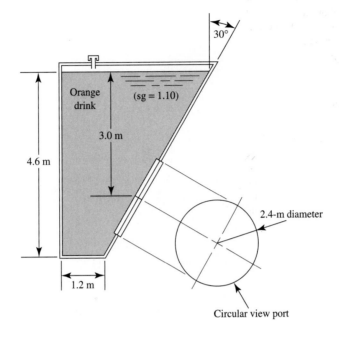

**4.21E** Refer to Fig. 4.33.

**FIGURE 4.33** Problem 4.21.

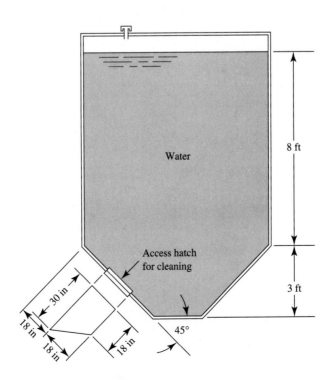

**4.22E** Refer to Fig. 4.34.

**FIGURE 4.34**   Problem 4.22.

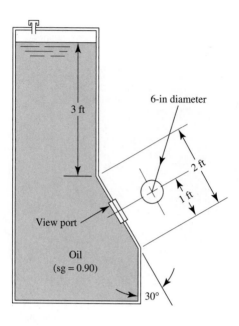

**4.23M** *Refer to Fig. 4.35.*

**FIGURE 4.35**   Problems 4.23,
4.38, and 4.39.

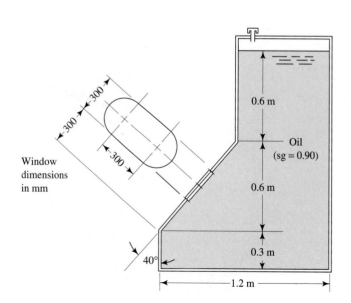

**4.24E** Refer to Fig. 4.36.

**FIGURE 4.36** Problem 4.24.

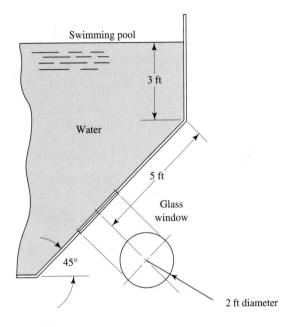

**4.25M** *Refer to Fig. 4.37.*

**FIGURE 4.37** Problem 4.25.

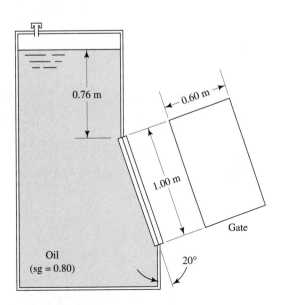

**4.26E** Refer to Fig. 4.38.

**FIGURE 4.38** Problems 4.26 and 4.45.

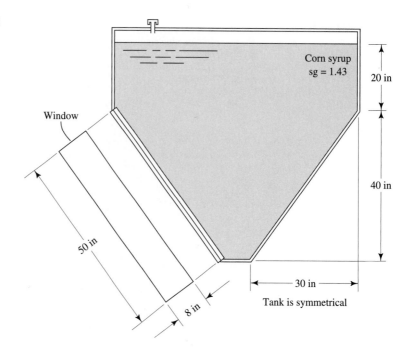

**4.27M** *Refer to Fig. 4.39.*

**FIGURE 4.39** Problem 4.27.

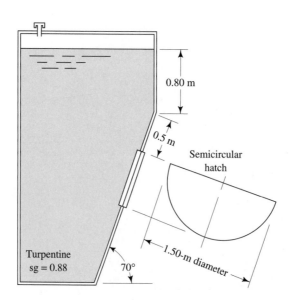

**4.28E** Refer to Fig. 4.40.

**FIGURE 4.40**  Problems 4.28 and 4.46.

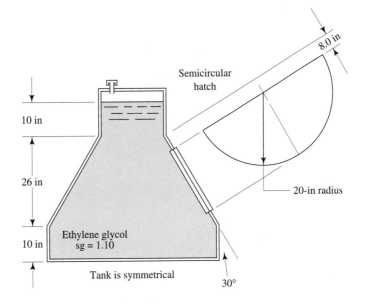

**4.29E** Refer to Fig. 4.41.

**FIGURE 4.41**  Problem 4.29.

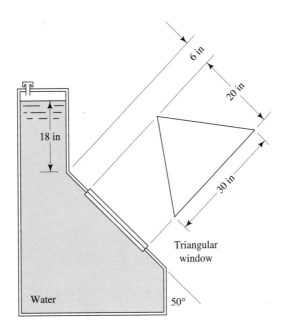

**4.30M** *Figure 4.42 shows a gasoline tank filled into the filler pipe. The gasoline has a specific gravity of 0.67. Calculate the total force on each flat end of the tank and determine the location of the center of pressure.*

**4.31M** *If the tank in Fig. 4.42 is filled just to the bottom of the filler pipe with gasoline (sg = 0.67). calculate the magnitude and location of the resultant force on the flat end.*

**4.32M** *If the tank in Fig. 4.42 is only half full of gasoline (sg = 0.67), calculate the magnitude and location of the resultant force on the flat end.*

**FIGURE 4.42**    Problems 4.30 to 4.32.

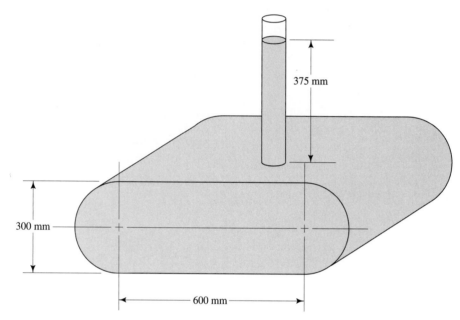

**4.33E** For the water tank shown in Fig. 4.43, compute the magnitude and location of the total force on the vertical back wall.

**FIGURE 4.43**    Problems 4.33 to 4.35.

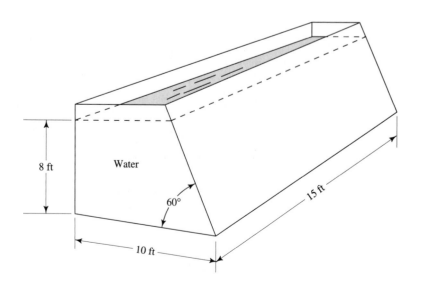

**4.34E** For the water tank shown in Fig. 4.43, compute the magnitude and location of the total force on each vertical end wall.

**4.35E** For the water tank shown in Fig. 4.43, compute the magnitude and location of the total force on the inclined wall.

**4.36M** *For the orange-drink tank shown in Fig. 4.32, compute the magnitude and location of the total force on each vertical end wall. The tank is 3.0 m long.*

**4.37M** *For the orange-drink tank shown in Fig. 4.32, compute the magnitude and location of the total force on the vertical back wall. The tank is 3.0 m long.*

**4.38M** *For the oil tank shown in Fig. 4.35, compute the magnitude and location of the total force on each vertical end wall. The tank is 1.2 m long.*

**4.39M** *For the oil tank shown in Fig. 4.35, compute the magnitude and location of the total force on the vertical back wall. The tank is 1.2 m long.*

**4.40E** Figure 4.44 shows a rectangular gate holding water behind it. If the water is 6.00 ft deep, compute the magnitude and location of the resultant force on the gate. Then, compute the forces on the hinge at the top and on the stop at the bottom.

**FIGURE 4.44**   Problem 4.40.

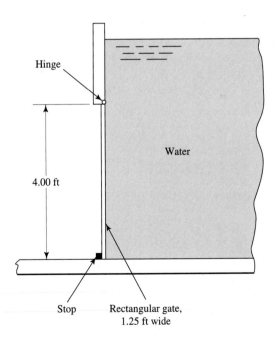

Hinge

Water

4.00 ft

Stop        Rectangular gate,
            1.25 ft wide

**4.41M** *Figure 4.45 shows a gate hinged at its bottom and held by a simple support at its top. The gate separates two fluids. Compute the net force on the gate due to the fluid on each side. Then, compute the force on the hinge and on the support.*

**FIGURE 4.45**   Problem 4.41.

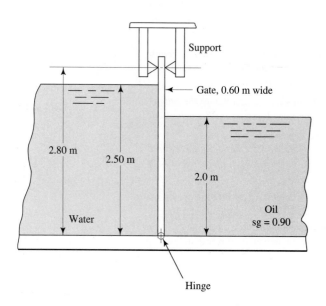

**4.42E** Figure 4.46 shows a tank of water with a circular pipe connected to its bottom. A circular gate seals the pipe opening to prohibit flow. To drain the tank, a winch is used to pull the gate open. Compute the amount of force that the winch cable must exert to open the gate.

**FIGURE 4.46**   Problem 4.42.

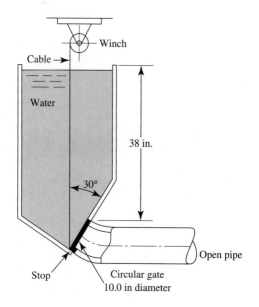

## Piezometric Head

**4.43M** *Repeat Problem 4.19M (Fig. 4.31), except that the tank is now sealed at the top with a pressure of 13.8 kPa above the oil.*

**4.44M** *Repeat Problem 4.20M (Fig. 4.32), except that the tank is now sealed at the top with a pressure of 25.0 kPa above the fluid.*

**4.45E** Repeat Problem 4.26E (Fig. 4.38), except that the tank is now sealed at the top with a pressure of 2.50 psig above the fluid.

**4.46E** Repeat Problem 4.28E (Fig. 4.40), except that the tank is now sealed at the top with a pressure of 4.0 psig above the fluid.

## Forces on Curved Surfaces

*General note for Problems 4.47 through 4.54.* For each problem, one curved surface is shown restraining a body of static fluid. Compute the magnitude of the horizontal component of the force, and compute the vertical component of the force exerted by the fluid on that surface. Then, compute the magnitude of the resultant force and its direction. Show the resultant force acting on the curved surface. In each case the surface of interest is a portion of a cylinder with the length of the surface given in the problem statement.

**4.47M** *Use Fig. 4.47. The surface is 2.00 m long.*

**4.48M** *Use Fig. 4.48. The surface is 2.50 m long.*

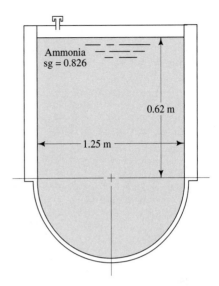

**FIGURE 4.48** Problems 4.48 and 4.56.

**4.49E** Use Fig. 4.49. The surface is 5.00 ft long.

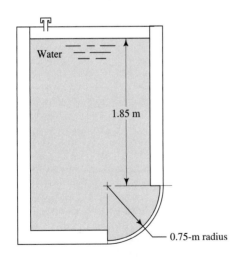

**FIGURE 4.47** Problems 4.47 and 4.55.

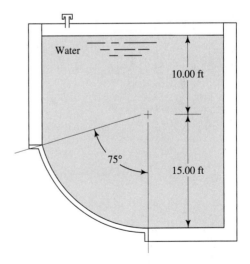

**FIGURE 4.49** Problem 4.49.

**4.50E**  Use Fig. 4.50. The surface is 4.50 ft long.

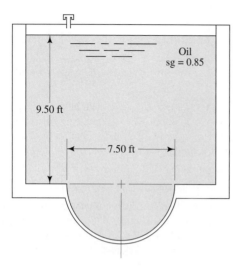

**FIGURE 4.50**   Problem 4.50.

**4.51M**  *Use Fig. 4.51. The surface is 4.00 m long.*

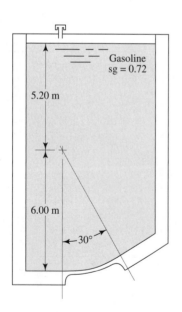

**FIGURE 4.51**   Problem 4.51.

**4.52M**  *Use Fig. 4.52. The surface is 1.50 m long.*

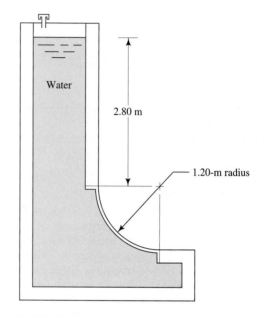

**FIGURE 4.52**   Problem 4.52.

**4.53M**  *Use Fig. 4.53. The surface is 1.50 m long.*

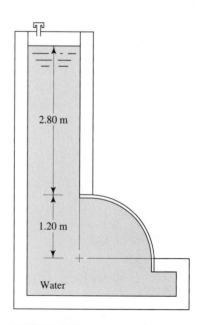

**FIGURE 4.53**   Problem 4.53.

**4.54E** Use Fig. 4.54. The surface is 60 in long.

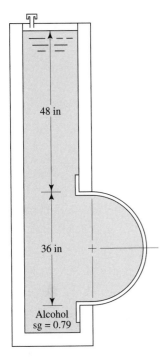

48 in

36 in

Alcohol
sg = 0.79

**FIGURE 4.54**    Problem 4.54.

**4.55M** *Repeat Problem 4.47 using Fig. 4.47, except that there is now 7.50 kPa air pressure above the fluid.*

**4.56M** *Repeat Problem 4.48 using Fig. 4.48, except that there is now 4.65 kPa air pressure above the fluid.*

**4.57E** Figure 4.55 shows a solid cylinder sitting on the bottom of a tank holding a static volume of water. Compute the

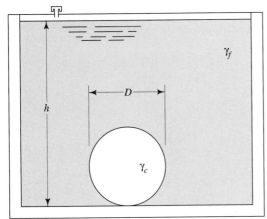

Cylinder length = L

**FIGURE 4.55**    Problems 4.57 to 4.63.

force exerted by the cylinder on the bottom of the tank for the following data: $D = 6.00$ in; $L = 10.00$ in; $\gamma_c = 0.284$ lb/in$^3$ (steel); $\gamma_f = 62.4$ lb/ft$^3$ (water); $h = 30$ in.

**4.58E** Repeat Problem 4.57, except use $\gamma_c = 0.100$ lb/in$^3$ (aluminum).

**4.59E** Repeat Problem 4.57, except use $\gamma_c = 30.00$ lb/ft$^3$ (wood).

**4.60** For the situation described in Problem 4.57, specify the required relationship between the specific weight of the cylinder and that of the fluid so that no force at all is exerted on the bottom of the tank.

**4.61E** Repeat Problem 4.57 for a depth of $h = 10.00$ in.

**4.62E** Repeat Problem 4.57 for a depth of $h = 5.00$ in.

**4.63E** For the situation described in Problem 4.57, compute the force exerted on the bottom of the tank for varying depths of fluid from $h = 30$ in to $h = 0$. Use any convenient increments of change in depth that will produce a well-defined curve of force versus depth.

**4.64** The tank in Figure 4.56 has a view port in the inclined side. Compute the magnitude of the resultant force on the panel. Show the resultant force on the door clearly and dimension its location.

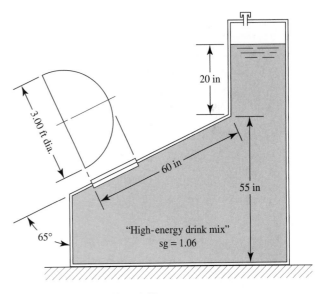

20 in

3.00 ft dia.

60 in

55 in

65°

"High-energy drink mix"
sg = 1.06

**FIGURE 4.56**    Problem 4.64.

# COMPUTER PROGRAMMING ASSIGNMENTS

**1.** Write a program to solve Problem 4.41 with any combination of data for the variables in Fig. 4.45, including the depth on either side of the gate and the specific gravity of the fluids.

2. Write a program to solve Problem 4.42 (Fig. 4.46) with any combination of data, including the size of the gate, the depth of the fluid, the specific gravity of the fluid, and the angle of inclination of the gate.

3. Write a program to solve curved surface problems of the type shown in Fig. 4.47 through 4.51 for any combination of variables, including the depth of the fluid, the angular size of the curved segment, the specific gravity of the fluid, and the radius of the surface.

4. Write a program to solve Problem 4.57 with any combination of data for the variables shown in Fig. 4.55.

5. For Program 1 above, cause the depth $h$ to vary over some specified range, giving output for each value.

# 5  Buoyancy and Stability

**Discussion Map**

Whenever an object is floating in a fluid or when it is completely submerged in the fluid, it is subjected to a force called the *buoyant force*. Usually, you will also be concerned with the *stability* of the body in the fluid.

□ *Buoyancy* is the tendency of a fluid to exert a supporting force on a body placed in the fluid.

□ *Stability* refers to the ability of a body to return to its original position after being tilted about a horizontal axis.

**Discover**

*Where have you experienced objects floating in water or other fluids?*

*Where have you seen objects completely submerged in a fluid?*

*Take a moment to write down at least five situations where you observed or felt the tendency of a fluid to support something.*

This chapter will provide the fundamental principles of both buoyancy and stability to help you develop the ability to analyze and design devices that will operate while floating or submerged.

Were your items on this list? See also the examples sketched in Figure 5.1.

- Sailing in a boat — anything from a toy sailboat, canoe, or rowboat to a sleek racing yacht, high-speed sport boat, aircraft carrier, coal barge, oil tanker, or majestic passenger liner
- Sitting on a floating dock or diving from a raft into a lake
- Floating on an inflatable raft or tube
- Swimming, exercising in a pool, or lying in a deep bath
- Playing water polo, surfboarding, windsurfing
- Snorkeling and SCUBA diving
- Fish, dolphins, whales, sea turtles, polar bears
- Cereal floating in milk for breakfast or ice floating in a cold drink or bobbing for apples
- Riding in a hot-air balloon (or looking up at one and wishing you *were* riding in it)
- Watching an advertising blimp at a sporting event
- Submarines, undersea research vehicles, and test equipment that must operate while submerged
- A floating bridge rapidly installed in an emergency
- A floating crane to assist in building a permanent bridge
- Navigation buoys and the bobber that holds a fishhook at just the right depth
- Floats to control the depth of fluid in a tank, perhaps in your car's fuel tank, a toilet tank, or in a factory using liquids from tanks in a process
- The float in the gasoline chamber of a carburetor for an engine

The two related phenomena, *buoyancy* and *stability,* are at work in these situations and in the countless others.

Anything placed in a fluid experiences a buoyant force that tends to lift it upward,

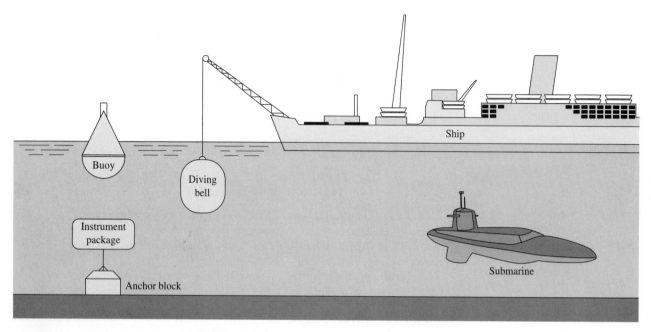

**FIGURE 5.1**   Examples of buoyancy problem types.

helping to support it. When you lie still in a swimming pool, you will float even though you are almost completely submerged. Wearing a life-vest or holding a buoyant cushion helps. How can you calculate the amount of force exerted on your body by the fluid? How can you apply that principle to some of the other applications listed above or others that were on your list?

The objects shown in Figure 5.1 show different floating tendencies. The buoy and the ship would obviously be designed to float. The diving bell would tend to sink unless supported by the cable from the crane on the ship. The instrument package tends to float and must be restrained by the cable attached to a heavy anchor block on the sea bottom. However, the submarine is designed to be able to adjust its ballast to hover at any depth (a condition called *neutral buoyancy*), or dive deeper, or rise to the surface and float.

Consider any kind of boat, raft, or other floating object that is expected to maintain a particular orientation when placed in a fluid. How can it be designed to ensure that it will be stable when given some angular displacement? Why is a canoe more likely to tip over than a large boat with a broad beam when you stand up or move around in it?

**5.2**
**OBJECTIVES**

After completing this chapter, you should be able to:

1. Write the equation for the buoyant force.
2. Analyze the case of bodies floating on a fluid.
3. Use the principle of static equilibrium to solve for the forces involved in buoyancy problems.
4. Define the conditions that must be met for a body to be stable when completely submerged in a fluid.
5. Define the conditions that must be met for a body to be stable when floating on a fluid.
6. Define the term *metacenter* and compute its location.

**5.3
BUOYANCY**

*A body in a fluid, whether floating or submerged, is buoyed up by a force equal to the weight of the fluid displaced.*

The buoyant force acts vertically upward through the centroid of the displaced volume and can be defined mathematically by Archimedes' principle as stated below.

⇨ **BUOYANT FORCE**

$$F_b = \gamma_f V_d \qquad (5\text{--}1)$$

where
$F_b$ = buoyant force
$\gamma_f$ = specific weight of the fluid
$V_d$ = displaced volume of the fluid

When a body is floating freely, it displaces a sufficient volume of fluid to just balance its own weight.

The analysis of problems dealing with buoyancy requires the application of the equation of static equilibrium in the vertical direction, $\Sigma F_v = 0$, assuming the object is at rest in the fluid. The following procedure is recommended for all problems, whether they involve floating or submerged bodies:

**PROCEDURE FOR SOLVING BUOYANCY PROBLEMS**

1. Determine the objective of the problem solution. Are you to find a force, weight, volume, or specific weight?
2. Draw a free body diagram of the object in the fluid. Show all forces that act on the free body in the vertical direction, including the weight of the body, the buoyant force, and all external forces. If the direction of some force is not known, assume the most probable direction and show it on the free body.
3. Write the equation of static equilibrium in the vertical direction, $\Sigma F_v = 0$, assuming the positive direction to be upward.
4. Solve for the desired force, weight, volume, or specific weight, remembering the following concepts:
   a. The buoyant force is calculated from $F_b = \gamma_f V_d$.
   b. The weight of a solid object is the product of its total volume and its specific weight; that is, $w = \gamma V$.
   c. An object with an average specific weight less than that of the fluid will tend to float because $w < F_b$ with the object submerged.
   d. An object with an average specific weight greater than that of the fluid will tend to sink because $w > F_b$ with the object submerged.
   e. *Neutral buoyancy* occurs when a body stays in a given position wherever it is submerged in a fluid. An object whose average specific weight is equal to that of the fluid would be neutrally buoyant.

## PROGRAMMED EXAMPLE PROBLEMS

☐ **EXAMPLE PROBLEM 5.1**   A cube, 0.50 m on a side, is made of bronze having a specific weight of 86.9 kN/m³. Determine the magnitude and direction of the force required to hold the cube in equilibrium completely submerged (a) in water and (b) in mercury. The specific gravity of mercury is 13.54.

*Solution*   Consider part (a) first. Imagine the cube of bronze submerged in water. Now do Step 1 of the procedure.

Assuming that the bronze cube will not stay in equilibrium by itself, some external force is required. The objective is to find the magnitude of this force and the direction in which it would act—that is, up or down.

Now do Step 2 of the procedure before looking at the next panel.

---

The free body is simply the cube itself. There are three forces acting on the cube in the vertical direction, as shown in Fig. 5.2: the weight of the cube $w$, acting downward through its center of gravity; the buoyant force $F_b$, acting upward through the centroid of the displaced volume; and the externally applied supporting force $F_e$.

**FIGURE 5.2**   Free body diagram of cube.

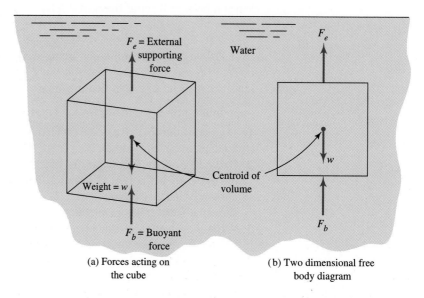

(a) Forces acting on the cube

(b) Two dimensional free body diagram

Part (a) of Fig. 5.2 shows the cube as a three-dimensional object with the three forces acting along a vertical line through the centroid of the volume. This is the preferred visualization of the free body diagram. However, for most problems it is suitable to use a simplified two-dimensional sketch as shown in part (b).

How do we know to draw the force $F_e$ in the upward direction?

---

We really do not know for certain. However, experience should indicate that without an external force the solid bronze cube would tend to sink in water. Therefore, an upward force seems to be required to hold the cube in equilibrium. If our choice is wrong, the final result will indicate that to us.

Now, assuming that the forces are as shown in Fig. 5.2, go on to Step 3.

---

The equation should look like this. (Assume that positive forces act upward.)

$$\Sigma F_v = 0$$

$$F_b + F_e - w = 0 \tag{5–2}$$

As a part of Step 4, solve this equation algebraically for the desired term.

---

You should now have

$$F_e = w - F_b \qquad\qquad (5\text{--}3)$$

since the objective is to find the external force.

How do we calculate the weight of the cube $w$?

---

Item b under Step 4 of the procedure indicates that $w = \gamma_B V$, where $\gamma_B$ is the specific weight of the bronze cube and $V$ is its total volume. For the cube, since each side is 0.50 m, we have

$$V = (0.50 \text{ m})^3 = 0.125 \text{ m}^3$$

and

$$w = \gamma_B\, V = (86.9 \text{ kN/m}^3)(0.125 \text{ m}^3) = 10.86 \text{ kN}$$

There is another unknown on the right side of Eq. (5–3). How do we calculate $F_b$?

---

Check item a under Step 4 of the procedure if you have forgotten.

$$F_b = \gamma_f V_d$$

In this case $\gamma_f$ is the specific weight of the water (9.81 kN/m³), and the displaced volume $V_d$ is equal to the total volume of the cube that we already know to be 0.125 m³. Then, we have

$$F_b = \gamma_f V_d = (9.81 \text{ kN/m}^3)(0.125 \text{ m}^3) = 1.23 \text{ kN}$$

Now we can complete our solution for $F_e$.

---

The solution is

$$F_e = w - F_b = 10.86 \text{ kN} - 1.23 \text{ kN} = 9.63 \text{ kN}$$

**Result Part a**     Notice that the result is positive. This means that our assumed direction for $F_e$ was correct. Then the solution to the problem is that an upward force of 9.63 kN is required to hold the block of bronze in equilibrium under water.

What about part (b) of the problem, where the cube is submerged in mercury? Our objective is the same as before—to determine the magnitude and direction of the force required to hold the cube in equilibrium.

Now do Step 2 of the procedure.

---

Either of two free body diagrams is correct as shown in Fig. 5.3, depending on the assumed direction for the external force $F_e$. The solution for the two diagrams will be carried out simultaneously so you can check your work regardless of which diagram looks like yours and to demonstrate that either approach will yield the correct answer.

Now do Step 3 of the procedure.

---

These are the correct equations of equilibrium. Notice the differences and relate them to the figures.

$$F_b + F_e - w = 0 \qquad | \qquad F_b - F_e - w = 0$$

Now, solve for $F_e$.

---

**FIGURE 5.3**   Two possible free
body diagrams.

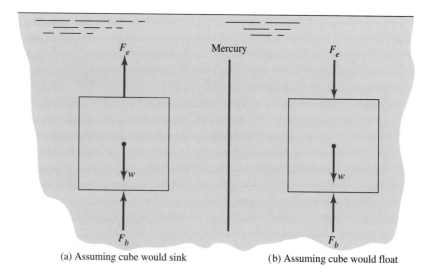

(a) Assuming cube would sink                    (b) Assuming cube would float

You should now have

$$F_e = w - F_b \qquad | \qquad F_e = F_b - w$$

Since the magnitude of $w$ and $F_b$ are the same for each equation, they can now be calculated.

___

As in part (a) of the problem, the weight of the cube is

$$w = \gamma_B V = (86.9 \text{ kN/m}^3)(0.125 \text{ m}^3) = 10.86 \text{ kN}$$

For the buoyant force $F_b$, you should have

$$F_b = \gamma_m V = (sg)_m(9.81 \text{ kN/m}^3)(V)$$

where the subscript $m$ refers to mercury. We then have

$$F_b = (13.54)(9.81 \text{ kN/m}^3)(0.125 \text{ m}^3) = 16.60 \text{ kN}$$

Now go on with the solution for $F_e$.

___

The correct answers are

$$F_e = w - F_b \qquad\qquad F_e = F_b - w$$
$$= 10.86 \text{ kN} - 16.60 \text{ kN} \qquad = 16.60 \text{ kN} - 10.86 \text{ kN}$$
$$= -5.74 \text{ kN} \qquad\qquad = +5.74 \text{ kN}$$

Notice that both solutions yield the same numerical value, but they have opposite signs. The negative sign for the solution on the left means that the assumed direction for $F_e$ in Fig. 5.3(a) was wrong. Therefore, both approaches give the same result.

**Result Part b**        The required external force is a downward force of 5.74 kN.
How could you have reasoned from the start that a downward force would be required?

___

Items c and d of Step 4 of the procedure suggest that the specific weight of the cube and the fluid be compared. In this case:

$$\text{For the bronze cube, } \gamma_B = 86.9 \text{ kN/m}^3$$
$$\text{For the fluid (mercury); } \gamma_m = (13.54)(9.81 \text{ kN/m}^3)$$
$$= 132.8 \text{ kN/m}^3$$

**Comment**  Since the specific weight of the cube is less than that of the mercury, it would tend to float without an external force. Therefore, a downward force, as pictured in Fig. 5.3(b), would be required to hold it in equilibrium under the surface of the mercury.

This example problem is concluded.

☐ **EXAMPLE PROBLEM 5.2**  A certain solid metal object has such an irregular shape that it is difficult to calculate its volume by geometry. Use the principle of buoyancy to calculate its volume.

**Solution**  First, the weight of the object is determined in the normal manner to be 60 lb. Then, using a setup similar to that in Fig. 5.4, we find its apparent weight while submerged in water to be 46.5 lb. Using these data and the procedure for analyzing buoyancy problems, we can find the volume of the object.

**FIGURE 5.4**  Metal object suspended in a fluid.

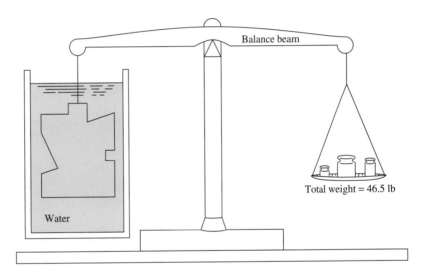

Now draw the free body diagram of the object while it is suspended in the water.

The free body diagram of the object while it is suspended in the water should look like Fig. 5.5. In this figure, what are the two forces $F_e$ and $w$?

From the problem statement we should know that $w = 60$ lb, the weight of the object in air, and $F_e = 46.5$ lb, the supporting force exerted by the balance shown in Fig. 5.4.

Now do Step 3 of the procedure.

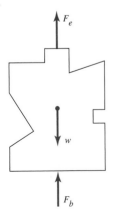

**FIGURE 5.5**  Free body diagram.

Using $\Sigma F_v = 0$, we get

$$F_b + F_e - w = 0$$

Our objective is to find the total volume $V$ of the object. How can we get $V$ from this equation?

---

We use this equation:

$$F_b = \gamma_f V$$

where $\gamma_f$ is the specific weight of the water, 62.4 lb/ft$^3$.

Substitute this into the preceding equation and solve for $V$.

---

You should now have

$$F_b + F_e - w = 0$$
$$\gamma_f V + F_e - w = 0$$
$$\gamma_f V = w - F_e$$
$$V = \frac{w - F_e}{\gamma_f}$$

Now we can put in the known values and calculate $V$.

---

**Result**   The result is $V = 0.216$ ft$^3$. This is how it is done:

$$V = \frac{w - F_e}{\gamma_f} = (60 - 46.5)\text{lb}\left(\frac{\text{ft}^3}{62.4\ \text{lb}}\right) = \frac{13.5\ \text{ft}^3}{62.4} = 0.216\ \text{ft}^3$$

**Comment**   Now that the volume of the object is known, the specific weight of the material can be found.

---

$$\gamma = \frac{w}{V} = \frac{60\ \text{lb}}{0.216\ \text{ft}^3} = 278\ \text{lb/ft}^3$$

This is approximately the specific weight of a titanium alloy.

■

---

The next two problems are worked out in detail and should serve to check your ability to solve buoyancy problems. After reading the problem statement, you should complete the solution yourself before reading the panel on which a correct solution is given. Be sure to read the problem carefully and use the proper units in your calculations. Although there is more than one way to solve some problems, it is possible to get the correct answer by the wrong method. If your method is different from that given, be sure yours is based on sound principles before assuming it is correct.

---

☐ **EXAMPLE PROBLEM 5.3**   A cube, 80 mm on a side, is made of a rigid foam material and floats in water with 60 mm below the surface. Calculate the magnitude and direction of the force required to hold it completely submerged in glycerine, which has a specific gravity of 1.26.

Complete the solution before looking at the next panel.

*Solution*   First calculate the weight of the cube, then the force required to hold the cube submerged in glycerine. Use the free body diagrams in Fig. 5.6: (a) cube floating on water and (b) cube submerged in glycerine.

**FIGURE 5.6**   Free body diagrams.

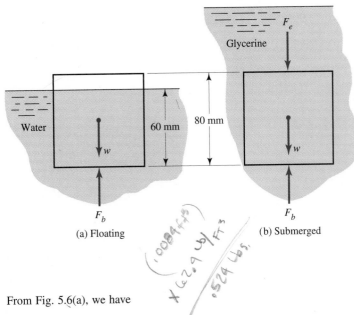

(a) Floating          (b) Submerged

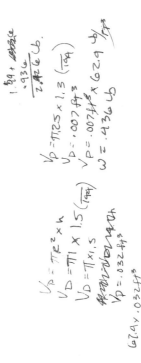

From Fig. 5.6(a), we have

$$\Sigma F_v = 0$$

$$F_b - w = 0$$

$$w = F_b = \gamma_f V_d$$

$$V_d = (80\text{ mm})(80\text{ mm})(60\text{ mm}) = 384 \times 10^3\text{ mm}^3$$

(submerged volume of cube)

$$w = \left(\frac{9.81 \times 10^3\text{ N}}{\text{m}^3}\right)(384 \times 10^3\text{ mm}^3)\left(\frac{1\text{ m}^3}{(10^3\text{ mm})^3}\right)$$

$$= 3.77\text{ N}$$

From Fig. 5.6(b), we have

$$\Sigma F_v = 0$$

$$F_b - F_e - w = 0$$

$$F_e = F_b - w = \gamma_f V_d - 3.77\text{ N}$$

$$V_d = (80\text{ mm})^3 = 512 \times 10^3\text{ mm}^3$$

(total volume of cube)

$$\gamma_f = (1.26)(9.81\text{ kN/m}^3) = 12.36\text{ kN/m}^3$$

$$F_e = \gamma_f V_d - 3.77\text{ N}$$

$$= \left(\frac{12.36 \times 10^3\text{ N}}{\text{m}^3}\right)(512 \times 10^3\text{ mm}^3)\left(\frac{1\text{ m}^3}{(10^3\text{ mm})^3}\right) - 3.77\text{ N}$$

*Result*

$$F_e = 6.33\text{ N} - 3.77\text{ N} = 2.56\text{ N}$$

A downward force of 2.56 N is required to hold the cube submerged in glycerine.

☐ **EXAMPLE PROBLEM 5.4**   A brass cube, 6 in on a side, weighs 67 lb. We want to hold this cube in equilibrium under water by attaching a light foam buoy to it. If the foam weighs 4.5 lb/ft³, what is the minimum required volume of the buoy?

Complete the solution yourself before looking at the next panel.

---

*Solution*   Calculate the minimum volume of foam to hold the brass cube in equilibrium.

Notice that the foam and brass in Fig. 5.7 are considered as parts of a single system and that there is a buoyant force on each. The subscript $F$ refers to the foam, and the subscript $B$ refers to the brass. No external force is required.

**FIGURE 5.7**   Free body diagram for brass and foam together.

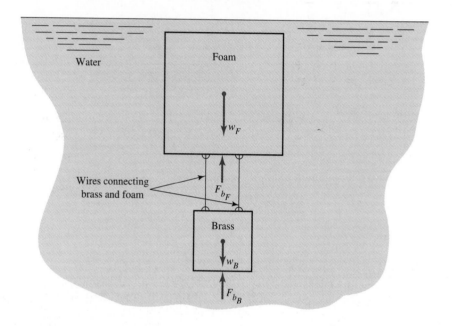

The equilibrium equation is

$$\Sigma F_v = 0$$

$$0 = F_{b_B} + F_{b_F} - w_B - w_F \tag{5-4}$$

$$w_B = 67 \text{ lb} \quad \text{(given)}$$

$$F_{b_B} = \gamma_f V_{d_B} = \left(\frac{62.4 \text{ lb}}{\text{ft}^3}\right)(6 \text{ in})^3\left(\frac{\text{ft}^3}{1728 \text{ in}^3}\right) = 7.8 \text{ lb}$$

$$w_F = \gamma_F V_F$$

$$F_{b_F} = \gamma_f V_F$$

Substitute these quantities into Eq. (5-4):

$$F_{b_B} + F_{b_F} - w_B - w_F = 0$$

$$7.8 \text{ lb} + \gamma_f V_F - 67 \text{ lb} - \gamma_F V_F = 0$$

Solve for $V_F$, using $\gamma_f = 62.4 \text{ lb/ft}^3$ and $\gamma_F = 4.5 \text{ lb/ft}^3$:

$$\gamma_f V_F - \gamma_F V_F = 67\ \text{lb} - 7.8\ \text{lb} = 59.2\ \text{lb}$$
$$V_F(\gamma_f - \gamma_F) = 59.2\ \text{lb}$$
$$V_F = \frac{59.2\ \text{lb}}{\gamma_f - \gamma_F} = \frac{59.2\ \text{lb ft}^3}{(62.4 - 4.5)\ \text{lb}}$$
$$V_F = 1.02\ \text{ft}^3$$

**Result**     This means that if 1.02 ft$^3$ of foam were attached to the brass cube, the combination would be in equilibrium in water without any external force. It would be neutrally buoyant.

This completes the programmed example problems.

∎

**5.4**
**STABILITY OF**
**COMPLETELY**
**SUBMERGED BODIES**

A body in a fluid is considered stable if it will return to its original position after being rotated a small amount about a horizontal axis. Two familiar examples of bodies completely submerged in a fluid are submarines and weather balloons. It is important for these kinds of objects to remain in a specific orientation despite the action of currents, winds, or maneuvering forces.

⇨ CONDITION OF STABILITY FOR
SUBMERGED BODIES

*The condition for stability of bodies completely submerged in a fluid is that the center of gravity of the body must be below the center of buoyancy.*

The center of buoyancy of a body is at the centroid of the displaced volume of fluid, and it is through this point that the buoyant force acts in a vertical direction. The weight of the body acts vertically downward through the center of gravity.

The undersea research vehicle shown in Fig. 5.8 has a stable configuration due to its shape and the location of equipment within the hull. Figure 5.9 shows the

**FIGURE 5.8** Grumman-BEN FRANKLIN, undersea research vessel. (Source of photo: Grumman Aerospace Corp. Ocean Systems Department, Bethpage, NY)

**FIGURE 5.9**   Stability of submerged body.

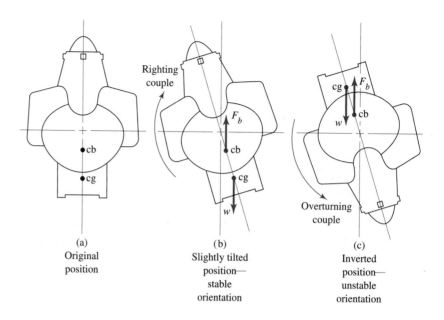

(a)
Original
position

(b)
Slightly tilted
position—
stable
orientation

(c)
Inverted
position—
unstable
orientation

approximate cross-sectional shape of the vessel. The circular section is a hollow cylinder providing a manned cabin and housing for delicate instruments and life-support systems. The rectangular section at the bottom contains heavy batteries and other durable equipment. With this distribution of weight and volume, the center of gravity (cg) and the center of buoyancy (cb) are located approximately as shown in Fig. 5.9(a).

Figure 5.9(b) shows the action of the buoyant force and the weight to produce a couple that tends to rotate the vessel back to its original position after being rotated slightly. Thus, the body is stable.

Contrast this with Fig. 5.9(c), which shows what would happen if the configuration were upside down from that in Fig. 5.9(a). When this body is rotated a small amount, the weight and the buoyant force produce a couple that tends to overturn it. Therefore, this orientation of the body is unstable.

If the center of gravity and center of buoyancy of a body coincide, as with a solid object, the weight and buoyant force act through the same point, producing no couple. In this case the body would have *neutral stability* and would remain in any orientation in which it is placed.

## 5.5 STABILITY OF FLOATING BODIES

The condition for stability of floating bodies is different from that for completely submerged bodies; the reason is illustrated in Fig. 5.10, which shows the approximate cross section of a ship's hull. In part (a) of the figure, the floating body is at its equilibrium orientation and the center of gravity (cg) is above the center of buoyancy (cb). A vertical line through these points will be called the *vertical axis* of the body. Figure 5.10(b) shows that if the body is rotated slightly, the center of buoyancy shifts to a new position because the geometry of the displaced volume has changed. The buoyant force and the weight now produce a righting couple that tends to return the body to its original orientation. Thus, the body is stable.

In order to state the condition for stability of a floating body, we must define a new term, *metacenter*. The metacenter (mc) is defined as the intersection of the

**FIGURE 5.10**   Method of finding
the metacenter.

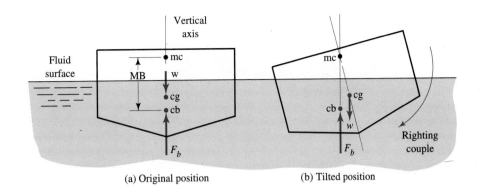

(a) Original position                    (b) Tilted position

vertical axis of a body when in its equilibrium position and a vertical line through the new position of the center of buoyancy when the body is rotated slightly. This is illustrated in Fig. 5.10(b).

CONDITION OF STABILITY FOR
FLOATING BODIES

*A floating body is stable if its center of gravity is below the metacenter.*

It is possible to determine analytically if a floating body is stable by calculating the location of its metacenter. The distance to the *me*tacenter from the center of *b*uoyancy is called MB and is calculated from

$$\text{MB} = I/V_d \qquad (5-5)$$

In this equation, $V_d$ is the displaced volume of fluid and $I$ is the *least* moment of inertia of a horizontal section of the body taken at the surface of the fluid. *If the distance MB places the metacenter above the center of gravity, the body is stable.*

**PROCEDURE FOR EVALUATING THE STABILITY OF FLOATING BODIES**

1. Determine the position of the floating body, using the principles of buoyancy.
2. Locate the center of buoyancy, cb; and compute the distance from some reference axis to cb, called $y_{cb}$. Usually, the bottom of the object is taken as the reference axis.
3. Locate the center of gravity, cg; and compute $y_{cg}$ measured from the same reference axis.
4. Determine the shape of the area at the fluid surface, and compute the *smallest* moment of inertia, $I$, for that shape.
5. Compute the displaced volume, $V_d$.
6. Compute $\text{MB} = I/V_d$.
7. Compute $y_{mc} = y_{cb} + \text{MB}$.
8. If $y_{mc} > y_{cg}$, the body is stable.
9. If $y_{mc} < y_{cg}$, the body is unstable.

## PROGRAMMED EXAMPLE PROBLEMS

☐ **EXAMPLE PROBLEM 5.5**    Figure 5.11(a) shows a flatboat hull that, when fully loaded, weighs 150 kN. Parts (b), (c), and (d) show the top, front, and side views of the boat. Note the location of the center of gravity, cg. Determine if the boat is stable in fresh water.

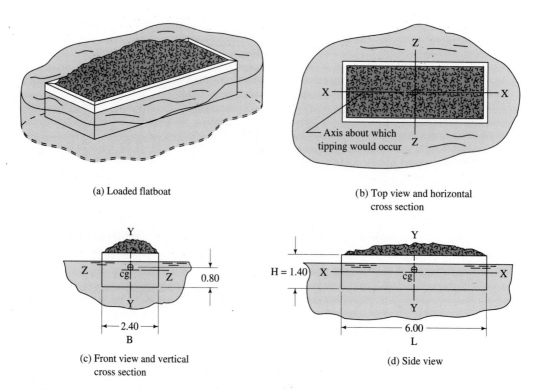

(a) Loaded flatboat

(b) Top view and horizontal cross section

(c) Front view and vertical cross section

(d) Side view

**FIGURE 5.11**    Shape of hull for flatboat for Example Problem 5.5.

***Solution***    First, find out if the boat will float.

This is done by finding how far the boat will sink into the water, using the principles of buoyancy stated in Section 5.2. Complete that calculation before going to the next panel.

The depth of submergence or draft of the boat is 1.06 m, as shown in Fig. 5.12, found by the method described below.

Equation of equilibrium:  $\sum F_v = 0 = F_b - w$

$$w = F_b$$

**FIGURE 5.12**    Free body diagram.

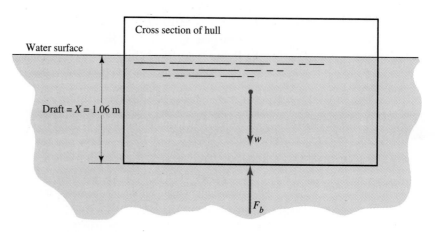

Submerged volume: $V_d = B \times L \times X$

Buoyant force: $F_b = \gamma_f V_d = \gamma_f \times B \times L \times X$

Then, we have

$$w = F_b = \gamma_f \times B \times L \times X$$

$$X = \frac{w}{B \times L \times \gamma_f} = \frac{150 \text{ kN}}{(2.4 \text{ m})(6.0 \text{ m})} \times \frac{\text{m}^3}{(9.81 \text{ kN})} = 1.06 \text{ m}$$

It floats with 1.06 m submerged. Where is the center of buoyancy?

---

It is at the center of the displaced volume of water. In this case, as shown in Fig. 5.13, it is on the vertical axis of the boat at a distance of 0.53 m from the bottom. That is half of the draft, $X$. Then $y_{cb} = 0.53$ m.

**FIGURE 5.13**  Location of center of buoyancy and center of gravity.

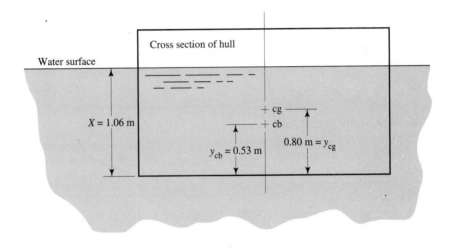

Since the center of gravity is above the center of buoyancy, we must locate the metacenter to determine if the boat is stable. Using Eq. (5–5), calculate the distance MB and show it on the sketch.

---

The result is MB = 0.45 m, as shown in Fig. 5.14. Here is how it is done.

$$MB = I/V_d$$
$$V_d = L \times B \times X = (6.0 \text{ m})(2.4 \text{ m})(1.06 \text{ m}) = 15.26 \text{ m}^3$$

The moment of inertia, $I$, is determined about the axis $X$-$X$ in Fig. 5.11(b), since this would yield the smallest value for $I$:

$$I = \frac{LB^3}{12} = \frac{(6.0 \text{ m})(2.4 \text{ m})^3}{12} = 6.91 \text{ m}^4$$

Then, the distance from the center of buoyancy to the metacenter is

$$MB = I/V_d = 6.91 \text{ m}^4/15.26 \text{ m}^3 = 0.45 \text{ m}$$

**FIGURE 5.14**  Location of meta-center.

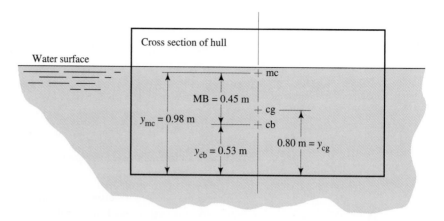

and

$$y_{mc} = y_{cb} + MB = 0.53 \text{ m} + 0.45 \text{ m} = 0.98 \text{ m}$$

Is the boat stable:

***Result***  Yes, it is. Since the metacenter is above the center of gravity, as shown in Fig. 5.14, the boat is stable. That is, $y_{mc} > y_{cg}$.

Now, read the next panel for another problem.

☐ **EXAMPLE PROBLEM 5.6**  A solid cylinder is 3.0 ft in diameter, is 6.0 ft high, and weighs 1550 lb. If the cylinder is placed in oil (sg = 0.90), with its axis vertical, would it be stable?

The complete solution is shown in the next panel. Do this problem and then look at the solution.

***Solution***  Position of cylinder in oil (Fig. 5.15):

$$V_d = \text{submerged volume} = AX = \frac{\pi D^2}{4}(X)$$

Equilibrium equation: $\Sigma F_v = 0$

$$w = F_b = \gamma_o V_d = \gamma_o \frac{\pi D^2}{4}(X)$$

$$X = \frac{4w}{\pi D^2 \gamma_o} = \frac{(4)(1550 \text{ lb}) \text{ ft}^3}{(\pi)(3.0 \text{ ft})^2(0.90)(62.4 \text{ lb})} = 3.90 \text{ ft}$$

The center of buoyancy, cb, is at a distance $X/2$ from the bottom of the cylinder:

$$y_{cb} = X/2 = 3.90 \text{ ft}/2 = 1.95 \text{ ft}$$

The center of gravity, cg, is at $H/2 = 3$ ft from the bottom of the cylinder, assuming the material of the cylinder is of uniform specific weight. The position of the metacenter, mc, using Eq. (5–5), is

$$MB = I/V_d$$

$$I = \frac{\pi D^4}{64} = \frac{\pi (3 \text{ ft})^4}{64} = 3.98 \text{ ft}^4$$

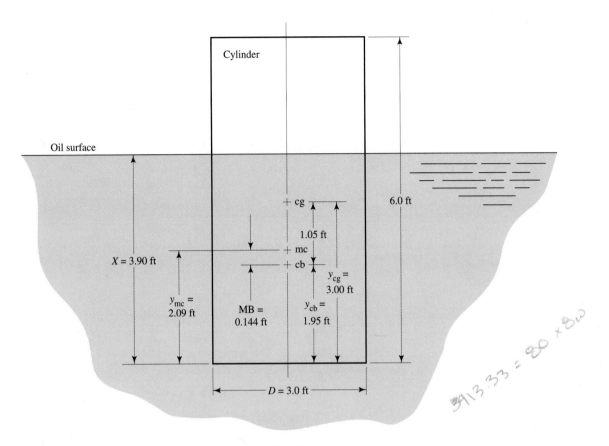

**FIGURE 5.15**   Complete solution for Example Problem 5.6.

$$V_d = AX = \frac{\pi D^2}{4}(X) = \frac{\pi(3 \text{ ft})^2}{4}(3.90 \text{ ft}) = 27.6 \text{ ft}^3$$

$$\text{MB} = I/V_d = 3.98 \text{ ft}^4/27.6 \text{ ft}^3 = 0.144 \text{ ft}$$

$$y_{mc} = y_{cb} + \text{MB} = 1.95 \text{ ft} + 0.14 \text{ ft} = 2.09 \text{ ft}$$

**Result**   Since this places the metacenter below the center of gravity, as shown in Fig. 5.15, the cylinder is not stable. That is, $y_{mc} < y_{cg}$.

This completes the programmed instruction.

∎

The conditions for stability of bodies in a fluid can be summarized as follows.

- *Completely submerged bodies are stable if the center of gravity is below the center of buoyancy.*
- *Floating bodies are stable if the center of gravity is below the metacenter.*

**5.6**
**DEGREE OF STABILITY**   Although the limiting case of stability has been stated as any design for which the metacenter is above the center of gravity, some objects can be more stable than others. One measure of relative stability is called the *metacentric height* defined as the distance to the metacenter from the center of gravity.

**FIGURE 5.16** Degree of stability as indicated by metacentric height and righting arm.

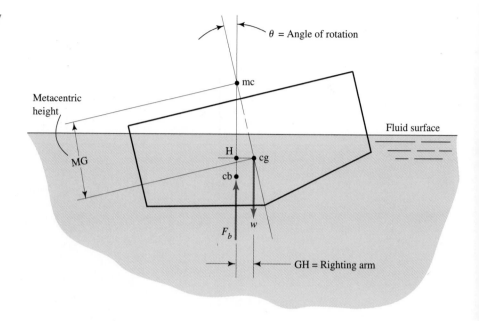

Refer to Fig. 5.16. The metacentric height is labeled MG. Using the procedures discussed in this chapter, we can compute MG from

$$MG = y_{mc} - y_{cg} \qquad (5\text{–}6)$$

Reference 1 states that small seagoing vessels should have a minimum value of MG of 1.5 ft (0.46 m). Large ships should have MG > 3.5 ft (1.07 m). But, the metacentric height should not be too large because the ship may then have the uncomfortable rocking motions that cause seasickness.

☐ **EXAMPLE PROBLEM 5.7**    Compute the metacentric height for the flatboat hull described in Example Problem 5.5.

**Solution**    From the results of Example Problem 5.5,

$$y_{mc} = 0.98 \text{ m from the bottom of the hull}$$

$$y_{cg} = 0.80 \text{ m}$$

Then, the metacentric height is

$$MG = y_{mc} - y_{cg} = 0.98 \text{ m} - 0.80 \text{ m} = 0.18 \text{ m}$$

■

**5.6.1**
**Static Stability Curve**

Another measure of the stability of a floating object is the amount of offset between the line of action of the weight of the object acting through the center of gravity and that of the buoyant force acting through the center of buoyancy. Earlier, in Fig. 5.10, it was shown that the product of one of these forces and the amount of the offset produces the righting couple that causes the object to return to its original position and thus to be stable.

Figure 5.16 shows a sketch of a boat hull in a rotated position with the weight and the buoyant force shown. A horizontal line drawn through the center of gravity intersects the line of action of the buoyant force at point *H*. The horizontal distance, GH, is called the *righting arm* and is a measure of the magnitude of the righting couple. The distance GH varies as the angle of rotation varies, and Fig. 5.17 shows a characteristic plot of the righting arm versus the angle of rotation for a ship. Such a plot is called a *static stability curve*. As long as the value of GH remains positive, the ship is stable. Conversely, when GH becomes negative, the ship is unstable and it will overturn.

**FIGURE 5.17**  Static stability curve for a floating body.

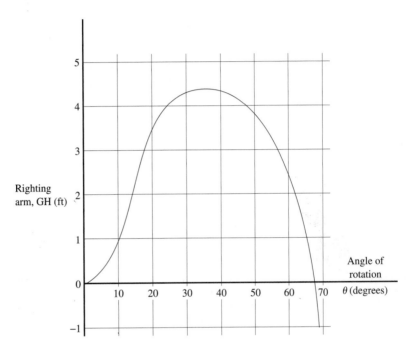

## REFERENCE

1. Avallone, Eugene A., and Theodore Baumeister III, eds. 1996. *Marks' Standard Handbook for Mechanical Engineers*. 10th ed. New York: McGraw-Hill.

## PRACTICE PROBLEMS

### Buoyancy

**5.1M** *The instrument package shown in Fig. 5.18 weighs 258 N. Calculate the tension in the cable if the package is completely submerged in seawater having a specific weight of 10.05 kN/m³.*

**5.2M** *A 1.0-m-diameter hollow sphere weighing 200 N is attached to a solid concrete block weighing 4.1 kN. If the concrete has a specific weight of 23.6 kN/m³, will the two objects together float or sink in water?*

**5.3M** *A certain standard steel pipe has an outside diameter of 168 mm, and a 1-m length of the pipe weighs 277 N.*

*Would the pipe float or sink in glycerine (sg = 1.26) if its ends are sealed?*

**5.4E** *A cylindrical float has a 10-in diameter and is 12 in long. What should be the specific weight of the float material if it is to have nine-tenths of its volume below the surface of a fluid with a specific gravity of 1.10?*

**5.5M** *A buoy is a solid cylinder 0.3 m in diameter and 1.2 m long. It is made of a material with a specific weight of 7.9 kN/m³. If it floats upright, how much of its length is above the water?*

**FIGURE 5.18**    Problem 5.1.

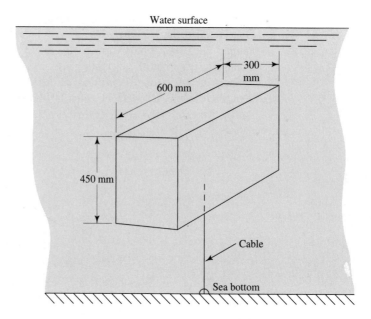

**5.6M** *A float to be used as a level indicator is being designed to float in oil, which has a specific gravity of 0.90. It is to be a cube, 100 mm on a side, and is to have 75 mm submerged in the oil. Calculate the required specific weight of the float material.*

**5.7M** *A concrete block with a specific weight of 23.6 kN/m³ is suspended by a rope in a solution with a specific gravity of 1.15. What is the volume of the concrete block if the tension in the rope is 2.67 kN?*

**5.8E** Figure 5.19 shows a pump partially submerged in oil (sg = 0.90) and supported by springs. If the total weight

of the pump is 14.6 lb and the submerged volume is 40 in³, calculate the supporting force exerted by the springs.

**5.9M** *A steel cube 100 mm on a side weighs 80 N. We want to hold the cube in equilibrium under water by attaching a light foam buoy to it. If the foam weighs 470 N/m³, what is the minimum required volume of the buoy?*

**5.10E** A cylindrical drum is 2 ft in diameter, is 3 ft long, and weighs 30 lb when empty. Aluminum weights are to be placed inside the drum in order to make it neutrally buoyant in fresh water. What volume of aluminum will be required if it weighs 0.100 lb/in³?

**5.11E** If the aluminum weights described in Problem 5.10 are placed outside the drum, what volume will be required?

**5.12** Figure 5.20 shows a cube floating in a fluid. Derive an expression relating the submerged depth $X$, the specific weight of the cube, and the specific weight of the fluid.

**5.13E** A hydrometer is a device for indicating the specific gravity of liquids. Figure 5.21 shows the design for a hydrometer in which the bottom part is a hollow cylinder with a 1.00-in diameter, and the top is a tube with a 0.25-in diameter. The empty hydrometer weighs 0.020 lb. What weight of steel balls should be added to make the hydrometer float in the position shown in fresh water? (Note that this is for a specific gravity of 1.00.)

**5.14E** For the hydrometer designed in Problem 5.13, what will be the specific gravity of the fluid in which the hydrometer would float at the top mark?

**5.15E** For the hydrometer designed in Problem 5.13, what will be the specific gravity of the fluid in which the hydrometer would float at the bottom mark?

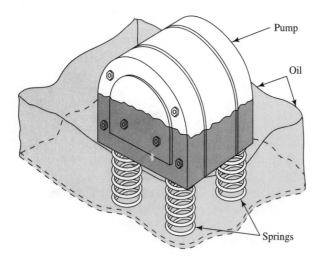

**FIGURE 5.19**    Problem 5.8.

**FIGURE 5.20**   Problems 5.12 and 5.60.

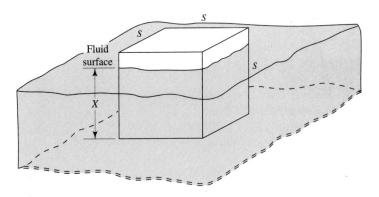

**5.16E** A buoy is to support a cone-shaped instrument package, as shown in Fig. 5.22. The buoy is made from a uniform material having a specific weight of 8.00 lb/ft³. At least 1.50 ft of the buoy must be above the surface of the seawater for safety and visibility. Calculate the maximum allowable weight of the instrument package.

**5.17E** A cube has side dimensions of 18.00 in. It is made of steel having a specific weight of 491 lb/ft³. What force is required to hold it in equilibrium under fresh water?

**5.18E** A cube has side dimensions of 18.00 in. It is made of steel having a specific weight of 491 lb/ft³. What force is required to hold it in equilibrium under mercury?

**5.19M** *A ship has a mass of 292 Mg. Compute the volume of sea water it will displace when floating.*

**5.20M** *An iceberg has a specific weight of 8.72 kN/m³. What portion of its volume is above the surface when in seawater?*

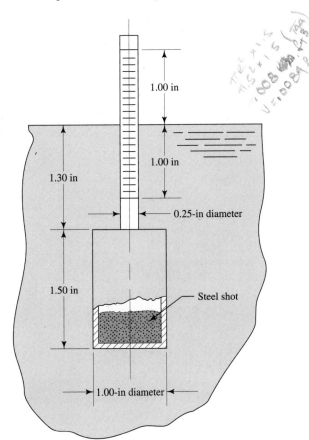

**FIGURE 5.21**   Hydrometer for Problems 5.13, 5.14, and 5.15.

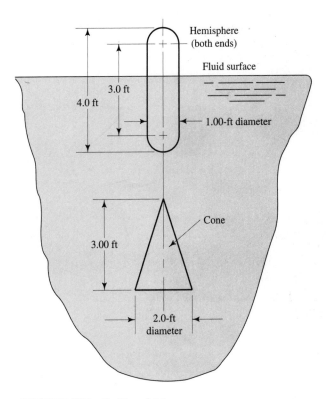

**FIGURE 5.22**   Problem 5.16.

**5.21M** *A cylindrical log has a diameter of 450 mm and a length of 6.75 m. When the log is floating in fresh water with its long axis horizontal, 110 mm of its diameter is above the surface. What is the specific weight of the wood in the log?*

**5.22M** *The cylinder shown in Fig. 5.23 is made from a uniform material. What is its specific weight?*

**5.23M** *If the cylinder from Problem 5.22 is placed in fresh water at 95°C, how much of its height would be above the surface?*

**5.24M** *A brass weight is to be attached to the bottom of the cylinder described in Problems 5.22 and 5.23, so that the cylinder will be completely submerged and neutrally buoyant in water at 95°C. The brass is to be a cylinder with the same diameter as the original cylinder shown in Fig. 5.24. What is the required thickness of the brass?*

**5.25M** *For the cylinder with the added brass (described in Problem 5.24), what will happen if the water were cooled to 15°C?*

**5.26M** *For the composite cylinder shown in Fig. 5.25, what thickness of brass is required to cause the cylinder to float in the position shown in carbon tetrachloride at 25°C?*

**5.27M** *A vessel for a special experiment has a hollow cylinder for its upper part and a solid hemisphere for its lower part, as shown in Fig. 5.26. What must be the total weight of the vessel if it is to sit upright, submerged to a depth of 0.75 m, in a fluid having a specific gravity of 1.16?*

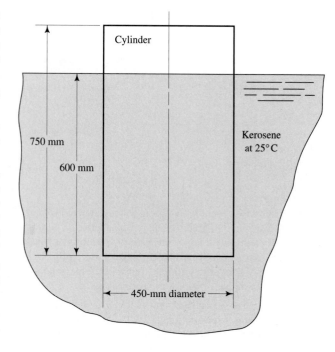

**FIGURE 5.23**    Problems 5.22, 5.23, 5.24, 5.25, and 5.52.

**FIGURE 5.24**    Problems 5.24 and 5.25.

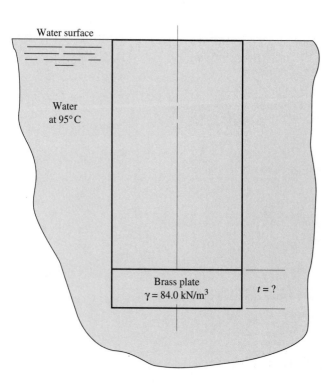

**FIGURE 5.25** Problems 5.26 and 5.53.

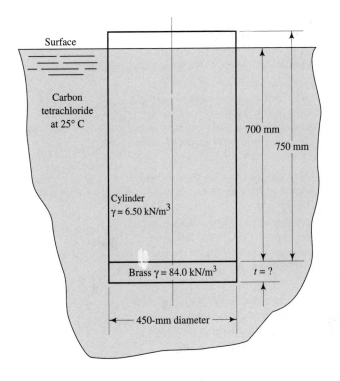

Surface

Carbon tetrachloride at 25° C

Cylinder
$\gamma = 6.50$ kN/m$^3$

Brass $\gamma = 84.0$ kN/m$^3$

700 mm

750 mm

$t = ?$

450-mm diameter

**FIGURE 5.26** Problems 5.27 and 5.48.

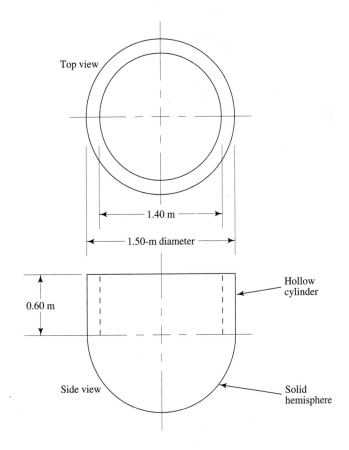

Top view

1.40 m

1.50-m diameter

0.60 m

Hollow cylinder

Side view

Solid hemisphere

**5.28M** *A light foam cup similar to a disposable coffee cup has a weight of 0.05 N, a uniform diameter of 82.0 mm, and a length of 150 mm. How much of its height would be submerged if placed in water?*

**5.29M** *A light foam cup similar to a disposable coffee cup has a weight of 0.05 N. A steel bar is placed inside the cup. The bar has a specific weight of 76.8 kN/m³, a diameter of 38.0 mm, and a length of 80.0 mm. How much of the height of the cup will be submerged if it is placed in water? The cup has a uniform diameter of 82.0 mm.*

**5.30M** *Repeat Problem 5.29, but consider that the steel bar is fastened outside the bottom of the cup instead of being placed inside.*

**5.31E** Figure 5.27 shows a raft made of four hollow drums supporting a platform. Each drum weighs 30 lb. How much total weight of the platform and anything placed on it can the raft support when the drums are completely submerged in fresh water?

**5.32E** Figure 5.28 shows the construction of the platform for the raft described in Problem 5.31. Compute its weight if it is made of wood from a specific weight of 40.0 lb/ft³.

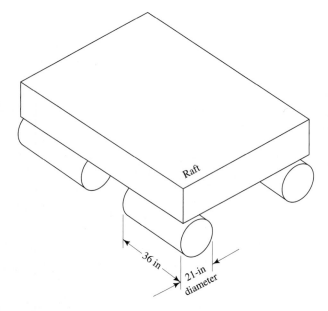

**FIGURE 5.27**   Problems 5.31, 5.33, and 5.34.

**FIGURE 5.28**   Raft construction for Problems 5.32 and 5.34.

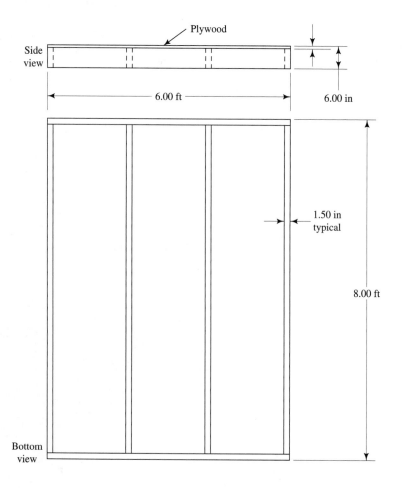

**5.33E** For the raft shown in Fig. 5.27, how much of the drums will be submerged when only the platform is being supported? Refer to Problems 5.31 and 5.32 for data.

**5.34E** For the raft and platform shown in Figs. 5.27 and 5.28 and described in Problems 5.31 and 5.32, what extra weight will cause all the drums and the platform itself to be submerged? Assume that no air is trapped beneath the platform.

**5.35E** A float in an ocean harbor is made from a uniform foam having a specific weight of 12.00 lb/ft³. It is made in the shape of a rectangular solid, 18.00 in square and 48.00 in long. A concrete (specific weight = 150 lb/ft³) block weighing 600 lb in air is attached to the float by a cable. The length of the cable is adjusted so that 14.00 in of the height of the float is above the surface with the long axis vertical. Compute the tension in the cable.

**5.36E** Describe how the situation described in Problem 5.35 will change if the water level rises by 18 in during high tide.

**5.37E** A cube, 6.00 in on a side, is made from aluminum having a specific weight of 0.100 lb/in³. If the cube is suspended on a wire with half its volume in water and the other half in oil (sg = 0.85), what is the tension in the string?

**5.38E** Figure 4.55 (Chapter 4) shows a solid cylinder sitting on the bottom of a tank holding a static volume of fluid. Compute the force exerted by the cylinder on the bottom of the tank for the following data: $D = 6.00$ in;

$L = 10.00$ in; $\gamma_c = 0.284$ lb/in³ (steel); $\gamma_f = 62.4$ lb/ft³; $h = 30.00$ in.

**Stability**

**5.39M** *A cylindrical block of wood is 1.00 m in diameter and 1.00 m long and has a specific weight of 8.00 kN/m³. Will it float in a stable manner in water with its axis vertical?*

**5.40E** A container for an emergency beacon is a rectangular shape 30.0 in wide, 40.0 in long, and 22.0 in high. Its center of gravity is 10.50 in above its base. The container weighs 250 lb. Will the box be stable with the 30 × 40 in side parallel to the surface in plain water?

**5.41E** The large platform shown in Fig. 5.29 carries equipment and supplies to offshore installations. The total weight of the system is 450 000 lb, and its center of gravity is even with the top of the platform, 8.00 ft from the bottom. Will the platform be stable in seawater in the position shown?

**5.42E** Will the cylindrical float described in Problem 5.4 be stable if placed in the fluid with its axis vertical?

**5.43M** *Will the buoy described in Problem 5.5 be stable if placed in the water with its axis vertical?*

**5.44M** *Will the float described in Problem 5.6 be stable if placed in the oil with its top surface horizontal?*

**5.45E** A closed, hollow, empty drum has a diameter of 24.0 in, a length of 48.0 in, and a weight of 70.0 lb. Will it float stably if placed upright in water?

**FIGURE 5.29**  Problem 5.41.

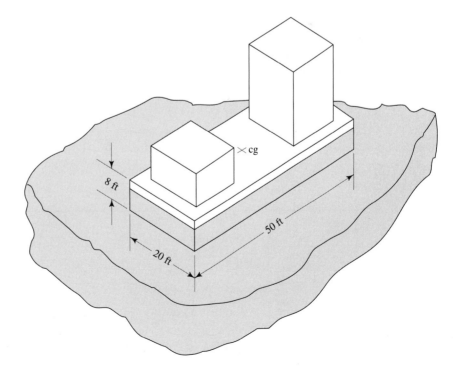

**5.46E** Figure 5.30 shows a river scow used to carry bulk materials. Assume that the scow's center of gravity is at its centroid and that it floats with 8.00 ft submerged. Determine the minimum width that will ensure stability in seawater.

**5.47E** Repeat Problem 5.46, except assume that crushed coal is added to the scow so that the scow is submerged to a depth of 16.0 ft and its center of gravity is raised to 13.50 ft from the bottom. Determine the minimum width for stability.

**5.48M** *For the vessel shown in Fig. 5.26 and described in Problem 5.27, assume that it floats with just the entire hemisphere submerged and that its center of gravity is 0.65 m from the top. Is it stable in the position shown?*

**5.49M** *For the foam cup described in Problem 5.28, will it float stably if placed in the water with its axis vertical?*

**5.50M** *Referring to Problem 5.29, assume that the steel bar is placed inside the cup with its long axis vertical. Will the cup float stably?*

**5.51M** *Referring to Problem 5.30, assume that the steel bar is fastened to the bottom of the cup with the long axis of the bar horizontal. Will the cup float stably?*

**5.52M** *Will the cylinder shown in Fig. 5.23 and described in Problem 5.22 be stable in the position shown?*

**5.53M** *Will the cylinder together with the brass plate shown in Fig. 5.25 and described in Problem 5.26 be stable in the position shown?*

**5.54E** A proposed design for a part of a seawall consists of a rectangular solid weighing 3840 lb with dimensions of 8.00 ft by 4.00 ft by 2.00 ft. The 8.00-ft side is to be vertical. Will this object float stably in seawater?

**5.55E** A platform is being designed to support some water pollution equipment. As shown in Fig. 5.31, its base is 36.00 in wide, 48.00 in long, and 12.00 in high. The entire system weighs 130 lb, and its center of gravity is 34.0 in

above the top surface of the platform. Is the proposed system stable when floating in seawater?

**5.56E** A block of wood with a specific weight of 32 lb/ft$^3$ is 6 by 6 by 12 in. If it is placed in oil (sg = 0.90) with the 6 by 12-inch surface parallel to the surface, would it be stable?

**5.57E** A barge is 60 ft long, 20 ft wide, and 8 ft deep. When empty, it weighs 210 000 lb, and its center of gravity is 1.5 ft above the bottom. Is it stable when floating in water?

**5.58E** If the barge in Problem 5.57 is loaded with 240 000 lb of loose coal having an average density of 45 lb/ft$^3$, how much of the barge would be below the water? Is it stable?

**5.59M** *A piece of cork having a specific weight of 2.36 kN/m$^3$ is shaped as shown in Fig. 5.32. (a) To what depth will it sink in turpentine (sg = 0.87) if placed in the orientation shown? (b) Is it stable in this position?*

**5.60M** *Figure 5.20 shows a cube floating in a fluid. (a) Derive an expression for the depth of submergence X that would ensure that the cube is stable in the position shown. (b) Using the expression derived in (a), determine the required distance X for a cube 75 mm on a side.*

**5.61M** *A boat has the cross section shown in Fig. 5.33(a). Its geometry at the water line is shown in the top view, Fig. 5.33(b). The hull is solid. Is the boat stable?*

**5.62E** (a) If the cone shown in Fig. 5.34 is made of pine wood with a specific weight of 30 lb/ft$^3$, will it be stable in the position shown floating in water? (b) Will it be stable if it is made of teak wood with a specific weight of 55 lb/ft$^3$?

**5.63M** *Refer to Figure 5.35. The vessel shown is to be used for a special experiment in which it will float in a fluid having a specific gravity of 1.16. It is required that the top surface of the vessel is 0.25 m above the fluid surface.*
**(a)** *What should be the total weight of the vessel and its contents?*

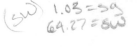
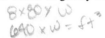

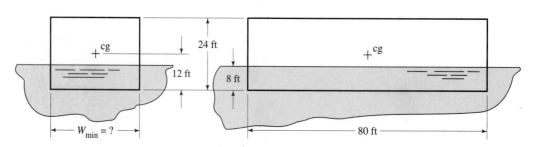

**FIGURE 5.30**   Problems 5.46 and 5.47.

**FIGURE 5.31**    Problem 5.55.

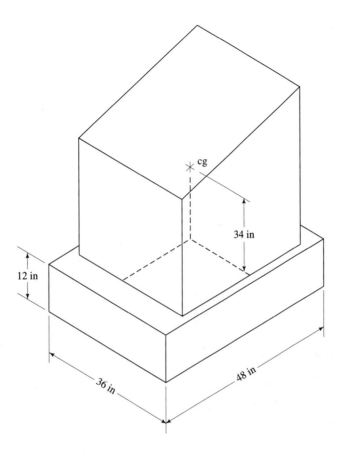

**(b)** *If the contents of the vessel have a weight of 5.0 kN, determine the required specific weight of the material from which the vessel is made.*

**(c)** *The center of gravity for the vessel and its contents is 0.40 m down from the rim of the open top of the cylinder. Is the vessel stable?*

**5.64E** A golf club head is made from aluminum having a specific weight of $0.100 \, \text{lb/in}^3$. In air it weighs 0.500 lb. What would be its apparent weight when suspended in cool water?

**FIGURE 5.32**    Problem 5.59.

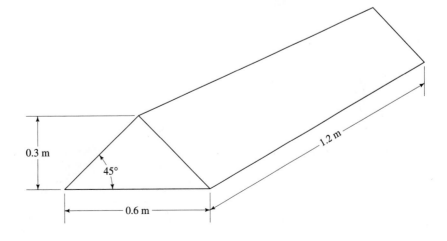

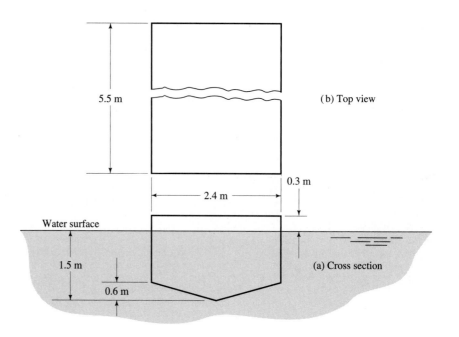

**FIGURE 5.33** Problem 5.61.

5.5 m

(b) Top view

0.3 m

2.4 m

Water surface

1.5 m

0.6 m

(a) Cross section

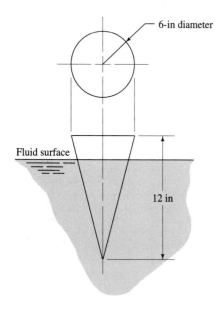

6-in diameter

Fluid surface

12 in

**FIGURE 5.34** Problem 5.62.

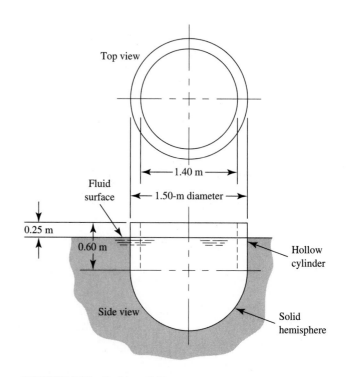

Top view

1.40 m

Fluid
surface

1.50-m diameter

0.25 m

0.60 m

Hollow
cylinder

Side view

Solid
hemisphere

**FIGURE 5.35** Problem 5.63.

## COMPUTER PROGRAMMING ASSIGNMENTS

1. Write a program to evaluate the stability of a circular cylinder placed in a fluid with its axis vertical. Call for input data for diameter, length, weight (or specific weight) of the cylinder; location of the center of gravity; and specific weight of the fluid. Solve for the position of the cylinder when it is floating, the location of the center of buoyancy, and the metacenter. Compare the location of the metacenter with the center of gravity to evaluate stability.

2. For any cylinder of a uniform density, floating in any fluid and containing a specified volume, vary the diameter from a small value to larger values in selected increments. Then, compute the required height of the cylinder to obtain the specified volume. Finally, evaluate the stability of the cylinder if it is placed in the fluid with its axis vertical.

3. For the results found in Programming Assignment 2, compute the metacentric height (as described in Section 5.6). Plot the metacentric height versus the diameter of the cylinder.

4. Write a program to evaluate the stability of a rectangular block placed in a fluid in a specified orientation. Call for input data for length, width, height, weight (or specific weight) of the block; location of the center of gravity; and specific weight of the fluid. Solve for the position of the block when it is floating, the location of the center of buoyancy, and the metacenter. Compare the location of the metacenter with the center of gravity to evaluate stability.

5. Write a program to determine the stability of a rectangular block with a given length and height as the width varies. Call for input data for length, height, weight (or specific weight), and fluid specific weight. Vary the width in selected increments from small values to larger values, and compute the range of widths for which the metacentric height is positive, that is, for which the design would be stable. Plot a graph of metacentric height versus width.

# 6 Flow of Fluids and Bernoulli's Equation

## 6.1
## The
## Big
## Picture

### Discussion Map

- You must learn the fundamental principles of fluid flow to help you develop the ability to analyze and design fluid systems to perform many useful functions.

- Three important parameters describe the flow of fluids:

  1. *Volume flow rate, Q*: Volume of fluid flowing past a given section per unit time.
  2. *Weight flow rate, W*: Weight of fluid flowing past a given section per unit time.
  3. *Mass flow rate, M*: Mass of fluid flowing past a given section per unit time.

- You will need to evaluate these terms at various points in a system using *the principle of continuity*.

- You must also learn to consider *kinetic energy, potential energy,* and *flow energy* when analyzing a fluid flow system. *Bernoulli's equation* is the tool for this and it is based on the principle of conservation of energy.

### Discover

- Where have you seen fluids being transported through pipes or tubes? Describe the system. What kind of fluid was involved? What was the fluid used for? How big was the pipe or tube? What material was it made from?
- Were there any devices in the pipe line that caused the fluid to change direction?
- Were there any places where the size of the pipe or tube changed? Did they get larger or smaller? Was the transition abrupt and sudden or was the fluid guided through the transition by a gradual, tapered path?
- Were there any control devices such as valves to adjust the amount of fluid that flows in the system or to permit the flow to be shut off completely?

*Take time to write descriptions of at least five situations where you observed fluids flowing in pipes and tubes.*

You will learn to apply these concepts as you study this chapter.

Were any of these items on your list of fluid flow systems?

- The water distribution system in a home consisting of copper or plastic pipe receiving water from a central source such as a municipal water supply system and delivering it to the kitchen sink, the bathroom, the laundry, and to the outside faucets.
- The municipal water supply system itself.
- The system that delivers washing and rinsing fluids in a car wash.

- The water pumping unit of the fire fighting brigade.
- An automatic lawn sprinkling system.
- A sprinkler system to protect offices, factories, and commercial buildings from damaging fires.
- The cooling system for an internal combustion engine in your car or truck.
- A system to deliver various processing fluids to manufacturing systems.
- The piping system for a fluid power automation system.
- The piping system for a fluid power system for construction of agricultural equipment.
- A pressure washer that uses a high-velocity stream of water or washing fluid to remove dirt from decks, houses, boats, machinery, or other surfaces.

Developing your ability to analyze and design such fluid flow systems is at the core of your study of fluid mechanics. You should be able to describe the way fluids flow in pipes, tubes, and noncircular flow conduits.

You should also be able to predict how the values of volume flow rate $Q$, weight flow rate $W$, and mass flow rate $M$ change as the size of the flow path changes. This is called the principle of continuity and it leads to the development and use of the continuity equation.

Next you will learn to account for three important kinds of energy contained by the fluid at any point in the system and to determine how the energy changes form as the fluid moves from one point to another. This requires the application of one of the classic relationships of physics, the principle of conservation of energy. The kinds of energy to be considered are:

**1.** Kinetic energy due to the velocity of the fluid at a given section of interest.
**2.** Potential energy due to the elevation of the fluid at the centerline of the section of interest.
**3.** Flow energy or pressure energy that is related to the pressure in the fluid at the section of interest.

You will bring all of these principles together in *Bernoulli's Equation* which states that the sum of these three forms of energy is the same at any point in a fluid flow system provided there are no energy additions or losses between the two points. But the distribution of energy among the three kinds may change. See Figure 6.1.

**FIGURE 6.1**   Portion of a fluid distribution system showing variations in velocity, pressure, and elevation.

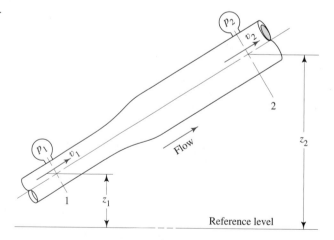

Bernoulli's equation provides the foundation for the work to come in the next seven chapters. Bear in mind that the material in these chapters is cumulative. One topic builds on another, to help you develop the ability to analyze and design complex fluid flow systems.

**6.2
OBJECTIVES**

After completing this chapter, you should be able to:

1. Define *volume flow rate*, *weight flow rate*, and *mass flow rate* and their units.
2. Define *steady flow* and the *principle of continuity*.
3. Write the continuity equation, and use it to relate the volume flow rate, area, and velocity of flow between two points in a fluid flow system.
4. Describe four types of commercially available pipe and tubing: steel pipe, ductile iron pipe, steel tubing, and copper tubing.
5. Specify the desired size of pipe or tubing to carry a given flow rate of fluid at a specified velocity.
6. State recommended velocities of flow and typical volume flow rates for various types of systems.
7. Define *potential energy*, *kinetic energy*, and *flow energy* as they relate to fluid flow systems.
8. Apply the principle of conservation of energy to develop *Bernoulli's equation*, and state the restrictions on its use.
9. Define the terms *pressure head*, *elevation head*, *velocity head*, and *total head*.
10. Apply Bernoulli's equation to fluid flow systems.
11. Define *Torricelli's theorem*, and apply it to compute the flow rate of fluid from a tank and the time required to empty a tank.

**6.3
FLUID FLOW RATE AND
THE CONTINUITY
EQUATION**

The quantity of fluid flowing in a system per unit time can be expressed by three different terms as defined below.

Q   The *volume flow rate* is the volume of fluid flowing past a section per unit time.

W   The *weight flow rate* is the weight of fluid flowing past a section per unit time.

M   The *mass flow rate* is the mass of fluid flowing past a section per unit time.

The most fundamental of these three terms is the volume flow rate $Q$, which is calculated from

▷ VOLUME FLOW RATE

$$Q = Av \qquad (6-1)$$

where $A$ is the area of the section and $v$ is the average velocity of flow. The units of $Q$ can be derived as follows, using SI units for illustration:

$$Q = Av = m^2 \times m/s = m^3/s$$

The weight flow rate $W$ is related to $Q$ by

▷ WEIGHT FLOW RATE

$$W = \gamma Q \qquad (6-2)$$

where $\gamma$ is the specific weight of the fluid. The units of $W$ are then

$$W = \gamma Q = N/m^3 \times m^3/s = N/s$$

The mass flow rate $M$ is related to $Q$ by

▷ MASS FLOW RATE

$$M = \rho Q \qquad (6-3)$$

where $\rho$ is the density of the fluid. The units of $M$ are then

$$M = \rho Q = kg/m^3 \times m^3/s = kg/s$$

**TABLE 6.1**   Flow rates

| Symbol | Name | Definition | SI Units | U.S. Customary System Units |
|--------|------|------------|----------|-----------------------------|
| $Q$ | Volume flow rate | $Q = Av$ | m³/s | ft³/s |
| $W$ | Weight flow rate | $W = \gamma Q$ | N/s | lb/s |
|   |   | $W = \gamma Av$ |   |   |
| $M$ | Mass flow rate | $M = \rho Q$ | kg/s | slugs/s |
|   |   | $M = \rho Av$ |   |   |

Table 6.1 summarizes these three types of fluid flow rates and gives the standard units in both the SI system and the U.S. Customary System. Because both the cubic meter per second and cubic foot per second are very large flow rates, other units are frequently used, such as liters per minute (L/min)*, m³/hr, and gallons per minute (gal/min or gpm; this text will use gal/min). Useful conversions are

CONVERSION FACTORS FOR VOLUME FLOW RATES

$$1.0 \text{ L/min} = 0.06 \text{ m}^3/\text{hr}$$
$$1.0 \text{ m}^3/\text{s} = 60\,000 \text{ L/min}$$
$$1.0 \text{ gal/min} = 3.785 \text{ L/min}$$
$$1.0 \text{ gal/min} = 0.2271 \text{ m}^3/\text{hr}$$
$$1.0 \text{ ft}^3/\text{s} = 449 \text{ gal/min}$$

Table 6.2 lists typical volume flow rates for different kinds of systems.

---

* The international symbol for liter is the lowercase "l," which can easily be confused with the numeral "1." The American National Metric Council, on the basis of a recommendation from the Department of Commerce, has therefore suggested the symbol "L" for United States use.

**TABLE 6.2**   Typical volume flow rates

| Flow rate m³/hr | (L/min) |  | Flow rate (gal/min) |
|-----------------|---------|--|---------------------|
| 0.9–7.5 | 15–125 | Reciprocating pumps handling heavy fluids and slurries | 4–33 |
| 0.60–6.0 | 10–100 | Industrial oil hydraulic systems | 3–30 |
| 6.0–36 | 100–600 | Hydraulic systems for mobile equipment | 30–150 |
| 2.4–270 | 40–4500 | Centrifugal pumps in chemical processes | 10–1200 |
| 12–240 | 200–4000 | Flood control and drainage pumps | 50–1000 |
| 2.4–900 | 40–15 000 | Centrifugal pumps handling mine wastes | 10–4000 |
| 108–570 | 1800–9500 | Centrifugal fire fighting pumps | 500–2500 |

☐ **EXAMPLE PROBLEM 6.1**    Convert a flow rate of 30 gal/min to ft³/s.

*Solution*    The flow rate is

$$Q = 30 \text{ gal/min} \left( \frac{1.0 \text{ ft}^3/\text{s}}{449 \text{ gal/min}} \right) = 6.68 \times 10^{-2} \text{ ft}^3/\text{s}$$

☐ **EXAMPLE PROBLEM 6.2**    Convert a flow rate of 600 L/min to m³/s.

*Solution*

$$Q = 600 \text{ L/min} \left( \frac{1.0 \text{ m}^3/\text{s}}{60\,000 \text{ L/min}} \right) = 0.010 \text{ m}^3/\text{s}$$

☐ **EXAMPLE PROBLEM 6.3**    Convert a flow rate of 30 gal/min to L/min.

*Solution*

$$Q = 30 \text{ gal/min} \left( \frac{3.785 \text{ L/min}}{1.0 \text{ gal/min}} \right) = 113.6 \text{ L/min}$$

The method of calculating the velocity of flow of a fluid in a closed pipe system depends on the *principle of continuity*. Consider the pipe in Fig. 6.1. A fluid is flowing from section 1 to section 2 at a constant rate. That is, the quantity of fluid flowing past any section in a given amount of time is constant. This is referred to as *steady flow*. Now if there is no fluid added, stored, or removed between section 1 and section 2, then the mass of fluid flowing past section 2 in a given amount of time must be the same as that flowing past section 1. This can be expressed in terms of the mass flow rate as

$$M_1 = M_2$$

or, since $M = \rho A v$, we have

$$\rho_1 A_1 v_1 = \rho_2 A_2 v_2 \tag{6-4}$$

**CONTINUITY EQUATION FOR ANY FLUID**

Equation (6–4) is a mathematical statement of the principle of continuity and is called the *continuity equation*. It is used to relate the fluid density, flow area, and velocity of flow at two sections of the system in which there is steady flow. It is valid for all fluids, whether gas or liquid.

If the fluid in the pipe in Fig. 6.1 is a liquid which can be considered incompressible, then the terms $\rho_1$ and $\rho_2$ in Eq. (6–4) are equal. The equation then becomes

**CONTINUITY EQUATION FOR LIQUIDS**

$$A_1 v_1 = A_2 v_2 \tag{6-5}$$

or, since $Q = Av$, we have

$$Q_1 = Q_2$$

Equation (6–5) is the continuity equation as applied to liquids; it states that for steady flow the volume flow rate is the same at any section. It can also be used for gases at low velocity, i.e., less than 100 m/s, with little error.

☐ **EXAMPLE PROBLEM 6.4**   In Fig. 6.1 the inside diameters of the pipe at sections 1 and 2 are 50 mm and 100 mm, respectively. Water at 70°C is flowing with an average velocity of 8 m/s at section 1. Calculate the following:
(a) Velocity at section 2
(b) Volume flow rate
(c) Weight flow rate
(d) Mass flow rate

*Solution*    (a) Velocity at section 2
From Eq. (6–5) we have

$$A_1 v_1 = A_2 v_2$$

$$v_2 = v_1 \left( \frac{A_1}{A_2} \right)$$

$$A_1 = \frac{\pi D_1^2}{4} = \frac{\pi (50 \text{ mm})^2}{4} = 1963 \text{ mm}^2$$

$$A_2 = \frac{\pi D_2^2}{4} = \frac{\pi (100 \text{ mm})^2}{4} = 7854 \text{ mm}^2$$

Then the velocity at section 2 is

$$v_2 = v_1 \left( \frac{A_1}{A_2} \right) = \frac{8.0 \text{ m}}{\text{s}} \times \frac{1963 \text{ mm}^2}{7854 \text{ mm}^2} = 2.0 \text{ m/s}$$

Notice that for steady flow of a liquid, as the flow area increases, the velocity decreases. This is independent of pressure and elevation.

(b) Volume flow rate, $Q$
From Table 6.1, $Q = Av$. Because of the principle of continuity we could use the conditions either at section 1 or at section 2 to calculate $Q$. At section 1 we have

$$Q = A_1 v_1 = 1963 \text{ mm}^2 \times \frac{8.0 \text{ m}}{\text{s}} \times \frac{1 \text{ m}^2}{(10^3 \text{ mm})^2} = 0.0157 \text{ m}^3/\text{s}$$

(c) Weight flow rate, $W$
From Table 6.1, $W = \gamma Q$. At 70°C the specific weight of water is 9.59 kN/m³.
Then the weight flow rate is

$$W = \gamma Q = \frac{9.59 \text{ kN}}{\text{m}^3} \times \frac{0.0157 \text{ m}^3}{\text{s}} = 0.151 \text{ kN/s}$$

(d) Mass flow rate, $M$
From Table 6.1, $M = \rho Q$. At 70°C the density of water is 978 kg/m³. Then the mass flow rate is

$$M = \rho Q = \frac{978 \text{ kg}}{\text{m}^3} \times \frac{0.0157 \text{ m}^3}{\text{s}} = 15.36 \text{ kg/s}$$

☐ **EXAMPLE PROBLEM 6.5**   At one section in an air distribution system, air at 14.7 psia and 100°F has an average velocity of 1200 ft/min and the duct is 12 in square. At another section, the duct is round with a diameter of 18 in, and the velocity is measured to be 900 ft/min. Calculate (a) the density of the air in the round section and (b) the weight flow rate of air in pounds per hour. At 14.7 psia and 100°F, the density of air is $2.20 \times 10^{-3}$ slugs/ft³ and the specific weight is $7.09 \times 10^{-2}$ lb/ft³.

**Solution** According to the continuity equation for gases, Eq. (6–4), we have

$$\rho_1 A_1 v_1 = \rho_2 A_2 v_2$$

Then, we can calculate the area of the two sections and solve for $\rho_2$.

$$\rho_2 = \rho_1 \left(\frac{A_1}{A_2}\right)\left(\frac{v_1}{v_2}\right)$$

$$A_1 = (12 \text{ in})(12 \text{ in}) = 144 \text{ in}^2$$

$$A_2 = \frac{\pi D_2^2}{4} = \frac{\pi(18 \text{ in})^2}{4} = 254 \text{ in}^2$$

**(a)** Then, the density of the air in the round section is

$$\rho_2 = (2.20 \times 10^{-3} \text{ slugs/ft}^3)\left(\frac{144 \text{ in}^2}{254 \text{ in}^2}\right)\left(\frac{1200 \text{ ft/min}}{900 \text{ ft/min}}\right)$$

$$\rho_2 = 1.66 \times 10^{-3} \text{ slugs/ft}^3$$

**(b)** The weight flow rate can be found at section 1 from $W = \gamma_1 A_1 v_1$. Then, the weight flow rate is

$$W = \gamma_1 A_1 v_1$$

$$W = (7.09 \times 10^{-2} \text{ lb/ft}^3)(144 \text{ in}^2)\left(\frac{1200 \text{ ft}}{\text{min}}\right)\left(\frac{1 \text{ ft}^2}{144 \text{ in}^2}\right)\left(\frac{60 \text{ min}}{\text{h}}\right)$$

$$W = 5100 \text{ lb/h}$$

■

---

**6.4**
**COMMERCIALLY**
**AVAILABLE PIPE**
**AND TUBING**

The actual outside and inside diameters of standard commercially available pipe and tubing may be quite different from the nominal size given. We will describe several widely used types of standard pipe and tubing in this section. Data are given in the appendices for outside diameter, inside diameter, wall thickness, and flow area for some of these types. The nominal sizes for commercially available pipe still refer to an "inch" size even though the transition to the SI system is an international trend. Since the nominal size is used solely to designate a certain pipe or tube, the standard conventional size will be used in this book. However, as you can see in Appendices F, G, H, and I, the dimensions are listed in millimeters (mm) for outside diameter, inside diameter, and wall thickness. The flow area is listed in square meters (m²) to help you maintain consistent units in calculations. Data are also given in the U.S. Customary System.

**6.4.1**
**Steel Pipe**

General-purpose pipe lines are often constructed of steel pipe. Standard pipe sizes are designated by the nominal size and schedule number. Schedule numbers are related to the permissible operating pressure of the pipe and to the allowable stress of the steel in the pipe. The range of schedule numbers is from 10 to 160, with the higher numbers indicating a heavier wall thickness. Since all schedules of pipe of a given nominal size have the same outside diameter, the higher schedules have a smaller inside diameter. The most complete series of steel pipe available are Schedules 40 and 80. Data for these two schedules are given in SI units and in U.S. Customary System units in Appendix F. Refer to ANSI/ASME Standard B31.1-1998 *Power Piping* for a method of computing the minimum acceptable wall thickness for pipes. See Reference 1.

**6.4.2**
**Steel Tubing**

Standard steel tubing is used in fluid power systems, condensers, heat exchangers, engine fuel systems, and industrial fluid processing systems. Sizes are designated by outside diameter and wall thickness. Standard sizes from $\frac{1}{8}$ in to 2 in for several wall thickness gauges are tabulated in Appendix G. Other wall thicknesses are available.

**6.4.3**
**Copper Tubing**

Household plumbing, refrigerant lines, and compressed air lines often use copper tubing manufactured as Type K or Type L. Type K has the greater wall thickness and is recommended for underground service. Type L is suitable for general-purpose interior plumbing. The nominal size of copper tubing is $\frac{1}{8}$ in less than the actual outside diameter of the tube. Data for wall thickness, inside diameter, and flow area are given in Appendix H for Type K tubing.

**6.4.4**
**Ductile Iron Pipe**

Water, gas, and sewage lines are often made of ductile iron pipe because of its strength, ductility, and relative ease of handling. It has replaced cast iron in many applications. Standard fittings are supplied with the pipe for convenient installation above or below ground. Several classes of ductile iron pipe are available for use in systems with a range of pressures. Appendix I lists the dimensions of Class 150 pipe for 150 psi (1.03 MPa) service in nominal sizes from 3 to 24 inches. Actual inside and outside diameters are larger than nominal sizes.

**6.4.5**
**Other Types of
Pipe and Tubing**

Brass pipe is used in corrosive fluids, as is stainless steel. Other materials used are aluminum, lead, tin, vitrified clay, concrete, and many types of plastics, such as polyethylene, nylon, and polyvinyl chloride. No specific data are included in this book for these types of pipes and tubing.

**6.4.6**
**Flow Areas**

When problems in this book identify a particular size and type of pipe or tube, look in the corresponding appendix table for the necessary diameters and flow areas. When actual diameters are given, you may compute the flow area from $A = \pi D^2/4$.

**6.5
RECOMMENDED
VELOCITY OF FLOW IN
PIPE AND TUBING**

Factors affecting the selection of a satisfactory velocity of flow in fluid systems are numerous. Some of the important ones are the type of fluid, the length of the flow system, the type of pipe or tube, the pressure drop that can be tolerated, the devices (such as pumps, valves, etc.) that may be connected to the pipe or tube, temperature, pressure, and noise.

When we discussed the continuity equation in Section 6.3, we learned that the velocity of flow increases as the area of the flow path decreases. Therefore, smaller tubes will cause high velocities, and larger tubes will provide low velocities. Later we will explain that energy losses and the corresponding pressure drop increase dramatically as the flow velocity increases. For this reason it is desirable to keep the velocities low. But, since larger pipes and tubes are more costly, some limits are necessary.

Table 6.3 provides some guidelines for specifying pipe and tube sizes for typical pumped fluid distribution systems. In general, the flow velocity is kept lower in suction lines providing flow into a pump to aid in the process of filling the suction inlet passages. Also, the pressure at the inlet to a pump must be kept relatively high to ensure that pure liquid enters the pump. Lower pressures than those specified by the pump manufacturer may cause a damaging condition called *cavitation* to occur. This is discussed in more detail in Chapter 13 on pump selection and application. Cavitation causes excessive noise and may lead to rapid erosion of pump surfaces in the inlet passages and the impeller.

**TABLE 6.3** Pipe size selection based on flow velocity

| Suction Lines Flowing to the Inlet to a Pump Recommended Range of Velocity: 3 to 20 ft/s | | | | | | |
|---|---|---|---|---|---|---|
| | | Volume flow rate (gal/min) for given velocity | | | | |
| Pipe size Sch 40 | Flow area (ft²) | Velocity (ft/s) | | | | |
| | | 3 | 7.5 | 10 | 15 | 20 |
| 1/2 | 0.00211 | 2.84 | 7.11 | 9.47 | 14.2 | 18.9 |
| 3/4 | 0.00370 | 4.98 | 12.5 | 16.6 | 24.9 | 33.2 |
| 1 | 0.00600 | 8.08 | 20.2 | 26.9 | 40.4 | 53.9 |
| 1 1/4 | 0.01039 | 14.0 | 35.0 | 46.7 | 70.0 | 93.3 |
| 1 1/2 | 0.01414 | 19.0 | 47.6 | 63.5 | 95.2 | 127 |
| 2 | 0.02333 | 31.4 | 78.6 | 105 | 157 | 210 |
| 2 1/2 | 0.03326 | 44.8 | 112 | 149 | 224 | 299 |
| 3 | 0.05132 | 69.1 | 173 | 230 | 346 | 461 |
| 3 1/2 | 0.06868 | 92.5 | 231 | 308 | 463 | 617 |
| 4 | 0.08840 | 119 | 298 | 397 | 595 | 794 |
| 5 | 0.1390 | 187 | 468 | 624 | 936 | 1248 |
| 6 | 0.2006 | 270 | 676 | 901 | 1351 | 1801 |
| 8 | 0.3472 | 468 | 1,169 | 1,559 | 2,338 | 3,118 |
| 10 | 0.5479 | 738 | 1,845 | 2,460 | 3,690 | 4,920 |

| Discharge Lines Flowing from the Outlet of a Pump Recommended Range of Velocity: 10 to 30 ft/s | | | | | | |
|---|---|---|---|---|---|---|
| | | Volume flow rate (gal/min) for given velocity | | | | |
| Pipe size Sch 40 | Flow area (ft²) | Velocity (ft/s) | | | | |
| | | 10 | 15 | 20 | 25 | 30 |
| 1/2 | 0.00211 | 9.47 | 14.21 | 18.9 | 23.7 | 28.4 |
| 3/4 | 0.00370 | 16.61 | 24.9 | 33.2 | 41.5 | 49.8 |
| 1 | 0.00600 | 26.9 | 40.4 | 53.9 | 67.4 | 80.8 |
| 1 1/4 | 0.01039 | 46.7 | 70.0 | 93.3 | 116.6 | 140 |
| 1 1/2 | 0.01414 | 63.5 | 95.2 | 127.0 | 159 | 190 |
| 2 | 0.02333 | 104.8 | 157 | 210 | 262 | 314 |
| 2 1/2 | 0.03326 | 149.3 | 224 | 299 | 373 | 448 |
| 3 | 0.05132 | 230 | 346 | 461 | 576 | 691 |
| 3 1/2 | 0.06868 | 308 | 463 | 617 | 771 | 925 |
| 4 | 0.08840 | 397 | 595 | 794 | 992 | 1191 |
| 5 | 0.1390 | 624 | 936 | 1,248 | 1,560 | 1,872 |
| 6 | 0.2006 | 901 | 1,351 | 1,801 | 2,252 | 2,702 |
| 8 | 0.3472 | 1,559 | 2,338 | 3,118 | 3,897 | 4,677 |
| 10 | 0.5479 | 2,460 | 3,690 | 4,920 | 6,150 | 7,380 |

**TABLE 6.3** (*continued*)

| | | Suction Lines Flowing to the Inlet to a Pump: Recommended Range of Velocity: 1.0 to 6.0 m/s | | | | |
|---|---|---|---|---|---|---|
| | Flow | Volume flow rate (m³/hr) for given velocity | | | | |
| Pipe size Sch 40 | area (m²) | Velocity (m/s) | | | | |
| | | 1 | 2.25 | 3 | 4.5 | 6 |
| 1/2 | 1.960E-04 | 0.71 | 1.59 | 2.12 | 3.18 | 4.23 |
| 3/4 | 3.437E-04 | 1.24 | 2.78 | 3.71 | 5.57 | 7.42 |
| 1 | 5.574E-04 | 2.01 | 4.52 | 6.02 | 9.03 | 12.0 |
| 1 1/4 | 9.653E-04 | 3.47 | 7.82 | 10.4 | 15.6 | 20.8 |
| 1 1/2 | 1.314E-03 | 4.73 | 10.6 | 14.2 | 21.3 | 28.4 |
| 2 | 2.167E-03 | 7.80 | 17.6 | 23 | 35.1 | 46.8 |
| 2 1/2 | 3.090E-03 | 11.1 | 25.0 | 33 | 50.1 | 66.7 |
| 3 | 4.768E-03 | 17.2 | 38.6 | 51 | 77.2 | 103 |
| 3 1/2 | 6.381E-03 | 23.0 | 51.7 | 69 | 103 | 138 |
| 4 | 8.213E-03 | 29.6 | 66.5 | 89 | 133 | 177 |
| 5 | 0.01291 | 46.5 | 105 | 139 | 209 | 279 |
| 6 | 0.01864 | 67.1 | 151 | 201 | 302 | 403 |
| 8 | 0.03226 | 116 | 261 | 348 | 523 | 697 |
| 10 | 0.05090 | 183 | 412 | 550 | 825 | 1099 |

| | | Discharge Lines Flowing from the Outlet of a Pump: Recommended Range of Velocity: 3.0 to 9.0 m/s | | | | |
|---|---|---|---|---|---|---|
| | Flow | Volume flow rate (m³/hr) for given velocity | | | | |
| Pipe size Sch 40 | area (m²) | Velocity (m/s) | | | | |
| | | 3 | 4.5 | 6 | 7.5 | 9 |
| 1/2 | 1.960E-04 | 2.12 | 3.18 | 4.23 | 5.29 | 6.35 |
| 3/4 | 3.437E-04 | 3.71 | 5.57 | 7.42 | 9.28 | 11.1 |
| 1 | 5.574E-04 | 6.02 | 9.03 | 12.0 | 15.1 | 18.1 |
| 1 1/4 | 9.653E-04 | 10.4 | 15.6 | 20.8 | 26.1 | 31.3 |
| 1 1/2 | 1.314E-03 | 14.2 | 21.3 | 28.4 | 35.5 | 42.6 |
| 2 | 2.167E-03 | 23.4 | 35.1 | 46.8 | 58.5 | 70.2 |
| 2 1/2 | 3.090E-03 | 33.4 | 50.1 | 66.7 | 83.4 | 100 |
| 3 | 4.768E-03 | 51.5 | 77.2 | 103 | 129 | 154 |
| 3 1/2 | 6.381E-03 | 68.9 | 103 | 138 | 172 | 207 |
| 4 | 8.213E-03 | 88.7 | 133 | 177 | 222 | 266 |
| 5 | 0.01291 | 139 | 209 | 279 | 349 | 418 |
| 6 | 0.01864 | 201 | 302 | 403 | 503 | 604 |
| 8 | 0.03226 | 348 | 523 | 697 | 871 | 1,045 |
| 10 | 0.05090 | 550 | 825 | 1,099 | 1,374 | 1,649 |

Discharge line velocities at the outlet of a pump can generally be higher than those in the suction line. Most commercially available pumps used for the transfer of fluids in water systems and process operations specify the required size of pipe for both the suction and discharge connections and the suction line size is larger than the discharge line size. Typically, then, the pipes leading to and from the pump have these same sizes. However, it is your responsibility as the system designer to verify that energy losses are within acceptable limits to ensure that an adequate flow rate is maintained and that pressures are adequate at functional elements such as sprinkler heads, nozzles, and within process equipment.

The total energy loss in a system must be overcome by the pump so it is desirable to minimize such losses for optimum energy use. This may require that larger pipes than the standard pump connection sizes are used in some sections of the system to produce lower velocities. Then appropriate enlargements and reducers can be used to facilitate connections to the pump. This topic will be discussed frequently in future chapters where methods of computing energy losses are presented.

You will note in Chapter 13 on pump selection and application that a given size pump can deliver a wide range of flow rates by changing its speed of operation, by using variable displacement techniques in the pump, and by changing the size of impellers in centrifugal pumps. Therefore, the nominal size pipe corresponding to the suction and discharge port sizes may not be suitable for the entire range.

It is recommended that the smaller pipes typically operate at lower velocities than larger pipes so the darker shaded areas of Table 6.3 are preferred for use in this book. The final specifications of pipe sizes should consider the entire life cycle of the system and the compatibility of all elements. Certainly, some successful systems use velocities different from those shown in Table 6.3.

Data for volume flow rate in Table 6.3 are given in gal/min for the U.S. Customary System and in $m^3/hr$ for the SI system because most manufacturers rate their pumps in such units. Conversions to the standard units of $ft^3/s$ and $m^3/s$ must be done before using the flow rates in calculations in this book.

---

☐ **EXAMPLE PROBLEM 6.6**

Determine the maximum allowable volume flow rate in L/min that can be carried through a standard steel tube with an outside diameter of $1\frac{1}{4}$ in and a 0.065 in wall thickness if the maximum velocity is to be 3.0 m/s.

**Solution**

Using the definition of volume flow rate, we have

$$Q = Av$$
$$A = 6.356 \times 10^{-4} \, m^2 \quad \text{(from Appendix G)}$$

Then we find the flow rate:

$$Q = (6.356 \times 10^{-4} \, m^2)(3.0 \, m/s) = 1.907 \times 10^{-3} \, m^3/s$$

Converting to L/min, we have

$$Q = 1.907 \times 10^{-3} \, m^3/s \left( \frac{60\,000 \, \text{L/min}}{1.0 \, m^3/s} \right) = 114 \, \text{L/min}$$

---

☐ **EXAMPLE PROBLEM 6.7**

Determine the required size standard Schedule 40 steel pipe to carry 192 $m^3/hr$ of water with a maximum velocity of 6.0 m/s.

*Solution*   Since $Q$ and $v$ are known, the required area can be found from

$$Q = Av$$

$$A = Q/v$$

First, we must convert the volume flow rate to the units of $m^3/s$:

$$Q = 192 \text{ m}^3/\text{hr} \ (1 \text{ hr}/3600 \text{ s}) = 0.0533 \text{ m}^3/\text{s}$$

Then, we have

$$A = \frac{Q}{v} = \frac{0.0533 \text{ m}^3/\text{s}}{6.0 \text{ m/s}} = 0.008 \ 88 \text{ m}^2 = 8.88 \times 10^{-3} \text{ m}^2$$

This must be interpreted as the *minimum* allowable area since any smaller area would produce a velocity higher than 6.0 m/s. Therefore, we must look in Appendix F for a standard pipe with a flow area just larger than $8.88 \times 10^{-3}$ m². A standard 5-in Schedule 40 steel pipe, with a flow area of $1.291 \times 10^{-2}$ m², is required. The actual velocity of flow when this pipe carries 0.0533 m³/s of water is

$$v = \frac{Q}{A} = \frac{0.0533 \text{ m}^3/\text{s}}{1.291 \times 10^{-2} \text{ m}^2} = 4.13 \text{ m/s}$$

If the next smaller pipe (a 4-in Schedule 40 pipe) were used, the velocity would have been

$$v = \frac{Q}{A} = \frac{0.0533 \text{ m}^3/\text{s}}{8.213 \times 10^{-3} \text{ m}^2} = 6.49 \text{ m/s} \quad \text{(too high)}$$

*Comment*   Table 6.3 could also be used to solve the problem.

■

**6.6**
**FLOW IN**
**NONCIRCULAR**
**SECTIONS**

The continuity equation applies equally well to flow in noncircular sections as it does in circular pipes and tubes. In the formula for volume flow rate, $Q = Av$, the area, $A$, is the *net flow area* and $v$ is the average velocity of flow when carrying the given volume flow rate $Q$.

☐ **EXAMPLE PROBLEM 6.8**   Figure 6.2 shows a heat exchanger used to transfer heat from the fluid flowing inside the inner tube to that flowing in the space between the outside of the tube and the inside of the square shell that surrounds the tube. Such a device is often called a *shell and tube heat exchanger*. Compute the volume flow rate in gal/min that would produce a velocity of 8.0 ft/s both inside the tube and in the shell.

*Solution*   We use the formula for volume flow rate, $Q = Av$, for each part.
(a) Inside the $\frac{1}{2}$-in Type K copper tube:
    From Appendix H, we can read:

$$OD = 0.625 \text{ in}$$

$$ID = 0.527 \text{ in}$$

$$\text{Wall thickness} = 0.049 \text{ in}$$

$$A_t = 1.515 \times 10^{-3} \text{ ft}^2$$

Then, the volume flow rate inside the tube is

$$Q_t = A_t v = (1.515 \times 10^{-3} \text{ ft}^2)(8.0 \text{ ft/s}) = 0.01212 \text{ ft}^3/\text{s}$$

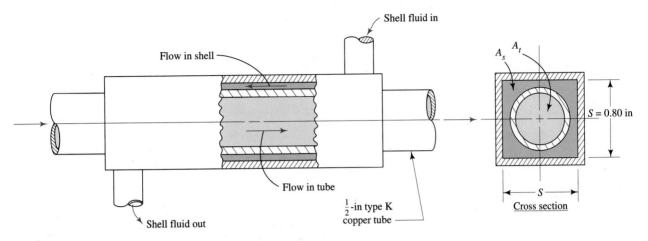

**FIGURE 6.2** Shell and tube heat exchanger for Example Problem 6.8.

Converting to gal/min gives

$$Q_t = 0.01212 \text{ ft}^3/\text{s} \; \frac{449 \text{ gal/min}}{1.0 \text{ ft}^3/\text{s}} = 5.44 \text{ gal/min}$$

**(b)** In the shell:

The net flow area is the difference between the area *inside* the square shell and the *outside* of the tube. Then,

$$A_s = S^2 - \pi OD^2/4$$
$$A_s = (0.80 \text{ in})^2 - \pi(0.625 \text{ in})^2/4 = 0.3332 \text{ in}^2$$

Converting to $\text{ft}^2$ gives

$$A_s = 0.3332 \text{ in}^2 \; \frac{1.0 \text{ ft}^2}{144 \text{ in}^2} = 2.314 \times 10^{-3} \text{ ft}^2$$

The required volume flow rate is then

$$Q_s = A_s v = (2.314 \times 10^{-3} \text{ ft}^2)(8.0 \text{ ft/s}) = 0.01851 \text{ ft}^3/\text{s}$$
$$Q_s = 0.01851 \text{ ft}^3/\text{s} \; \frac{449 \text{ gal/min}}{1.0 \text{ ft}^3/\text{s}} = 8.31 \text{ gal/min}$$

The ratio of the flow in the shell to the flow in the tube is

$$\text{Ratio} = Q_s/Q_t = 8.31/5.44 = 1.53$$

■

**6.7
CONSERVATION
OF ENERGY—
BERNOULLI'S EQUATION**

The analysis of a pipe line problem such as that illustrated in Fig. 6.1 accounts for all the energy within the system. In physics you learned that energy can be neither created nor destroyed, but it can be transformed from one form to another. This is a statement of the law of *conservation of energy*.

There are three forms of energy which are always considered when analyzing a pipe flow problem. Consider an element of fluid, as shown in Fig. 6.3, that may be inside a pipe in a flow system. It would be located at a certain elevation $z$, have a certain velocity $v$, and have a pressure $p$. The element of fluid would possess the following forms of energy:

**FIGURE 6.3**   Element of a fluid in a pipe.

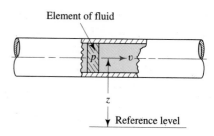

1. *Potential Energy.* Due to its elevation, the potential energy of the element relative to some reference level is

$$PE = wz \qquad (6\text{--}6)$$

where $w$ is the weight of the element.

2. *Kinetic Energy.* Due to its velocity, the kinetic energy of the element is

$$KE = wv^2/2g \qquad (6\text{--}7)$$

3. *Flow Energy.* Sometimes called *pressure energy* or *flow work*, this represents the amount of work necessary to move the element of fluid across a certain section against the pressure $p$. Flow energy is abbreviated FE and is calculated from

$$FE = wp/\gamma \qquad (6\text{--}8)$$

Equation (6–8) can be derived as follows. Figure 6.4 shows the element of fluid in the pipe being moved across a section. The force on the element is $pA$, where $p$ is the pressure at the section and $A$ is the area of the section. In moving the element across the section, the force moves a distance $L$ equal to the length of the element. Therefore, the work done is

$$\text{Work} = pAL = pV$$

where $V$ is the volume of the element. The weight of the element $w$ is

$$w = \gamma V$$

where $\gamma$ is the specific weight of the fluid. Then, the volume of the element is

$$V = w/\gamma$$

and we have

$$\text{Work} = pV = pw/\gamma$$

which is called flow energy in Eq. (6–8).

**FIGURE 6.4**   Flow energy.

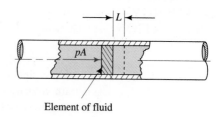

The total amount of energy of these three forms possessed by the element of fluid would be the sum, called $E$:

$$E = FE + PE + KE$$
$$E = wp/\gamma + wz + wv^2/2g$$

Each of these terms is expressed in units of energy which are newton-meters (N·m) in the SI unit system or foot-pounds (ft-lb) in the U.S. Customary System.

Now consider the element of fluid in Fig. 6.5, which moves from a section 1 to a section 2. The values for $p$, $z$, and $v$ will be different at the two sections. At section 1, the total energy is

$$E_1 = \frac{wp_1}{\gamma} + wz_1 + \frac{wv_1^2}{2g}$$

**FIGURE 6.5**  Fluid elements used in Bernoulli's equation.

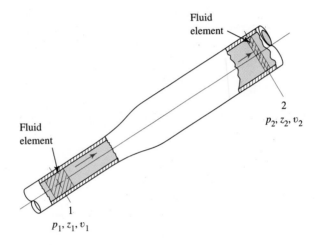

At section 2, the total energy is

$$E_2 = \frac{wp_2}{\gamma} + wz_2 + \frac{wv_2^2}{2g}$$

If no energy is added to the fluid or lost between sections 1 and 2, then the principle of conservation of energy requires that

$$E_1 = E_2$$
$$\frac{wp_1}{\gamma} + wz_1 + \frac{wv_1^2}{2g} = \frac{wp_2}{\gamma} + wz_2 + \frac{wv_2^2}{2g}$$

The weight of the element $w$ is common to all terms and can be divided out. The equation then becomes

$$\frac{p_1}{\gamma} + z_1 + \frac{v_1^2}{2g} = \frac{p_2}{\gamma} + z_2 + \frac{v_2^2}{2g} \tag{6–9}$$

⇨  BERNOULLI'S EQUATION

This is referred to as *Bernoulli's equation.*

**6.8**
**INTERPRETATION OF**
**BERNOULLI'S EQUATION**

Each term in Bernoulli's equation (Eq. 6–9) resulted from dividing an expression for energy by the weight of an element of the fluid. Therefore,

*Each term in Bernoulli's equation is one form of the energy possessed by the fluid per unit weight of fluid flowing in the system.*

The units for each term are "energy per unit weight." In the SI system the units are N·m/N and in the U.S. Customary System the units are lb·ft/lb.

But notice that the force (or weight) unit appears in both the numerator and the denominator and it can be cancelled. The resulting unit is simply the meter (m) or foot (ft) and it can be interpreted to be a height. In fluid flow analysis, the terms are typically expressed as "head" referring to a height above a reference level. Specifically,

> $p/\gamma$ is called the *pressure head*.
>
> *z* is called the *elevation head*.
>
> $v^2/2g$ is called the *velocity head*.
>
> *The sum of these three terms is called the total head.*

Because each term in Bernoulli's equation represents a height, a diagram similar to that shown in Fig. 6.6 is helpful to visualize the relationship among the three types of energy. As the fluid moves from point 1 to point 2, the magnitude of each term may change in value. But, if no energy is lost from or added to the fluid, the total head remains at a constant level. Bernoulli's equation is used to determine how the values of pressure head, elevation head, and velocity head change as the fluid moves through the system.

In Fig. 6.6 you should see that the velocity head at section 2 will be less than that at section 1. This can be shown by the continuity equation,

$$A_1 v_1 = A_2 v_2$$
$$v_2 = v_1 (A_1/A_2)$$

**FIGURE 6.6** Pressure head, elevation head, velocity head, and total head.

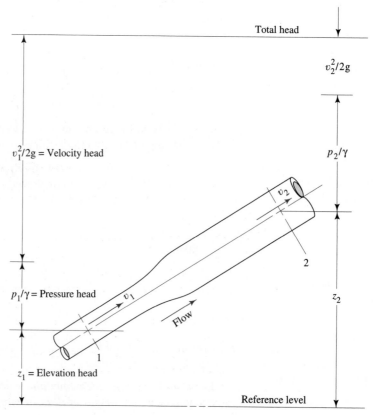

Total head

$v_2^2/2g$

$v_1^2/2g$ = Velocity head

$p_2/\gamma$

$p_1/\gamma$ = Pressure head

$v_1$

Flow

2

$v_2$

$z_2$

$z_1$ = Elevation head

Reference level

Because $A_1 < A_2$, $v_2$ must be less than $v_1$. And because the velocity is squared in the velocity head term, $v_2^2/2g$ is much less than $v_1^2/2g$.

Typically, when the section size expands as it does in Fig. 6.6, the pressure head increases because the velocity head decreases. That is how Fig. 6.6 is constructed. However, the actual change is also affected by the change in the elevation head.

In summary,

> *Bernoulli's equation accounts for the changes in elevation head, pressure head, and velocity head between two points in a fluid flow system. It is assumed that there are no energy losses or additions between the two points so the total head remains constant.*

When writing Bernoulli's equation, it is essential that the pressures at the two reference points both be expressed as absolute pressures or both as gage pressures. That is, they must both have the same reference pressure. In most problems it will be convenient to use gage pressure because parts of the fluid system exposed to the atmosphere will then have zero pressure. Also, most pressures are measured by a gage relative to the local atmospheric pressure.

**6.9**
**RESTRICTIONS ON BERNOULLI'S EQUATION**

Although Bernoulli's equation is applicable to a larger number of practical problems, there are several limitations that must be understood in order to apply it properly.

1. It is valid only for incompressible fluids since the specific weight of the fluid is assumed to be the same at the two sections of interest.
2. There can be no mechanical devices between the two sections of interest that would add energy to or remove energy from the system, since the equation states that the total energy in the fluid is constant.
3. There can be no heat transferred into or out of the fluid.
4. There can be no energy lost due to friction.

In reality no system satisfies all these restrictions. However, there are many systems for which only a negligible error will result when Bernoulli's equation is used. Also, the use of this equation may allow a fast estimate of a result when that is all that is required. In Chapter 7, limitations 2 and 4 will be eliminated by expanding Bernoulli's equation into the *general energy equation*.

**6.10**
**APPLICATIONS OF BERNOULLI'S EQUATION**

We will present several programmed example problems to illustrate the use of Bernoulli's equation. Although it is not possible to cover all types of problems with a certain solution method, the general approach to problems of fluid flow is described here.

**PROCEDURE FOR APPLYING BERNOULLI'S EQUATION**

1. Decide which items are known and what is to be found.
2. Decide which two sections in the system will be used when writing Bernoulli's equation. One section is chosen for which much data is known. The second is usually the section at which something is to be calculated.
3. Write Bernoulli's equation for the two selected sections in the system. It is important that the equation is written *in the direction of flow*. That is, the flow must proceed *from* the section on the left side of the equation *to* that on the right side.
4. Be explicit when labeling the subscripts for the pressure head, elevation head, and velocity head terms in Bernoulli's equation. You should note where the reference points are on a sketch of the system.

5. Simplify the equation, if possible, by cancelling terms that are zero or those that are equal on both sides of the equation.
6. Solve the equation algebraically for the desired term.
7. Substitute known quantities and calculate the result, being careful that consistent units are used throughout the calculation.

## PROGRAMMED EXAMPLE PROBLEM

☐ **EXAMPLE PROBLEM 6.9**    In Fig. 6.6, water at 10°C is flowing from section 1 to section 2. At section 1, which is 25 mm in diameter, the gage pressure is 345 kPa and the velocity of flow is 3.0 m/s. Section 2, which is 50 mm in diameter, is 2.0 m above section 1. Assuming there are no energy losses in the system, calculate the pressure $p_2$.

List the items that are known from the problem statement before looking at the next panel.

$$D_1 = 25 \text{ mm} \qquad v_1 = 3.0 \text{ m/s} \qquad z_2 - z_1 = 2.0 \text{ m}$$
$$D_2 = 50 \text{ mm} \qquad p_1 = 345 \text{ kPa(gage)}$$

The pressure $p_2$ is to be found. In other words, we are asked to calculate the pressure at section 2, which is different from the pressure at section 1 because there is a change in elevation and flow area between the two sections.

We are going to use Bernoulli's equation to solve the problem. Which two sections should be used when writing the equation?

In this case, sections 1 and 2 are the obvious choices. At section 1, we know $p_1$, $v_1$, and $z_1$. The unknown pressure $p_2$ is at section 2.

Now write Bernoulli's equation. [See Eq. (6–9).]

It should look like this:

$$\frac{p_1}{\gamma} + z_1 + \frac{v_1^2}{2g} = \frac{p_2}{\gamma} + z_2 + \frac{v_2^2}{2g}$$

The three terms on the left refer to section 1, while the three on the right refer to section 2.

Solve for $p_2$ in terms of the other variables.

The algebraic solution for $p_2$ could look like this:

$$\frac{p_1}{\gamma} + z_1 + \frac{v_2^2}{2g} = \frac{p_2}{\gamma} + z_2 + \frac{v_2^2}{2g}$$
$$\frac{p_2}{\gamma} = \frac{p_1}{\gamma} + z_1 + \frac{v_1^2}{2g} - z_2 - \frac{v_2^2}{2g}$$
$$p_2 = \gamma\left(\frac{p_1}{\gamma} + z_1 + \frac{v_1^2}{2g} - z_2 - \frac{v_2^2}{2g}\right) .$$

This is correct. However, it is convenient to group the elevation heads and velocity heads together. Also, since $\gamma(p_1/\gamma) = p_1$, the final solution for $p_2$ should be

$$p_2 = p_1 + \gamma\left(z_1 - z_2 + \frac{v_1^2 - v_2^2}{2g}\right) \tag{6-10}$$

Are the values of all the terms on the right side of this equation known?

---

Everything was given except $\gamma$, $v_2$, and $g$. Of course, $g = 9.81$ m/s$^2$. Since water at 10°C is flowing in the system, $\gamma = 9.81$ kN/m$^3$. How can $v_2$ be determined?

---

The continuity equation is used:

$$A_1 v_1 = A_2 v_2$$
$$v_2 = v_1(A_1/A_2)$$

Calculate $v_2$ now.

---

You should have $v_2 = 0.75$ m/s. This is found from

$$A_1 = \pi D_1^2/4 = \pi(25 \text{ mm})^2/4 = 491 \text{ mm}^2$$
$$A_2 = \pi D_2^2/4 = \pi(50 \text{ mm})^2/4 = 1963 \text{ mm}^2$$
$$v_2 = v_1(A_1/A_2) = 3.0 \text{ m/s}(491 \text{ mm}^2/1963 \text{ mm}^2) = 0.75 \text{ m/s}$$

Now substitute the known values into Eq. (6–10).

---

$$p_2 = 345 \text{ kPa} + \frac{9.81 \text{ kN}}{\text{m}^3}\left(-2.0 \text{ m} + \frac{(3.0 \text{ m/s})^2 - (0.75 \text{ m/s})^2}{2(9.81 \text{ m/s}^2)}\right)$$

Notice that $z_1 - z_2 = -2.0$ m. Neither $z_1$ nor $z_2$ is known, but it is known that $z_2$ is 2.0 m greater than $z_1$. Therefore, the difference $z_1 - z_2$ must be negative.
Now complete the calculation for $p_2$.

---

The final answer is $p_2 = 329.6$ kPa. This is 15.4 kPa less than $p_1$. The details of the solution are

$$p_2 = 345 \text{ kPa} + \frac{9.81 \text{ kN}}{\text{m}^3}\left(-2.0 \text{ m} + \frac{(9.0 - 0.563)\text{m}^2/\text{s}^2}{2(9.81)\text{m/s}^2}\right)$$
$$p_2 = 345 \text{ kPa} + \frac{9.81 \text{ kN}}{\text{m}^3}(-2.0 \text{ m} + 0.43 \text{ m})$$
$$p_2 = 345 \text{ kPa} - 15.4 \text{ kN/m}^2 = 345 \text{ kPa} - 15.4 \text{ kPa}$$
$$p_2 = 329.6 \text{ kPa}$$

The pressure $p_2$ is a gage pressure because it was computed relative to $p_1$, which was also a gage pressure. In later problem solutions, we will assume the pressures to be gage unless otherwise stated.

■

---

**6.10.1**
**Tanks, Reservoirs, and Nozzles Exposed to the Atmosphere**

Figure 6.7 shows a fluid flow system in which a siphon draws fluid from a tank or reservoir and delivers it through a nozzle at the end of the pipe. Note that the surface of the tank (point A) and the free stream of fluid exiting the nozzle (section F) are not confined by solid boundaries and are exposed to the prevailing atmosphere.

**FIGURE 6.7**   Siphon for Example
Problem 6.10.

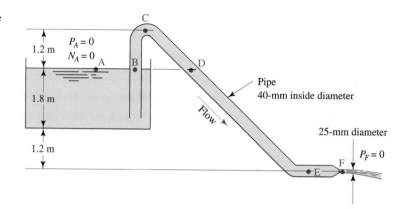

Therefore, the pressure at those sections is zero gage pressure. We then use the following rule:

> *When the fluid at a reference point is exposed to the atmosphere, the pressure is zero and the pressure head term can be cancelled from Bernoulli's equation.*

The tank from which the fluid is being drawn can be assumed to be quite large compared to the size of the flow area inside the pipe. Now, because $v = Q/A$, the velocity at the surface of such a tank will be very small. Furthermore, when we use the velocity to compute the velocity head, $v^2/2g$, we *square* the velocity. The process of squaring a small number produces an even smaller number. For these reasons, we adopt the following rule:

> *The velocity head at the surface of a tank or reservoir is considered to be zero and it can be cancelled from Bernoulli's equation.*

**6.10.2**

**When Both Reference Points Are in the Same Pipe**

Also notice in Fig. 6.7 that several points of interest (points B, C, D, and E) are inside the pipe that has a uniform flow area. Under the conditions of steady flow assumed for these problems, the velocity will be the same throughout the pipe. Then the following rule applies when steady flow occurs:

> *When the two points of reference for Bernoulli's equation are both inside a pipe of the same size, the velocity head terms on both sides of the equation are equal and can be cancelled.*

**6.10.3**

**When Elevations Are Equal at Both Reference Points**

Similarly, the following rule applies when the reference points are on the same level:

> *When the two points of reference for Bernoulli's equation are both at the same elevation, the elevation head terms, $z_1$ and $z_2$, are equal and can be cancelled.*

The four observations made in sections 6.10.1–3 enable the simplification of Bernoulli's equation and make the algebraic manipulations easier. Example Problem 6.10 uses these observations.

## PROGRAMMED EXAMPLE PROBLEM

☐ **EXAMPLE PROBLEM 6.10**   Figure 6.7 shows a siphon that is used to draw water from a swimming pool. The pipe that makes up the siphon has an inside diameter of 40 mm and terminates with a 25-mm diameter nozzle. Assuming that there are no energy losses in the system, calculate the volume flow rate through the siphon and the pressure at points B, C, D, and E.

The first step in this problem solution is to calculate the volume flow rate $Q$, using Bernoulli's equation. The two most convenient points to use for this calculation are A and F. What is known about point A?

---

Point A is the free surface of the water in the pool. Therefore, $p_A = 0$ Pa. Also, since the surface area of the pool is very large, the velocity of the water at the surface is very nearly zero. Therefore, we will assume $v_A = 0$.

What do we know about point F?

---

Point F is in the free stream of water outside the nozzle. Since the stream is exposed to atmospheric pressure, the pressure $p_F = 0$ Pa. We also know that point F is 3.0 m below point A.

Now write Bernoulli's equation for points A and F.

---

You should have

$$\frac{p_A}{\gamma} + z_A + \frac{v_A^2}{2g} = \frac{p_F}{\gamma} + z_F + \frac{v_F^2}{2g}$$

Taking into account the information in the previous two panels, how can we simplify this equation?

---

Since $p_A = 0$ Pa, $p_F = 0$ Pa, and $v_A$ is approximately zero, we can cancel them from the equation. What remains is this:

$$\frac{p_A}{\gamma}^{0} + z_A + \frac{v_A^2}{2g}^{0} = \frac{p_F}{\gamma}^{0} + z_F + \frac{v_F^2}{2g}$$

$$z_A = z_F + \frac{v_F^2}{2g}$$

The objective is to calculate the volume flow rate, which depends on the velocity. Solve for $v_F$ now.

---

You should have

$$v_F = \sqrt{(z_A - z_F)2g}$$

What is $z_A - z_F$?

---

From Fig. 6.7 we see that $z_A - z_F = 3.0$ m. Notice that the difference is positive since $z_A$ is greater than $z_F$. We can now solve for the value of $v_F$.

---

The result is

$$v_F = \sqrt{(3.0 \text{ m})(2)(9.81 \text{ m/s}^2)} = \sqrt{58.9} \text{ m/s} = 7.67 \text{ m/s}$$

Now, how can $Q$ be calculated?

---

Using the continuing equation $Q = Av$, compute the volume flow rate.

The result is

$$Q = A_F v_F$$

$$v_F = 7.67 \text{ m/s}$$

$$A_F = \pi(25 \text{ mm})^2/4 = 491 \text{ mm}^2$$

$$Q = 491 \text{ mm}^2 \left(\frac{7.67 \text{ m}}{\text{s}}\right)\left(\frac{1 \text{ m}^2}{10^6 \text{ mm}^2}\right) = 3.77 \times 10^{-3} \text{ m}^3/\text{s}$$

The first part of the problem is now complete. Now use Bernoulli's equation to determine $p_B$. What two points should be used?

---

Points A and B are the best. As shown in the previous panels, using point A allows the equation to be simplified greatly. And since we are looking for $p_B$, we must choose point B. Write Bernoulli's equation for points A and B, simplify it as before and solve for $p_B$.

---

Here is one possible solution procedure:

$$\cancel{\frac{p_A}{\gamma}}^{0} + z_A + \cancel{\frac{v_A^2}{2g}}^{0} = \frac{p_B}{\gamma} + z_B + \frac{v_B^2}{2g}$$

Since $p_A = 0$ Pa and $v_A = 0$, we have

$$z_A = \frac{p_B}{\gamma} + z_B + \frac{v_B^2}{2g}$$

$$p_B = \gamma[(z_A - z_B) - v_B^2/2g] \qquad (6\text{–}11)$$

What is $z_A - z_B$?

---

It is zero. Since the two points are on the same level, their evaluations are the same. Can you find $v_B$?

---

We can calculate $v_B$ by using the continuity equation:

$$Q = A_B v_B$$

$$v_B = Q/A_B$$

The area of a 40-mm diameter pipe can be found in Appendix J. Complete the calculation for $v_B$.

---

The result is

$$v_B = Q/A_B$$

$$Q = 3.77 \times 10^{-3} \text{ m}^3/\text{s}$$

$$A_B = 1.257 \times 10^{-3} \text{ m}^2$$

$$v_B = \frac{3.77 \times 10^{-3} \text{ m}^3}{\text{s}} \times \frac{1}{1.257 \times 10^{-3} \text{ m}^2} = 3.00 \text{ m/s}$$

We now have all the data we need to calculate $p_B$ from Eq. (6–11).

The pressure at point B is

$$p_B = \gamma[(z_A - z_B) - v_B^2/2g]$$

$$\frac{v_B^2}{2g} = \frac{(3.00)^2 \, m^2}{s^2} \times \frac{s^2}{(2)(9.81) \, m} = 0.459 \, m$$

$$p_B = (9.81 \, kN/m^3)(0 - 0.459 \, m)$$

$$p_B = -4.50 \, kN/m^2$$

$$p_B = -4.50 \, kPa$$

The negative sign indicates that $p_B$ is 4.50 kPa below atmospheric pressure. Notice that when we deal with fluids in motion, the concept that points on the same level have the same pressure does *not* apply as it does with fluids at rest.

The next three panels present the solutions for the pressure $p_C$, $p_D$, and $p_E$, which can be found in a manner very similar to that used for $p_B$. Complete the solution for $p_C$ yourself before looking at the next panel.

---

The answer is $p_C = -16.27$ kPa. We use Bernoulli's equation.

$$\frac{\cancel{p_A}^{\,0}}{\gamma} + z_A + \frac{\cancel{v_A^2}^{\,0}}{2g} = \frac{p_C}{\gamma} + z_C + \frac{v_C^2}{2g}$$

Since $p_A = 0$ and $v_A = 0$, the pressure at point C is

$$z_A = \frac{p_C}{\gamma} + z_C + \frac{v_C^2}{2g}$$

$$P_C = \gamma[(z_A - z_C) - v_C^2/2g]$$

$$z_A - z_C = -1.2 \, m \quad \text{(negative, since } z_C \text{ is greater than } z_A)$$

$$v_C = v_B = 3.00 \, m/s \quad \text{(since } A_C = A_B)$$

$$\frac{v_C^2}{2g} = \frac{v_B^2}{2g} = 0.459 \, m$$

$$p_C = (9.81 \, kN/m^3)(-1.2 \, m - 0.459 \, m)$$

$$p_C = -16.27 \, kN/m^2$$

$$p_C = -16.27 \, kPa$$

Complete the calculation for $p_D$ before looking at the next panel.

---

The answer is $p_D = -4.50$ kPa. This is the same as $p_B$ because the elevation and the velocity at points B and D are equal. Solution by Bernoulli's equation would prove this. Now find $p_E$.

---

The pressure at point E is 24.93 kPa. We use Bernoulli's equation:

$$\frac{\cancel{p_A}^{\,0}}{\gamma} + z_A + \frac{\cancel{v_A^2}^{\,0}}{2g} = \frac{p_E}{\gamma} + z_E + \frac{v_E^2}{2g}$$

Since $p_A = 0$ and $v_A = 0$, we have

$$z_A = \frac{p_E}{\gamma} + z_E + \frac{v_E^2}{2g}$$

$$P_E = \gamma[(z_A - z_E) - v_E^2/2g]$$

$$z_A - z_E = +3.0 \text{ m}$$

$$v_E = v_B = 3.00 \text{ m/s}$$

$$\frac{v_E^2}{2g} = \frac{v_B^2}{2g} = 0.459 \text{ m}$$

$$p_E = (9.81 \text{ kN/m}^3)(3.0 \text{ m} - 0.459 \text{ m})$$

$$p_E = 24.93 \text{ kN/m}^2$$

$$p_E = 24.93 \text{ kPa}$$

**SUMMARY OF THE RESULTS OF EXAMPLE PROBLEM 6.10**

1. The velocity of flow from the nozzle, and therefore the volume flow rate delivered by the siphon, depends on the elevation difference between the free surface of the fluid and the outlet of the nozzle.
2. The pressure at point B is below atmospheric pressure even though it is on the same level as point A, which is exposed to the atmosphere. In Eq. (6–11), Bernoulli's equation shows that the pressure head at B is decreased by the amount of the velocity head. That is, some of the energy is converted to kinetic energy, resulting in a lower pressure at B.
3. The velocity of flow is the same at all points where the pipe size is the same, when steady flow exists.
4. The pressure at point C is the lowest in the system because point C is at the highest elevation.
5. The pressure at point D is the same as that at point B, because both are on the same elevation and the velocity head at both points is the same.
6. The pressure at point E is the highest in the system because point E is at the lowest elevation.

■

### 6.10.4
### Venturi Meters and Other Closed Systems with Unknown Velocities

Figure 6.8 shows a device called a *venturi meter* that can be used to measure the velocity of flow in a fluid flow system. A more complete description of the venturi meter is given in Chapter 15. However, the analysis of such a device is based on the application of Bernoulli's equation. The reduced diameter section at B causes the velocity of flow to increase there with a corresponding decrease in the pressure. It will be shown that the velocity of flow is dependent on the *difference* in pressure between points A and B. Therefore, a differential manometer as shown is convenient to use.

We will also show in the solution to the following problem that we must combine the continuity equation with Bernoulli's equation to solve for desired velocity of flow.

☐ **EXAMPLE PROBLEM 6.11**

The venturi meter shown in Fig. 6.8 carries water at 60°C. The specific gravity of the gage fluid in the manometer is 1.25. Calculate the velocity of flow at section A and the volume flow rate of water.

*Solution*

The problem solution will be shown in the steps outlined at the beginning of this section but the programmed technique will not be used.

**FIGURE 6.8**  Venturi meter system
for Example Problem 6.11.

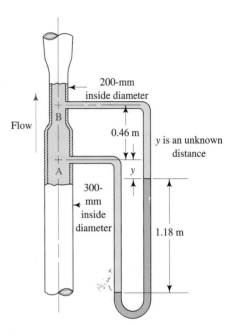

1. Decide what items are known and what is to be found. The elevation difference between points A and B is known. The manometer allows the determination of the difference in pressure between points A and B. The sizes of the sections at A and B are known.

   The velocity is not known at any point in the system and the velocity at point A was specifically requested.
2. Decide on sections of interest. Points A and B are the obvious choices.
3. Write Bernoulli's equation between points A and B.

$$\frac{p_A}{\gamma} + z_A + \frac{v_A^2}{2g} = \frac{p_B}{\gamma} + z_B + \frac{v_B^2}{2g}$$

   The specific weight $\gamma$ is for water at 60°C, which is 9.65 kN/m³.
4. Simplify the equation, if possible, by eliminating terms that are zero or terms that are equal on both sides of the equation. No simplification can be done here.
5. Solve the equation algebraically for the desired term. This step will require significant effort. First, note that *both* of the velocities are unknown. But we can find the difference in pressures between A and B and the elevation difference is known. Therefore, it is convenient to bring both pressure terms and both elevation terms onto the left side of the equation in the form of differences. Then the two velocity terms can be moved to the right side. The result is

$$\frac{p_A - p_B}{\gamma} + (z_A - z_B) = \frac{v_B^2 - v_A^2}{2g} \qquad (6\text{--}12)$$

6. Calculate the result. Several steps are required. The elevation difference is

$$z_A - z_B = -0.46 \text{ m} \qquad (6\text{--}13)$$

The value is negative because B is higher than A. This value will be used in Eq. (6–12) later.

   The pressure head difference term can be evaluated by writing the equation for the manometer. We will use $\gamma_g$ for the specific weight of the gage fluid where

$$\gamma_g = 1.25(\gamma_w \text{ at } 4°C) = 1.25(9.81 \text{ kN/m}^3) = 12.26 \text{ kN/m}^3$$

A new problem occurs here because the data in Fig. 6.8 do not include the vertical distance from point A to the level of the gage fluid in the right leg of the manometer. We will show that this problem will be eliminated by simply calling this unknown distance $y$ or any other variable name.

Now we can write the manometer equation starting at A.

$$p_A + \gamma(y) + \gamma(1.18\ m) - \gamma_g(1.18\ m) - \gamma(y) - \gamma(0.46\ m) = p_B$$

Note that the two terms containing the unknown $y$ variable can be cancelled out.

Solving for the pressure difference $p_A - p_B$,

$$p_A - p_B = \gamma(0.46\ m - 1.18\ m) + \gamma_g(1.18\ m)$$
$$p_A - p_B = \gamma(-0.72\ m) + \gamma_g(1.18\ m)$$

But notice in Eq. 6–12 above, we really need $(p_A - p_B)/\gamma$. If we divide both sides of the above equation by $\gamma$ we get the desired term.

$$\frac{p_A - p_B}{\gamma} = -0.72\ m + \frac{\gamma_g(1.18\ m)}{\gamma} = -0.72\ m + \frac{12.26\ kN/m^3\,(1.18\ m)}{9.65\ kN/m^3}$$

$$(p_A - p_B)/\gamma = -0.72\ m + 1.50\ m = 0.78\ m \qquad \textbf{(6–14)}$$

The entire left side of Eq. 6–12 has now been evaluated. But note that there are still *two* unknowns on the right side, $v_A$ and $v_B$. We can eliminate one unknown by finding another independent equation that relates those two variables. A convenient equation is the *continuity equation*.

$$A_A v_A = A_B v_B$$

Solving for $v_B$ in terms of $v_A$,

$$v_B = v_A(A_A/A_B)$$

The areas for the 200 mm and 300 mm diameter sections can be found in Appendix J. Then,

$$v_B = v_A(7.069 \times 10^{-2}/3.142 \times 10^{-2}) = 2.25\ v_A$$

But we need $v_B^2$.

$$v_B^2 = 5.06\, v_A^2$$

Then,

$$v_B^2 - v_A^2 = 5.06\, v_A^2 - v_A^2 = 4.06\, v_A^2 \qquad \textbf{(6–15)}$$

We can now take this result, the elevation head difference [Eq. (6–13)] and the pressure head difference [Eq. (6–14)] back into Eq. (6–12). We can complete the solution now. Equation (6–12) becomes

$$0.78\ m - 0.46\ m = 4.06\, v_A^2/2g$$

Solving for $v_A$ gives

$$v_A = \sqrt{\frac{2g(0.32\ m)}{4.06}} = \sqrt{\frac{2(9.81\ m/s^2)(0.32\ m)}{4.06}}$$

$$v_A = 1.24\ m/s$$

The problem statement also asked for the volume flow rate which can be computed from

$$Q = A_A v_A = (7.069 \times 10^{-2}\,\text{m}^2)(1.24\,\text{m/s}) = 8.77 \times 10^{-2}\,\text{m}^3\text{/s}$$

This example problem is completed.

■

**6.11**
**TORRICELLI'S**
**THEOREM**

In the siphon analyzed in Example Problem 6.10, it was observed that the velocity of flow from the siphon depends on the elevation difference between the free surface of the fluid and the outlet of the siphon. A classic application of this observation is shown in Fig. 6.9. Fluid is flowing from the side of a tank through a smooth, rounded nozzle. To determine the velocity of flow from the nozzle, write Bernoulli's equation between a reference point on the fluid surface and a point in the jet issuing from the nozzle:

$$\frac{p_1}{\gamma} + z_1 + \frac{v_1^2}{2g} = \frac{p_2}{\gamma} + z_2 + \frac{v_2^2}{2g}$$

But $p_1 = p_2 = 0$, and $v_1$ is approximately zero.

$$\frac{\cancel{p_1}^{\,0}}{\gamma} + z_1 + \frac{\cancel{v_1^2}^{\,0}}{2g} = \frac{\cancel{p_2}^{\,0}}{\gamma} + z_2 + \frac{v_2^2}{2g}$$

**FIGURE 6.9**   Flow from a tank.

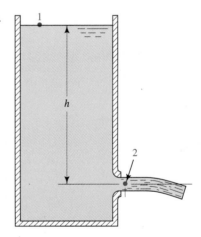

Then, solving for $v_2$ gives

$$v_2 = \sqrt{2g(z_1 - z_2)}$$

Letting $h = (z_1 - z_2)$, we have

TORRICELLI'S THEOREM

$$v_2 = \sqrt{2gh} \qquad (6\text{–}16)$$

Equation (6–16) is called *Torricelli's theorem* in honor of Evangelista Torricelli, who discovered it in approximately 1645.

☐ **EXAMPLE PROBLEM 6.12**

For the tank shown in Fig. 6.9, compute the velocity of flow from the nozzle for a fluid depth $h$ of 3.00 m.

*Solution*   This is a direct application of Torricelli's theorem:

$$v_2 = \sqrt{2gh} = \sqrt{(2)(9.81 \text{ m/s}^2)(3.0 \text{ m})} = 7.67 \text{ m/s}$$

☐ **EXAMPLE PROBLEM 6.13**   For the tank shown in Fig. 6.9, compute the velocity of flow from the nozzle and the volume flow rate for a range of depth from 3.0 m to 0.50 m in steps of 0.50 m. The diameter of the jet at the nozzle is 50 mm.

*Solution*   The same procedure used in Example Problem 6.12 can be used to determine the velocity at any depth. So, at $h = 3.0$ m, $v_2 = 7.67$ m/s. The volume flow rate is computed by multiplying this velocity by the area of the jet:

$$A_j = 1.963 \times 10^{-3} \text{ m}^2 \quad \text{from Appendix J.}$$

Then,

$$Q = A_j v_2 = (1.963 \times 10^{-3} \text{ m}^2)(7.67 \text{ m/s}) = 1.51 \times 10^{-2} \text{ m}^3/2$$

Using the same procedure, we compute the following data:

| Depth, $h$ (m) | $v_2$ (m/s) | $Q$ (m³/s) |
|---|---|---|
| 3.0 | 7.67 | $1.51 \times 10^{-2}$ |
| 2.5 | 7.00 | $1.38 \times 10^{-2}$ |
| 2.0 | 6.26 | $1.23 \times 10^{-2}$ |
| 1.5 | 5.42 | $1.07 \times 10^{-2}$ |
| 1.0 | 4.43 | $0.87 \times 10^{-2}$ |
| 0.5 | 3.13 | $0.61 \times 10^{-2}$ |

Figure 6.10 is a plot of velocity and volume flow rate versus depth.

**FIGURE 6.10**   Jet velocity and volume flow rate vs. fluid depth.

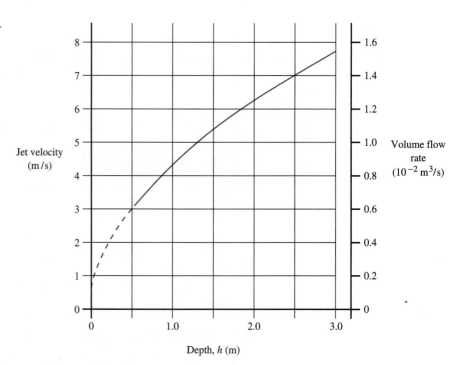

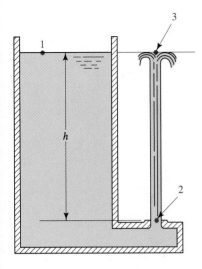

**FIGURE 6.11** Vertical jet.

Another interesting application of Torricelli's theorem is shown in Fig. 6.11, in which a jet of fluid is shooting upward. If no energy losses occur, the jet will reach a height equal to the elevation of the free surface of the fluid in the tank. And, of course, at this height the velocity in the stream is zero. This can be demonstrated using Bernoulli's equation. First obtain an expression for the velocity of the jet at point 2:

$$\cancelto{0}{\frac{p_1}{\gamma}} + z_1 + \cancelto{0}{\frac{v_1^2}{2g}} = \cancelto{0}{\frac{p_2}{\gamma}} + z_2 + \frac{v_2^2}{2g}$$

This is an identical situation to that encountered in the initial development of Torricelli's theorem. Then, as in Eq. (6–16),

$$v_2 = \sqrt{2gh}$$

Now, write Bernoulli's equation between point 2 and point 3 at the level of the free surface of the fluid but in the fluid stream.

$$\cancelto{0}{\frac{p_2}{\gamma}} + z_2 + \frac{v_2^2}{2g} = \cancelto{0}{\frac{p_3}{\gamma}} + z_3 + \frac{v_3^2}{2g}$$

But, $p_2 = p_3 = 0$. Then, solving for $v_3$, we have

$$v_3 = \sqrt{v_2^2 + 2g(z_2 - z_3)}$$

From Eq. (6–16), $v_2^2 = 2gh$. Also, $(z_2 - z_3) = -h$. Then,

$$v_3 = \sqrt{2gh + 2g(-h)} = 0$$

This result verifies that the stream just reaches the height of the free surface of the fluid in the tank.

To make a jet go higher (as with some decorative fountains, for example), a greater pressure can be developed above the fluid in the reservoir, or a pump can be used to develop a higher pressure.

---

☐ **EXAMPLE PROBLEM 6.14**

Using a system similar to that shown in Fig. 6.12, compute the required air pressure above the water to cause the jet to rise 40.0 ft from the nozzle. The depth $h = 6.0$ ft.

**Solution**

First, use Bernoulli's equation to obtain an expression for the velocity of flow from the nozzle as a function of the air pressure.

$$\frac{p_1}{\gamma} + z_1 + \cancelto{0}{\frac{v_1^2}{2g}} = \cancelto{0}{\frac{p_2}{\gamma}} + z_2 + \frac{v_2^2}{2g}$$

Here, we can see $v_1 = 0$ and $p_2 = 0$. Solving for $v_2$ gives

$$v_2 = \sqrt{2g[(p_1/\gamma) + (z_1 - z_2)]}$$

As before, letting $h = (z_1 - z_2)$, we have

$$v_2 = \sqrt{2g[(p_1/\gamma) + h]} \qquad\qquad \textbf{(6–17)}$$

This is similar to Torricelli's theorem. It was shown above that for $v = \sqrt{2gh}$, the jet rises to a height $h$. By analogy, the pressurized system would cause the jet to rise to a height of $[(p_1/\gamma) + h]$. Then, in this problem, if we want a height of 40.0 ft and $h = 6.0$ ft,

$$p_1/\gamma + h = 40.0 \text{ ft}$$
$$p_1/\gamma = 40.0 \text{ ft} - h = 40.0 \text{ ft} - 6.0 \text{ ft} = 34.0 \text{ ft}$$

**FIGURE 6.12**   Pressurized tank delivering a vertical jet. Also for Problems 6.93 and 6.94.

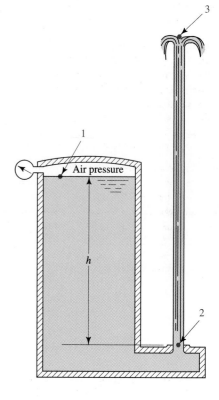

And,

$$p_1 = \gamma(34.0 \text{ ft})$$

$$p_1 = (62.4 \text{ lb/ft}^3)(34.0 \text{ ft})(1 \text{ ft}^2)/(144 \text{ in}^2)$$

$$p_1 = 14.73 \text{ psig}$$

In Chapter 4, we defined the pressure head, $p/\gamma$, in such applications as the *piezometric head*. Then the *total head* above the nozzle is $p_1/\gamma + h$.

■

**6.12**
**FLOW DUE TO A**
**FALLING HEAD**

As stated earlier, most problems considered in this book are for situations in which the flow rate is constant. However, in Section 6.11, it was shown that the flow rate depends on the pressure head available to cause the flow. The results of Example Problem 6.13, plotted in Fig. 6.10, show that the velocity and volume flow rate issuing from an orifice in a tank decrease in a nonlinear manner as the fluid flows from the tank and the depth of the fluid decreases.

In this section, we will develop a method for computing the time required to empty a tank, considering the variation of velocity as the depth decreases. Figure 6.13 shows a tank with a smooth, well-rounded nozzle in the bottom through which fluid is discharging. For a given depth of fluid $h$, Torricelli's theorem tells us that the velocity of flow in the jet is

$$v_j = \sqrt{2gh}$$

**FIGURE 6.13** Flow from a tank with falling head. Also for Problems 6.95–6.106.

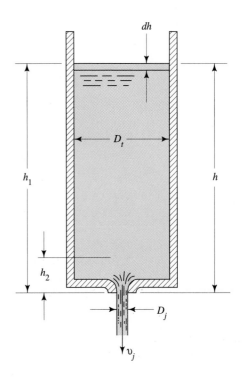

The volume flow rate through the nozzle is $Q = A_j v_j$ in such units as cubic meters per second (m³/s) or cubic feet per second (ft³/s). In a small amount of time, $dt$, the volume of fluid flowing through the nozzle is

$$\text{Volume flowing} = Q(dt) = A_j v_j (dt) \qquad (6\text{–}18)$$

Meanwhile, because fluid is leaving the tank, the fluid level is decreasing. During the small time increment, $dt$, the fluid level drops a small distance, $dh$. Then, the volume of fluid removed from the tank is

$$\text{Volume removed} = -A_t dh \qquad (6\text{–}19)$$

These two volumes must be equal. Then,

$$A_j v_j (dt) = -A_t dh \qquad (6\text{–}20)$$

Solving for the time $dt$, we have

$$dt = \frac{-(A_t/A_j)}{v_j} dh \qquad (6\text{–}21)$$

From Torricelli's theorem, we can substitute $v_j = \sqrt{2gh}$. Then,

$$dt = \frac{-(A_t/A_j)}{\sqrt{2gh}} dh \qquad (6\text{–}22)$$

Rewriting to separate the terms involving $h$ gives

$$dt = \frac{-(A_t/A_j)}{\sqrt{2g}} h^{-1/2} dh \qquad (6\text{–}23)$$

The time required for the fluid level to fall from one depth, $h_1$, to another depth $h_2$, can be found by integrating Eq. (6–23):

$$\int_{t_1}^{t_2} dt = \frac{-(A_t/A_j)}{\sqrt{2g}} \int_{h_1}^{h_2} h^{-1/2} dh \tag{6–24}$$

$$t_2 - t_1 = \frac{-(A_t/A_j)}{\sqrt{2g}} \frac{(h_2^{1/2} - h_1^{1/2})}{\frac{1}{2}} \tag{6–25}$$

We can reverse the two terms involving $h$ and remove the minus sign. At the same time, clearing the $\frac{1}{2}$ from the denominator, we get

⇨ TIME REQUIRED TO DRAIN A TANK

$$t_2 - t_1 = \frac{2(A_t/A_j)}{\sqrt{2g}} (h_1^{1/2} - h_2^{1/2}) \tag{6–26}$$

Equation (6–26) can be used to compute the time required to drain a tank from $h_1$ to $h_2$.

---

☐ **EXAMPLE PROBLEM 6.15**

For the tank shown in Fig. 6.13, find the time required to drain the tank from a level of 3.0 m to 0.50 m. The tank has a diameter of 1.50 m, and the nozzle has a diameter of 50 mm.

**Solution**

To use Eq. (6–26), the required areas are

$$A_t = \pi(1.50 \text{ m})^2/4 = 1.767 \text{ m}^2$$
$$A_j = \pi(0.05 \text{ m})^2/4 = 0.001963 \text{ m}^2$$

The ratio of these two areas is required:

$$\frac{A_t}{A_j} = \frac{1.767 \text{ m}^2}{0.001963 \text{ m}^2} = 900$$

Now, in Eq. (6–26),

$$t_2 - t_1 = \frac{2(A_t/A_j)}{\sqrt{2g}} (h_1^{1/2} - h_2^{1/2})$$
$$t_2 - t_1 = \frac{2(900)}{\sqrt{2(9.81 \text{ m/s}^2)}} [(3.0 \text{ m})^{1/2} - (0.5 \text{ m})^{1/2})]$$
$$t_2 - t_1 = 417 \text{ s}$$

This is equivalent to 6 min and 57 s.

■

**6.12.1**
**Draining a Pressurized Tank**

If the tank in Fig. 6.13 is sealed with a pressure above the fluid, the piezometric head, $p/\gamma$, should be added to the actual liquid depth before completing the calculations called for in Eq. (6–25).

**6.12.2**
**Effect of the Type of Nozzle**

The development of Eq. (6–26) assumes that the diameter of the jet of fluid flowing from the nozzle is the same as the diameter of the nozzle itself. This is very nearly true for the well-rounded nozzles depicted in Figs. 6.9, 6.11, and 6.13. However, if the nozzle is sharper, the minimum diameter of the jet is significantly smaller than the diameter of the opening. For example, Fig. 6.14 shows the flow from a tank through a sharp-edged orifice. The proper area to use for $A_j$ in Eq. (6–26) is that at the smallest diameter. This point, called the *vena contracta*, occurs slightly outside the orifice. For this sharp-edged orifice, $A_j = 0.62 A_o$ is a good approximation.

**FIGURE 6.14** Flow through a sharp-edged orifice.

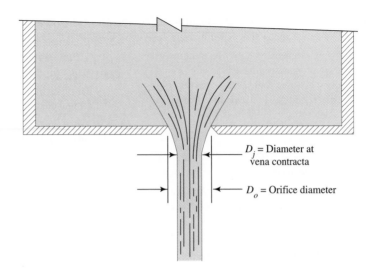

$D_j$ = Diameter at vena contracta

$D_o$ = Orifice diameter

# REFERENCE

1. American Society of Mechanical Engineers. 1992. *ANSI/ASME Standard B31.1-1998 Power Piping*. New York: Author.

# PRACTICE PROBLEMS

## Conversion Factors

**6.1** Convert a volume flow rate of 3.0 gal/min to $m^3/s$.

**6.2** Convert 459 gal/min to $m^3/s$.

**6.3** Convert 8720 gal/min to $m^3/s$.

**6.4** Convert 84.3 gal/min to $m^3/s$.

**6.5** Convert a volume flow rate of 125 L/min to $m^3/s$.

**6.6** Convert 4500 L/min to $m^3/s$.

**6.7** Convert 15 000 L/min to $m^3/s$.

**6.8** Convert 459 gal/min to L/min.

**6.9** Convert 8720 gal/min to L/min.

**6.10** Convert 23.5 $cm^3/s$ to $m^3/s$.

**6.11** Convert 0.296 $cm^3/s$ to $m^3/s$.

**6.12** Convert 0.105 $m^3/s$ to L/min.

**6.13** Convert $3.58 \times 10^{-3}$ $m^3/s$ to L/min.

**6.14** Convert $5.26 \times 10^{-6}$ $m^3/s$ to L/min.

**6.15** Convert 459 gal/min to $ft^3/s$.

**6.16** Convert 20 gal/min to $ft^3/s$.

**6.17** Convert 2500 gal/min to $ft^3/s$.

**6.18** Convert 2.50 gal/min to $ft^3/s$.

**6.19** Convert 125 $ft^3/s$ to gal/min.

**6.20** Convert 0.060 $ft^3/s$ to gal/min.

**6.21** Convert 7.50 $ft^3/s$ to gal/min.

**6.22** Convert 0.008 $ft^3/s$ to gal/min.

**6.23** Table 6.2 lists the range of typical volume flow rates for centrifugal fire-fighting pumps to be 500 to 2500 gal/min. Express this range in the units of $ft^3/s$ and $m^3/s$.

**6.24** Table 6.2 lists the range of typical volume flow rates for industrial oil hydraulic systems pumps to be 3 to 30 gal/min. Express this range in the units of $ft^3/s$ and $m^3/s$.

**6.25** A certain deep well pump for a residence is rated to deliver 745 gal/h of water. Express this flow rate in $ft^3/s$.

**6.26** A small pump delivers 0.85 gal/h of liquid fertilizer. Express this flow rate in $ft^3/s$.

**6.27** A small metering pump delivers 11.4 gal of a water treatment chemical per 24 h. Express this flow rate in $ft^3/s$.

**6.28** A small metering pump delivers 19.5 mL/min of water to dilute a waste stream. Express this flow rate in $m^3/s$.

*General Note:* In the following problems you may be required to refer to the appendix for fluid properties, dimensions of pipe and tubing, or conversion factors. Assume that there are no energy losses in all problems. Unless otherwise stated, the pipe sizes given are actual inside diameters.

## Fluid Flow Rates

**6.29M** *Water at 10°C is flowing at 0.075 $m^3/s$. Calculate the weight flow rate and the mass flow rate.*

**6.30M** *Oil for a hydraulic system (sg = 0.90) is flowing at 2.35 $\times$ $10^{-3}$ $m^3/s$. Calculate the weight flow rate and mass flow rate.*

**6.31M** *A liquid refrigerant (sg = 1.08) is flowing at a weight flow rate of 28.5 N/h. Calculate the volume flow rate and the mass flow rate.*

**6.32M** *After the refrigerant from Problem 6.31 flashes into a vapor, its specific weight is 12.50 $N/m^3$. If the weight flow rate remains at 28.5 N/h, compute the volume flow rate.*

**6.33C** A fan delivers 640 $ft^3$/min (CFM) of air. If the density of the air is 1.20 $kg/m^3$, compute the mass flow rate in slugs/s and the weight flow rate in lb/h.

**6.34E** A large blower for a furnace delivers 47 000 $ft^3$/min (CFM) of air having a specific weight of 0.075 $lb/ft^3$. Calculate the weight flow rate and mass flow rate.

**6.35E** A furnace requires 1200 lb/h of air for efficient combustion. If the air has a specific weight of 0.062 $lb/ft^3$, compute the required volume flow rate.

**6.36E** If a pump removes 1.65 gal/min of water from a tank, how long will it take to empty the tank if it contains 7425 lb of water?

## Continuity Equation

**6.37E** Calculate the diameter of a pipe that would carry 75.0 $ft^3$/s of a liquid at an average velocity of 10.0 ft/s.

**6.38E** If the velocity of a liquid is 1.65 ft/s in a 12-in diameter pipe, what is the velocity in a 3-in diameter jet exiting from a nozzle attached to the pipe?

**6.39M** *When 2000 L/min of water flow through a 300-mm diameter pipe that later reduces to a 150-mm diameter pipe, calculate the average velocity of flow in each pipe.*

**6.40M** *Water flows at 1.20 m/s in a 150-mm diameter pipe. Calculate the velocity of flow in a 300-mm pipe connected to it.*

**6.41M** *A 150-mm diameter pipe carries 0.072 $m^3/s$ of water. The pipe branches into two pipes, as shown in Fig. 6.15. If the velocity in the 50-mm pipe is 12.0 m/s, what is the velocity in the 100-mm pipe?*

**6.42E** A standard Schedule 40 steel pipe is to be selected to carry 10 gal/min of water with a maximum velocity of 1.0 ft/s. What size pipe should be used?

**6.43E** If water at 180°F is flowing with a velocity of 4.50 ft/s in a standard 6-in Schedule 40 pipe, calculate the weight flow rate in lb/h.

**6.44M** *A standard 1-in OD (outside diameter) steel tube (0.065-in wall thickness) is carrying 19.7 L/min of oil. Calculate the velocity of flow.*

**6.45E** The recommended velocity of flow in the discharge line of an oil hydraulic system is in the range of 8.0 to 25.0 ft/s. If the pump delivers 30 gal/min of oil, specify the smallest and largest suitable sizes of steel tubing.

**6.46E** Repeat Problem 6.45, except specify suitable sizes for the suction lines to maintain the velocity between 2.0 and 7.0 ft/s for 30 gal/min of flow.

**FIGURE 6.15**    Problem 6.41.

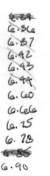

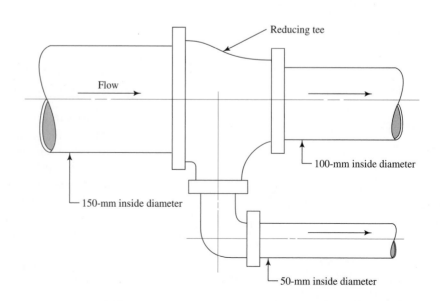

**FIGURE 6.16** Shell-and-tube heat exchanger for Problem 6.55.

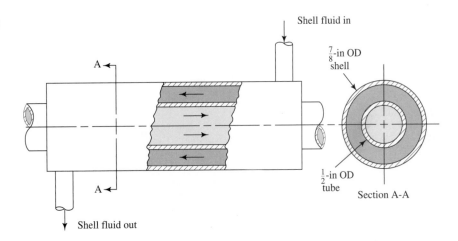

**6.47M** *Table 6.2 shows the typical volume flow rate for centrifugal firefighting pumps is in the range of 1800 to 9500 L/min. Specify the smallest suitable size of Schedule 40 steel pipe for each flow rate that will maintain the maximum velocity of flow at 2.0 m/s.*

**6.48M** *Repeat Problem 6.47, but use Schedule 80 pipe.*

**6.49M** *Compute the resulting velocity of flow if 400 L/min of fluid flow through a 2-in Schedule 40 pipe.*

**6.50M** Repeat Problem 6.49 for a 2-in Schedule 80 pipe.

**6.51E** Compute the resulting velocity of flow if 400 gal/min of fluid flow through a 4-in Schedule 40 pipe.

**6.52E** Repeat Problem 6.51 for a 4-in Schedule 80 pipe.

**6.53M** *From the list of standard steel tubing in Appendix G, select the smallest size that would carry 2.80 L/min of oil with a maximum velocity of 0.30 m/s.*

**6.54E** A standard 6-in Schedule 40 steel pipe is carrying 95 gal/min of water. The pipe then branches into two standard 3-in pipes. If the flow divides evenly between the branches, calculate the velocity of flow in all three pipes.

**Noncircular Sections, Problems 6.55–6.57**

**6.55E** A shell and tube heat exchanger is made of two standard steel tubes, as shown in Fig. 6.16. Each tube has a wall thickness of 0.049 in. Calculate the required ratio of the volume flow rate in the shell to that in the tube if the average velocity of flow is to be the same in each.

**6.56M** *Figure 6.17 shows a heat exchanger in which each of two 6-in Schedule 40 pipes carries 450 L/min of water. The pipes are inside a rectangular duct whose inside dimensions are 200 mm by 400 mm. Compute the velocity of flow in the pipes. Then, compute the required volume flow rate of water in the duct to obtain the same average velocity.*

**FIGURE 6.17** Problem 6.56.

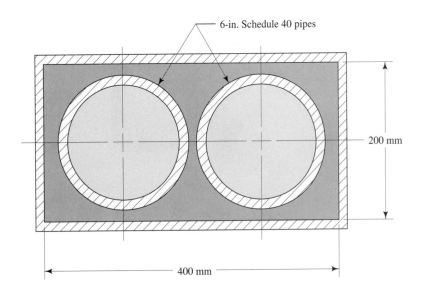

**6.57E** Figure 6.18 shows the cross section of a shell-and-tube heat exchanger. Compute the volume flow rate required in each small pipe and in the shell to obtain an average velocity of flow of 25 ft/s in all parts.

**FIGURE 6.18**   Problem 6.57.

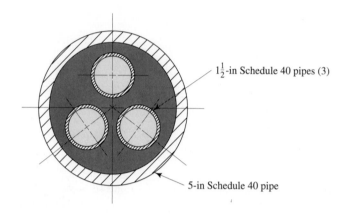

$1\frac{1}{2}$-in Schedule 40 pipes (3)

5-in Schedule 40 pipe

**6.58M** *A venturi meter is a device that uses a constriction in a flow system to measure the velocity of flow. Figure 6.19 illustrates one type of design. If the main pipe section is a standard 4-in Type K copper tube, compute the volume flow rate when the velocity there is 3.0 m/s. Then, for that volume flow rate, specify the required size of the throat section that would make the velocity there at least 15.0 m/s.*

**FIGURE 6.19**   Venturi meter for Problem 6.58.

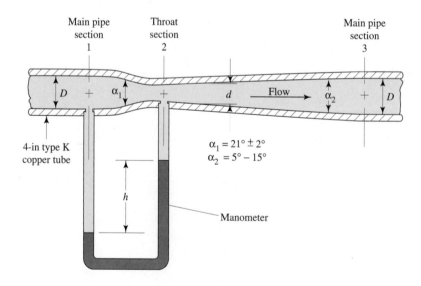

Main pipe section 1

Throat section 2

Main pipe section 3

$D$  $\alpha_1$  $d$  Flow  $\alpha_2$  $D$

4-in type K copper tube

$\alpha_1 = 21° \pm 2°$
$\alpha_2 = 5° - 15°$

$h$

Manometer

**6.59E** A flow nozzle, shown in Fig. 6.20, is used to measure the velocity of flow. If the nozzle is installed inside a 14-in Schedule 40 pipe and the nozzle diameter is 2.75 in, compute the velocity of flow at section 1 and the throat of the nozzle when 7.50 ft³/s of water flow through the system.

**FIGURE 6.20** Nozzle meter for Problem 6.59.

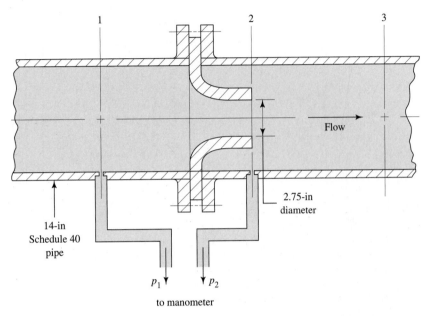

**Bernoulli's Equation**

**6.60M** *Gasoline (sg = 0.67) is flowing at 0.11 m³/s in the pipe shown in Fig. 6.21. If the pressure before the reduction is 415 kPa, calculate the pressure in the 75-mm diameter pipe.*

**FIGURE 6.21** Problem 6.60.

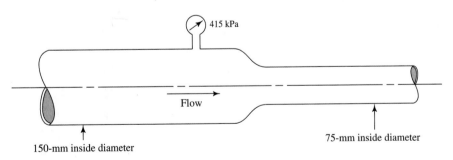

**6.61M** *Water at 10°C is flowing from point A to point B through the pipe shown in Fig. 6.22 at the rate of 0.37 m³/s. If the pressure at A is 66.2 kPa, calculate the pressure at B.*

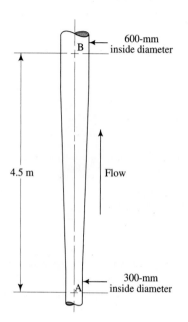

**FIGURE 6.22**    Problem 6.61.

**6.62M** *Calculate the volume flow rate of water at 5°C through the system shown in Fig. 6.23.*

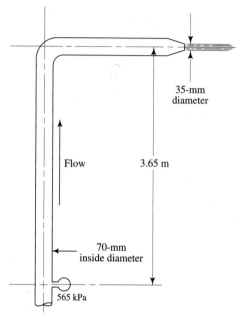

**FIGURE 6.23**    Problem 6.62.

**6.63E** Calculate the pressure required in the pipe just ahead of the nozzle in Fig. 6.24 to produce a jet velocity of 75 ft/s. The fluid is water at 180°F.

**FIGURE 6.24**    Problem 6.63.

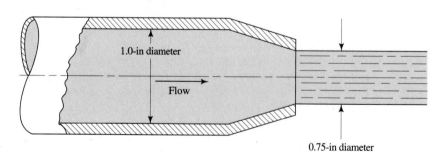

**6.64E** Kerosene with a specific weight of 50.0 lb/ft³ is flowing at 10 gal/min from a standard 1-in Schedule 40 steel pipe to a standard 2-in Schedule 40 steel pipe. Calculate the difference in pressure in the two pipes.

**6.65M** *For the system shown in Fig. 6.25, calculate (a) the volume flow rate of water from the nozzle and (b) the pressure at point A.*

**FIGURE 6.25**   Problem 6.65.

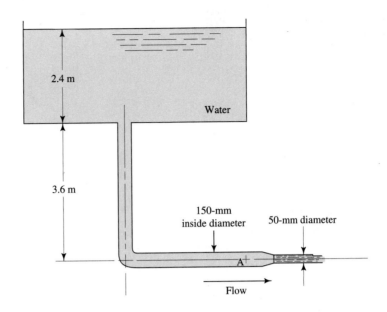

**6.66M** *For the system shown in Fig. 6.26, calculate (a) the volume flow rate of oil from the nozzle and (b) the pressures at A and B.*

**6.67E** For the tank shown in Fig. 6.27, calculate the volume flow rate of water from the nozzle. The tank is sealed with a pressure of 20 psig above the water. The depth, *h*, is 8 ft.

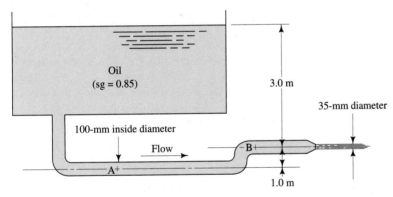

**FIGURE 6.26**   Problem 6.66.

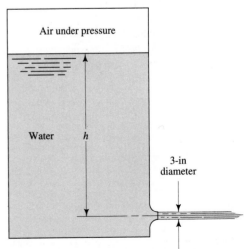

**FIGURE 6.27**   Problems 6.67 and 6.68.

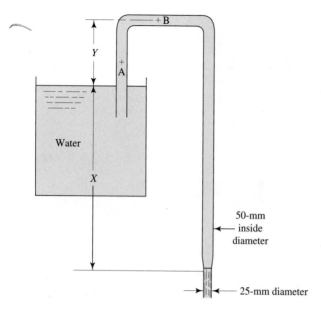

**FIGURE 6.28**    Problems 6.69, 6.70, and 6.71.

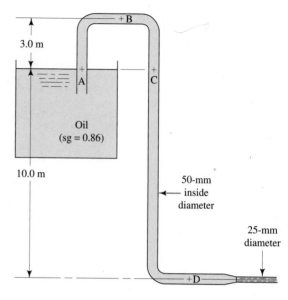

**FIGURE 6.29**    Problems 6.72 and 6.83.

**6.68E**    Calculate the pressure of the air in the sealed tank shown in Fig. 6.27 that would cause the velocity of flow to be 20 ft/s from the nozzle. The depth, *h*, is 10 ft.

**6.69M**    *For the siphon in Fig. 6.28, calculate (a) the volume flow rate of water through the nozzle and (b) the pressure at points A and B. The distance X = 4.6 m and Y = 0.90 m.*

**6.70M**    *For the siphon in Fig. 6.28, calculate the distance X required to obtain a volume flow rate of $7.1 \times 10^{-3}$ $m^3/s$.*

**6.71M**    *For the siphon in Fig. 6.28, assume that the volume flow rate is $5.6 \times 10^{-3}$ $m^3/s$. Determine the maximum allowable distance Y if the minimum allowable pressure in the system is −18 kPa (gage).*

**6.72M**    *For the siphon shown in Fig. 6.29, calculate (a) the volume flow rate of oil from the tank and (b) the pressures at points A, B, C, and D.*

**6.73E**    For the pipe reducer shown in Fig. 6.30, the pressure at A is 50.0 psig and the pressure at B is 42.0 psig. Calculate the velocity of flow of water at point B.

**6.74E**    In the enlargement shown in Fig. 6.31, the pressure at A is 25.6 psig and the pressure at B is 28.2 psig. Calculate the volume flow rate of oil (sg = 0.90).

**6.75M**    *Fig. 6.32 shows a manometer being used to indicate the pressure difference between two points in a pipe system. Calculate the volume flow rate of water in the system if the manometer deflection, h, is 250 mm. (This arrangement is called a venturi meter, which is often used for flow measurement.)*

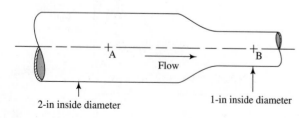

**FIGURE 6.30**    Problems 6.73 and 6.84.

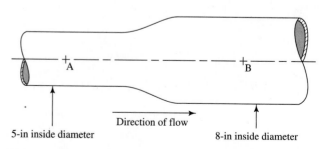

**FIGURE 6.31**    Problem 6.74.

**6.76M** *For the venturi meter shown in Fig. 6.32, calculate the manometer deflection, h, if the velocity of flow of water in the 25-mm diameter section is 10 m/s.*

**FIGURE 6.32** Problems 6.75 and 6.76.

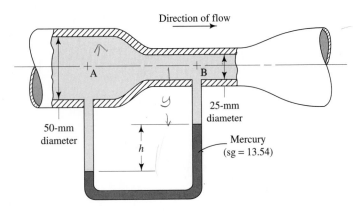

**6.77M** *Oil with a specific weight of 8.64 kN/m³ flows from A to B through the system shown in Fig. 6.33. Calculate the volume flow rate of oil.*

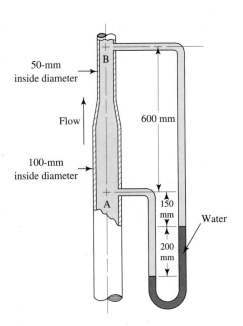

**FIGURE 6.33** Problem 6.77.

**6.78M** *The venturi meter shown in Fig. 6.34 carries oil (sg = 0.90). The specific gravity of the gage fluid in the manometer is 1.40. Calculate the volume flow rate of oil.*

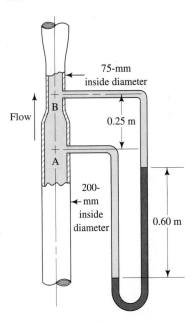

**FIGURE 6.34** Problem 6.78.

**6.79E** Oil with a specific gravity of 0.90 is flowing downward through the venturi meter shown in Fig. 6.35. If the manometer deflection $h$ is 28 in, calculate the volume flow rate of oil.

**6.80E** Oil with a specific gravity of 0.90 is flowing downward through the venturi meter shown in Fig. 6.35. If the velocity of flow in the 2-in diameter section is 10.0 ft/s, calculate the deflection, $h$, of the manometer.

**FIGURE 6.35**   Problems 6.79 and 6.80.

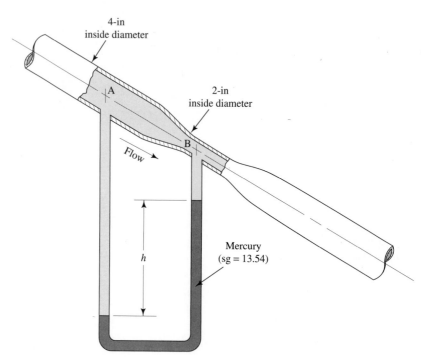

4-in inside diameter

2-in inside diameter

A

B

Flow

$h$

Mercury (sg = 13.54)

**6.81E** Gasoline (sg = 0.67) is flowing at 4.0 ft³/s in the pipe shown in Fig. 6.36. If the pressure before the reduction is 60 psig, calculate the pressure in the 3-in diameter pipe.

**6.82E** Oil with a specific weight of 55.0 lb/ft³ flows from A to B through the system shown in Fig. 6.37. Calculate the volume flow rate of the oil.

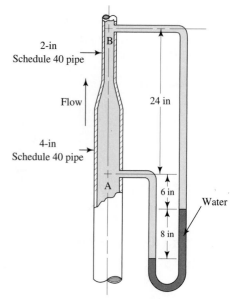

2-in Schedule 40 pipe

B

Flow

24 in

4-in Schedule 40 pipe

A

6 in

Water

8 in

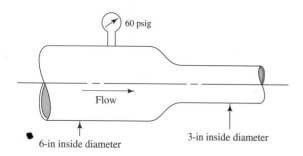

60 psig

Flow

6-in inside diameter

3-in inside diameter

**FIGURE 6.36**   Problem 6.81.

**FIGURE 6.37**   Problem 6.82.

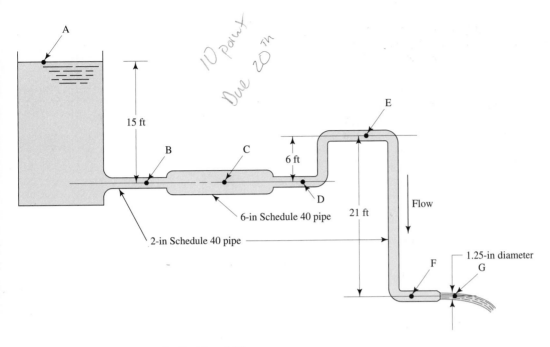

**FIGURE 6.38**   Flow system for Problem 6.85.

**6.83M** *Draw a plot of elevation head, pressure head, velocity head, and total head for the siphon system shown in Fig. 6.29 and analyzed in Problem 6.70.*

**6.84E** Draw a plot of elevation head, pressure head, velocity head, and total head for the system shown in Fig. 6.30 and analyzed in Problem 6.73.

**6.85E** Fig. 6.38 shows a system in which water flows from a tank through a pipe system having several sizes and el-

evations. For points A through G, compute the elevation head, the pressure head, the velocity head, and the total head. Plot these values on a sketch similar to that shown in Fig. 6.6.

**6.86M** *Fig. 6.39 shows a venturi meter with a U-tube manometer to measure the velocity of flow. When no flow occurs, the mercury column is balanced and its top is 300 mm below the throat. Compute the volume flow rate*

**FIGURE 6.39**   Venturi meter for Problem 6.86.

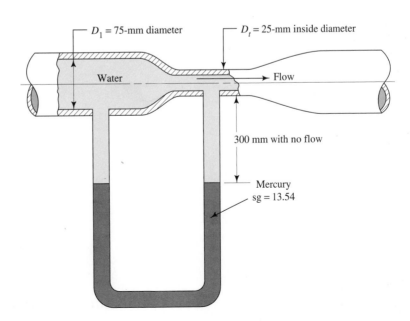

*through the meter that will cause the mercury to flow into the throat. Note that for a given manometer deflection, h, the left side will move down h/2 while the right side would rise h/2.*

**6.87E** For the tank shown in Fig. 6.40, compute the velocity of flow from the outlet nozzle at varying depths from 10.0 ft to 2.0 ft in 2.0-ft increments. Then, use increments of 0.5 ft to zero. Plot the velocity versus depth.

**6.88E** What depth of fluid above the outlet nozzle is required to deliver 200 gal/min of water from the tank shown in Fig. 6.40? The nozzle has a 3.00-in diameter.

## Torricelli's Theorem

**6.89** Derive Torricelli's theorem for the velocity of flow from a tank through an orifice opening into the atmosphere under a given depth of fluid.

**6.90E** Solve Problem 6.88 using the direct application of Torricelli's theorem.

**6.91M** *To what height will the jet of fluid rise for the conditions shown in Fig. 6.41?*

**6.92E** To what height will the jet of water rise for the conditions shown in Fig. 6.42?

**6.93E** What pressure is required above the water in Fig. 6.12 to cause the jet to rise to 28.0 ft? The water depth is 4.50 ft.

**6.94M** *What pressure is required above the water in Fig. 6.12 to cause the jet to rise to 9.50 m? The water depth is 1.50 m.*

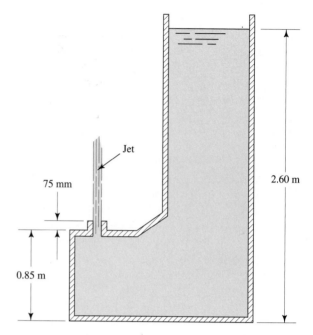

**FIGURE 6.41**    Problem 6.91

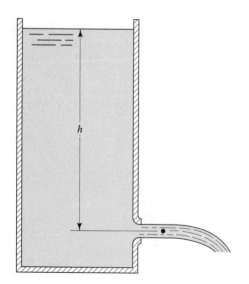

**FIGURE 6.40**    Tank for Problems 6.87–6.88.

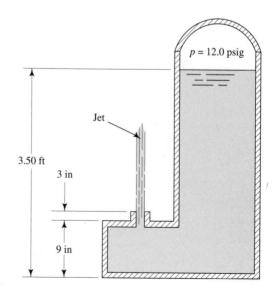

**FIGURE 6.42**    Problem 6.92.

## Flow Due to Falling Head

**6.95M** *Compute the time required to empty the tank shown in Fig. 6.13 if the original depth is 2.68 m. The tank diameter is 3.00 m, and the orifice diameter is 150 mm.*

**6.96M** *Compute the time required to empty the tank shown in Fig. 6.13 if the original depth is 55 mm. The tank diameter is 300 mm, and the orifice diameter is 20 mm.*

**6.97E** Compute the time required to empty the tank shown in Fig. 6.13 if the original depth is 15 ft. The tank diameter is 12.0 ft, and the orifice diameter is 6.00 in.

**6.98E** Compute the time required to empty the tank shown in Fig. 6.13 if the original depth is 18.5 in. The tank diameter is 22.0 in, and the orifice diameter is 0.50 in.

**6.99M** *Compute the time required to reduce the depth in the tank shown in Fig. 6.13 by 1.50 m if the original depth is 2.68 m. The tank diameter is 2.25 m, and the orifice diameter is 50 mm.*

**6.100M** *Compute the time required to reduce the depth in the tank shown in Fig. 6.13 by 225 mm if the original depth*

*is 1.38 m. The tank diameter is 1.25 m, and the orifice diameter is 25 mm.*

**6.101E** Compute the time required to reduce the depth in the tank shown in Fig. 6.13 by 12.5 in if the original depth is 38 in. The tank diameter is 6.25 ft, and the orifice diameter is 0.625 in.

**6.102E** Compute the time required to reduce the depth in the tank shown in Fig. 6.13 by 21.0 ft if the original depth is 23.0 ft. The tank diameter is 46.5 ft, and the orifice diameter is 8.75 in.

**6.103E** Repeat Problem 6.97 if the tank is sealed and a pressure of 5.0 psig is above the water in the tank.

**6.104E** Repeat Problem 6.101 if the tank is sealed and a pressure of 2.8 psig is above the water in the tank.

**6.105M** *Repeat Problem 6.96 if the tank is sealed and a pressure of 20 kPa(gage) is above the water in the tank.*

**6.106M** *Repeat Problem 6.100 if the tank is sealed and a pressure of 35 kPa(gage) is above the water in the tank.*

## COMPUTER PROGRAMMING ASSIGNMENTS

1. Create a spreadsheet to compute the values of the pressure head, the velocity head, the elevation head, and the total head for given values for pressure, velocity, and elevation.

2. Enhance the spreadsheet in Assignment 1 by causing it to list side-by-side in several combinations the various head components, in order to compare one with another as done when using Bernoulli's equation.

3. In the spreadsheet in Assignment 1, include the ability to compute the velocity of flow from given data for volume flow rate and pipe size.

4. Create a spreadsheet to compute, using Eq. (6–26), the time required to decrease the fluid level in a tank between two values for any combination of tank size and nozzle diameter. Apply it to Problems 6.95–6.102.

5. Add the ability to pressurize the system to the spreadsheet in Assignment 4. Apply it to Problems 6.103–6.106.

6. Create a spreadsheet to compute the velocity of flow from an orifice, using Torricelli's theorem for any depth of fluid and

any amount of pressure above the fluid. Apply it to Problems 6.90–6.94.

7. Prepare a spreadsheet that presents the data from Table 6.3 differently. Let the primary variable in the left column be the volume flow rate in gal/min. Then determine the flow area of a pipe or tube that would produce the recommended velocity of flow from the shaded area of the table. Finally, list a standard pipe (or tube) size that will produce approximately that velocity of flow. Because there is a range of recommended velocities, more than one pipe (or tube) size could be listed. For the listed pipe (or tube) size, compute and list the actual velocity of flow in ft/s that will result from a particular size for the listed volume flow rate. Note that different lists are required for suction lines and for discharge lines.

8. Repeat Assignment 7 but use metric data. The volume flow rate should be in m³/hr (or m³/s or L/s) and the velocity of flow should be in m/s.

# 7 General Energy Equation

<table>
<tr><td>

## 7.1
## The
## Big
## Picture

</td><td>

### Discussion Map

□ You will now expand your ability to analyze the energy in fluid flow systems by adding terms to Bernoulli's equation.

□ Considering energy added, energy removed, and energy lost from a system transforms the Bernoulli equation into the general energy equation.

□ Using the general energy equation eliminates many of the restrictions that were identified for the Bernoulli equation.

</td><td>

### Discover

■ Think again about the fluid flow situations that were discussed in the Big Picture part of Chapter 6. Remember the water distribution system in your house, the lawn sprinkling system, the piping for a fluid power system, and the fluid distribution systems used in so many places in manufacturing systems.

■ In what ways do those systems fail to conform to the restrictions listed above for the use of Bernoulli's equation?

■ Do some of those systems contain pumps to provide the energy that causes flow and increases the pressure in the fluid?

■ Are there any kinds of flow control devices such as valves in those systems?

■ Does the fluid make any changes in direction as it travels through the system?

■ Are there places where the size of the flow system changes, either getting larger or smaller?

■ Will there be pressure losses as the fluid flows through the pipes and tubes?

</td></tr>
</table>

In this chapter, you will learn how to apply the general energy equation to real systems with pumps, fluid motors, turbines, and energy losses from friction, valves, and fittings.

You should now have a basic understanding of how to analyze fluid flow systems from your work in Chapter 6. You should be able to compute volume flow rate, weight flow rate, and mass flow rate. You should be comfortable with various uses of the principle of continuity that states that the mass flow rate throughout a steady flow system is the same. We use the following form of the continuity equation involving volume flow rate most often when liquids are flowing in the system.

$$Q_1 = Q_2$$

But since $Q = Av$,

$$A_1 v_1 = A_2 v_2$$

These relationships allow you to determine the velocity of flow at any point in a system if you know the volume flow rate and the areas of the pipes at the sections of interest.

You should also be familiar with the terms that express energy possessed by a fluid per unit weight of the fluid flowing in the system.

$p/\gamma$ is the pressure head.

$z$ is the elevation head.

$v^2/2g$ is the velocity head.

The sum of these three terms is called the total head.

All of this came together in Bernoulli's equation.

$$\frac{p_1}{\gamma} + z_1 + \frac{v_1^2}{2g} = \frac{p_2}{\gamma} + z_2 + \frac{v_2^2}{2g}$$

But there are several restrictions on the use of Bernoulli's equation as you learned in Section 6.9.

1. It is valid only for incompressible fluids.
2. There can be no mechanical devices such as pumps, fluid motors, or turbines between the two sections of interest.
3. There can be no energy lost due to friction or to the turbulence created by valves and fittings in the flow system.
4. There can be no heat transferred into or out of the fluid.

In reality, no system satisfies all these restrictions.

Refer now to Figure 7.1 that shows a portion of an industrial fluid distribution system. The pump adds energy to the fluid. Several valves are in the piping system to control the flow or, in fact, to shut it off. There are tees, elbows, and several sizes of pipe used. Each of these devices causes energy to be lost from the fluid as it flows through them. Also, even as the fluid flows through straight lengths of pipe, energy is lost and that causes the fluid pressure to decrease. So Bernoulli's equation is not adequate to analyze such systems.

We can overcome these restrictions by making a few modest additions to Bernoulli's equation to produce an expanded form we will call *the general energy equation*. This chap-

**FIGURE 7.1** Typical pipeline installation, showing a pump, valves, tees, and other fittings. (Source of photo: Ingersoll-Rand Co., Phillipsburg, NJ)

ter introduces the general energy equation and demonstrates how to use it to compute pressures, and to determine the primary parameters for the application of pumps.

We also discuss fluid motors and turbines that take the energy from a fluid and deliver useful work such as driving a shaft to operate a piece of construction machinery, some manufacturing equipment, or an electrical generator.

The power required to drive a pump and the power delivered from a fluid motor are also important parts of what you will learn in this chapter.

**7.2**
**OBJECTIVES**

After completing this chapter, you should be able to:

1. Identify the conditions under which energy losses occur in fluid flow systems.
2. Identify the means by which energy can be added to a fluid flow system.
3. Identify the means by which energy can be removed from a fluid flow system.
4. Expand Bernoulli's equation to form the general energy equation by considering energy losses, energy additions, and energy removals.
5. Apply the general energy equation to a variety of practical problems.
6. Compute the power added to a fluid by pumps.
7. Define the *efficiency of pumps*.
8. Compute the power required to drive pumps.
9. Compute the power delivered by a fluid to a fluid motor.
10. Define the *efficiency of fluid motors*.
11. Compute the power output from a fluid motor.

**7.3**
**ENERGY LOSSES AND ADDITIONS**

The objective of this section is to describe, in general terms, the various types of devices and components of fluid flow systems. They occur in most fluid flow systems and they either add energy to the fluid, remove energy from the fluid, or cause undesirable losses of energy from the fluid.

At this time we are only describing these devices in conceptual terms. We discuss pumps, fluid motors, friction losses as fluid flows in pipes and tubes, energy losses from changes in the size of the flow path, and energy losses from valves and fittings.

In later chapters you will learn more details about how to compute the amount of energy losses in pipes and specific types of valves and fittings. You will learn the method of using performance curves for pumps to apply them properly.

**7.3.1**
**Pumps**

A pump is a common example of a mechanical device that adds energy to a fluid. An electric motor or some other prime power device drives a rotating shaft in the pump. The pump then takes this kinetic energy and delivers it to the fluid, resulting in fluid flow and increased fluid pressure.

Many configurations are used in pump designs. The system pictured in Fig. 7.1 contains a centrifugal pump mounted inline with the process piping. Figures 7.2 and 7.3 show two types of fluid power pumps capable of producing very high pressures in the range from 1500 to 5000 psi (10.3 to 34.5 MPa). Chapter 15 gives an extensive discussion of these and several other styles of pumps along with their selection and application.

**7.3.2**
**Fluid Motors**

Fluid motors, turbines, rotary actuators, and linear actuators are examples of devices that take energy from a fluid and deliver it in the form of work, causing the rotation of a shaft or the linear movement of a piston.

Many fluid motors have the same basic configurations as the pumps shown in Figs. 7.2 and 7.3. The major difference between a pump and a fluid motor is that,

**FIGURE 7.2**   Gear pump. (Source of photo: Danfoss Fluid Power, a division of Danfoss, Inc., Racine, WI; source of drawing: *Machine Design Magazine*)

(a) Cutaway

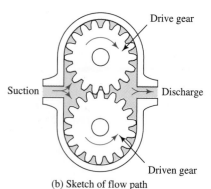

(b) Sketch of flow path

when acting as a motor, the fluid drives the rotating elements of the device. The reverse is true for pumps. For some designs, such as the gear-on-gear type in Fig. 7.2, a pump could act as a motor by forcing a flow of fluid through the device. In other types, a change in the valve arrangement or in the configuration of the rotating elements would be required.

The hydraulic motor shown in Fig. 7.4 is often used as a drive for the wheels of construction equipment and trucks, and for rotating components of material transfer systems, conveyors, agricultural equipment, special machines, and automation equipment. The design incorporates a stationary internal gear with a special shape. The rotating component is like an external gear, sometimes called a *gerotor*, that has one fewer teeth than the internal gear. The external gear rotates in a circular orbit around the center of the internal gear. High pressure fluid entering the cavity between the two gears acts on the rotor and develops a torque that rotates the output shaft. The magnitude of the output torque depends on the pressure difference between the input and output sides of the rotating gear. The speed of rotation is a function of the displacement of the motor (volume per revolution) and the volume flow rate of fluid through the motor.

Fig. 7.5 shows a photograph of a cutaway model of a fluid power cylinder or linear actuator.

**FIGURE 7.3**   Piston pump. (Source of photo: Danfoss Fluid Power, a division of Danfoss, Inc., Racine, WI: Source of drawing: *Machine Design Magazine*)

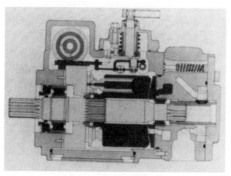

(a) Cutaway

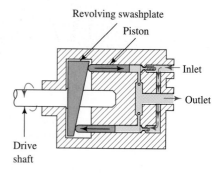

(b) Sketch of flow path

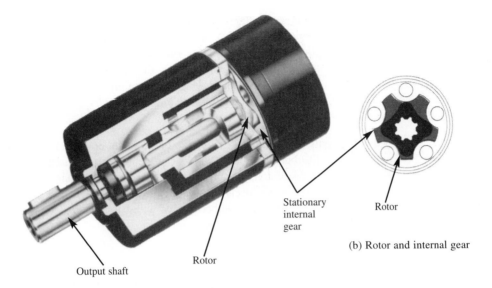

Stationary
internal
gear

Rotor

(b) Rotor and internal gear

Output shaft

Rotor

(a) Cutaway

**FIGURE 7.4**   Hydraulic motor. (Source: Danfoss Fluid Power, a division of
Danfoss, Inc., Racine, WI)

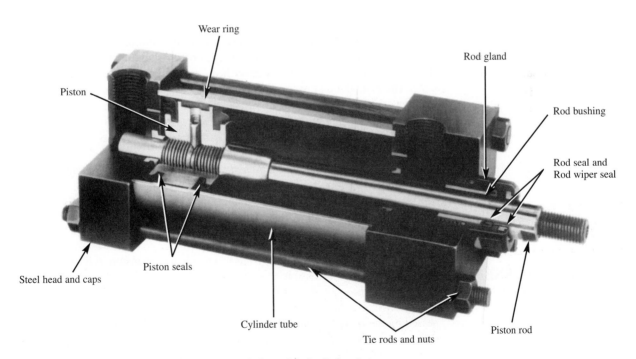

Wear ring

Rod gland

Piston

Rod bushing

Rod seal and
Rod wiper seal

Steel head and caps

Piston seals

Cylinder tube

Tie rods and nuts

Piston rod

**FIGURE 7.5**   Fluid power cylinder. (Source of photo: Mosier Industries,
(Brookville, OH)

**7.3.3**
**Fluid Friction**

A fluid in motion offers frictional resistance to flow. Part of the energy in the system is converted into *thermal energy* (heat), which is dissipated through the walls of the pipe in which the fluid is flowing. The magnitude of the energy loss is dependent on the properties of the fluid, the flow velocity, the pipe size, and smoothness of the pipe wall, and the length of the pipe. We will develop methods of calculating this frictional energy loss in later chapters.

**7.3.4**
**Valves and Fittings**

Elements that control the direction or flow rate of a fluid in a system typically set up local turbulence in the fluid, causing energy to be dissipated as heat. Whenever there is a restriction, a change in flow velocity, or a change in the direction of flow, these energy losses occur. In a large system the magnitude of losses due to valves and fittings is usually small compared with frictional losses in the pipes. Therefore, such losses are referred to as *minor losses*.

**7.4**
**NOMENCLATURE OF ENERGY LOSSES AND ADDITIONS**

We will account for energy losses and additions in a system in terms of energy per unit weight of fluid flowing in the system. This is also known as "head," as described in Chapter 6. As an abbreviation for head we will use the symbol $h$ for energy losses and additions. Specifically, we will use the following terms throughout the next several chapters:

$h_A$ = *Energy added* to the fluid with a mechanical device such as a pump. This is often referred to as the *total head* on the pump.

$h_R$ = *Energy removed* from the fluid by a mechanical device such as a fluid motor.

$h_L$ = *Energy losses* from the system due to friction in pipes or minor losses due to valves and fittings.

We will not consider the effects of heat transferred into or out of the fluid at this time because they are negligible in the types of problems with which we are dealing. Courses in thermodynamics cover heat energy.

The magnitude of energy losses produced by fluid friction, valves, and fittings is directly proportional to the velocity head of the fluid. This can be expressed mathematically as

$$h_L = K(v^2/2g)$$

The term $K$ is the *resistance coefficient*. You will learn how to determine the value of $K$ for fluid friction in Chapter 9 using the Darcy equation. In Chapter 10, you will see methods of finding $K$ for many kinds of valves, fittings, and changes in flow cross section and direction. Most of these are found from experimental data.

**7.5**
**GENERAL ENERGY EQUATION**

The general energy equation as used in this text is an expansion of Bernoulli's equation, which makes it possible to solve problems in which energy losses and additions occur. The logical interpretation of the energy equation can be seen in Fig. 7.6, which represents a flow system. The terms $E_1'$ and $E_2'$ denote the energy possessed by the fluid per unit weight at sections 1 and 2, respectively. The energy additions, removals, and losses, $h_A$, $h_R$, and $h_L$, are shown. For such a system the expression of the principle of conservation of energy is

$$E_1' + h_A - h_R - h_L = E_2' \tag{7-1}$$

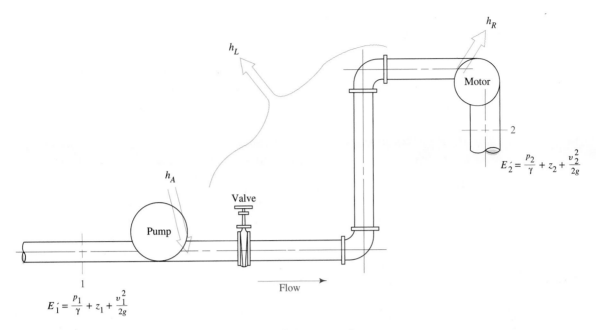

$$E_2' = \frac{p_2}{\gamma} + z_2 + \frac{v_2^2}{2g}$$

$$E_1' = \frac{p_1}{\gamma} + z_1 + \frac{v_1^2}{2g}$$

**FIGURE 7.6** Fluid flow system illustrating the general energy equation.

The energy possessed by the fluid per unit weight is

$$E' = \frac{p}{\gamma} + z + \frac{v^2}{2g} \tag{7-2}$$

Equation (7–1) then becomes

 GENERAL ENERGY EQUATION

$$\frac{p_1}{\gamma} + z_1 + \frac{v_1^2}{2g} + h_A - h_R - h_L = \frac{p_2}{\gamma} + z_2 + \frac{v_2^2}{2g} \tag{7-3}$$

This is the form of the energy equation that we will use most often in this book. As with Bernoulli's equation, each term in Eq. (7–3) represents a quantity of energy per unit weight of fluid flowing in the system. Typical SI units are N·m/N, or meters. U.S. Customary System units are lb-ft/lb or ft.

It is essential that the general energy equation be written *in the direction of flow*, that is, *from* the reference point on the left side of the equation *to* that on the right side. Algebraic signs are critical because the left side of Equation (7–3) states that an element of fluid having a certain amount of energy per unit weight at section 1 may have energy added $(+h_A)$, energy removed $(-h_R)$, or energy lost $(-h_L)$ from it before it reaches section 2. There it contains a different amount of energy per unit weight as indicated by the terms on the right side of the equation.

For example, in Fig. 7.6, reference points are shown to be points 1 and 2 with the pressure head, elevation head, and velocity head indicated at each point. After the fluid leaves point 1 it enters the pump where energy is added. A prime mover such as an electric motor drives the pump and the impeller of the pump transfers the energy to the fluid $(+h_A)$. Then the fluid flows through a piping system composed of a valve, elbows, and the lengths of pipe in which energy is dissipated from the fluid and is lost $(-h_L)$. Before reaching point 2, the fluid flows through a fluid motor which removes some of the energy to drive an external device $(-h_R)$. The general energy equation accounts for all of these energies.

In a particular problem, it is possible that not all of the terms in the general energy equation will be required. For example, if there is no mechanical device between the sections of interest, the terms $h_A$ and $h_R$ will be zero and can be left out of the equation. If energy losses are so small that they can be neglected, the term $h_L$ can be left out. If both of these conditions exist, it can be seen that Eq. (7–3) reduces to Bernoulli's equation.

## PROGRAMMED EXAMPLE PROBLEMS

☐ **EXAMPLE PROBLEM 7.1**

Water flows from a large reservoir at the rate of 1.20 ft³/s through a pipe system as shown in Fig. 7.7. Calculate the total amount of energy lost from the system because of the valve, the elbows, the pipe entrance, and fluid friction.

**FIGURE 7.7** Pipe system for Example Problem 7.1.

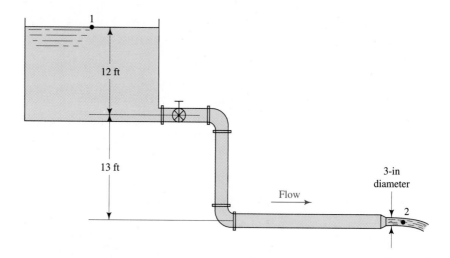

Using an approach similar to that used with Bernoulli's equation, select two sections of interest and write the general energy equation before looking at the next panel.

---

The sections at which we know the most information about pressure, velocity, and elevation are the surface of the reservoir and the free stream of fluid at the exit of the pipe. Call these section 1 and section 2, respectively. Then, the complete general energy equation [Eq. (7–3)] is

$$\frac{p_1}{\gamma} + z_1 + \frac{v_1^2}{2g} + h_A - h_R - h_L = \frac{p_2}{\gamma} + z_2 + \frac{v_2^2}{2g}$$

The value of some of these terms is zero. Determine which are zero and simplify the energy equation accordingly.

---

The value of the following terms is zero

$$p_1 = 0 \qquad \text{Surface of reservoir exposed to the atmosphere}$$
$$p_2 = 0 \qquad \text{Free stream of fluid exposed to the atmosphere}$$
$$v_1 = 0 \qquad \text{(Approximately) Surface area of reservoir is large}$$
$$h_A = h_R = 0 \qquad \text{No mechanical device in the system}$$

Then, the energy equation becomes

$$\cancel{\frac{p_1}{\gamma}}^{0} + z_2 + \cancel{\frac{v_1^2}{2g}}^{0} + \cancel{h_A}^{0} - \cancel{h_R}^{0} - h_L = \cancel{\frac{p_2}{\gamma}}^{0} + z_2 + \frac{v_2^2}{2g}$$

$$z_1 - h_L = z_2 + v_2^2/2g$$

Since we are looking for the total energy lost from the system, solve this equation for $h_L$.

---

You should have

$$h_L = (z_1 - z_2) - v_2^2/2g$$

Now evaluate the terms on the right side of the equation to determine $h_L$ in the units of lb-ft/lb.

---

The answer is $h_L = 15.75$ lb-ft/lb. Here is how it is done:

$$z_1 - z_2 = +25 \text{ ft}$$

$$v_2 = Q/A_2$$

Since $Q$ was given as 1.20 ft³/s and the area of a 3-in diameter jet is 0.0491 ft², we have

$$v_2 = \frac{Q}{A_2} = \frac{1.20 \text{ ft}^3}{s} \times \frac{1}{0.0491 \text{ ft}^2} = 24.4 \text{ ft/s}$$

$$\frac{v_2^2}{2g} = \frac{(24.4)^2 \text{ ft}^2}{s^2} \times \frac{s^2}{(2)(32.2) \text{ ft}} = 9.25 \text{ ft}$$

Then the total amount of energy lost from the system is

$$h_L = (z_1 - z_2) - v_2^2/2g = 25 \text{ ft} - 9.25 \text{ ft}$$

$$h_L = 15.75 \text{ ft, or } 15.75 \text{ lb-ft/lb}$$

---

☐ **EXAMPLE PROBLEM 7.2**  The volume flow rate through the pump shown in Fig. 7.8 is 0.014 m³/s. The fluid being pumped is oil with a specific gravity of 0.86. Calculate the energy delivered by the pump to

**FIGURE 7.8**  Pump system for Example Problem 7.2.

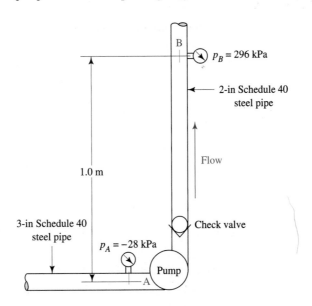

the oil per unit weight of oil flowing in the system. Energy losses in the system are caused by the check valve and friction losses as the fluid flows through the piping. The magnitude of such losses has been determined to be 1.86 N·m/N.

Using the sections where the pressure gages are located as the sections of interest, write the energy equation for the system, including only the necessary terms.

---

You should have

$$\frac{p_A}{\gamma} + z_A + \frac{v_A^2}{2g} + h_A - h_L = \frac{p_B}{\gamma} + z_B + \frac{v_B^2}{2g}$$

Notice that the term $h_R$ has been left out of the general energy equation.

The objective of the problem is to calculate the energy added to the oil by the pump. Solve for $h_A$ before looking at the next panel.

---

One correct solution is

$$h_A = \frac{p_B - p_A}{\gamma} + (z_B - z_A) + \frac{v_B^2 - v_A^2}{2g} + h_L \qquad (7\text{–}4)$$

Notice that similar terms have been grouped. This will be convenient when performing the calculations.

Equation 7–4 should be studied well, It indicates that the total head on the pump, $h_A$, is a measure of all of the tasks the pump is required to do in a system. It must increase the pressure from that at point A at the inlet to the pump to the pressure at point B. It must raise the fluid by the amount of the elevation difference between points A and B. It must supply the energy to increase the velocity of the fluid from that in the larger pipe at the pump inlet (called the suction pipe) to the velocity in the smaller pipe at the pump outlet (called the discharge pipe). And it must overcome any energy losses that occur in the system such as those due to the check valve and friction in the discharge pipe.

We recommend that you evaluate each of the terms in Equation 7–4 separately and then combine them at the end. The first term is the difference between the pressure head at point A and that at point B. What is the value of $\gamma$?

---

Remember that the specific weight of the fluid being pumped must be used. In this case, the specific weight of the oil is

$$\gamma = (sg)(\gamma_w) = (0.86)(9.81 \text{ kN/m}^3) = 8.44 \text{ kN/m}^3$$

Now complete the evaluation of $(p_B - p_A)/\gamma$.

---

Since $p_B = 296$ kPa and $p_A = -28$ kPa, we have

$$\frac{p_B - p_A}{\gamma} = \frac{[296 - (-28)] \text{ kN}}{\text{m}^2} \times \frac{\text{m}^3}{8.44 \text{ kN}} = 38.4 \text{ m}$$

Now evaluate the elevation difference, $z_B - z_A$.

---

You should have $z_B - z_A = 1.0$ m. Notice that point B is at a higher elevation than point A and, therefore, $z_B > z_A$. The result is that $z_B - z_A$ is a positive number.

Now compute the velocity head difference term, $(v_B^2 - v_A^2)/2g$.

---

We can use the definition of volume flow rate and the continuity equation to determine each velocity.

$$Q = Av = A_A v_A = A_B v_B$$

Then, solving for the velocities and using the flow areas for the suction and discharge pipes from Appendix F gives,

$$v_A = Q/A_A = (0.014 \text{ m}^3/\text{s})/(4.768 \times 10^{-3} \text{ m}^2) = 2.94 \text{ m/s}$$
$$v_B = Q/A_B = (0.014 \text{ m}^3/\text{s})/(2.168 \times 10^{-3} \text{ m}^2) = 6.46 \text{ m/s}$$

Finally,

$$\frac{(v_B^2 - v_A^2)}{2g} = \frac{[(6.46)^2 - (2.94)^2] \text{ m}^2/\text{s}^2}{2(9.81 \text{ m/s}^2)} = 1.69 \text{ m}$$

The only remaining term in Equation 7–4 is the energy loss, $h_L$, that is given to be 1.86 N·m/N or 1.86 m. We can now combine all of these terms and complete the calculation of $h_A$.

---

The energy added to the system is

$$h_A = 38.4 \text{ m} + 1.0 \text{ m} + 1.69 \text{ m} + 1.86 \text{ m} = 42.9 \text{ m, or } 42.9 \text{ N·m/N}$$

That is, the pump delivers 42.9 N·m of energy to each newton of oil flowing through it.
This completes the programmed instruction.

■

---

**7.6**
**POWER REQUIRED**
**BY PUMPS**

Power is defined as the rate of doing work. In fluid mechanics we can modify this statement and consider that power is the rate at which energy is being transferred.

We first develop the basic concept of power in SI units. Then we show the units for the U.S. Customary System. The unit for power in the SI system is the watt (W), which is equivalent to 1.0 N·m/s or 1.0 joule/s.

In Example Problem 7.2 we found that the pump was delivering 42.9 N·m of energy to each newton of oil as it flowed through the pump. In order to calculate the power delivered to the oil, we must determine how many newtons of oil are flowing through the pump in a given amount of time. This is called the *weight flow rate W*, which we defined in Chapter 6, and is expressed in units of N/s. Power is calculated by multiplying the energy transferred per newton of fluid by the weight flow rate. This is

$$P_A = h_A W$$

But since $W = \gamma Q$, we can also write

⇨ **POWER ADDED TO A FLUID BY A PUMP**

$$P_A = h_A \gamma Q \qquad (7\text{--}5)$$

where $P_A$ denotes power added to the fluid, $\gamma$ is the specific weight of the fluid flowing through the pump, and $Q$ is the volume flow rate of the fluid.

By using the data of Example Problem 7.2, we can find the power delivered by the pump to the oil as follows:

$$P_A = h_A \gamma Q$$

We know from Example Problem 7.2 that

$$h_A = 42.9 \text{ N·m/N}$$
$$\gamma = 8.44 \text{ kN/m}^3 = 8.44 \times 10^3 \text{ N/m}^3$$
$$Q = 0.014 \text{ m}^3/\text{s}$$

Substituting these values into Eq. (7–5), we get

$$P_A = \frac{42.9 \text{ N·m}}{\text{N}} \times \frac{8.44 \times 10^3 \text{ N}}{\text{m}^3} \times \frac{0.014 \text{ m}^3}{\text{s}} = 5069 \text{ N·m/s}$$

Because $1.0 \text{ W} = 1.0 \text{ N·m/s}$, express the result in watts.

$$P_A = 5069 \text{ W} = 5.07 \text{ kW}$$

### 7.6.1 Power in the U.S. Customary System

The unit for power in the U.S. Customary System is lb-ft/s. Since it is common practice to refer to power in horsepower (hp), the conversion factor required is

$$1 \text{ hp} = 550 \text{ lb-ft/s}$$

In Eq. (7–5) the energy added, $h_A$, is expressed in feet of the fluid flowing in the system. Then, expressing the specific weight of the fluid in lb/ft³ and the volume flow rate in ft³/s would yield the weight flow rate $\gamma Q$ in lb/s. Finally, in the power equation $P_A = h_A \gamma Q$, power would be expressed in lb-ft/s.

To convert these units to the SI system we use the factors

$$1 \text{ lb-ft/s} = 1.356 \text{ W}$$

$$1 \text{ hp} = 745.7 \text{ W}$$

### 7.6.2 Mechanical Efficiency of Pumps

The term *efficiency* is used to denote the ratio of the power delivered by the pump to the fluid to the power supplied to the pump. Because of energy losses due to mechanical friction in pump components, fluid friction in the pump, and excessive fluid turbulence in the pump, not all of the input power is delivered to the fluid. Then, using the symbol $e_M$ for mechanical efficiency, we have

 PUMP EFFICIENCY

$$e_M = \frac{\text{Power delivered to fluid}}{\text{Power put into pump}} = \frac{P_A}{P_I} \qquad (7\text{–}6)$$

The value of $e_M$ will always be less than 1.0.

Continuing with the data of Example Problem 7.2, we could calculate the power input to the pump if $e_M$ is known. For commercially available pumps the value of $e_M$ is published as part of the performance data. If we assume that for the pump in this problem the efficiency is 82 percent, then

$$P_I = P_A / e_M = 4.66/0.82 = 5.68 \text{ kW}$$

The value of the mechanical efficiency of pumps depends not only on the design of the pump but also on the conditions under which it is operating, particularly the total head and the flow rate. For pumps used in hydraulic systems, such as those shown in Figs. 7.2 and 7.3, efficiencies range from about 70 to 90 percent. For centrifugal pumps used primarily to transfer or circulate liquids, the efficiencies range from about 50 to 85 percent.

See Chapter 13 for more data and discussion of pump performance. Efficiency values for positive displacement fluid power pumps are reported differently from those for centrifugal pumps. Three values are often used: *overall efficiency, $e_o$, volumetric efficiency, $e_v$,* and *torsional efficiency, $e_T$.* More is said in Chapter 13 about the details of these efficiencies. But in general the overall efficiency is analogous to the mechanical efficiency discussed for other types of pumps in this section. Volumetric efficiency is a measure of the actual delivery from the pump compared with the ideal delivery found from the displacement per revolution times the rotational speed of the pump. A high volumetric efficiency is desired because the operation of the fluid power system depends on a nearly uniform flow rate of fluid through all operating conditions. Torsional efficiency is a measure of the ratio of the ideal torque required to drive the pump against the pressure it is developing to the actual torque.

The following programmed example problem illustrates a possible setup for measuring pump efficiency.

## PROGRAMMED EXAMPLE PROBLEM

☐ **EXAMPLE PROBLEM 7.3**     For the pump test arrangement shown in Fig. 7.9, determine the mechanical efficiency of the pump if the power input is measured to be 3.85 hp when pumping 500 gal/min of oil ($\gamma = 56.0$ lb/ft$^3$).

**FIGURE 7.9**   Pump test system for Example Problem 7.3.

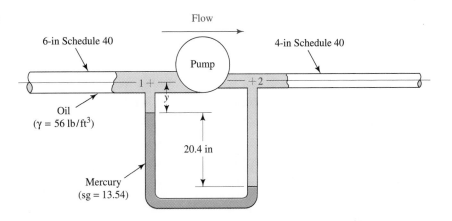

To begin, write the energy equation for this system.

---

Using the points identified as 1 and 2 in Fig. 7.9, we have

$$\frac{p_1}{\gamma} + z_1 + \frac{v_1^2}{2g} + h_A = \frac{p_2}{\gamma} + z_2 + \frac{v_2^2}{2g}$$

Because we must find the power delivered by the pump to the fluid, we should now solve for $h_A$.

---

This equation is used:

$$h_A = \frac{p_2 - p_1}{\gamma} + (z_2 - z_1) + \frac{v_2^2 - v_1^2}{2g} \qquad (7\text{–}7)$$

It is convenient to solve for each term individually and then combine the results. The manometer enables us to calculate $(p_2 - p_1)/\gamma$ because it measures the pressure difference. Using the procedure outlined in Chapter 3, write the manometer equation between points 1 and 2.

---

Starting at point 1, we have

$$p_1 + \gamma_o y + \gamma_m(20.4 \text{ in}) - \gamma_o(20.4 \text{ in}) - \gamma_o y = p_2$$

where $y$ is the unknown distance from point 1 to the top of the mercury column in the left leg of the manometer. The terms involving $y$ cancel out. Also, in this equation $\gamma_o$ is the specific weight of the oil and $\gamma_m$ is the specific weight of the mercury gage fluid.

The desired result for use in Eq. (7–7) is $(p_2 - p_1)/\gamma_o$. Solve for this now and compute the result.

---

The correct solution is $(p_2 - p_1)/\gamma_o = 24.0$ ft. Here is one way to do it:

$$\gamma_m = (13.54)(\gamma_w) = (13.54)(62.4 \text{ lb/ft}^3) = 844.9 \text{ lb/ft}^3$$

$$p_2 = p_1 + \gamma_m(20.4 \text{ in}) - \gamma_o(20.4 \text{ in})$$

$$p_2 - p_1 = \gamma_m(20.4 \text{ in}) - \gamma_o(20.4 \text{ in})$$

$$\frac{p_2 - p_1}{\gamma_o} = \frac{\gamma_m(20.4 \text{ in})}{\gamma_o} - 20.4 \text{ in} = \left(\frac{\gamma_m}{\gamma_o} - 1\right)20.4 \text{ in}$$

$$= \left(\frac{844.9 \text{ lb/ft}^3}{56.0 \text{ lb/ft}^3} - 1\right)20.4 \text{ in} = (15.1 - 1)(20.4 \text{ in})$$

$$\frac{p_1 - p_2}{\gamma_o} = (14.1)(20.4 \text{ in})\left(\frac{1 \text{ ft}}{12 \text{ in}}\right) = 24.0 \text{ ft}$$

The next term in Eq. (7–7) is $z_2 - z_1$. What is its value?

---

It is zero. Both points are on the same elevation. These terms could have been cancelled from the original equation. Now find $(v_2^2 - v_1^2)/2g$.

---

You should have $(v_2^2 - v_1^2)/2g = 1.99$ ft.

$$Q = 500 \text{ gal/min}\left(\frac{1 \text{ ft}^3/s}{449 \text{ gal/min}}\right) = 1.11 \text{ ft}^3/s$$

Using $A_1 = 0.2006 \text{ ft}^2$ and $A_2 = 0.0884 \text{ ft}^2$ from Appendix F, we get

$$v_1 = \frac{Q}{A_1} = \frac{1.11 \text{ ft}^3}{s} \times \frac{1}{0.2006 \text{ ft}^2} = 5.55 \text{ ft/s}$$

$$v_2 = \frac{Q}{A_2} = \frac{1.11 \text{ ft}^3}{s} \times \frac{1}{0.0884 \text{ ft}^2} = 12.6 \text{ ft/s}$$

$$\frac{v_2^2 - v_1^2}{2g} = \frac{(12.6)^2 - (5.55)^2}{(2)(32.2)}\frac{\text{ft}^2}{s^2}\frac{s^2}{\text{ft}} = 1.99 \text{ ft}$$

Now place these results into Eq. (7–7) and solve for $h_A$.

---

Solving for $h_A$, we get

$$h_A = 24.0 \text{ ft} + 0 + 1.99 \text{ ft} = 25.99 \text{ ft}$$

We can now calculate the power delivered to the oil, $P_A$.

---

The result is $P_A = 2.95$ hp.

$$P_A = h_a \gamma Q = 25.99 \text{ ft} \left( \frac{56.0 \text{ lb}}{\text{ft}^3} \right) \left( \frac{1.11 \text{ ft}^3}{\text{s}} \right)$$

$$P_A = 1620 \text{ lb-ft/s} \left( \frac{1 \text{ hp}}{550 \text{ lb-ft/s}} \right) = 2.95 \text{ hp}$$

The final step is to calculate $e_M$, the mechanical efficiency of the pump.

---

From Eq. (7–6) we get

$$e_M = P_A/P_I = 2.95/3.85 = 0.77$$

Expressed as a percentage, the pump is 77 percent efficient at the stated conditions.
This completes the programmed instruction.

■

---

**7.7**
**POWER DELIVERED**
**TO FLUID MOTORS**

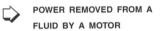
POWER REMOVED FROM A
FLUID BY A MOTOR

The energy delivered by the fluid to a mechanical device such as a fluid motor or turbine is denoted in the general energy equation by the term $h_R$. This is a measure of the energy delivered by each unit weight of fluid as it passes through the device. We find the power delivered by multiplying $h_R$ by the weight flow rate $W$:

$$P_R = h_R W = h_R \gamma Q \tag{7–8}$$

where $P_R$ is the power delivered by the fluid to the fluid motor.

**7.7.1**
**Mechanical Efficiency of**
**Fluid Motors**

⇨ MOTOR EFFICIENCY

As was described for pumps, energy losses in a fluid motor are produced by mechanical and fluid friction. Therefore, not all the power delivered to the motor is ultimately converted to power output from the device. Mechanical efficiency is then defined as

$$e_M = \frac{\text{Power output from motor}}{\text{Power delivered by fluid}} = \frac{P_O}{P_R} \tag{7–9}$$

Here again, the value of $e_M$ is always less than 1.0.
Refer to Section 7.6 for power units.

---

## PROGRAMMED EXAMPLE PROBLEM

☐ **EXAMPLE PROBLEM 7.4**

Water at 10°C is flowing at a rate of 115 L/min through the fluid motor shown in Fig. 7.10. The pressure at A is 700 kPa and the pressure at B is 125 kPa. It is estimated that due to friction in the piping there is an energy loss of 4.0 N·m/N of water flowing. (a) Calculate the power delivered to the fluid motor by the water. (b) If the mechanical efficiency of the fluid motor is 85 percent, calculate the power output.

Start the solution by writing the energy equation.

---

Choosing points A and B as our reference points, we get

$$\frac{p_A}{\gamma} + z_A + \frac{v_A^2}{2g} - h_R - h_L = \frac{p_B}{\gamma} + z_B + \frac{v_B^2}{2g}$$

**FIGURE 7.10**   Fluid motor for
Example Problem 7.4.

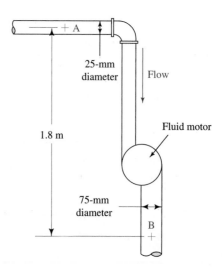

The value of $h_R$ is needed to determine the power output. Solve the energy equation for this
term.

---

Compare this equation with your result:

$$h_R = \frac{p_A - p_B}{\gamma} + (z_A - z_B) + \frac{v_A^2 - v_B^2}{2g} - h_L \qquad (7\text{--}10)$$

Before looking at the next panel, solve for the value of each term in this equation using the
unit of N·m/N or m.

---

The correct results are

**1.** $\dfrac{p_A - p_B}{\gamma} = \dfrac{(700 - 125)(10^3)\text{N}}{\text{m}^2} \times \dfrac{\text{m}^3}{9.81 \times 10^3 \text{ N}} = 58.6 \text{ m}$

**2.** $z_A - z_B = 1.8 \text{ m}$

**3.** Solving for $(v_A^2 - v_B^2)/2g$,

$$Q = 115 \text{ L/min} \times \frac{1.0 \text{ m}^3/\text{s}}{60\,000 \text{ L/min}} = 1.92 \times 10^{-3} \text{ m}^3/\text{s}$$

$$v_A = \frac{Q}{A_A} = \frac{1.92 \times 10^{-3} \text{ m}^3}{\text{s}} \times \frac{1}{4.909 \times 10^{-4} \text{ m}^2} = 3.91 \text{ m/s}$$

$$v_B = \frac{Q}{A_B} = \frac{1.92 \times 10^{-3} \text{ m}^3}{\text{s}} \times \frac{1}{4.418 \times 10^{-3} \text{ m}^2} = 0.43 \text{ m/s}$$

$$\frac{v_A^2 - v_B^2}{2g} = \frac{(3.91)^2 - (0.43)^2}{(2)(9.81)} \frac{\text{m}^2}{\text{s}^2} \frac{\text{s}^2}{\text{m}} = 0.77 \text{ m}$$

**4.** $h_L = 4.0 \text{ m}$ (given)

Complete the solution of Eq. (7–10) for $h_R$ now.

---

The energy delivered by the water to the turbine is

$$h_R = (58.6 + 1.8 + 0.77 - 4.0) \text{ m} = 57.2 \text{ m}$$

To complete part (a) of the problem, calculate $P_R$.

---

Substituting the known values into Eq. (7–8), we get

$$P_R = h_R \gamma Q$$

$$P_R = 57.2 \text{ m} \times \frac{9.81 \times 10^3 \text{ N}}{\text{m}^3} \times \frac{1.92 \times 10^{-3} \text{ m}^3}{\text{s}} = 1080 \text{ N·m/s}$$

$$P_R = 1.08 \text{ kW}$$

This is the power delivered to the fluid motor by the water. How much useful power can be expected to be put out by the motor?

---

Because the efficiency of the motor is 85 percent, we get 0.92 kW of power out. Using Eq. (7–9), $e_M = P_O/P_R$, we get

$$P_O = e_M P_R$$

$$= (0.85)(1.08 \text{ kW})$$

$$P_O = 0.92 \text{ kW}$$

This completes the programmed example problem.

■

---

# PRACTICE PROBLEMS

It may be necessary to refer to the appendix for data concerning the dimensions of pipes or the properties of fluids. Assume there are no energy losses unless stated otherwise.

**7.1E** A horizontal pipe carries oil with a specific gravity of 0.83. If two pressure gages along the pipe read 74.6 psig and 62.2 psig, respectively, calculate the energy loss between the two gages.

**7.2E** Water at 40°F is flowing downward through the pipe shown in Fig. 7.11. At point A the velocity is 10 ft/s and the pressure is 60 psig. The energy loss between points A and B is 25 lb-ft/lb. Calculate the pressure at point B.

**7.3M** *Find the volume flow rate of water exiting from the tank shown in Fig. 7.12. The tank is sealed with a pressure of 140 kPa above the water. There is an energy loss of 2.0 N·m/N as the water flows through the nozzle.*

**7.4M** *A long 6-in Schedule 40 steel pipe discharges 0.085 m³/s of water from a reservoir into the atmosphere as shown in Fig. 7.13. Calculate the energy loss in the pipe.*

**7.5E** Figure 7.14 shows a setup to determine the energy loss due to a certain piece of apparatus. The inlet is through a 2-in Schedule 40 pipe and the outlet is a 4-in Schedule 40 pipe. Calculate the energy loss between points A and B if water is flowing upward at 0.20 ft³/s. The gage fluid is mercury (sg = 13.54).

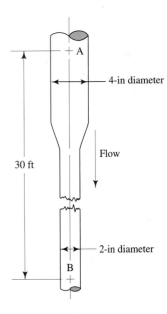

**FIGURE 7.11**  Problem 7.2.

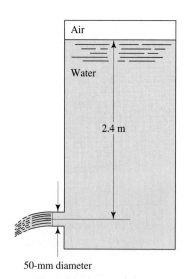

FIGURE 7.12 Problem 7.3.

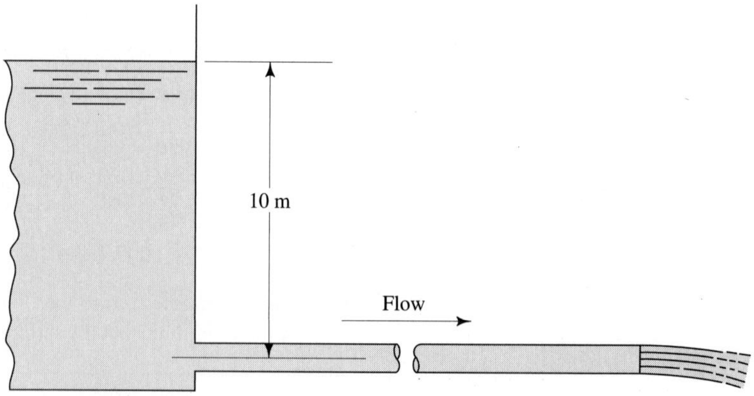

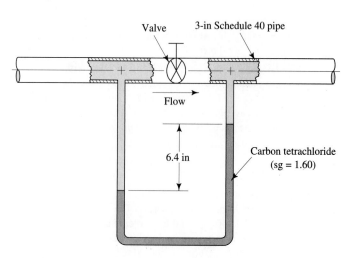

**FIGURE 7.13** Problem 7.4.

**7.6E** A test setup to determine the energy loss as water flows through a valve is shown in Fig. 7.15. Calculate the energy loss if 0.10 ft³/s of water at 40°F is flowing. Also calculate the *resistance* coefficient $K$ if the energy loss is expressed as $K(v^2/2g)$.

**7.7M** *The setup shown in Fig. 7.16 is being used to measure the energy loss across a valve. The velocity of flow of the oil is 1.2 m/s. Calculate the value of K if the energy loss is expressed as $K(v^2/2g)$.*

**7.8M** *A pump is being used to transfer water from an open tank to one that has air at 500 kPa above the water, as shown in Fig. 7.17. If 2250 L/min are being pumped, calculate the power delivered by the pump to the water. Assume that the level of the surface in each tank is the same.*

**7.9M** *In Problem 7.8 (Fig. 7.17), if the left-hand tank were also sealed and air pressure above the water is 68 kPa, calculate the pump power.*

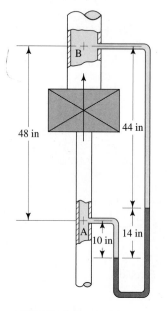

**FIGURE 7.14** Problem 7.5.

**FIGURE 7.15** Problem 7.6.

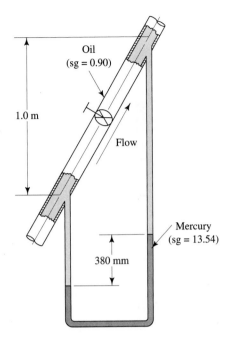

**FIGURE 7.16**    Problem 7.7.

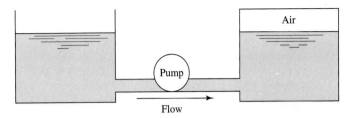

**FIGURE 7.17**    Problems 7.8 and 7.9.

**7.10E**  A commercially available sump pump is capable of delivering 2800 gal/h of water through a vertical lift of
20 ft. The inlet to the pump is just below the water surface and the discharge is to the atmosphere through a
1¼-in Schedule 40 pipe. (a) Calculate the power deliv-

ered by the pump to the water. (b) If the pump draws
0.5 hp, calculate its efficiency.

**7.11E**  A submersible deep-well pump delivers 745 gal/h of water through a 1-in Schedule 40 pipe when operating in
the system sketched in Fig. 7.18. An energy loss of
10.5 lb-ft/lb occurs in the piping system. (a) Calculate the
power delivered by the pump to the water. (b) If the pump
draws 1 hp, calculate its efficiency.

**7.12M**  *In a pump test the suction pressure at the pump inlet is
30 kPa below atmospheric pressure. The discharge pressure at a point 750 mm above the inlet is 520 kPa. Both
pipes are 75 mm in diameter. If the volume flow rate of
water is 75 L/min, calculate the power delivered by the
pump to the water.*

**7.13M**  *The pump shown in Fig. 7.19 is delivering hydraulic oil
with a specific gravity of 0.85 at a rate of 75 L/min. The
pressure at A is −20 kPa while the pressure at B is
275 kPa. The energy loss in the system is 2.5 times the
velocity head in the discharge pipe. Calculate the power
delivered by the pump to the oil.*

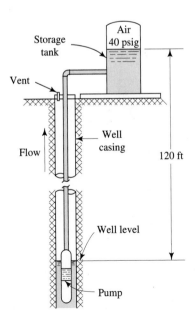

**FIGURE 7.18**    Problem 7.11.

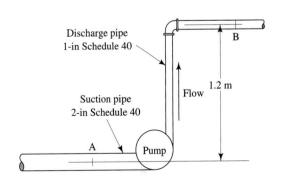

**FIGURE 7.19**    Problem 7.13.

**7.14E** The pump in Fig. 7.20 delivers water from the lower to the upper reservoir at the rate of 2.0 ft³/s. The energy loss between the suction pipe inlet and the pump is 6 lb-ft/lb and between the pump outlet and the upper reservoir is 12 lb-ft/lb. Both pipes are 6-in Schedule 40 steel pipe. Calculate (a) the pressure at the pump inlet, (b) the pressure at the pump outlet, (c) the total head on the pump, and (d) the power delivered by the pump to the water.

**7.15E** Repeat Problem 7.14, but assume that the level of the lower reservoir is 10 ft above the pump instead of below it. All other data remain the same.

**7.16M** *Figure 7.21 shows a pump delivering 840 L/min of crude oil (sg = 0.85) from an underground storage drum to the first stage of a processing system. (a) If the total energy loss in the system is 4.2 N·m/N of oil flowing, calculate the power delivered by the pump. (b) If the energy loss in the suction pipe is 1.4 N·m/N of oil flowing, calculate the pressure at the pump inlet.*

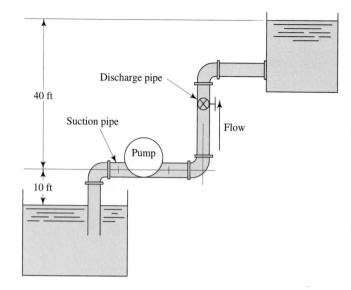

**FIGURE 7.20**    Problems 7.14 and 7.15.

**FIGURE 7.21**    Problem 7.16.

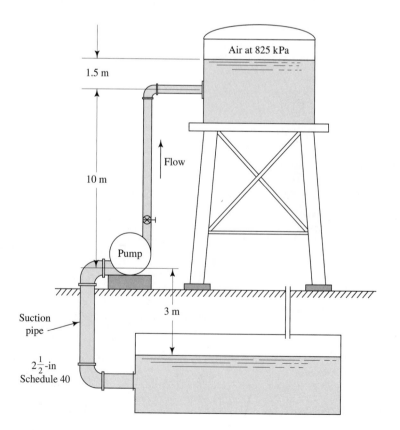

**7.17M** *Figure 7.22 shows a submersible pump being used to circulate 60 L/min of a water-based coolant (sg = 0.95) to the cutter of a milling machine. The outlet is through a $\frac{3}{4}$-in Schedule 40 steel pipe. Assuming a total energy loss due to the piping of 3.0 N·m/N, calculate the total head developed by the pump and the power delivered to the coolant.*

**FIGURE 7.22**    Problem 7.17.

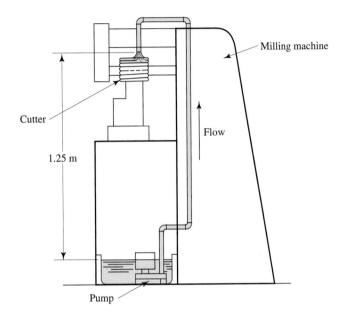

**7.18M** *Figure 7.23 shows a small pump in an automatic washer discharging into a laundry sink. The washer tub is 525 mm in diameter and 250 mm deep. The average head above the pump is 375 mm as shown. The discharge hose has an inside diameter of 18 mm. The energy loss in the hose system is 0.22 N·m/N. If the pump empties the tub in 90 s, calculate the average total head on the pump.*

**FIGURE 7.23**    Problem 7.18.

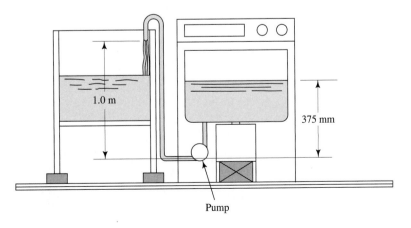

**7.19E** The water being pumped in the system shown in Fig. 7.24 discharges into a tank which is being weighed. It is found that 556 lb of water are collected in 10 s. If the pressure at A is 2.0 psi below atmospheric pressure, calculate the horsepower delivered by the pump to the water.

**FIGURE 7.24**   Problem 7.19.

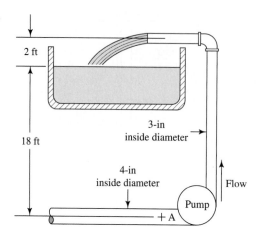

**7.20E** A manufacturer's rating for a gear pump states that 0.85 hp is required to drive the pump when it is pumping 9.1 gal/min of oil (sg = 0.90) with a total head of 257 ft. Calculate the mechanical efficiency of the pump.

**7.21M** *The specifications for an automobile fuel pump state that it should pump 1 L of gasoline in 40s with a suction pressure of 150 mm of mercury vacuum and a discharge pressure of 30 kPa. Assuming that the pump efficiency is 60 percent, calculate the power drawn from the engine. See Fig. 7.25. The suction and discharge lines are the same size.*

**FIGURE 7.25**   Automobile fuel pump for Problem 7.21.

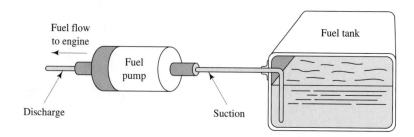

**7.22E** Figure 7.26 shows the arrangement of a circuit for a hydraulic system. The pump draws oil with a specific gravity of 0.90 from a reservoir and delivers it to the hydraulic cylinder. The cylinder has an inside diameter of 5.0 in, and in 15 s the piston must travel 20 in while exerting a force of 11 000 lb. It is estimated that there are energy losses of 11.5 lb-ft/lb in the suction pipe and 35.0 lb-ft/lb in the discharge pipe. Both pipes are ³/₈-in Schedule 80 steel pipes. Calculate

**a.** The volume flow rate through the pump.
**b.** The pressure at the cylinder.
**c.** The pressure at the outlet of the pump.
**d.** The pressure at the inlet to the pump.
**e.** The power delivered to the oil by the pump.

**FIGURE 7.26**   Problem 7.22.

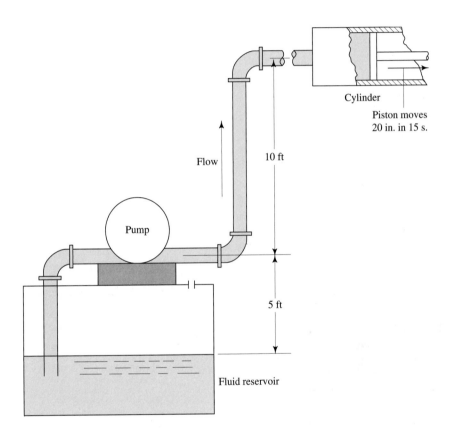

**7.23M** *Calculate the power delivered to the hydraulic motor in Fig. 7.27 if the pressure at A is 6.8 MPa and the pressure at B is 3.4 MPa. The motor inlet is a 1-in steel tube (0.065-in wall thickness) and the outlet is a 2-in steel tube (0.065-in wall thickness). The fluid is oil (sg = 0.90) and the velocity of flow is 1.5 m/s at point B.*

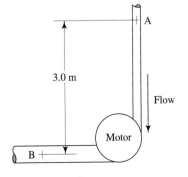

**FIGURE 7.27**   Problem 7.23.

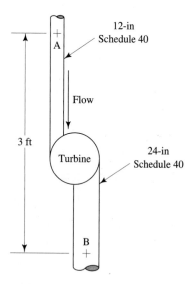

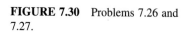

**FIGURE 7.28**    Problem 7.24.

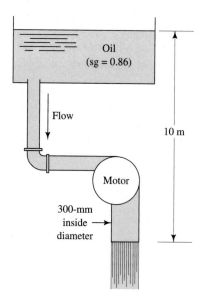

**FIGURE 7.29**    Problem 7.25.

**7.24E** Water flows through the turbine shown in Fig. 7.28 at a rate of 3400 gal/min when the pressure at A is 21.4 psig and the pressure at B is −5 psig. The friction energy loss between A and B is twice the velocity head in the 12-in pipe. Determine the power delivered by the water to the turbine.

**7.25M** *Calculate the power delivered by the oil to the fluid motor shown in Fig. 7.29 if the volume flow rate is 0.25 m³/s. There is an energy loss of 1.4 N·m/N in the piping sys-*

*tem. If the motor has an efficiency of 75 percent, calculate the power output.*

**7.26E** What hp must the pump shown in Fig. 7.30 deliver to a fluid having a specific weight of 60.0 lb/ft³ if energy losses of 3.40 lb-ft/lb occur between points 1 and 2? The pump delivers 40 gal/min of fluid.

**7.27E** If the pump in Problem 7.26 operates with an efficiency of 75 percent, what is the power input to the pump?

**FIGURE 7.30**    Problems 7.26 and 7.27.

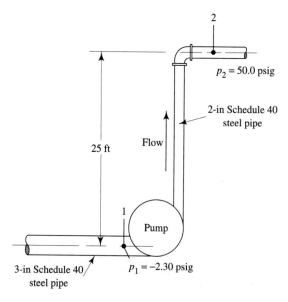

**7.28M** *The system shown in Fig. 7.31 delivers 600 L/min of water. The outlet is directly into the atmosphere. Determine the energy losses in the system.*

**FIGURE 7.31** Problem 7.28.

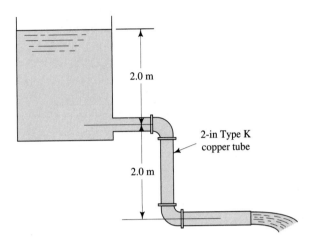

2.0 m

2-in Type K copper tube

2.0 m

**7.29M** *Kerosene (sg = 0.823) flows at 0.060 m³/s in the pipe shown in Fig. 7.32. Compute the pressure at B if the total energy loss in the system is 4.60 N·m/N.*

**FIGURE 7.32** Problem 7.29.

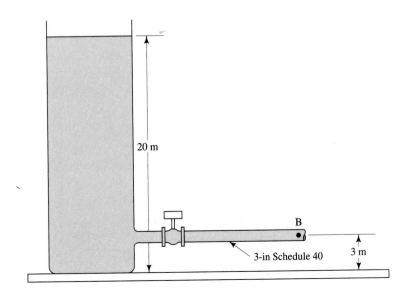

20 m

B

3-in Schedule 40

3 m

**7.30E** Water at 60°F flows from a large reservoir through a turbine at the rate of 1000 gal/min in the system shown in Fig. 7.33. If the turbine removes 37 hp from the fluid, calculate the energy losses in the system.

**FIGURE 7.33**    Problem 7.30.

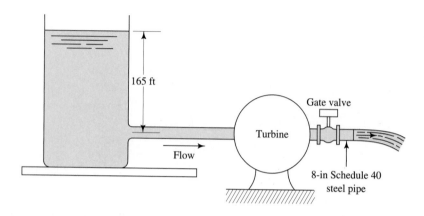

**7.31E** Figure 7.34 shows a portion of a fire protection system in which a pump draws 1500 gal/min of water at 50°F from a reservoir and delivers it to point B. The energy loss between the reservoir and point A at the inlet to the pump is 0.65 lb-ft/lb. Specify the required depth $h$ to maintain at least 5.0 psig pressure at point A.

**7.32E** For the conditions of Problem 7.31, and if we assume that the pressure at point A is 5.0 psig, calculate the power delivered by the pump to the water to maintain a pressure of 85 psig at point B. Energy losses between the pump and point B total 28.0 lb-ft/lb.

**7.33M** *In Fig. 7.35, kerosene at 25°C is flowing at 500 L/min from the lower tank to the upper tank through 2-in Type K copper tubing and a valve. If the pressure above the fluid is 15.0 psig, how much energy loss occurs in the system?*

**7.34M** *For the system shown in Fig. 7.35 and analyzed in Problem 7.33, assume that the energy loss is proportional to the velocity head in the tubing. Compute the pressure in the tank required to cause a flow of 1000 L/min.*

**General Data for Problems 7.35E Through 7.40E**

Figure 7.36 shows a diagram of a fluid power system for a hydraulic press used to extrude rubber parts. The following data are known:

1. The fluid is oil (sg = 0.93).
2. Volume flow rate is 175 gal/min.
3. Power input to the pump is 28.4 hp.
4. Pump efficiency is 80 percent.
5. Energy loss from point 1 to 2 is 2.80 lb-ft/lb.
6. Energy loss from point 3 to 4 is 28.50 lb-ft/lb.
7. Energy loss from point 5 to 6 is 3.50 lb-ft/lb.

**FIGURE 7.34**    Problems 7.31 and 7.32.

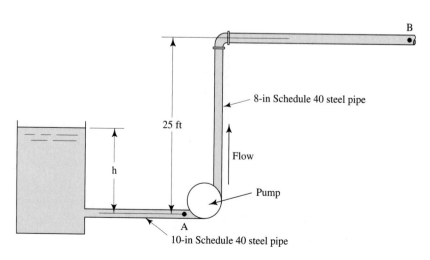

**FIGURE 7.35** Problems 7.33 and 7.34.

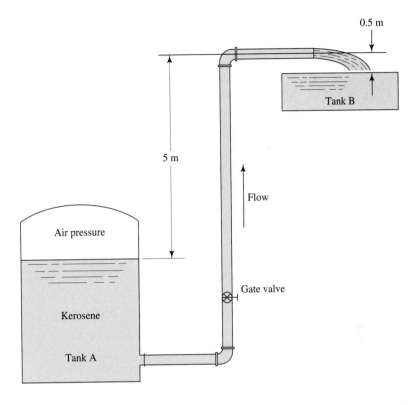

**7.35E** Compute the power removed from the fluid by the press.

**7.36E** Compute the pressure at point 2 at the pump inlet.

**7.37E** Compute the pressure at point 3 at the pump outlet.

**7.38E** Compute the pressure at point 4 at the press inlet.

**7.39E** Compute the pressure at point 5 at the press outlet.

**7.40E** Compare the velocity of flow in the suction line and in the discharge line of the system, using the recommendations listed in Table 6.3. Are they satisfactory? If not, specify suitable pipe sizes.

**7.41E** The portable, pressurized fuel can shown in Fig. 7.37 is used to deliver fuel to a race car during a pit stop. What pressure must be above the fuel to deliver 40 gal in 8.0 s? The specific gravity of the fuel is 0.76. An energy loss of 4.75 lb-ft/lb occurs at the nozzle.

**7.42E** Professor Crocker is building a cabin on a hillside and has proposed the water system shown in Fig. 7.38. The distribution tank in the cabin maintains a pressure of 30.0 psig above the water. There is an energy loss of 15.5 lb-ft/lb in the piping. When the pump is delivering

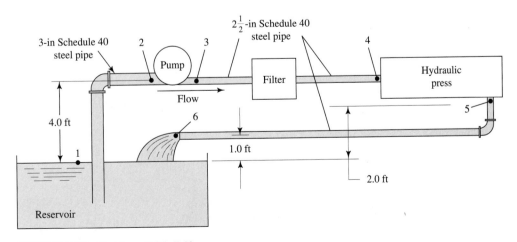

**FIGURE 7.36** Problems 7.35–7.40.

**FIGURE 7.37**   Problem 7.41.

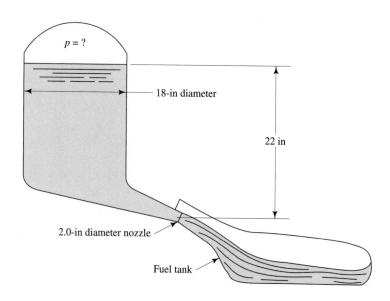

**FIGURE 7.38**   Problems 7.42 and 7.43.

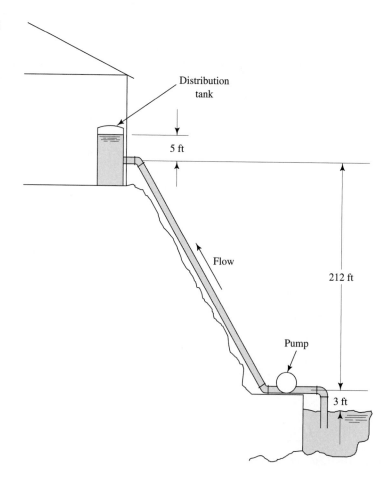

**FIGURE 7.39** Problems 7.44 and 7.45.

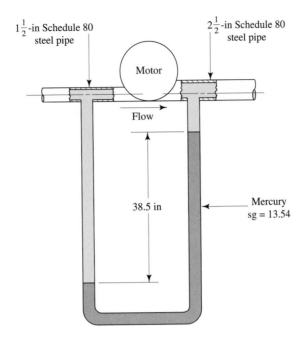

$1\frac{1}{2}$-in Schedule 80 steel pipe

$2\frac{1}{2}$-in Schedule 80 steel pipe

Motor

Flow

38.5 in

Mercury sg = 13.54

40 gal/min of water, compute the horsepower delivered by the pump to the water.

**7.43E** If Professor Crocker's pump, described in Problem 7.42, has an efficiency of 72 percent, what size motor is required to drive the pump?

**7.44E** The test setup in Fig. 7.39 measures the pressure difference between the inlet and outlet of the fluid motor. The flow rate of hydraulic oil (sg = 0.90) is 135 gal/min. Compute the power removed from the fluid by the motor.

**7.45E** If the fluid motor in Problem 7.44 has an efficiency of 78 percent, how much power is delivered by the motor?

# 8 Reynolds Number, Lamina Flow, and Turbulent Flow

■ ■ ■ ■

## 8.1 The Big Picture

### Discussion Map

▢ In this chapter you begin the development of your skills in analyzing the energy losses that occur as fluids flow in real pipe line systems.

▢ To analyze such energy losses, you must be able to use the Reynolds number to characterize the nature of the flow to determine if it is laminar or turbulent.

▢ Review Chapter 2 for the important concepts related to viscosity now.

### Discover

■ Describe what you observe as you turn a water faucet on to a very slow flow rate. What does the stream look like?
■ Now, slowly open the faucet more and more and observe how the character of the flow stream changes.
■ Now close the faucet slowly and carefully.
■ What did you observe?
■ Think about other fluid flow systems where you could observe the character of the flow. Consider fountains, waterfalls, open streams, and transparent tubes.

You will learn in this chapter that we can characterize the flow by comparing a dimensionless number, the *Reynolds number*, relating the important variables for the flow: velocity, size and shape of the flow path, fluid density, and viscosity.

As the water flows from a faucet at a very low velocity, the flow appears to be smooth and steady. The stream has a fairly uniform diameter and there is little or no evidence of mixing of the various parts of the stream. This is called *laminar flow*, a term derived from the word *layer*, because the fluid appears to be flowing in continuous layers with little or no mixing from one layer to the adjacent layers.

When the faucet is nearly fully open, the water has a rather high velocity. The elements of fluid appear to be mixing chaotically within the stream. This is a general description of *turbulent flow*.

Let's go back to when you observed laminar flow and then continued to open the faucet slowly. As you increased the velocity of flow, did you notice that the stream became less smooth with ripples developing along its length? The cross section of the flow stream might have appeared to oscillate in and out, even when the flow was generally smooth. This region of flow is called the *transition zone* in which the flow is changing from laminar to turbulent. Higher velocities produced more of these oscillations until the flow eventually became fully turbulent.

The example of the flow of water from a faucet illustrates the importance of the flow velocity on the character of the flow. There is another parameter that is also important.

Review the discussion from Chapter 2 in which you studied *fluid viscosity*. Both dynamic viscosity, $\mu$, and kinematic viscosity, $v$, were defined. Recall that $v = \mu/\rho$, where $\rho$ is the density of the fluid. One general observation you made is that fluids with low viscosity flow more easily than those with higher viscosity. To aid in your review, consider the following questions.

- What are some fluids that have a relatively low viscosity?
- What are some fluids that have a rather high viscosity?
- What happens with regard to the ease with which a high viscosity fluid flows when the temperature is increased?
- What happens when the temperature of a high viscosity fluid is decreased?

Heating a high viscosity fluid such as an engine lubricating oil, lowers its viscosity and allows it to flow more easily. Conversely, reducing its temperature increases the viscosity and the oil flows more slowly. This illustrates the concept that the character of the flow is also dependent on fluid viscosity. The flow of heavy viscous fluids like cold oil is more likely to be laminar. The flow of low viscosity fluids like water is more likely to be turbulent.

You will also see in this chapter that the size and shape of the flow path affects the character of the flow. Much of our work will deal with fluid flow through circular pipes and tubes as discussed in Chapter 6. The inside flow diameter of the pipe plays an important role in characterizing the flow. But, you also studied flow paths that were non-circular in Chapter 6. For these types of conduits, we use the *hydraulic radius* to indicate the size of the flow path.

Figure 8.1 shows one way of visualizing laminar flow in a circular pipe. Concentric rings of fluid are flowing in a straight, smooth path. There is little or no mixing of the fluid across the "boundaries" of each layer as the fluid flows along in the pipe. Of course, in real fluids an infinite number of layers makes up the flow.

Another way to visualize laminar flow is depicted in Figure 8.2, which shows a transparent fluid such as water flowing in a clear glass tube. When a stream of a dark fluid such as a dye is injected into the flow, the stream remains intact as long as the flow remains laminar. The dye stream will not mix with the bulk of the fluid.

In contrast to laminar flow, turbulent flow appears chaotic and rough with much intermixing of the fluid. Figure 8.3 shows that when a dye stream is introduced into turbulent flow, it immediately dissipates throughout the primary fluid.

Indeed, an important reason for creating turbulent flow is to promote mixing in such applications as:

**1.** Blending two or more fluids.
**2.** Hastening chemical reactions.
**3.** Increasing heat transfer into or out of a fluid.

Open channel flow is the type in which one surface of the fluid is exposed to the atmosphere. Figure 8.4 shows a reservoir discharging fluid into an open channel that eventually allows the stream to fall into a lower pool. Have you seen fountains that have this feature?

**FIGURE 8.1**    Illustration of laminar flow in a circular pipe.

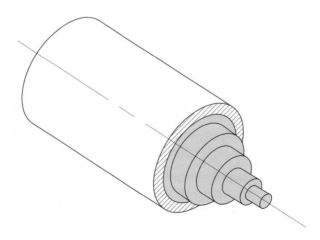

**FIGURE 8.2** Dye stream in laminar flow.

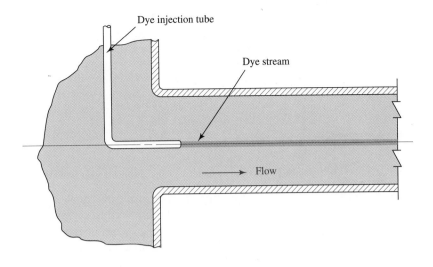

Dye injection tube

Dye stream

Flow

**FIGURE 8.3** Dye stream mixing with turbulent flow.

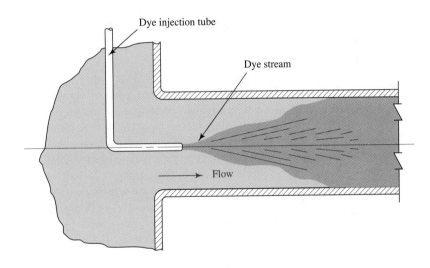

Dye injection tube

Dye stream

Flow

**FIGURE 8.4** Tranquil (laminar) flow over a wall.

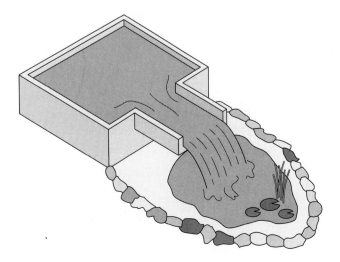

Here, as with the flow in a circular pipe, laminar flow would appear to be smooth and layered. The discharge from the channel into the pool would be like a smooth sheet. Turbulent flow would appear to be chaotic. Have you seen Niagara Falls or some other fast falling water?

You will learn in this chapter that we can characterize the flow by computing a dimensionless number, the *Reynolds number*, relating the important variables for the flow: velocity, size and shape of the flow path, fluid density, and viscosity. These concepts will be used extensively in Chapters 9 to 13.

**8.2**
**OBJECTIVES**

After completing this chapter, you should be able to:

1. Describe the appearance of laminar flow and turbulent flow.
2. State the relationship used to compute the Reynolds number.
3. Identify the limiting values of the Reynolds number by which you can predict whether flow is laminar or turbulent.
4. Compute the Reynolds number for the flow of fluids in round pipes and tubes.
5. Define the term *hydraulic radius* as it applies to the description of the size of noncircular flow paths.
6. Compute the Reynolds number for flow in noncircular flow paths.
7. Describe the velocity profile for laminar and turbulent flow.
8. Describe the laminar boundary layer as it occurs in turbulent flow.

**8.3**
**REYNOLDS NUMBER**

The behavior of a fluid, particularly with regard to energy losses, is quite dependent on whether the flow is laminar or turbulent, as will be demonstrated in Chapter 9. For this reason we want to have a means of predicting the type of flow without actually observing it. Indeed, direct observation is impossible for fluids in opaque pipes. It can be shown experimentally and verified analytically that the character of flow in a round pipe depends on four variables: fluid density $\rho$, fluid viscosity $\mu$, pipe diameter $D$, and average velocity of flow $v$. Osborne Reynolds was the first to demonstrate that laminar or turbulent flow can be predicted if the magnitude of a dimensionless number, now called the Reynolds number $(N_R)$, is known. Equation (8–1) shows the basic definition of the Reynolds number.

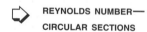 REYNOLDS NUMBER—
CIRCULAR SECTIONS

$$N_R = \frac{vD\rho}{\mu} = \frac{vD}{\nu} \qquad (8\text{–}1)$$

These two forms of the equation are equivalent because $\nu = \mu/\rho$ as discussed in Chapter 2.

You must use a consistent set of units to ensure that the Reynolds number is dimensionless. Table 8.1 lists the required units in both the SI metric unit system

**TABLE 8.1** Standard Units for Quantities Used in the Calculation of Reynolds Number to Ensure That It Is Dimensionless

| Quantity | SI Units | U.S. Customary Units |
|---|---|---|
| Velocity | m/s | ft/s |
| Diameter | m | |
| Density | kg/m$^3$ or N·s$^2$/m$^4$ | slugs/ft$^3$ or lb·s$^2$/ft$^4$ |
| Dynamic Viscosity | N·s/m$^2$ or Pa·s or kg/m·s | lb·s/ft$^2$ or slugs/ft·s |
| Kinematic Viscosity | m$^2$/s | ft$^2$/s |

and the U.S. Customary unit system. Converting to these standard units prior to entering data into the calculation for $N_R$ is recommended. But, of course, you could enter the given data with units into the calculation and perform the appropriate conversions as the calculation is being finalized. Review Sections 2.2 and 2.3 in Chapter 2 for the discussion about viscosity. Consult Appendix K for conversion factors.

We can demonstrate that the Reynolds number is dimensionless by substituting standard SI units into Eq. (8–1):

$$N_R = \frac{vD\rho}{\mu} = v \times D \times \rho \times \frac{1}{\mu}$$

$$N_R = \frac{m}{s} \times m \times \frac{kg}{m^3} \times \frac{m \cdot s}{kg}$$

Because all units can be cancelled, $N_R$ is dimensionless.

Reynolds number is one of several dimensionless numbers useful in the study of fluid mechanics and heat transfer. The process called *dimensional analysis* can be used to determine dimensionless numbers (see Reference 2).

Reynolds number is the ratio of the inertia force on an element of fluid to the viscous force. The inertia force is developed from Newton's second law of motion, $F = ma$. As discussed in Chapter 2, the viscous force is related to the product of the shear stress times area.

Flows having large Reynolds numbers, typically because of high velocity and/or low viscosity, tend to be turbulent. Those fluids having high viscosity and/or moving at low velocities will have low Reynolds numbers and will tend to be laminar. The following section gives some quantitative data with which to predict whether a given flow system will be laminar or turbulent.

The formula for Reynolds number takes a different form for noncircular cross sections, open channels, and for the flow of fluid around immersed bodies. These situations are discussed elsewhere in this book.

**8.4**
**CRITICAL REYNOLDS**
**NUMBERS**

For practical applications in pipe flow we find that if the Reynolds number for the flow is less than 2000, the flow will be laminar. Also, if the Reynolds number is greater than 4000, the flow can be assumed to be turbulent. In the range of Reynolds numbers between 2000 and 4000, it is impossible to predict which type of flow exists; therefore this range is called the *critical region*. Typical applications involve flows that are well within the laminar flow range or well within the turbulent flow range, so the existence of this region of uncertainty does not cause great difficulty. If the flow in a system is found to be in the critical region, the usual practice is to change the flow rate or pipe diameter to cause the flow to be definitely laminar or turbulent. More precise analysis is then possible.

By carefully minimizing external disturbances, it is possible to maintain laminar flow for Reynolds numbers as high as 50 000. However, when $N_R$ is greater than about 4000, a minor disturbance of the flow stream will cause the flow to suddenly change from laminar to turbulent. For this reason, and because we are dealing with practical applications in this book, we will assume the following:

If $N_R < 2000$, the flow is laminar.

If $N_R > 4000$, the flow is turbulent.

☐ **EXAMPLE PROBLEM 8.1**    Determine whether the flow is laminar or turbulent if glycerine at 25°C flows in a pipe with a 150-mm inside diameter. The average velocity of flow is 3.6 m/s.

**Solution**    We must first evaluate the Reynolds number using Eq. (8–1):

$$N_R = vD\rho/\mu$$

$$v = 3.6 \text{ m/s}$$

$$D = 0.15 \text{ m}$$

$$\rho = 1258 \text{ kg/m}^3 \quad \text{(from Appendix B)}$$

$$\mu = 9.60 \times 10^{-1} \text{ Pa·s} \quad \text{(from Appendix B)}$$

Then we have

$$N_R = \frac{(3.6)(0.15)(1258)}{9.60 \times 10^{-1}} = 708$$

Because $N_R = 708$, which is less than 2000, the flow is laminar. Notice that each term was expressed in consistent SI units before $N_R$ was evaluated.

---

☐ **EXAMPLE PROBLEM 8.2**    Determine if the flow is laminar or turbulent if water at 70°C flows in a 1-in Type K copper tube with a flow rate of 285 L/min.

**Solution**    Evaluate the Reynolds number, using Eq. (8–1):

$$N_R = \frac{vD\rho}{\mu} = \frac{vD}{\nu}$$

For a 1-in Type K copper tube, $D = 0.02527$ m and $A = 5.017 \times 10^{-4}$ m² (from Appendix H). Then we have

$$v = \frac{Q}{A} = \frac{285 \text{ L/min}}{5.017 \times 10^{-4} \text{ m}^2} \times \frac{1 \text{ m}^3/\text{s}}{60\,000 \text{ L/min}} = 9.47 \text{ m/s}$$

$$\nu = 4.11 \times 10^{-7} \text{ m}^2/\text{s} \quad \text{(from Appendix A)}$$

$$N_R = \frac{(9.47)(0.02527)}{4.11 \times 10^{-7}} = 5.82 \times 10^5$$

Because the Reynolds number is greater than 4000, the flow is turbulent.

---

☐ **EXAMPLE PROBLEM 8.3**    Determine the range of average velocity of flow for which the flow would be in the critical region if SAE 10 oil at 60°F is flowing in a 2-in Schedule 40 steel pipe. The oil has a specific gravity of 0.89.

**Solution**    The flow would be in the critical region if $2000 < N_R < 4000$. First, we use the Reynolds number and solve for velocity:

$$N_R = \frac{vD\rho}{\mu}$$

$$v = \frac{N_R\mu}{D\rho} \tag{8–2}$$

Then we find the values for $\mu$, $D$, and $\rho$:

$$D = 0.1723 \text{ ft} \qquad \text{(from Appendix F)}$$

$$\mu = 2.10 \times 10^{-3} \text{ lb-s/ft}^2 \qquad \text{(from Appendix D)}$$

$$\rho = (\text{sg})(1.94 \text{ slugs/ft}^3) = (0.89)(1.94 \text{ slugs/ft}^3) = 1.73 \text{ slugs/ft}^3$$

Substituting these values into Eq. (8–2), we get

$$v = \frac{N_R(2.10 \times 10^{-3})}{(0.1723)(1.73)} = (7.05 \times 10^{-3})N_R$$

For $N_R = 2000$, we have

$$v = (7.05 \times 10^{-3})(2 \times 10^3) = 14.1 \text{ ft/s}$$

For $N_R = 4000$, we have

$$v = (7.05 \times 10^{-3})(4 \times 10^3) = 28.2 \text{ ft/s}$$

Therefore, if $14.1 < v < 28.2$ ft/s, the flow will be in the critical region.

∎

**8.5
VELOCITY PROFILES**

Unless otherwise stated, we assume that the term *velocity* indicates the average velocity of flow found from the continuity equation $v = Q/A$. However, in some cases we must determine the fluid velocity at a point within the flow stream. The magnitude of velocity is by no means uniform across a particular section of a pipe, and the manner in which the velocity varies with position is dependent on the type of flow that exists, as shown in Fig. 8.5. We observed in Chapter 2 (Section 2.2) that the velocity of a fluid in contact with a stationary solid boundary is zero. The maximum velocity for any type of flow occurs at the center of the pipe. The reason for the different shapes of the velocity profiles is that, because of the rather chaotic motion and violent mixing of fluid molecules in turbulent flow, there is a transfer of momentum between molecules resulting in a more uniform velocity distribution than in the case of laminar flow. Because laminar flow is essentially made up of layers of fluid, the momentum transfer between molecules is less and the velocity profile becomes parabolic.

**FIGURE 8.5**   Velocity profiles for pipe flow.

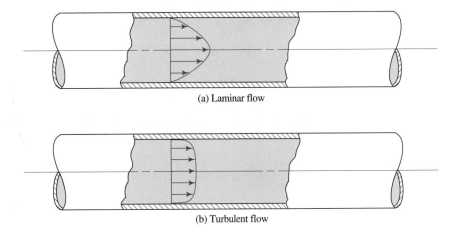

(a) Laminar flow

(b) Turbulent flow

Note in Fig. 8.5(b) that, even though the flow as a whole is turbulent, there exists a thin layer of fluid near the pipe wall where the velocity is quite small and in which the flow is actually laminar. This is known as the *boundary layer*. The actual thickness of the boundary layer and the velocity distribution within it are very important in analyzing heat transfer into the fluid or in determining the drag on bodies submerged in a fluid.

Because of the regularity of the velocity profile in laminar flow, we can define an equation for the local velocity at any point within the flow path. If we call the local velocity $U$ at a radius $r$, the maximum radius $r_o$, and the average velocity $v$,

$$U = 2v[1 - (r/r_o)^2] \qquad (8\text{--}3)$$

---

☐ **EXAMPLE PROBLEM 8.4**   Using the data from Example Problem 8.1, compute points on the velocity profile from the pipe wall to the middle of the pipe at increments of 15 mm. Plot the data for $U$ versus $r$. Also, show the average velocity on the plot.

**Solution**   From Example Problem 8.1, we find the Reynolds number to be 708, which indicates laminar flow. The average velocity of flow was given to be 3.60 m/s. Then, we compute $r_o$:

$$r_o = D/2 = 150/2 = 75 \text{ mm}$$

At $r = 75$ mm $= r_o$ at the pipe wall, $r/r_o = 1$ and $U = 0$ from Eq. (8–3). This is consistent with the observation that the velocity of a fluid at a solid boundary is equal to the velocity of that boundary.

At $r = 60$ mm,

$$U = 2(3.6 \text{ m/s})[1 - (60/75)^2] = 2.59 \text{ m/s}$$

If we use a similar technique, we can compute the following values:

| $r$ (mm) | $r/r_o$ | $U$ (m/s) | |
|---|---|---|---|
| 75 | 1.0 | 0 | (at the pipe wall) |
| 60 | 0.8 | 2.59 | |
| 45 | 0.6 | 4.61 | |
| 30 | 0.4 | 6.05 | |
| 15 | 0.2 | 6.91 | |
| 0 | 0.0 | 7.20 | (middle of the pipe) |

Notice that the local velocity at the middle of the pipe is 2.0 times the average velocity. Figure 8.6 shows the plot of $U$ versus $r$.

---

☐ **EXAMPLE PROBLEM 8.5**   Compute the radius at which the local velocity $U$ would equal the average velocity $v$ for laminar flow.

**Solution**   In Eq. (8–3), for the condition that $U = v$, we can first divide by $U$ to obtain

$$1 = 2[1 - (r/r_o)^2]$$

Now, solving for $r$ gives

$$r = \sqrt{0.5} \; r_o = 0.707 r_o \qquad (8\text{--}4)$$

**FIGURE 8.6**   Results of Example Problems 8.4 and 8.5. Velocity profile for laminar flow.

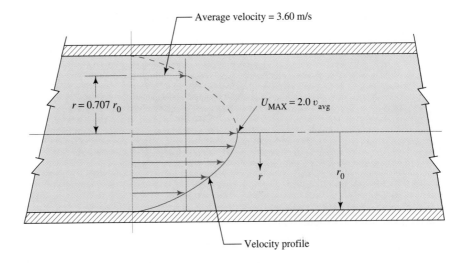

For the data from Example Problem 8.1, the local velocity is equal to the average velocity 3.6 m/s at

$$r = 0.707(75 \text{ mm}) = 53.0 \text{ mm}$$

Computing the local velocity for points within turbulent flow requires knowledge of the *friction factor, f,* which is developed in Chapter 9. Further discussion of the velocity profile for turbulent flow will be presented there.

**8.6**
**HYDRAULIC RADIUS FOR NONCIRCULAR CROSS SECTIONS**

All equations and examples up to this point for computing the Reynolds number have been for fluid flowing in a full circular pipe. In such cases, the characteristic dimension of the flow path is the inside diameter. However, many practical fluid mechanics problems involve flow in noncircular cross sections.

Noncircular cross sections can be either closed conduits running full or open channels, one surface of which is exposed to the local atmosphere. One type is quite different from the other. Here, we will work with only the sections running full, because the analysis of energy losses presented in the next chapter is quite similar for either circular pipes or closed, noncircular cross sections running full. Open channel flow is discussed in Chapter 14, and a revised form of the Reynolds number will be presented there.

Examples of typical closed, noncircular cross sections are shown in Fig. 8.7. The sections shown could represent (a) a shell-and-tube heat exchanger, (b) and (c) air distribution ducts, and (d) a flow path inside a machine. See also Section 6.6 for another example.

The characteristic dimension of noncircular cross sections is called the *hydraulic radius, R,* defined as the ratio of the net cross-sectional area of a flow stream to the wetted perimeter of the section. That is

⇨ **HYDRAULIC RADIUS**

$$R = \frac{A}{WP} = \frac{\text{area}}{\text{wetted perimeter}} \qquad (8–5)$$

**FIGURE 8.7**   Examples of closed noncircular cross sections.

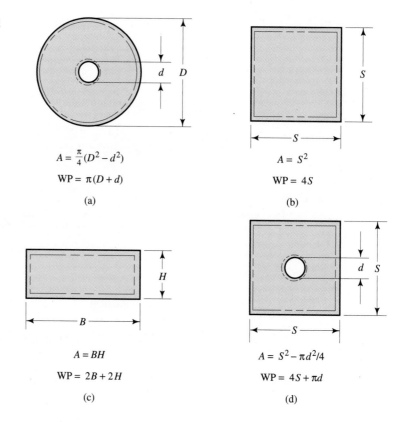

$$A = \frac{\pi}{4}(D^2 - d^2)$$

$$WP = \pi(D + d)$$

(a)

$$A = S^2$$

$$WP = 4S$$

(b)

$$A = BH$$

$$WP = 2B + 2H$$

(c)

$$A = S^2 - \pi d^2/4$$

$$WP = 4S + \pi d$$

(d)

The unit for $R$ is the meter in the SI unit system. In the U.S. Customary System, $R$ is expressed in feet.

In the calculation of the hydraulic radius, the net cross-sectional area should be evident from the geometry of the section.

> *The wetted perimeter is defined as the sum of the length of the boundaries of the section actually in contact with (that is, wetted by) the fluid.*

Expressions for the area $A$ and the wetted perimeter $WP$ are given in Fig. 8.7 for those sections illustrated. In each case, the fluid flows in the shaded portion of the section. A dashed line is shown adjacent to the boundaries that make up the wetted perimeter.

☐ **EXAMPLE PROBLEM 8.6**       Determine the hydraulic radius of the section shown in Fig. 8.7(d) if the inside dimension of each side of the square is 250 mm and the outside diameter of the tube is 150 mm.

*Solution*       The net flow area is the difference between the area of the square and the area of the circle:

$$A = S^2 - \pi d^2/4 = (250)^2 - \pi(150)^2/4 = 44\ 829 \text{ mm}^2$$

The wetted perimeter is the sum of the four sides of the square and the circumference of the circle:

$$WP = 4S + \pi d = 4(250) + \pi(150) = 1471 \text{ mm}$$

Then, the hydraulic radius $R$ is

$$R = \frac{A}{WP} = \frac{44\,829 \text{ mm}^2}{1471 \text{ mm}} = 30.5 \text{ mm} = 0.0305 \text{ m}$$

∎

**8.7**
**REYNOLDS NUMBER**
**FOR CLOSED**
**NONCIRCULAR CROSS**
**SECTIONS**

When the fluid completely fills the available cross-sectional area and is under pressure, the average velocity of flow is determined by using the volume flow rate and the net flow area in the familiar continuity equation. That is,

$$v = Q/A$$

Note that the area is the same as that used to compute the hydraulic radius.

The Reynolds number for flow in noncircular sections is computed in a very similar manner to that used for circular pipes and tubes. The only alteration to Eq. (8–1) is the replacement of the diameter $D$ with $4R$, four times the hydraulic radius. The result is

⇨ REYNOLDS NUMBER—
NONCIRCULAR SECTIONS

$$N_R = \frac{v(4R)\rho}{\mu} = \frac{v(4R)}{v} \tag{8–6}$$

The validity of this substitution can be demonstrated by calculating the hydraulic radius for a circular pipe:

$$R = \frac{A}{WP} = \frac{\pi D^2/4}{\pi D} = \frac{D}{4}$$

Then,

$$D = 4R$$

Therefore, $4R$ is equivalent to $D$ for the circular pipe. Thus, by analogy, the use of $4R$ as the characteristic dimension for noncircular cross sections is appropriate. This approach will give reasonable results as long as the cross section has an *aspect ratio* not much different from that of the circular cross section. In this context, aspect ratio is the ratio of the width of the section to its height. So, for a circular section, the aspect ratio is 1.0. In Fig. 8.7, all the examples shown have reasonable aspect ratios.

An example of a shape that has an unacceptable aspect ratio is a rectangle for which the width is more than four times the height. (See Reference 2.) For such shapes, the hydraulic radius is approximately one-half the height. (See Reference 1.) Some annular shapes, similar to that shown in Fig. 8.7(a), would have high aspect ratios if the space between the two pipes was small. However, general data are not readily available for what constitutes a "small" space or for how to determine the hydraulic radius. Performance testing of such sections is recommended.

☐ **EXAMPLE PROBLEM 8.7**

Compute the Reynolds number for the flow of ethylene glycol at 25°C through the section shown in Fig. 8.7(d). The volume flow rate is 0.16 m³/s. Use the dimensions given in Example Problem 8.6.

*Solution*

The result for the hydraulic radius for the section from Example Problem 8.6 can be used: $R = 0.0305$ m. Now the Reynolds number can be computed from Eq. (8–6). We can use

$\mu = 1.62 \times 10^{-2}$ Pa·s and $\rho = 1100$ kg/m³ (from Appendix B). The area must be converted to m².

$$A = (44\ 829\ \text{mm}^2)(1\ \text{m}^2/10^6\ \text{mm}^2) = 0.0448\ \text{m}^2$$

The average velocity of flow is

$$v = \frac{Q}{A} = \frac{0.16\ \text{m}^3/\text{s}}{0.0448\ \text{m}^2} = 3.57\ \text{m/s}$$

The Reynolds number can now be calculated:

$$N_R = \frac{v(4R)\rho}{\mu} = \frac{(3.57)(4)(0.0305)(1100)}{1.62 \times 10^{-2}}$$
$$N_R = 2.96 \times 10^4$$

■

## REFERENCES

1. Crane Co. 1988. *Flow of Fluids Through Valves, Fittings, and Pipe*. Technical Paper No. 410. Joliet, IL: Author.

2. Fox, Robert W., and Alan T. McDonald, 1992. *Introduction to Fluid Mechanics*. 4th ed. New York: John Wiley & Sons, Inc.

## PRACTICE PROBLEMS

The following problems require the use of the reference data listed below:

- Appendices A–C: Properties of liquids
- Appendix D: Dynamic viscosity of fluids
- Appendices F–J: Dimensions of pipe and tubing
- Appendix K: Conversion factors
- Appendix L: Properties of areas

### Reynolds Numbers

**8.1E** A 4-in diameter pipe carries 0.20 ft³/s of glycerine (sg = 1.26) at 100°F. Is the flow laminar or turbulent?

**8.2C** Calculate the minimum velocity of flow in ft/s and m/s of water at 160°F in a 2-in diameter pipe for which the flow is turbulent.

**8.3M** *Calculate the maximum volume flow rate of fuel oil at 45°C at which the flow will remain laminar in a 100-mm-diameter pipe. For the fuel oil, use sg = 0.895 and dynamic viscosity = $4.0 \times 10^{-2}$ Pa·s.*

**8.4E** Calculate the Reynolds number for the flow of each of the following fluids in a 2-in Schedule 40 steel pipe if the volume flow rate is 0.25 ft³/s: (a) water at 60°F, (b) acetone at 77°F, (c) castor oil at 77°F, and (d) SAE 10 oil at 210°F (sg = 0.87).

**8.5M** *Determine the smallest Type K copper tube size that will carry 4 L/min of the following fluids while maintaining laminar flow: (a) water at 40°C, (b) gasoline (sg = 0.68) at 25°C, (c) ethyl alcohol (sg = 0.79) at 0°C, and (d) heavy fuel oil at 25°C.*

**8.6M** *In an existing installation, SAE 10 oil (sg = 0.89) must be carried in a 3-in Schedule 40 steel pipe at the rate of 850 L/min. Efficient operation of a certain process requires that the Reynolds number of the flow be approximately $5 \times 10^4$. To what temperature must the oil be heated to accomplish this?*

**8.7E** From the data in Appendix C, we can see that automotive hydraulic oil and the medium machine tool hydraulic oil have nearly the same kinematic viscosity at 212°F. However, because of their different viscosity index, their viscosities at 104°F are quite different. Calculate the Reynolds number for the flow of each oil at each temperature in a 5-in Schedule 80 steel pipe at 10 ft/s velocity. Are the flows laminar or turbulent?

**8.8M** *Compute the Reynolds number for the flow of 325 L/min of water at 10°C in a standard 2-in steel tube with a wall thickness of 0.065 in. Is the flow laminar or turbulent?*

**8.9M** *Benzene (sg = 0.86) at 60°C is flowing at 25 L/min in a 1-in Schedule 80 steel pipe. Is the flow laminar or turbulent?*

**8.10M** *Hot water at 80°C is flowing to a dishwasher at a rate of 15.0 L/min through a ¹/₂-in Type K copper tube. Is the flow laminar or turbulent?*

**8.11E** A major water main is an 18-in ductile iron pipe. Compute the Reynolds number if the pipe caries 16.5 ft³/s of water at 50°F.

**8.12E** An engine crankcase contains SAE 10 motor oil (sg = 0.88). The oil is distributed to other parts of the engine by an oil pump through an $\frac{1}{8}$-in steel tube with a wall thickness of 0.032 in. The ease with which the oil is pumped is obviously affected by its viscosity. Compute the Reynolds number for the flow of 0.40 gal/h of the oil at 40°F.

**8.13E**  Repeat Problem 8.12 for an oil temperature of 160°F.

**8.14E**  At approximately what volume flow rate will propyl alcohol at 77°F become turbulent when flowing in a 3-in Type K copper tube?

**8.15M**  *SAE 30 oil (sg = 0.89) is flowing at 45 L/min through a $^7/8$-in steel tube with a wall thickness of 0.065 in. If the oil is at 110°C, is the flow laminar or turbulent?*

**8.16M**  *Repeat Problem 8.15 for an oil temperature of 0°C.*

**8.17M**  *Repeat Problem 8.15, except the tube is a 2-in size with a wall thickness of 0.065.*

**8.18M**  *Repeat Problem 8.17 for an oil temperature of 0°C.*

**8.19C**  The lubrication system for a punch press delivers 1.65 gal/min of a light lubricating oil (see Appendix C) through $^5/16$-in steel tubes having a wall thickness of 0.049 in. Shortly after the press is started, the oil temperature is 104°F. Compute the Reynolds number for the oil flow.

**8.20C**  After the press has run for some time, the lubricating oil described in Problem 8.19 heats to 212°F. Compute the Reynolds number for the oil flow at this temperature Discuss the possible operating difficulty as the oil heats up.

**8.21E**  A system is being designed to carry 500 gal/min of ethylene glycol at 77°F. Chapter 6 recommended that the maximum velocity should be approximately 10.0 ft/s. Specify the smallest standard Schedule 40 steel pipe to meet this condition. Then, for the selected pipe, compute the Reynolds number for the flow.

**8.22E**  The range of Reynolds numbers between 2000 and 4000 is described as the *critical region*, because it is not possible to predict whether the flow is laminar or turbulent. One should avoid operation of fluid flow systems in this range. Compute the range of volume flow rates in gal/min of water at 60°F for which the flow would be in the critical region in a $^3/4$-in Type K copper tube.

**8.23E**  The water line described in Problem 8.2 was a cold water distribution line. At another point in the system, the same size tube delivers water at 180°F. Compute the range of volume flow rates for which the flow would be in the critical region.

**8.24C**  In a dairy, milk at 100°F is reported to have a kinematic viscosity of 1.30 centistokes. Compute the Reynolds number for the flow of the milk at 45 gal/min through a $1\frac{1}{4}$-in steel tube with a wall thickness of 0.065 in.

**8.25C**  In a soft-drink bottling plant, the concentrated syrup used to make the drink has a kinematic viscosity of 17.0 centistokes at 80°F. Compute the Reynolds number for the flow of 215 L/min of the syrup through a 1-in Type K copper tube.

**8.26C**  A certain jet fuel has a kinematic viscosity of 1.20 centistokes. If the fuel is being delivered to the engine at 200 L/min through a 1-in steel tube with a wall thickness of 0.065 in, compute the Reynolds number for the flow.

### Velocity Profile

**8.27M**  *A small velocity probe is to be inserted through a pipe wall. If we measure from the outside of the 6-in Schedule 80 pipe, how far (in mm) should the probe be inserted to sense the average velocity if the flow in the pipe is laminar?*

**8.28M**  *If the accuracy of positioning the probe described in Problem 8.27 is plus or minus 5.0 mm, compute the possible error in measuring the average velocity.*

**8.29M**  *An alternative scheme for using the velocity probe described in Problem 8.27 is to place it in the middle of the pipe, where the velocity is expected to be 2.0 times the average velocity. Compute the amount of insertion required to center the probe. Then, if the accuracy of placement is again plus or minus 5.0 mm, compute the possible error in measuring the average velocity.*

**8.30M**  *An existing fixture inserts the velocity probe described in Problem 8.27 exactly 60.0 mm from the outside surface of the pipe. If the probe reads 2.48 m/s, compute the actual average velocity of flow, assuming the flow is laminar. Then, check to see if the flow actually is laminar if the fluid is a heavy fuel oil with a kinematic viscosity of 850 centistokes.*

### Noncircular Cross Sections

**8.31M**  *Air with a specific weight of 12.5 N/m$^3$ and a dynamic viscosity of $2.0 \times 10^{-5}$ Pa·s flows through the shaded portion of the duct shown in Fig. 8.8 at the rate of 150 m$^3$/h. Calculate the Reynolds number of the flow.*

**8.32E**  Carbon dioxide with a specific weight of 0.114 lb/ft$^3$ and a dynamic viscosity of $3.34 \times 10^{-7}$ lb-s/ft$^2$ flows in the

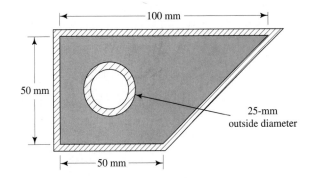

**FIGURE 8.8**  Problem 8.31.

shaded portion of the duct shown in Fig. 8.9. If the volume flow rate is 200 ft³/min, calculate the Reynolds number of the flow.

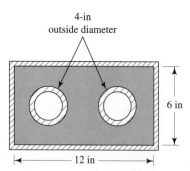

FIGURE 8.9   Problem 8.32.

**8.33E** Water at 90°F flows in the space between 6-in Schedule 40 steel pipe and a square duct with inside dimensions of 10.0 in. The duct is similar to that shown in Fig. 8.7(d). Compute the Reynolds number if the volume flow rate is 4.00 ft³/sec.

**8.34C** *Refer to the shell-and-tube heat exchanger shown in Fig. 6.16. Both tubes are standard steel tubes with 0.049-in. wall thicknesses. The inside tube carries 4.75 gal/min of water at 95°C, while the shell carries 30.0 gal/min of ethylene glycol at 25°C to carry heat away from the water. Compute the Reynolds number for the flow in both the tube and the shell.*

**8.35M** *Refer to Fig. 6.17, which shows two 6-in Schedule 40 pipes inside a rectangular duct. Each pipe carries 450 L/min of water at 20°C. Compute the Reynolds number for the flow of water. Then, for benzene (sg = 0.862) at 70°C flowing inside the duct, compute the volume flow rate required to produce the same Reynolds number.*

**8.36E** Refer to Fig. 6.18, which shows three pipes inside a larger pipe. The inside pipes carry water at 200°F, while the large pipe carries water at 60°F. The average velocity of flow is 25.0 ft/s in each pipe; compute the Reynolds number for each.

**8.37M** *Water at 10°C is flowing in the shell shown in Fig. 8.10 at the rate of 850 L/min. The shell is a 2-in Type K copper tube, while the tubes are ³/₈-in Type K copper tubes. Compute the Reynolds number for the flow.*

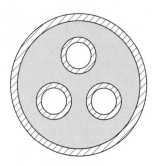

FIGURE 8.10   Problem 8.37.

**8.38E** Figure 8.11 shows the cross section of a heat exchanger used to cool a bank of electronic devices. Ethylene glycol at 77°F flows in the shaded area. Compute the volume flow rate required to produce a Reynolds number of 1500.

**8.39M** *Figure 8.12 shows a liquid-to-air heat exchanger in which air flows at 50 m³/h inside a rectangular passage and around a set of five vertical tubes. Each tube is a ½-in steel tube with a wall thickness of 0.049 in. The air has a density of 1.15 kg/m³ and a dynamic viscosity of $1.63 \times 10^{-5}$ Pa·s. Compute the Reynolds number for the air flow.*

FIGURE 8.11   Problem 8.38.

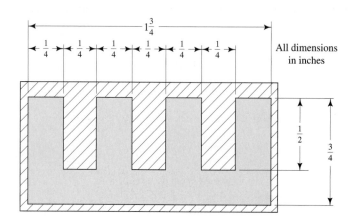

**FIGURE 8.12**  Problem 8.39.

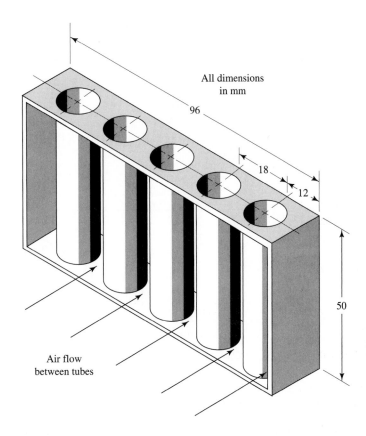

All dimensions
in mm

96

18

12

50

Air flow
between tubes

**8.40M** *Glycerine (sg = 1.26) at 40°C flows in the portion of the duct outside the square tubes shown in Fig. 8.13. Calculate the Reynolds number for a flow rate of 0.10 m³/s.*

**8.41M** *Each of the square tubes shown in Fig. 8.13 carries 0.75 m³/s of water at 90°C. The thickness of the walls of the tubes is 2.77 mm. Compute the Reynolds number of the flow of water.*

**8.42E** A heat sink for an electronic circuit is made by machining a pocket into a block of aluminum and then covering it with a flat plate to provide a passage for cooling wa-

ter as shown in Fig. 8.14. Compute the Reynolds number if the flow of water at 50°F is 78.0 gal/min.

**8.43E** Figure 8.15 shows the cross section of a cooling passage for an odd-shaped device. Compute the volume flow rate of water at 50°F that would produce a Reynolds number of $1.5 \times 10^5$.

**8.44E** Figure 8.16 shows the cross section of a flow path machined from a casting using a 3/4-in-diameter milling cutter. Considering all the fillets, compute the hydraulic radius for the passage, and then compute the volume flow rate of acetone at 77°F required to produce a Reynolds number for the flow of $2.6 \times 10^4$.

**FIGURE 8.13**  Problems 8.40 and 8.41.

Both 150 mm square outside

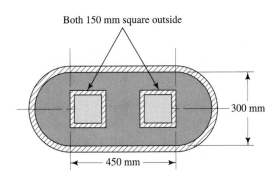

300 mm

450 mm

**FIGURE 8.14** Problem 8.42.

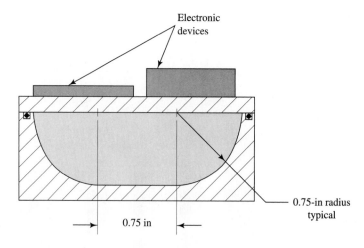

Electronic devices

0.75-in radius typical

0.75 in

**FIGURE 8.15** Problem 8.43.

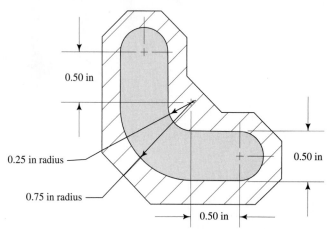

0.50 in

0.25 in radius

0.75 in radius

0.50 in

0.50 in

**FIGURE 8.16** Problem 8.44.

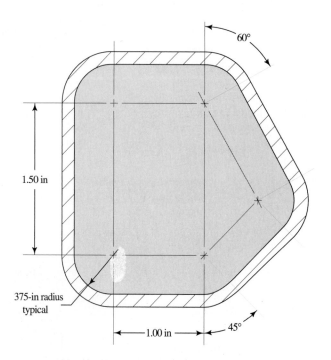

60°

1.50 in

375-in radius typical

1.00 in

45°

**FIGURE 8.17**   Problem 8.45.

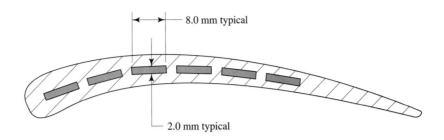

8.0 mm typical

2.0 mm typical

**8.45M** *The blade of a gas turbine engine contains internal cool-ing passages, as shown in Fig. 8.17. Compute the volume flow rate of air required to produce an average velocity of flow in each passage of 25.0 m/s. The air flow dis-tributes evenly to all six passages. Then, compute the Reynolds number if the air has a density of 1.20 kg/m³ and a dynamic viscosity of $1.50 \times 10^{-5}$ Pa·s.*

**8.46E**   A simple heat exchanger is made by welding one-half of a 1 3/4 in steel tube to a flat plate as shown in Figure 8.18. Water at 40°F flows in the open space and cools the plate. Compute the volume flow rate required so that the Reynolds number of the flow is $3.5 \times 10^4$.

**FIGURE 8.18**   Problem 8.46.

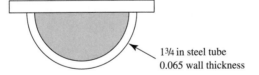

1 3/4 in steel tube
0.065 wall thickness

**8.47E**   Three surfaces of an instrument package are cooled by soldering half-sections of copper tubing to it as shown in Fig. 8.19. Compute the Reynolds number for each sec-tion if ethylene glycol at 77°F flows with an average ve-locity of 15 ft/s.

**FIGURE 8.19**   Problem 8.47.

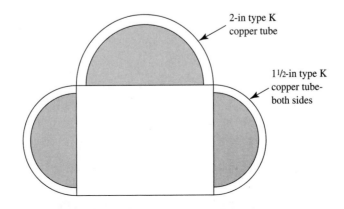

2-in type K
copper tube

1 1/2-in type K
copper tube-
both sides

**FIGURE 8.20**   Problem 8.48.

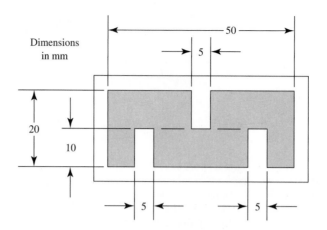

**8.48M** *Figure 8.20 shows a heat exchanger with internal fins. Compute the Reynolds number for the flow of brine (20% NaCl) at 0°C at a volume flow rate of 225 L/min inside the heat exchanger. The brine has a specific gravity of 1.10.*

# 9 Energy Losses Due to Friction

**9.1**
**The**
**Big**
**Picture**

## Discussion Map

- Your skill development continues here toward the goal of analyzing the flow of fluids in real pipe line systems.

- *Friction loss* is an important component of the total energy loss in fluid flow systems. It is the energy lost as the fluid flows along straight lengths of pipe and tubing.

- We call this *friction loss* because it results from internal friction within the fluid and interactions between the fluid and the interior walls of the pipe or tubing.

- The general effect of these energy losses is a decrease in pressure between two points of interest in the flow system.

- It also contributes to the amount of power that a pump must deliver to a fluid to cause the flow to move through the system against the frictional resistance.

### Discover

- Think about times when you may have observed pressure losses due to friction.
- Consider the case of water flowing through a long household garden hose.
- Consider the case of water flowing from the municipal water supply tower to homes and businesses miles away from the tower.
- Consider a pipeline that delivers oil from a well field in Texas to a refinery in Ohio, a distance of over 1000 miles.
- Consider the many pipeline systems that exist in a typical industrial plant delivering water, fuels, and process liquids to widely separated points within the plant.

In this chapter you will learn how to determine analytically the amount of energy loss due to friction and to predict the effect of such losses on pressure drop and on the amount of power required to pump the fluid through the system.

In any flow system, the fluid enters the system at a certain volume flow rate and velocity and with a certain pressure. At points farther down the system, the pressure will typically be noticeably lower than at the start of the system. Several mechanisms for energy loss could exist between the source and the destination point of the fluid. Here we are only concerned with the energy loss due to friction in the pipe, hose, or tubing.

For the garden hose, the water pressure at the faucet decreases to atmospheric pressure at the end of the hose where the water flows out onto the ground. This is caused by the frictional energy loss from the fluid as it flows along the hose.

In the municipal water system, the operator of the system is responsible for supplying a specified volume flow rate of water at a specified pressure at the inlet to the water distribution system. The pressure continually decreases as it flows along the network of piping. If your home is at the far end of the system, you would typically have much lower water pressure than your fellow citizens closer to the source of the water. If the piping system is very long, the municipality may also have installed booster pumps at intermediate points in the system to ensure an adequate pressure for all users.

The oil in a long pipeline experiences similar pressure drop. If it drops to such a low pressure that the flow of oil is impaired, the pipeline designers would install booster pumps at appropriate intervals to overcome the frictional energy losses.

Industrial fluid distribution systems often must be carefully designed to ensure that the pressure at a destination within the plant is adequate for the intended use. For example, if a washing fluid is to be sent through a nozzle or spray head in a washing booth, the effectiveness of the spray depends on having a certain pressure level at the inlet to the head. The same would apply to paint spray systems or to sprinkler systems used to suppress fires. It would be tragic if the fire suppression system did not function properly because of an excessive loss in pressure due to friction in the piping system.

In this chapter you will learn how to determine analytically the amount of energy loss due to friction and to predict the effect of such losses on pressure drop and on the amount of power required to pump the fluid through the system. Several approaches will be described. The one used most often in this book is called the *Darcy equation*. You will also learn about *the Hagen-Poiseuille equation*, and the *Hazen-Williams formula*.

**9.2**
**OBJECTIVES**

After completing this chapter, you should be able to:

1. State *Darcy's equation* for computing energy loss due to friction.
2. Define the *friction factor*, sometimes called the *coefficient of friction*.
3. State the *Hagen-Poiseuille equation* for computing energy loss due to friction in laminar flow.
4. Compute the equivalent friction factor for laminar flow for use in Darcy's equation.
5. Select design values for the pipe wall roughness.
6. Compute the relative roughness.
7. Determine the friction factor for turbulent flow using Moody's diagram.
8. Compute the magnitude of the energy loss for either laminar flow or turbulent flow in round pipes and tubes and use the energy loss in the general energy equation.
9. Use formulas for computing the magnitude of the friction factor.
10. Determine the friction factor and energy losses for flow in noncircular cross sections.
11. Use the Hazen-Williams formula to compute energy loss due to friction for the flow of water in pipes.

**9.3**
**DARCY'S EQUATION**

In the general energy equation

$$\frac{p_1}{\gamma} + z_1 + \frac{v_1^2}{2g} + h_A - h_R - h_L = \frac{p_2}{\gamma} + z_2 + \frac{v_2^2}{2g}$$

the term $h_L$ is defined as the energy loss from the system. One component of the energy loss is due to friction in the flowing fluid. Friction is proportional to the velocity head of the flow and to the ratio of the length to the diameter of the flow stream, for the case of flow in pipes and tubes. This is expressed mathematically as Darcy's equation:

▷ DARCY'S EQUATION FOR
ENERGY LOSS

$$h_L = f \times \frac{L}{D} \times \frac{v^2}{2g} \tag{9-1}$$

where $h_L$ = energy loss due to friction (N·m/N, m, lb-ft/lb, or ft)

$L$ = length of flow stream (m or ft)

$D$ = pipe diameter (m or ft)

$v$ = average velocity of flow (m/s or ft/s)

$f$ = friction factor (dimensionless)

Darcy's equation can be used to calculate the energy loss due to friction in long straight sections of round pipe for both laminar and turbulent flow. The difference between the two is in the evaluation of the dimensionless friction factor $f$.

## 9.4 FRICTION LOSS IN LAMINAR FLOW

When laminar flow exists, the fluid seems to flow as several layers, one on another. Because of the viscosity of the fluid, a shear stress is created between the layers of fluid. Energy is lost from the fluid by the action of overcoming the frictional forces produced by the shear stress. Since laminar flow is so regular and orderly, we can derive a relationship between the energy loss and the measurable parameters of the flow system. This relationship is known as the *Hagen-Poiseuille equation*:

 HAGEN-POISEUILLE EQUATION

$$h_L = \frac{32 \mu L v}{\gamma D^2} \qquad (9-2)$$

The parameters involved are the fluid properties of viscosity and specific weight, the geometrical features of length and pipe diameter, and the dynamics of the flow characterized by the average velocity. The Hagen-Poiseuille equation has been verified experimentally many times. You should observe from Eq. (9–2) that the energy loss in laminar flow is independent of the condition of the pipe surface. Viscous friction losses within the fluid govern the magnitude of the energy loss.

The Hagen-Poiseuille equation is valid only for laminar flow ($N_R < 2000$). However, we stated earlier that Darcy's equation, Eq. (9–1), could also be used to calculate the friction loss for laminar flow. If the two relationships for $h_L$ are set equal to each other, we can solve for the value of the friction factor:

$$f \times \frac{L}{D} \times \frac{v^2}{2g} = \frac{32 \mu L v}{\gamma D^2}$$

$$f = \frac{32 \mu L v}{\gamma D^2} \times \frac{D2g}{Lv^2} = \frac{64 \mu g}{v D \gamma}$$

Since $\rho = \gamma/g$, we get

$$f = \frac{64 \mu}{v D \rho}$$

The Reynolds number is defined as $N_R = v D \rho / \mu$. Then we have

FRICTION FACTOR FOR LAMINAR FLOW

$$f = \frac{64}{N_R} \qquad (9-3)$$

In summary, the energy loss due to friction in *laminar flow* can be calculated either from the Hagen-Poiseuille equation:

$$h_L = \frac{32 \mu L v}{\gamma D^2}$$

or from Darcy's equation:

$$h_L = f \times \frac{L}{D} \times \frac{v^2}{2g}$$

where $f = 64/N_R$.

---

☐ **EXAMPLE PROBLEM 9.1**     Determine the energy loss if glycerine at 25°C flows 30 m through a 150-mm diameter pipe with an average velocity of 4.0 m/s.

*Solution*     First, we must determine whether the flow is laminar or turbulent by evaluating the Reynolds number:

$$N_R = \frac{vD\rho}{\mu}$$

From Appendix B, we find that for glycerine at 25°C

$$\rho = 1258 \text{ kg/m}^3$$
$$\mu = 9.60 \times 10^{-1} \text{ Pa·s}$$

Then, we have

$$N_R = \frac{(4.0)(0.15)(1258)}{9.60 \times 10^{-1}} = 786$$

Because $N_R < 2000$, the flow is laminar.

Using Darcy's equation, we get

$$h_L = f \times \frac{L}{D} \times \frac{v^2}{2g}$$

$$f = \frac{64}{N_R} = \frac{64}{786} = 0.081$$

$$h_L = 0.081 \times \frac{30}{0.15} \times \frac{(4.0)^2}{2(9.81)} \text{ m} = 13.2 \text{ m}$$

Notice that each term in each equation is expressed in the units of the SI unit system. Therefore, the resulting units for $h_L$ are m or N·m/N. This means that 13.2 N·m of energy is lost by each newton of the glycerine as it flows along the 30 m of pipe.

■

---

**9.5**
**FRICTION LOSS IN**
**TURBULENT FLOW**

For turbulent flow of fluids in circular pipes it is most convenient to use Darcy's equation to calculate the energy loss due to friction. We cannot determine the friction factor $f$ by a simple calculation as we did for laminar flow because turbulent flow does not conform to regular predictable motions. It is rather chaotic and is constantly varying. For these reasons we must rely on experimental data to determine the value of $f$.

Tests have shown that the dimensionless number $f$ is dependent on two other dimensionless numbers, the Reynolds number and the relative roughness of the pipe. The relative roughness is the ratio of the pipe diameter $D$ to the average pipe wall roughness $\epsilon$ (Greek letter epsilon). Figure 9.1 illustrates pipe wall roughness (ex-

**FIGURE 9.1** Pipe wall roughness (exaggerated).

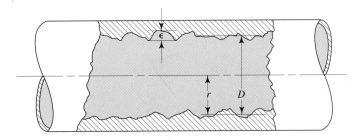

aggerated) as the height of the peaks of the surface irregularities. The condition of the pipe surface is very much dependent on the pipe material and the method of manufacture.

For commercially available pipe and tubing, the design value of the wall roughness $\epsilon$ has been determined as shown in Table 9.1. *These are only average values for new, clean pipe. Some variation should be expected. After a pipe has been in service for a time, the roughness could change due to the formation of deposits on the wall or due to corrosion.*

**TABLE 9.1** Pipe roughness—Design values

| Material | Roughness, $\epsilon$ (m) | Roughness, $\epsilon$ (ft) |
|---|---|---|
| Glass, plastic | Smooth | Smooth |
| Copper, brass, lead (tubing) | $1.5 \times 10^{-6}$ | $5 \times 10^{-6}$ |
| Commercial steel or welded steel | $4.6 \times 10^{-5}$ | $1.5 \times 10^{-4}$ |
| Wrought iron | $4.6 \times 10^{-5}$ | $1.5 \times 10^{-4}$ |
| Ductile iron—coated | $1.2 \times 10^{-4}$ | $4 \times 10^{-4}$ |
| Ductile iron—uncoated | $2.4 \times 10^{-4}$ | $8 \times 10^{-4}$ |
| Concrete | $1.2 \times 10^{-4}$ | $4 \times 10^{-4}$ |
| Riveted steel | $1.8 \times 10^{-3}$ | $6 \times 10^{-3}$ |

Refer to Chapter 6 for discussion of the sizes of standard pipe and tubing used in this book. Glass and plastic piping has an inside surface that is virtually hydraulically smooth. The value of surface roughness, $\epsilon$, is quite small and the resulting relative roughness, $D/\epsilon$, is very large. Copper and brass tubing is drawn to its final shape and size, leaving a fairly smooth surface. For standard steel pipe (such as Schedule 40 and Schedule 80) and steel tubing, we use the value for roughness listed for commercial steel or welded steel. There are grades of drawn steel tubing that may have lower roughness values. Ductile iron pipe is typically coated on the inside with a cement mortar for corrosion protection and to improve the surface roughness. In this book we use the roughness values for coated ductile iron unless stated otherwise. Some manufacturers have a smoother pipe surface, approaching that of steel. Well-made concrete pipe can have roughness values similar to the values for coated ductile iron. There are relatively fewer applications of lead tubing, wrought iron pipe, or riveted steel pipe in recent years.

**9.5.1**
**The Moody Diagram**

One of the most widely used methods for evaluating the friction factor employs the Moody diagram shown in Fig. 9.2. The diagram shows the friction factor $f$ plotted versus the Reynolds number $N_R$, with a series of parametric curves related to the

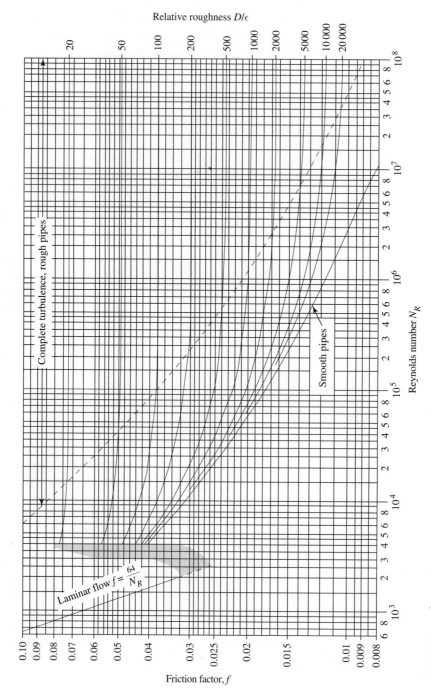

**FIGURE 9.2** Moody's diagram. (Source: Pao, R. H. F. 1961. *Fluid Mechanics*. New York: John Wiley & Sons, p. 284.)

relative roughness $D/\epsilon$. These curves were generated from experimental data by L. F. Moody (see Reference 2).

Both $f$ and $N_R$ are plotted on logarithmic scales because of the broad range of values encountered. At the left end of the chart, for Reynolds numbers less than 2000, the straight line shows the relationship $f = 64/N_R$ for laminar flow. For $2000 < N_R < 4000$, no curves are drawn since this is the critical zone between laminar and turbulent flow and it is not possible to predict the type of flow. The change from laminar to turbulent flow results in values for friction factors within the shaded band. Beyond $N_R = 4000$, the family of curves for different values of $D/\epsilon$ are plotted. Several important observations can be made from these curves.

1. For a given Reynolds number of flow, as the relative roughness $D/\epsilon$ is increased, the friction factor $f$ decreases.
2. For a given relative roughness $D/\epsilon$, the friction factor $f$ decreases with increasing Reynolds number until the zone of complete turbulence is reached.
3. Within the zone of complete turbulence, the Reynolds number has no effect on the friction factor.
4. As the relative roughness $D/\epsilon$ increases, the value of the Reynolds number at which the zone of complete turbulence begins also increases.

Figure 9.3 is a simplified sketch of Moody's diagram in which the various zones are identified. The *laminar zone* at the left has already been discussed. At the right of the dashed line downward across the diagram is the *zone of complete turbulence*. The lowest possible friction factor for a given Reynolds number in turbulent flow is indicated by the *smooth pipes* line.

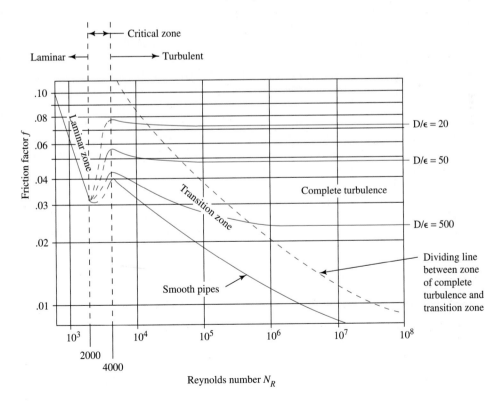

**FIGURE 9.3**   Explanation of parts of Moody's diagram.

Between the smooth pipes line and the line marking the start of the complete turbulence zone is the *transition zone*. Here, the various $D/\epsilon$ lines are curved, and care must be exercised to evaluate the friction factor properly. You can see, for example that the value of the friction factor for a relative roughness of 500 decreases from 0.0420 at $N_R = 4000$ to 0.0240 at $N_R = 6.0 \times 10^5$, where the zone of complete turbulence starts.

Check your ability to read the Moody diagram correctly now. Verify the following values for friction factors for the given values of Reynolds number and relative roughness, using Figure 9.2.

| $N_R$ | $D/\epsilon$ | $f$ |
|---|---|---|
| $6.7 \times 10^3$ | 150 | 0.0430 |
| $1.6 \times 10^4$ | 2000 | 0.0284 |
| $1.6 \times 10^6$ | 2000 | 0.0171 |
| $2.5 \times 10^5$ | 733 | 0.0223 |

As stated in Chapter 8, the critical zone between the Reynolds numbers of 2000 and 4000 is to be avoided if possible, because within this range the type of flow cannot be predicted. The dashed curves show how the friction factor could change according to the value of the relative roughness. For low values of $D/\epsilon$ (indicating large pipe wall roughness), the increase in friction factor is great as the flow changes from laminar to turbulent. For example, for flow in a pipe with $D/\epsilon = 20$, the friction factor would increase from 0.032 for $N_R = 2000$ at the end of the laminar range to approximately 0.077 at $N_R = 4000$ at the beginning of the turbulent range, an increase of 240 percent. Moreover, the value of the Reynolds number where this would occur cannot be predicted. Because the energy loss is directly proportional to the friction factor, changes of such magnitude are significant.

It should be noted that because relative roughness is defined as $D/\epsilon$, a high relative roughness indicates a low value of $\epsilon$, that is, a smooth pipe. In fact, the curve labeled *smooth pipes* is used for materials such as glass which have such a low roughness that $D/\epsilon$ would be an extremely large number.

Some texts and references use other conventions for reporting relative roughness, such as $\epsilon/D$, $\epsilon/r$, or $r/\epsilon$, where $r$ is the pipe radius. We feel that the convention used in this book makes calculations and interpolations easier.

**9.5.2**
**Use of the Moody Diagram**

The Moody diagram is used to help determine the value of the friction factor $f$ for turbulent flow. The value of the Reynolds number and the relative roughness must be known. Therefore, the basic data required are the pipe inside diameter, the pipe material, the flow velocity, and the kind of fluid and its temperature, from which the viscosity can be found. The following example problems illustrate the procedure for finding $f$.

☐ **EXAMPLE PROBLEM 9.2**

Determine the friction factor $f$ if water at 160°F is flowing at 30.0 ft/s in an uncoated ductile iron pipe having an inside diameter of 1 in.

*Solution*

The Reynolds number must first be evaluated to determine whether the flow is laminar or turbulent:

$$N_R = \frac{vD}{\nu}$$

But $D = 1$ in $= 0.0833$ ft, and $\nu = 4.38 \times 10^{-6}$ ft$^2$/s. We now have

$$N_R = \frac{(30.0)(0.0833)}{4.38 \times 10^{-6}} = 5.70 \times 10^5$$

Thus, the flow is turbulent. Now the relative roughness must be evaluated. From Table 9.1 we find $\epsilon = 8 \times 10^{-4}$ ft. Then, the relative roughness is

$$\frac{D}{\epsilon} = \frac{0.0833 \text{ ft}}{8 \times 10^{-4} \text{ ft}} = 1.04 \times 10^2 = 104$$

Notice that in order for $D/\epsilon$ to be a dimensionless ratio, both $D$ and $\epsilon$ must be in the same units.

The final steps in the procedure are

1. Locate the Reynolds number on the abscissa of the Moody diagram:

$$N_R = 5.70 \times 10^5$$

2. Project vertically until the curve for $D/\epsilon = 104$ is reached. Since 104 is so close to 100, that curve can be used.
3. Project horizontally to the left, and read $f = 0.038$.

---

☐ **EXAMPLE PROBLEM 9.3**  If the flow rate of water in Problem 9.2 was 0.45 ft/s with all other conditions being the same, determine the friction factor $f$.

*Solution*

$$N_R = \frac{\nu D}{\nu} = \frac{(0.45)(0.0833)}{4.38 \times 10^{-6}} = 8.55 \times 10^3$$

$$\frac{D}{\epsilon} = \frac{0.0833}{8 \times 10^{-4}} = 104$$

Then, from Fig. 9.2, $f = 0.044$. Notice that this is on the curved portion of the $D/\epsilon$ curve and that there is a significant increase in the friction factor over that in Example Problem 9.2.

---

☐ **EXAMPLE PROBLEM 9.4**  Determine the friction factor $f$ if ethyl alcohol at 25°C is flowing at 5.3 m/s in a standard 1¹/₂-in Schedule 80 steel pipe.

*Solution*  Evaluating the Reynolds number, we use the equation

$$N_R = \frac{\nu D \rho}{\mu}$$

From Appendix B, $\rho = 787$ kg/m$^3$ and $\mu = 1.00 \times 10^{-3}$ Pa·s. Also, for a 1¹/₂-in Schedule 80 pipe, $D = 0.0381$ m. Then we have

$$N_R = \frac{(5.3)(0.0381)(787)}{1.00 \times 10^{-3}} = 1.59 \times 10^5$$

Thus, the flow is turbulent. For a steel pipe, $\epsilon = 4.6 \times 10^{-5}$ m, so the relative roughness is

$$\frac{D}{\epsilon} = \frac{0.0381 \text{ m}}{4.6 \times 10^{-5} \text{ m}} = 828$$

From Fig. 9.2, $f = 0.0225$. You must interpolate on both $N_R$ and $D/\epsilon$ to determine this value, and you should expect some variation. However, you should be able to read the value of the friction factor $f$ within ±0.0005 in this portion of the graph.

■

The following is a programmed example problem illustrating a typical fluid piping situation. The energy loss due to friction must be calculated as a part of the solution.

## PROGRAMMED EXAMPLE PROBLEM

☐ **EXAMPLE PROBLEM 9.5**      See Figure 9.4. In a chemical processing plant, benzene at 50°C (sg = 0.86) must be delivered to point B with a pressure of 550 kPa. A pump is located at point A 21 m below point B, and the two points are connected by 240 m of plastic pipe having an inside diameter of 50 mm. If the volume flow rate is 110 L/min, calculate the required pressure at the outlet of the pump.

**FIGURE 9.4**   Example Problem 9.5.

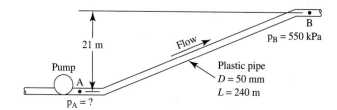

Write the energy equation between points A and B.

---

The relation is

$$\frac{p_A}{\gamma} + z_A + \frac{v_A^2}{2g} - h_L = \frac{p_B}{\gamma} + z_B + \frac{v_B^2}{2g}$$

The term $h_L$ is required because there is an energy loss due to friction between points A and B. Point A is at the outlet of the pump, and the objective of the problem is to calculate $p_A$.
    Can any terms be cancelled from this energy equation?

---

Yes, the velocity of flow is the same at points A and B. Therefore, the two velocity head terms may be canceled as shown below.

$$\frac{p_A}{\gamma} + z_A + \cancelto{0}{\frac{v_A^2}{2g}} - h_L = \frac{p_B}{\gamma} + z_B + \cancelto{0}{\frac{v_B^2}{2g}}$$

Solve algebraically for $p_A$ now.

---

The equation is

$$p_A = p_B + \gamma[(z_B - z_A) + h_L] \qquad (9\text{--}4)$$

What is the value of $z_B - z_A$?

---

We find that $z_B - z_A = +21$ m because point B is higher than point A.

This brings us to $h_L$, the energy loss due to friction between A and B. What is the first step?

---

The evaluation of the Reynolds number is the first step. The type of flow, laminar or turbulent, must be determined. Complete the calculation of the Reynolds number before looking at the next panel.

---

The correct value is $N_R = 9.54 \times 10^4$. Here is how it is found:

$$N_R = vD\rho/\mu$$

For a 50-mm pipe, $D = 0.050$ m and $A = 1.963 \times 10^{-3}$ m$^2$. Then, we have

$$Q = (110 \text{ L/min})\left(\frac{1 \text{ m}^3/\text{s}}{60\,000 \text{ L/min}}\right) = 1.83 \times 10^{-3} \text{ m}^3/\text{s}$$

$$v = \frac{Q}{A} = \frac{1.83 \times 10^{-3} \text{ m}^3/\text{s}}{1.963 \times 10^{-3} \text{ m}^2} = 0.932 \text{ m/s}$$

For benzene at 50°C with a specific gravity of 0.86, we find that

$$\rho = (0.86)(1000 \text{ kg/m}^3) = 860 \text{ kg/m}^3$$
$$\mu = 4.2 \times 10^{-4} \text{ Pa·s} \quad \text{(from Appendix D)}$$

Then we have

$$N_R = \frac{(0.932)(0.050)(860)}{4.2 \times 10^{-4}} = 9.54 \times 10^4$$

Therefore, the flow is turbulent. What relationship should be used to calculate $h_L$?

---

For turbulent flow, Darcy's equation should be used:

$$h_L = f \times \frac{L}{D} \times \frac{v^2}{2g}$$

In order to use the Moody diagram to find the value of $f$, is the value of the relative roughness $D/\epsilon$ needed?

---

Not in this case. Because the pipe is plastic, the inner surface is smooth and we can use the curve in the Moody diagram labeled *smooth pipes*. Evaluate $f$ now.

---

The result is $f = 0.018$. Now the calculation of $h_L$ can be completed.

---

The correct value is $h_L = 3.83$ m.

$$h_L = f \times \frac{L}{D} \times \frac{v^2}{2g} = 0.018 \times \frac{240}{0.050} \times \frac{(0.932)^2}{2(9.81)} \text{ m}$$

$$h_L = 3.83 \text{ m}$$

Going back to Eq. (9–4), you can now calculate $p_A$.

---

You should have $p_A = 759$ kPa.

$$p_A = p_B + \gamma[(z_B - z_A) + h_L]$$

$$p_A = 550 \text{ kPa} + \frac{(0.86)(9.81 \text{ kN})}{\text{m}^3}(21 \text{ m} + 3.83 \text{ m})$$

$$p_A = 550 \text{ kPa} + 209 \text{ kN/m}^2 = 550 \text{ kPa} + 209 \text{ kPa}$$

$$p_A = 759 \text{ kPa}$$

■

## 9.6
## EQUATIONS FOR
## FRICTION FACTOR

The Moody diagram in Fig. 9.2 is a convenient and sufficiently accurate means of determining the value of the friction factor when solving problems by manual calculations. However, if the calculations are to be automated for solution on a computer or a programmable calculator, we need equations for the friction factor.

The equations used in the work by Moody (1944) form the basis of the computational approach.* But those equations were cumbersome, requiring an iterative approach. We show here two equations that allow the direct solution for the friction factor. One covers laminar flow and the other is used for turbulent flow.

In the *laminar flow zone*, for values below 2000, $f$ can be found from Eq. (9–3):

⇨ FRICTION FACTOR FOR
LAMINAR FLOW

$$f = 64/N_R \qquad (9\text{–}3)$$

This relationship, developed in Section 9.2, plots in the Moody diagram as a straight line on the left side of the chart.

Of course, for Reynolds numbers from 2000 to 4000, the flow is in the critical range and it is impossible to predict the value of $f$.

The following equation, which allows the direct calculation of the value of the friction factor for turbulent flow, was developed by P. K. Swamee and A. K. Jain and is reported in Reference 4 (1976).

⇨ FRICTION FACTOR FOR
TURBULENT FLOW

$$f = \frac{0.25}{\left[\log\left(\dfrac{1}{3.7\,(D/\epsilon)} + \dfrac{5.74}{N_R^{0.9}}\right)\right]^2} \qquad (9\text{–}5)$$

Equation (9–5) produces values for $f$ that are within $\pm 1.0$ percent within the range of relative roughness, $D/\epsilon$, from 100 to $1 \times 10^6$ and for Reynolds numbers from $5 \times 10^3$ to $1 \times 10^8$. This is virtually the entire turbulent zone of the Moody diagram.

### Summary

To calculate the value of the friction factor, $f$, when the Reynolds number and relative roughness are known, use Eq. (9–3) for laminar flow and Eq. (9–5) for turbulent flow.

---

* Earlier work in developing the equations was done by several researchers, most notably C. F. Colebrook, L. Prandtl, H. Rouse, T. van Karman, and J. Nikuradse, whose papers are listed in the bibliography of Moody's ASME paper (see Reference 2).

☐ **EXAMPLE PROBLEM 9.6**   Compute the value for the friction factor if the Reynolds number for the flow is $1 \times 10^5$ and the relative roughness is 2000.

**Solution**   Because this is in the turbulent zone, Eq. (9–5) is used.

$$f = \frac{0.25}{\left[\log\left(\dfrac{1}{3.7(2000)} + \dfrac{5.74}{(1 \times 10^5)^{0.9}}\right)\right]^2}$$

$$f = 0.0204$$

This value compares closely with the value read from Fig. 9.2.

■

## 9.7 FRICTION LOSS IN NONCIRCULAR CROSS SECTIONS

Darcy's equation for friction loss can be used for noncircular cross sections if the geometry is represented by the hydraulic radius instead of the pipe diameter, as is used for circular sections. The definition of hydraulic radius and the method of computing its value were presented in Chapter 8.

After computing the hydraulic radius, we can compute the Reynolds number by substituting $4R$ for the diameter $D$. The equations then become

**REYNOLDS NUMBER FOR NONCIRCULAR SECTIONS**

$$N_R = \frac{v(4R)\rho}{\mu} = \frac{v(4R)}{\nu} \tag{9–6}$$

A similar substitution can be made for computing friction loss. In Darcy's equation, replacing $D$ with $4R$ gives

**DARCY'S EQUATION FOR NONCIRCULAR SECTIONS**

$$h_L = f\frac{L}{4R}\frac{v^2}{2g} \tag{9–7}$$

The relative roughness $D/\epsilon$ becomes $4R/\epsilon$. The friction factor can be found from the Moody diagram.

☐ **EXAMPLE PROBLEM 9.7**   Determine the pressure drop for a 50-m length of a duct with the cross section shown in Fig. 9.5. Ethylene glycol at 25°C is flowing at the rate of 0.16 m³/s. The inside dimension of the square is 250 mm, and the outside diameter of the tube is 150 mm. Use $\epsilon = 3 \times 10^{-5}$ m, somewhat smoother than commercial steel pipe.

**Solution**   The area, velocity, hydraulic radius, and Reynolds number were computed in Example Problems 8.6 and 8.7. The results are

$$A = 0.0448 \text{ m}^2$$
$$v = 3.57 \text{ m/s}$$
$$R = 0.0305 \text{ m}$$
$$N_R = 2.96 \times 10^4$$

The flow is turbulent, and Darcy's equation can be used to calculate the energy loss between two points 50 m apart. In order to determine the friction factor, we must first find the relative roughness:

$$4R/\epsilon = (4)(0.0305)/(3 \times 10^{-5}) = 4067$$

**FIGURE 9.5** Cross section for duct for Example Problem 9.7.

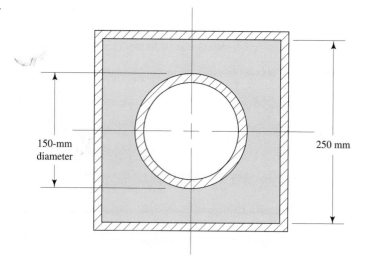

150-mm diameter

250 mm

From the Moody diagram, $f = 0.0245$, Then, we have

$$h_L = f \times \frac{L}{4R} \times \frac{v^2}{2g} = 0.0245 \times \frac{50}{(4)(0.0305)} \times \frac{(3.57)^2}{(2)(9.81)} \text{ m}$$

$$h_L = 6.52 \text{ m}$$

If the duct is horizontal,

$$h_L = \Delta p / \gamma$$

$$\Delta p = \gamma h_L$$

where $\Delta p$ is the pressure drop caused by the energy loss. Then, we have

$$\Delta p = \frac{10.79 \text{ kN}}{\text{m}^3} \times 6.52 \text{ m} = 70.4 \text{ kPa}$$

■

**9.8
VELOCITY PROFILE FOR
TURBULENT FLOW**

In Chapter 8 we showed that the velocity profile for laminar flow was a parabolic shape for which the local velocity at any point could be computed simply from Eq. (8–4). The only variables were the average velocity of flow and the radius of the pipe. Figure 8.6 showed a detailed plot of local velocity versus position within the fluid.

The velocity profile for turbulent flow is far different from the parabolic distribution for laminar flow. As shown in Fig. 9.6, the fluid velocity near the wall of the pipe changes rapidly from zero at the wall to a nearly uniform velocity distribution throughout the bulk of the cross section. The actual shape of the velocity profile varies with the friction factor, $f$, which in turn varies with the Reynolds number and the relative roughness of the pipe. The governing equation (from Reference 1) is

$$U = v[1 + 1.43\sqrt{f} + 2.15\sqrt{f}\log_{10}(1 - r/r_o)] \tag{9–8}$$

**FIGURE 9.6** General form of velocity profile for turbulent flow.

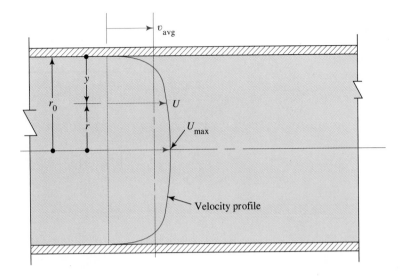

An alternate form of this equation can be developed by defining the distance from the wall of the pipe as $y = r_o - r$. Then, the argument of the logarithm term becomes

$$1 - \frac{r}{r_o} = \frac{r_o - r}{r_o} = \frac{y}{r_o}$$

Equation (9–8) is then

$$U = v[1 + 1.43\sqrt{f} + 2.15\sqrt{f}\log_{10}(y/r_o)] \qquad (9\text{–}9)$$

When evaluating Eq. (9–8) or (9–9), recall that the logarithm of zero is undefined. You may allow $r$ to *approach* $r_o$, but not equal it. Similarly, $y$ can only approach zero.

The maximum velocity occurs at the center of the pipe ($r = 0$ or $y = r_o$, and its value can be computed from

$$U_{max} = v(1 + 1.43\sqrt{f}) \qquad (9\text{–}10)$$

Figure 9.7 compares the velocity profiles for laminar flow and for turbulent flow at a variety of Reynolds numbers.

☐ **EXAMPLE PROBLEM 9.8**   For the data from Example Problem 9.5, compute the expected maximum velocity of flow, and compute several points on the velocity profile. Plot the velocity versus the distance from the pipe wall.

**Solution**   From Example Problem 9.5, the following data can be found:

$$D = 50 \text{ mm} = \text{pipe inside diameter}$$
$$v = 0.932 \text{ m/s} = \text{average velocity}$$
$$f = 0.018 = \text{friction factor}$$
$$N_R = 9.54 \times 10^4 \qquad \text{(turbulent)}$$

Now, from Eq. (9–10), we see that the maximum velocity of flow is

$$U_{max} = v(1 + 1.43\sqrt{f}) = (0.932 \text{ m/s})(1 + 1.43\sqrt{0.018})$$
$$U_{max} = 1.111 \text{ m/s at the center of the pipe}$$

**FIGURE 9.7** Velocity profiles in laminar and turbulent flow in a smooth pipe. (Source: Miller, R. W. 1983. *Flow Measurement Engineering Handbook.* New York: McGraw-Hill.)

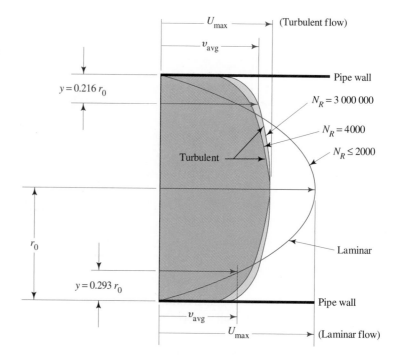

Equation (9–9) can be used to determine the points on the velocity profile. We know that the velocity equals zero at the pipe wall ($y = 0$). Also, the rate of change of velocity with position is greater near the wall than near the center of the pipe. Therefore, increments of 0.5 mm will be used from $y = 0.5$ to $y = 2.5$ mm. Then, increments of 2.5 mm will be used up to $y = 10$ mm. Finally, increments of 5.0 mm will provide sufficient definition of the profile near the center of the pipe. At $y = 1.0$ mm, and $r_o = 25$ mm,

$$U = v[1 + 1.43\sqrt{f} + 2.15\sqrt{f}\log_{10}(y/r_o)]$$
$$U = (0.932 \text{ m/s})[1 + 1.43\sqrt{0.018} + 2.15\sqrt{0.018}\log_{10}(1/25)]$$
$$U = 0.735 \text{ m/s}$$

Using similar calculations, we can compute the following values:

| $y$ (mm) | $y/r_o$ | $U$ (m/s) | |
|---|---|---|---|
| 0.5 | 0.02 | 0.654 | |
| 1.0 | 0.04 | 0.735 | |
| 1.5 | 0.06 | 0.782 | |
| 2.0 | 0.08 | 0.816 | |
| 2.5 | 0.10 | 0.842 | |
| 5.0 | 0.20 | 0.923 | |
| 7.5 | 0.30 | 0.970 | |
| 10.0 | 0.40 | 1.004 | |
| 15.0 | 0.60 | 1.051 | |
| 20.0 | 0.80 | 1.085 | |
| 25.0 | 1.00 | 1.111 | ($U_{max}$ at center of pipe) |

Figure 9.8 is the plot of $y$ versus velocity in the form in which the velocity profile is normally shown. Because the plot is symmetrical, only one-half of the profile is shown.

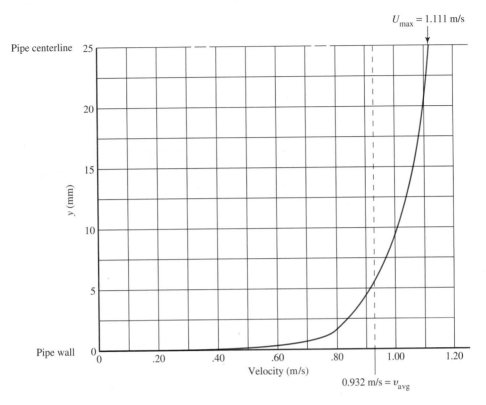

**FIGURE 9.8**    Velocity profile for turbulent flow for Example Problem 9.8.

■

**9.9**
**HAZEN-WILLIAMS**
**FORMULA FOR**
**WATER FLOW**

The Darcy equation presented in this chapter for calculating energy loss due to friction is applicable for any newtonian fluid. An alternate approach is convenient for the special case of the flow of water in pipe line systems.

The *Hazen-Williams formula* is one of the most popular formulas for the design and analysis of water systems. Its use is limited to the flow of water in pipes larger than 2.0 in and smaller than 6.0 ft in diameter. The velocity of flow should not exceed 10.0 ft/s. Also, it has been developed for water at 60°F. Use at temperatures much lower or higher would result in some error.

The Hazen-Williams formula is unit-specific. In the U.S. Customary unit system it takes the form

➪  **HAZEN-WILLIAMS FORMULA**
**U.S. CUSTOMARY UNITS**

$$v = 1.32\, C_h\, R^{0.63}\, s^{0.54} \tag{9–11}$$

where  $v$ = average velocity of flow (ft/s)

$C_h$ = Hazen-Williams coefficient (dimensionless)

$R$ = Hydraulic radius of flow conduit (ft)

$s$ = Ratio of $h_L/L$: energy loss/length of conduit (ft/ft)

The use of the hydraulic radius in the formula allows its application to noncircular sections as well as circular pipes. Recall that $R = D/4$ for circular pipes.

The coefficient $C_h$ is dependent only on the condition of the surface of the pipe or conduit. Table 9.2 gives typical values. Note that some are described as pipe in new, clean condition while the design value accounts for the accumulation of deposits that develop on the inside surfaces of pipe after time, even when clean water flows through them. Smoother pipes have higher values of $C_h$ as compared with rougher pipes.

**TABLE 9.2**   Hazen-Williams coefficient, $C_h$

| Type of Pipe | Average for New, Clean Pipe | Design Value |
|---|---|---|
| Steel, ductile iron, or cast iron with centrifugally applied cement or bituminous lining | 150 | 140 |
| Plastic, copper, brass, glass | 140 | 130 |
| Steel, uncoated | 130 | 100 |
| Concrete | 120 | 100 |
| Corrugated steel | 60 | 60 |

The Hazen-Williams formula for SI units is:

**HAZEN-WILLIAMS FORMULA SI UNITS**

$$v = 0.85\, C_h\, R^{0.63}\, s^{0.54} \tag{9-12}$$

where

$v$ = average velocity of flow (m/s)

$C_h$ = Hazen-Williams coefficient (dimensionless)

$R$ = hydraulic radius of flow conduit (m)

$s$ = ratio of $h_L/L$: energy loss/length of conduit (m/m)

As before, the volume flow rate can be computed from $Q = Av$.

□ **EXAMPLE PROBLEM 9.9**

For what velocity of flow of water in a new, clean 6-in Schedule 40 steel pipe would an energy loss of 20 ft of head occur over a length of 1000 ft? Compute the volume flow rate at that velocity. Then refigure the velocity using the design value of $C_h$ for steel pipe.

*Solution*

We can use Equation (9–11).

$$s = h_L/L = (20 \text{ ft})/(1000 \text{ ft}) = 0.02$$
$$R = D/4 = (0.5054 \text{ ft})/4 = 0.126 \text{ ft}$$
$$C_h = 130$$

Then,

$$v = 1.32\, C_h\, R^{0.63}\, s^{0.54}$$
$$v = (1.32)(130)(0.126)^{0.63}(0.02)^{0.54} = 5.64 \text{ ft/s}$$
$$Q = Av = (0.2006 \text{ ft}^2)(5.64 \text{ ft/s}) = 1.13 \text{ ft}^3/\text{s}$$

Now we can adjust the result for the design value of $C_h$.

Note that the velocity and volume flow rate are both directly proportional to the value of $C_h$. If the pipe degrades after use so the value of $C_h = 100$, the allowable volume flow rate to limit the energy loss to the same value of 20 ft per 1000 ft of pipe length would be

$$v = (5.64 \text{ ft/s})(100/130) = 4.34 \text{ ft/s}$$
$$Q = (1.13 \text{ ft}^3/\text{s})(100/130) = 0.869 \text{ ft}^3/\text{s}$$

$\blacksquare$

**9.10
OTHER FORMS OF THE
HAZEN-WILLIAMS
FORMULA**

Equations (9–11) and (9–12) allow the direct computation of the velocity of flow for a given type and size of flow conduit when the energy loss per unit length is known or specified. The volume flow rate can be simply calculated by using $Q = Av$. Other types of calculations that are often desired are:

1. To determine the required size of pipe to carry a given flow rate while limiting the energy loss to some specified value.
2. To determine the energy loss for a given flow rate through a given type and size of pipe of a known length.

Table 9.3 shows several forms of the Hazen-Williams formula that facilitate such calculations.

**TABLE 9.3**   Alternate forms of the Hazen-Williams formula

| U.S. Customary Units | SI Units |
|---|---|
| $v = 1.32 \, C_h R^{0.63} s^{0.54}$ | $v = 0.85 \, C_h R^{0.63} s^{0.54}$ |
| $Q = 1.32 \, A \, C_h R^{0.63} s^{0.54}$ | $Q = 0.85 \, A \, C_h R^{0.63} s^{0.54}$ |
| $h_L = L\left[\dfrac{Q}{1.32 \, A \, C_h R^{0.63}}\right]^{1.852}$ | $h_L = L\left[\dfrac{Q}{0.85 \, A \, C_h R^{0.63}}\right]^{1.852}$ |
| $D = \left[\dfrac{2.31 \, Q}{C_h s^{0.54}}\right]^{0.380}$ | $D = \left[\dfrac{3.59 \, Q}{C_h s^{0.54}}\right]^{0.380}$ |

*Note*: Units must be consistent:

| | |
|---|---|
| $v$ in ft/s | $v$ in m/s |
| $Q$ in ft$^3$/s | $Q$ in m$^3$/s |
| $A$ in ft$^2$ | $A$ in m$^2$ |
| $h_L$, $L$, $R$, and $D$ in ft | $h_L$, $L$, $R$, and $D$ in m |
| $s$ in ft/ft (dimensionless) | $s$ in m/m (dimensionless) |

**9.11
NOMOGRAPH
FOR SOLVING
HAZEN-WILLIAMS
FORMULA**

The nomograph shown in Figure 9.9 allows the solution of the Hazen-Williams formula to be done by simply aligning known quantities with a straight edge and reading the desired unknowns at the intersection of the straight edge with the appropriate vertical axis. *Note that this nomograph is constructed for the value of the Hazen-Williams coefficient of $C_h = 100$.* If the actual pipe condition warrants the use of a different value of $C_h$, the following formulas can be used to adjust the results. The subscript "100" refers to the value read from the nomograph for $C_h = 100$. The subscript "c" refers to the value for the given $C_h$.

| | | |
|---|---|---|
| $v_c = v_{100}(C_h/100)$ | [velocity] | (9–13) |
| $Q_c = Q_{100}(C_h/100)$ | [volume flow rate] | (9–14) |
| $D_c = D_{100}(100/C_h)^{0.38}$ | [pipe diameter] | (9–15) |
| $s_c = s_{100}(100/C_h)^{1.85}$ | [head loss/length] | (9–16) |

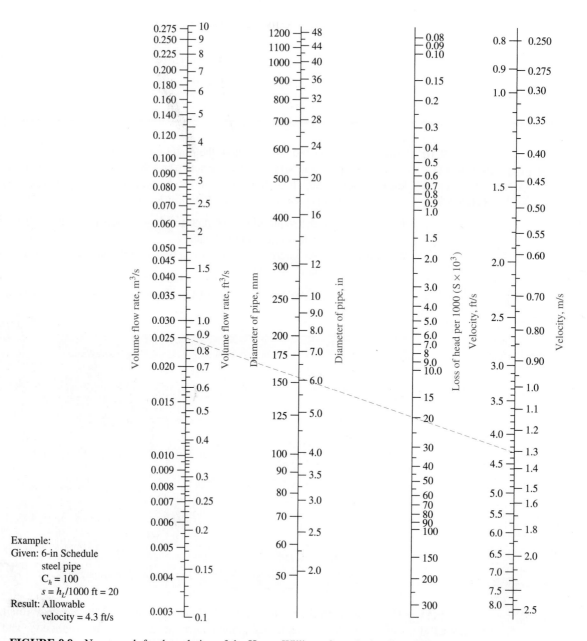

Example:
Given: 6-in Schedule
    steel pipe
    $C_h = 100$
    $s = h_L/1000 \text{ ft} = 20$
Result: Allowable
    velocity = 4.3 ft/s

**FIGURE 9.9**    Nomograph for the solution of the Hazen-Williams formula for $C_h = 100$.

The dashed line on the chart shows the use of the nomograph using data from Example Problem 9.9 for the case of $C_h = 100$.

One frequent use of a nomograph like that in Fig. 9.9 is to determine the required size of pipe to carry a given flow rate while limiting the energy loss to some specified value. Thus it is a convenient design tool.

☐ **EXAMPLE PROBLEM 9.10**    Specify the required size of Schedule 40 steel pipe to carry 1.20 ft³/s of water with no more than 4.0 ft of head loss over a 1000 ft length of pipe. Use the design value for $C_h$.

**Solution**   Table 9.2 suggests $C_h = 100$. Now, using Fig. 9.9, we can place a straight edge from $Q = 1.20$ ft$^3$/s on the volume flow rate line to the value of $s = (4.0$ ft$)/(1000$ ft$)$ on the energy loss line. The straight edge then intersects the pipe size line at approximately 9.7 in. The next larger standard pipe size listed in Appendix F is the nominal 10-in pipe with an inside diameter of 10.02 in.

Returning to the chart in Fig. 9.9 and slightly realigning $Q = 1.20$ ft$^3$/s with $D = 10.02$ in, we can read an average velocity of $v = 2.25$ ft/s. This is relatively low for a water distribution system and the pipe is quite large. If the pipeline is long, the cost for piping would be excessively large.

If we allow the velocity of flow to increase to approximately 6.0 ft/s for the same volume flow rate, we can use the chart to show that a 6-in pipe could be used with a head loss of approximately 37 ft per 1000 ft of pipe. The lower cost of the pipe compared with the 10-in pipe would have to be compared with the higher energy cost required to overcome the additional head loss.

---

# REFERENCES

1. Miller, R. W. 3rd ed. 1996. *Flow Measurement Engineering Handbook*. New York: McGraw-Hill.

2. Moody, L. F. 1944. Friction Factors for Pipe Flow. *Transactions of the ASME* 66(8):671–684. New York: American Society of Mechanical Engineers.

3. Steel, E. W., and Terence J. McGhee. 1979. *Water Supply and Sewerage*. New York: McGraw-Hill.

4. Swamee, P. K., and A. K. Jain. 1976. "Explicit Equations for Pipe-flow Problems." *Journal of the Hydraulics Division*, 102(HY5):657–664. New York: American Society of Civil Engineers.

# PRACTICE PROBLEMS

**9.1M** *Crude oil is flowing vertically downward through 60 m of 1-in Schedule 80 steel pipe at a velocity of 0.64 m/s. The oil has a specific gravity of 0.86 and is at 0°C. Calculate the pressure difference between the top and bottom of the pipe.*

**9.2M** *Water at 75°C is flowing in a $^1$/$_2$-in Type K copper tube at a rate of 12.9 L/min. Calculate the pressure difference between two points 45 m apart if the tube is horizontal.*

**9.3E** Fuel oil is flowing in a 4-in Schedule 40 steel pipe at the maximum rate for which the flow is laminar. If the oil has a specific gravity of 0.895 and a dynamic viscosity of $8.3 \times 10^{-4}$ lb-s/ft$^2$, calculate the energy loss per 100 ft of pipe.

**9.4E** A 3-in Schedule 40 steel pipe is 5000 ft long and carries a lubricating oil between two points A and B such that the Reynolds number is 800. Point B is 20 ft higher than A. The oil has a specific gravity of 0.90 and a dynamic viscosity of $4 \times 10^{-4}$ lb-s/ft$^2$. If the pressure at A is 50 psig, calculate the pressure at B.

**9.5M** *Benzene at 60°C is flowing in a 1-in Schedule 80 steel pipe at the rate of 20 L/min. The specific weight of the benzene is 8.62 kN/m$^3$. Calculate the pressure difference between two points 100 m apart if the pipe is horizontal.*

**9.6M** *As a test to determine the wall roughness of an existing pipe installation, water at 10°C is pumped through it at the rate of 225 L/min. The pipe is standard $1^1$/$_2$-in commercial steel tubing with a wall thickness of 0.083 in. Pressure gages located at 30 m apart in a horizontal run of the pipe read 1035 kPa and 669 kPa. Determine the pipe wall roughness.*

**9.7E** Water at 80°F flows from a storage tank through 550 ft of 6-in Schedule 40 steel pipe, as shown in Fig. 9.10. Taking the energy loss due to friction into account, calculate the required head $h$ above the pipe inlet to produce a volume flow rate of 2.50 ft$^3$/s.

**9.8E** A water main is an 18-in diameter concrete pressure pipe. Calculate the pressure drop over a 1-mi length due to pipe wall friction if the pipe carries 15 ft$^3$/s of water at 50°F.

**9.9E** Figure 9.11 shows a portion of a fire protection system in which a pump draws water at 60°F from a reservoir and delivers it to a point B at the flow rate of 1500 gal/min.
   **a.** Calculate the required height $h$ of the water level in the tank in order to maintain 5.0 psig pressure at point A.
   **b.** Assuming that the pressure at A is 5.0 psig, calculate the power delivered by the pump to the water in

**FIGURE 9.10**    Problem 9.7.

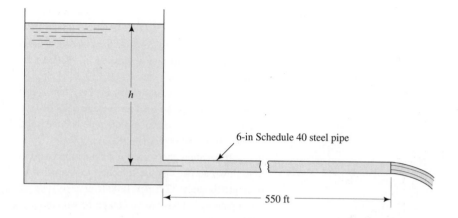

order to maintain the pressure at point B at 85 psig. Include energy losses due to friction but neglect any other energy losses.

**9.10E** A submersible deep-well pump delivers 745 gal/h of water at 60°F through a 1-in Schedule 40 steel pipe when operating in the system shown in Fig. 9.12. If the total length of pipe is 140 ft, calculate the power delivered by the pump to the water.

**9.11E** On a farm, water at 60°F is delivered from a pressurized storage tank to an animal watering trough through 300 ft of 1¹/₂-in Schedule 40 steel pipe as shown in Fig. 9.13. Calculate the required air pressure above the water in the tank to produce 75 gal/min of flow.

**9.12M** *Figure 9.14 shows a system for delivering lawn fertilizer in liquid form. The nozzle on the end of the hose requires 140 kPa of pressure to operate effectively. The hose is smooth plastic with an inside diameter of 25 mm. The fertilizer solution has a specific gravity of 1.10 and a dy-*

*namic viscosity of $2.0 \times 10^{-3}$ Pa·s. If the length of hose is 85 m, determine (a) the power delivered by the pump to the solution and (b) the pressure at the outlet of the pump. Neglect the energy losses on the suction side of the pump. The flow rate is 95 L/min.*

**9.13M** *A pipeline transporting crude oil (sg = 0.93) at 1200 L/min is made of 6-in Schedule 80 steel pipe. Pumping stations are spaced 3.2 km apart. If the oil is at 10°C, calculate (a) the pressure drop between stations and (b) the power required to maintain the same pressure at the inlet of each pump.*

**9.14M** *For the pipeline described in Problem 9.13, consider that the oil is to be heated to 100°C to decrease its viscosity.*
   **a.** *How does this affect the pump power requirement?*
   **b.** *At what distance apart could the pumps be placed with the same pressure drop as that from Problem 9.13?*

**9.15M** *Water at 10°C flows at the rate of 900 L/min from the reservoir and through the pipe shown in Fig. 9.15.*

**FIGURE 9.11**    Problem 9.9.

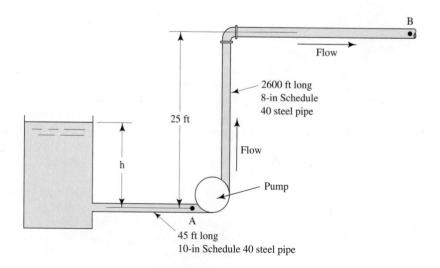

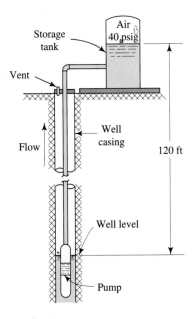

**FIGURE 9.12** Problem 9.10.

*Compute the pressure at point B, considering the energy loss due to friction, but neglecting other losses.*

**9.16E** For the system shown in Fig. 9.16, compute the power delivered by the pump to the water to pump 50 gal/min of water at 60°F to the tank. The air in the tank is at 40 psig. Consider the friction loss in the 225-ft-long discharge pipe, but neglect other losses. Then, redesign the system by using a larger pipe size to reduce the energy loss and reduce the power required to no more than 5.0 hp.

**9.17E** Fuel oil (sg = 0.94) is being delivered to a furnace at a rate of 60 gal/min through a 1½-in Schedule 40 steel pipe. Compute the pressure difference between two points 40.0 ft apart if the pipe is horizontal and the oil is at 85°F.

**9.18E** Figure 9.17 shows a system used to spray polluted water into the air to increase the water's oxygen content and to cause volatile solvents in the water to vaporize. The pressure at point B just ahead of the nozzle must be 25.0 psig for proper nozzle performance. The pressure at point A

**FIGURE 9.13** Problem 9.11.

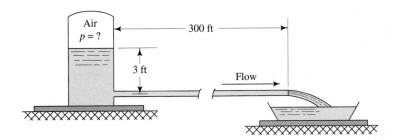

**FIGURE 9.14** Problem 9.12.

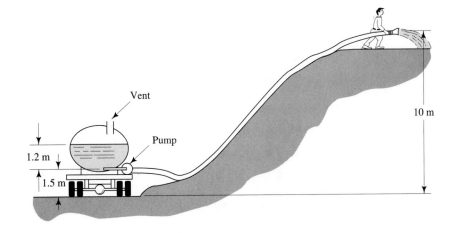

**FIGURE 9.15**   Problem 9.15.

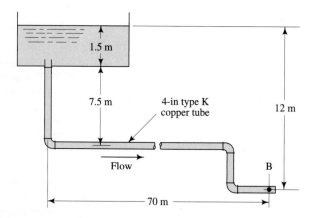

(the pump inlet) is −3.50 psig. The volume flow rate is 0.50 ft³/s. The dynamic viscosity of the fluid is $4.0 \times 10^{-5}$ lb·s/ft². The specific gravity of the fluid is 1.026. Compute the power delivered by the pump to the fluid, considering friction energy loss in the discharge line.

**9.19E** In a chemical processing system, the flow of glycerine at 60°F (sg = 1.24) in a copper tube must remain laminar with a Reynolds number approximately equal to but not exceeding 300. Specify the smallest standard Type K copper tube that will carry a flow rate of 0.90 ft³/s. Then, for a flow of 0.90 ft³/s in the tube you specified, compute the

**FIGURE 9.16**   Problem 9.16.

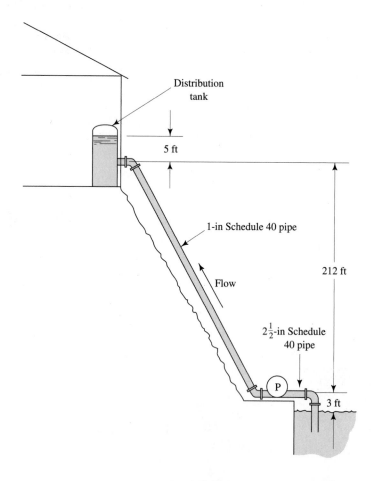

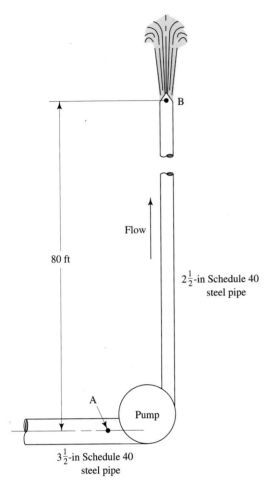

**FIGURE 9.17** Problem 9.18.

pressure drop between two points 55.0 ft apart if the pipe is horizontal.

**9.20E** Water at 60°F is being pumped from a stream to a reservoir whose surface is 210 ft above the pump. See Figure 9.18. The pipe from the pump to the reservoir is an 8-in Schedule 40 steel pipe 2500 ft long. If 4.00 ft³/s is being pumped, compute the pressure at the outlet of the pump. Consider the friction loss in the discharge line, but neglect other losses.

**9.21E** For the pump described in Problem 9.20, if the pressure at the pump inlet is −2.36 psig, compute the power delivered by the pump to the water.

**9.22E** Gasoline at 50°F flows from A to B through 3200 ft of standard 10-in Schedule 40 steel pipe at the rate of 4.25 ft³/s. Point B is 85 ft above point A and the pressure at B must be 40.0 psig. Considering the friction loss in the pipe, compute the required pressure at A.

**9.23E** Figure 9.19 shows a pump recirculating 300 gal/min of heavy machine tool lubricating oil at 104°F to test the oil's stability. The total length of 4-in pipe is 25.0 ft, and the total length of 3-in pipe is 75.0 ft. Compute the power delivered by the pump to the oil.

**9.24E** Linseed oil at 25°C flows at 3.65 m/s in a ³/₄-in Type K copper tube. Compute the pressure difference between two points in the tube 17.5 m apart if the first point is 1.88 m above the second point.

**9.25M** *Glycerine at 25°C flows through a straight copper tube (3-in type K) at a flow rate of 180 L/min. Compute the pressure difference between two points 25.8 m apart if the first point is 0.68 m below the second point.*

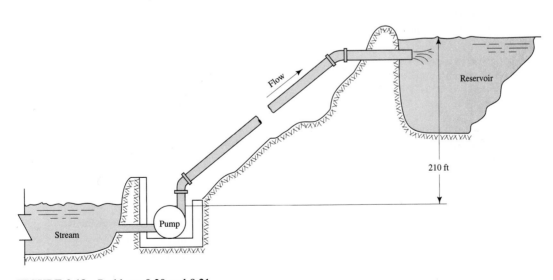

**FIGURE 9.18** Problems 9.20 and 9.21.

**FIGURE 9.19**   Problem 9.23.

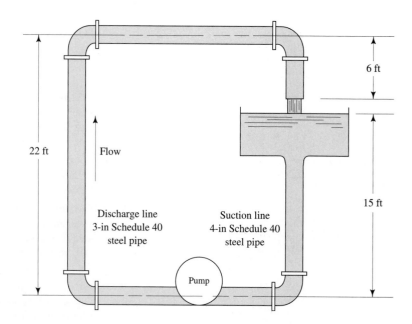

NOTE: For Problems 9.26 through 9.36, use the equations for friction factor from Section 9.6 to compute the friction factor.

**9.26M** *Water at 75°C flows in a ¹/₂-in Type K copper tube at a rate of 12.9 L/min.*

**9.27M** *Benzene (sg = 0.88) at 60°C flows in a 1-in Schedule 80 steel pipe at the rate of 20 L/min.*

**9.28E** Water at 80°F flows in a 6-in coated ductile iron pipe at a rate of 2.50 ft³/s.

**9.29E** Water at 50°F flows at 15.0 ft³/s in a concrete pipe with an inside diameter of 18.0 in.

**9.30E** Water at 60°F flows at 1500 gal/min in a 10-in Schedule 40 steel pipe.

**9.31M** *A liquid fertilizer solution (sg = 1.10) with a dynamic viscosity of 2.0 × 10⁻³ Pa·s flows at 95 L/min through a 25-mm diameter smooth plastic hose.*

**9.32M** *Crude oil (sg = 0.93) at 100°C flows at a rate of 1200 L/min in a 6-in Schedule 80 steel pipe.*

**9.33M** *Water at 65°C flows in a 1¹/₂-in Schedule 40 steel pipe at a rate of 10 m/s.*

**9.34M** *Propyl alcohol flows in a 3-in Type K copper tube at 25°C at a rate of 0.026 m³/s.*

**9.35E** Water at 70°F flows in a 12-in diameter concrete pipe at 3.0 ft³/s.

**9.36E** Heavy fuel oil at 77°F flows in a 6-in Schedule 40 steel pipe at 12 ft/s.

## Noncircular Cross Sections

**9.37E** For the system described in Problem 8.33, compute the pressure difference between two points 30.0 ft apart if the duct is horizontal. Use $\epsilon = 8.5 \times 10^{-5}$ ft.

**9.38M** *For the shell-and-tube heat exchanger described in Problem 8.34, compute the pressure difference for both fluids between two points 5.25 m apart if the heat exchanger is horizontal.*

**9.39M** *For the system described in Problem 8.35, compute the pressure drop for both fluids between two points 3.80 m apart if the duct is horizontal. Use the roughness for steel pipe for all surfaces.*

**9.40E** For the system described in Problem 8.36, compute the pressure difference in both the small pipes and the large pipe between two points 50.0 ft apart if the pipes are horizontal. Use the roughness for steel pipe for all surfaces.

**9.41M** *For the shell-and-tube heat exchanger described in Problem 8.37, compute the pressure drop for the flow of water in the shell. Use the roughness for copper for all surfaces. The length is 3.60 m.*

**9.42E** For the heat exchanger described in Problem 8.38, compute the pressure drop for a length of 57 in.

**9.43M** *For the glycerine described in Problem 8.40, compute the pressure drop for a horizontal duct 22.6 m long. All surfaces are copper.*

**9.44M** *For the flow of water in the square tubes described in Problems 8.41, compute the pressure drop over a length of 22.6 m. All surfaces are copper, and the duct is horizontal.*

**9.45E** If the heat sink described in Problem 8.42 is 105 in long, compute the pressure drop for the water. Use $\epsilon = 2.5 \times 10^{-5}$ ft for the aluminum.

**9.46E** Compute the energy loss for the flow of water in the cooling passage described in Problem 8.43 if its total length is 45 in. Use $\epsilon$ for steel. Also compute the pressure difference across the total length of the cooling passage.

**9.47E** In Fig. 9.20, ethylene glycol (sg = 1.10) at 77°F flows around the tubes and inside the rectangular passage. Calculate the volume flow rate of ethylene glycol in gal/min required for the flow to have a Reynolds number of 8000. Then, compute the energy loss over a length of 128 in. All surfaces are brass.

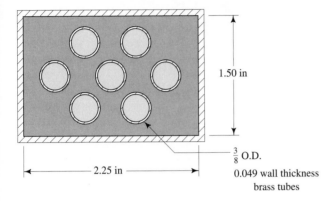

1.50 in

$\frac{3}{8}$ O.D.

2.25 in

0.049 wall thickness brass tubes

**FIGURE 9.20** Problem 9.47.

**9.48M** *Figure 9.21 shows a duct in which methyl alcohol at 25°C flows at the rate of 3000 L/min. Compute the energy loss over a 2.25-m length of the duct. All surfaces are smooth plastic.*

**9.49E** A furnace heat exchanger has a cross section like that shown in Fig. 9.22. The air flows around the three thin passages in which hot gases flow. The air is at 140°F and has a density of $2.06 \times 10^{-3}$ slugs/ft³ and a dynamic viscosity of $4.14 \times 10^{-7}$ lb·s/ft². Compute the Reynolds number for the flow if the velocity is 20 ft/sec.

**9.50E** Figure 9.23 shows a system in which methyl alcohol at 77°F flows outside the three tubes while ethyl alcohol at 0°F flows inside the tubes. Compute the volume flow rate of each fluid required to produce a Reynolds number of $3.5 \times 10^4$ in all parts of the system. Then, compute the pressure difference between two points 10.5 ft apart if the system is horizontal. All surfaces are copper.

## Velocity Profile for Turbulent Flow

**9.51M** *For the flow of 12.9 L/min of water at 75°C in a $\frac{1}{2}$-in Type K copper tube, compute the expected maximum velocity of flow from Eq. (9–10).*

**9.52M** *A large pipeline with a 1.200-m inside diameter carries oil similar to SAE 10 at 40°C (sg = 0.8.). Compute the volume flow rate required to produce a Reynolds number of $3.60 \times 10^4$. Then, if the pipe is clean steel, compute several points of the velocity profile and plot the data in a manner similar to that shown in Fig. 9.8.*

**9.53M** *Repeat Problem 9.52 if the oil is at 110°C but with the same flow rate. Discuss the differences in the velocity profile.*

**9.54** Using Eq. (9–9), compute the distance $y$ for which the local velocity $U$ is equal to the average velocity $v$.

**FIGURE 9.21** Problem 9.48.

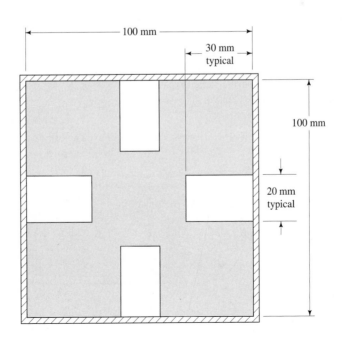

100 mm

30 mm typical

100 mm

20 mm typical

**FIGURE 9.22**   Problem 9.49.

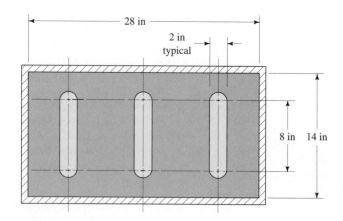

**9.55E**  The result for Problem 9.54 predicts that the average velocity for turbulent flow will be found at a distance of $0.216(r_o)$ from the wall of the pipe. Compute this distance for a 24-in Schedule 40 steel pipe. Then, if the pipe carries water at 50°F at a flow rate of 16.75 ft³/s, compute the velocity at points 0.50 in on either side of the average velocity point.

**9.56**  Using Eq. (9–10), compute the ratio of the average velocity to the maximum velocity of flow in smooth pipes with Reynolds numbers of 4000, $10^4$, $10^5$, and $10^6$.

**9.57**  Using Eq. (9–10), compute the ratio of the average velocity to the maximum velocity of flow for the flow of a liquid through a concrete pipe with an inside diameter of 8.00 in with Reynolds numbers of 4000, $10^4$, $10^5$, and $10^6$.

**9.58E**  Using Eq. (9–9), compute several points on the velocity profile for the flow of 400 gal/min of water at 50°F in a new, clean 4-in Schedule 40 steel pipe. Make a plot similar to Fig. 9.8 with a fairly large scale.

**9.59E**  Repeat Problem 9.58 for the same conditions, except that the inside of the pipe is roughened by age so that $\epsilon = 5.0 \times 10^{-3}$. Plot the results on the same graph as that used for the results of Problem 9.58.

**9.60E**  For both situations described in Problems 9.58 and 9.59, compute the pressure drop that would occur over a distance of 250 ft of horizontal pipe.

## Energy Loss Using the Hazen-Williams Formula

Use the design values for the coefficient $C_h$ from Table 9.2 unless stated otherwise. Use either of the various forms of the formula or the nomograph in Fig. 9.9 as assigned.

**9.61E**  Water flows at a rate of 1.50 ft³/s through 550 ft of 6-in cement-lined ductile iron pipe. Compute the energy loss.

**9.62M**  *Compute the energy loss as water flows in a 4-in type K copper tube at a rate of 1000 L/min over a length of 45 m.*

**9.63E**  A water main is an 18-in diameter concrete pressure pipe. Calculate the energy loss over a 1-mile length if it carries 7.50 ft³/s of water.

**9.64E**  A fire protection system includes 1500 ft of 10-in Schedule 40 steel pipe. Compute the energy loss in the pipe when it carries 1500 gal/min of water.

**9.65M**  *A 4-in type K copper tube carries 900 L/min of water over a length of 80 m. Compute the energy loss.*

**FIGURE 9.23**   Problem 9.50.

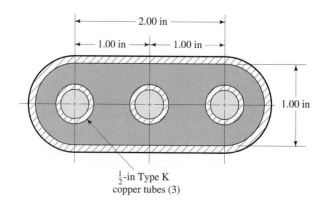

$\frac{1}{2}$-in Type K
copper tubes (3)

**9.66E** Compute the energy loss as 0.20 ft$^3$/s of water flows through a length of 80 ft of 2$^1$/$_2$-in Schedule 40 steel pipe.

**9.67E** It is desired to flow 2.0 ft$^3$/s of water through 2500 ft of 8-in pipe. Compute the head loss for both plain Schedule 40 steel pipe and ductile iron pipe coated with a centrifugally applied cement lining.

**9.68E** Specify a suitable size of new, clean Schedule 40 steel pipe that would carry 300 gal/min of water over a length of 1200 ft with no more than 10 ft of head loss. For the selected pipe, compute the actual expected head loss.

**9.69E** For the pipe selected in Problem 9.68, compute the head loss using the design value for $C_h$ rather than that for new, clean pipe.

**9.70E** Compare the head loss that would result from the flow of 100 gal/min of water through 1000 ft of new, clean Schedule 40 steel pipe for 2-in and 3-in sizes.

## COMPUTER PROGRAMMING ASSIGNMENTS

1. Write a program to compute the friction factor for the flow of any fluid through pipes and tube using Eqs. (9–3) and (9–5). The program must compute the Reynolds number and the relative roughness. Then, decisions must be made as follows:
   **a.** If $N_R < 2000$, use $f = 64/N_R$ [Eq. (9–3)].
   **b.** If $2000 < N_R < 4000$, the flow is in the critical range and no reliable value can be computed for $f$. Print a message to the user of the program.
   **c.** If $N_R > 4000$, the flow is turbulent. Use Eq. (9–5) to compute $f$.
   **d.** Print out $N_R$, $D/\epsilon$, and $f$.

2. Incorporate Program 1 into an enhanced program to compute the pressure drop for the flow of any fluid through a pipe of any size. The two points of interest can be any distance apart, and one end can be any elevation relative to the other. The program should be able to complete such analyses as required for Problems 9.1, 9.2, and 9.5. The program also can be set up to determine the energy loss only in order to solve problems such as 9.3.

3. Write a program or spreadsheet to compute the velocity profile for turbulent flow using Eqs. (9–8), (9–9), and (9–10). Specified increments of position within the pipe can be input by the operator. Program 1 could be incorporated.

4. Write a program to solve the Hazen-Williams formula in any of its forms listed in Table 9.3. Allow the program operator to specify the unit system to be used, which values are known, and which values are to be solved for.

5. Create a spreadsheet to solve the Hazen-Williams formula in any of its forms listed in Table 9.3. Different parts of the spreadsheet can compute different quantities: velocity, head loss, or pipe diameter. Provide for solutions in both U.S. customary and SI units.

# 10 Minor Losses

## 10.1 The Big Picture

### Discussion Map

- Here you continue to learn techniques for analyzing real pipe line problems in which several types of flow system components exist.
- You will build on the principles you learned in Chapters 6, 7, 8, and 9.
- Now you will develop your ability to determine the energy losses that occur as fluid flows through numerous kinds of valves and fittings commonly found in pipe line systems.

### Discover

- Study Figure 7.1, which shows a photograph of an industrial piping system delivering fluid from storage tanks to processes that use the fluid.
- List all of the components shown in Figure 7.1 that are used to control the flow or to direct it to specific destinations.
- Now, can you think of other fluid flow systems you have seen and identify other kinds of elements that could cause energy losses to occur?

In this chapter you will learn how to determine the amount of energy loss through such devices as enlargements in the size of the flow conduit, contractions, the entrance of fluid from a tank into a pipe, the exit of fluid from a pipe into a tank, elbows, tees, and valves.

Here you continue to learn techniques for analyzing real pipe line problems in which several types of flow system components exist. You are almost to the goal we set in Chapter 6, where Bernoulli's equation was introduced.

We said that in Chapters 6 to 11 you would continue to develop concepts related to the flow of fluids in pipe line systems. The goal is to put them all together to analyze the performance of such systems. You will do this in Chapter 11.

In Chapter 7, you developed the *general energy equation* that extended Bernoulli's equation to account for energy losses and additions that typically occur in real flow systems. Refer now to the first part of Chapter 7 where the general energy equation was introduced.

From your study of the industrial piping system in Figure 7.1, how does your list of fluid control components compare with this?

1. The fluid comes into the picture through a fairly large pipe at the left bringing fluid to the pump. This is called the *suction pipe* or the *suction line*.
2. Just before the fluid enters the pump through its suction flange, the pipe size is reduced through a *gradual reducer*. The reducer is needed because the suction pipe size is larger than the standard connection provided by the pump manufacturer. As a result, the fluid velocity increases somewhat as it moves from the pipe into the suction inlet of the pump.
3. The pump, driven by an electric motor, pulls the fluid from the suction line and adds energy to it as it moves the fluid into the *discharge pipe*, sometimes called the *discharge line*. The fluid in the discharge line now has a higher energy level resulting in a higher pressure head.

4. Connected to the discharge flange of the pump is an *enlargement* that increases the size of the flow conduit from the output of the pump to the full size of the discharge line. As the fluid moves through the enlargement, the flow velocity decreases.

5. Just to the right of the enlargement there is a *tee* in the pipe. This, of course, allows the operator of the system to direct the flow in either of two ways. The normal direction is to continue through the main discharge line. This would happen if the valve to the side of the tee is shut off. But, if that valve is opened, all or part of the flow would turn into the branch line through the tee and flow through the adjacent *valve*. It would then continue on through the branch line.

6. Now let's assume that the valve in the branch line is shut off. The fluid continues in the discharge line and encounters another *valve*. Normally, this valve is fully open, allowing the fluid to go on to its destination. The valve is there to permit the system to be shut down after the pump is stopped. This allows for the pump to be maintained or replaced without draining the piping system downstream from the pump. It is likely that a similar valve is in the suction line off to the left outside the bounds of the photograph.

7. After flowing through the valve in the discharge line, another *tee* allows the fluid to either go straight or to branch to the long pipe that proceeds toward the rear of the photograph. Let's assume that it does go into the branch line.

8. After leaving the tee through the branch line, the fluid immediately encounters an *elbow* that redirects it from a vertical to a horizontal direction.

9. After moving through a short length of pipe, another *valve* is in the line to control the flow to the rest of the system.

10. Beyond the valve the piping is covered with insulation, so it is difficult to see details of the piping system. But notice that there is a set of flanges just beyond the valve. Here there is a *flow meter* to permit the operator to measure how much fluid is flowing in the pipe.

11. After flowing through the meter, the fluid continues through the long pipe to the process that will use it.

12. Did you also notice the large *valve* at the base of the tank in the left background of the photograph just to the left of the motor? This allows the tank to be drained, perhaps into a truck that would remove residual fluid.

Look at the numerous control devices (shown italicized) in this list. Energy is lost from the system through each of these devices. When you design such a system, you will need to account for these energy losses.

Now, can you think of other fluid flow systems you have seen and identify other kinds of elements that could cause energy losses to occur?

- Consider the plumbing system in your home. Track how the water gets from the main supply point to the kitchen sink. Write down each element that causes an obstruction to the flow (such as a valve), that changes the direction of the flow, or that changes velocity of flow.
- Consider how the water gets to an outside faucet that can be used to water the lawn or garden. Track the flow all the way to the sprinkler head.
- How does the water get from the city supply wells or reservoir to your home?
- How does the cooling fluid in an automotive engine move from the radiator through the engine and back to the radiator?
- How does the windshield washing fluid get from the reservoir to the windshield?
- How does the gasoline in your car or a truck get from the fuel tank to the engine intake ports?
- How does fuel on an airplane get from its fuel tanks in the wings to the engines?
- How does the refrigerant in your car's air conditioning system flow from the compressor attached to the engine through the system that makes the car cool?
- How does the refrigerant in your refrigerator move through its cooling system?
- How does the water in a clothes washer get from the house piping system into the wash tub?

■ How does the wash water drain from the tub and get pumped into the sewer drain?
■ How does the water flow through a squirt toy?
■ Have you seen a high pressure washing system that can be used to remove heavy dirt from a deck, a driveway, or a boat? Track the flow of fluid through that kind of system.
■ How does water in an apartment building or hotel get from the city supply line to each apartment or hotel room?
■ How does the water flow from the city supply line through the sprinkler system in an office building or warehouse to protect the people, products, and equipment from a fire?
■ How does the oil in a fluid power system flow from the pump through the control valves, cylinders, and other fluid power devices to actuate industrial automation systems, construction equipment, agricultural machinery, or an aircraft landing gear?
■ How does engine oil get pumped from the oil pan to lubricate the moving parts of the engine?
■ How does the lubricating fluid in a complex piece of manufacturing equipment get distributed to critical moving parts?
■ How do liquid components of chemical processing systems move through those systems?
■ How does milk, juice, or soft drink mix flow through the systems that finally deliver it to the bottling station?

What others can you think of?

Now let's learn how to analyze the energy losses in these kinds of systems. In this chapter you will learn how to determine the amount of energy losses through such devices. Many of the losses through elbows, valves, tees, meters, and other control elements are small in comparison with the energy losses due to friction in long straight sections of pipe. For this reason they are often called *minor losses*. However, the combined effect of all such losses can be significant. Good system design practice calls for you to evaluate them.

**10.2**
**OBJECTIVES**

After completing this chapter, you should be able to:

1. Recognize the sources of minor losses.
2. Define *resistance coefficient*.
3. Determine the energy loss for flow through the following types of minor losses.
    a. Sudden enlargement of the flow path.
    b. Exit loss when fluid leaves a pipe and enters a static reservoir.
    c. Gradual enlargement of the flow path.
    d. Sudden contraction of the flow path.
    e. Gradual contraction of the flow path.
    f. Entrance loss when fluid enters a pipe from a static reservoir.
4. Define the term *vena contracta*.
5. Define and use the *equivalent length technique* for computing energy losses in valves, fittings, and pipe bends.

**10.3**
**RESISTANCE**
**COEFFICIENT**

MINOR LOSS USING
RESISTANCE COEFFICIENT

Energy losses are proportional to the velocity head of the fluid as it flows around an elbow, through an enlargement or contraction of the flow section, or through a valve. Experimental values for energy losses are usually reported in terms of a resistance coefficient, $K$, as follows:

$$h_L = K(v^2/2g) \qquad (10-1)$$

In Eq. (10–1), $h_L$ is the minor loss, $K$ is the resistance coefficient, and $v$ is the average velocity of flow in the pipe in the vicinity where the minor loss occurs. In some cases, there may be more than one velocity of flow, as with enlargements or contractions. It is most important for you to know which velocity is to be used with each resistance coefficient.

If the velocity head $v^2/2g$ in Eq. (10–1) is expressed in the units of meters, then the energy loss $h_L$ will also be in meters or N·m/N of fluid flowing. The resistance coefficient is unitless, as it represents a constant of proportionality between the energy loss and the velocity head. The magnitude of the resistance coefficient depends on the geometry of the device that causes the loss and sometimes on the velocity of flow. In the following sections, we will describe the process for determining the value of $K$ and for calculating the energy loss for many types of minor loss conditions.

**10.4**
**SUDDEN ENLARGEMENT**

As a fluid flows from a smaller pipe into a larger pipe through a sudden enlargement, its velocity abruptly decreases, causing turbulence that generates an energy loss (Fig. 10.1). The amount of turbulence, and therefore the amount of energy loss, is dependent on the ratio of the sizes of the two pipes.

**FIGURE 10.1**   Sudden enlargement.

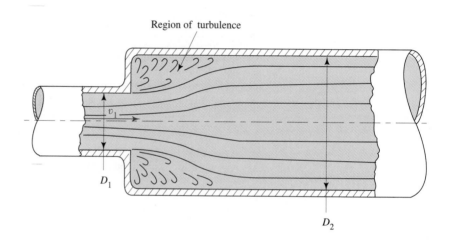

The minor loss is calculated from the equation

$$h_L = K(v_1^2/2g) \tag{10-2}$$

where $v_1$ is the average velocity of flow in the smaller pipe ahead of the enlargement. Tests have shown that the value of the loss coefficient $K$ is dependent on both the ratio of the sizes of the two pipes and the magnitude of the flow velocity. This is illustrated graphically in Fig. 10.2 and in tabular form in Table 10.1.

By making some simplifying assumptions about the character of the flow stream as it expands through the sudden enlargement, it is possible to analytically predict the value of $K$ from the following equation:

$$K = [1 - (A_1/A_2)]^2 = [1 - (D_1/D_2)^2]^2 \tag{10-3}$$

The subscripts 1 and 2 refer to the smaller and larger sections, respectively, as shown in Fig. 10.1. Values for $K$ from this equation agree well with experimental data when the velocity $v_1$ is approximately 1.2 m/s. At higher velocities, the actual values of $K$ are lower than the theoretical values. We recommend that experimental values be used if the velocity of flow is known.

**FIGURE 10.2**  Resistance coefficient—Sudden enlargement.

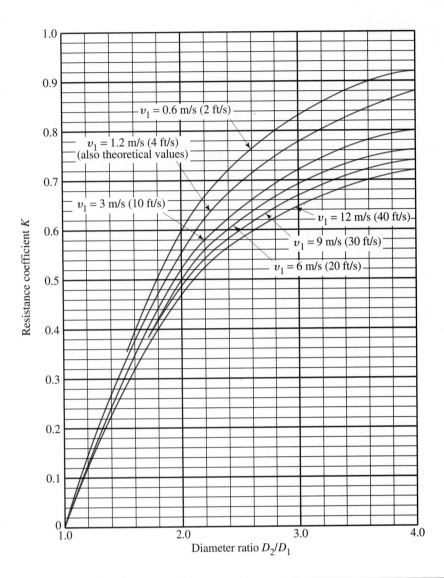

☐ **EXAMPLE PROBLEM 10.1**   Determine the energy loss that will occur as 100 L/min of water flows through a sudden enlargement from a 1-in copper tube (Type K) to a 3-in tube (Type K). See Appendix H for tube dimensions.

*Solution*   Using the subscript 1 for the section just ahead of the enlargement and 2 for the section downstream from the enlargement, we get

$$D_1 = 25.3 \text{ mm} = 0.0253 \text{ m}$$

$$A_1 = 5.017 \times 10^{-4} \text{ m}^2$$

$$D_2 = 73.8 \text{ mm} = 0.0738 \text{ m}$$

$$A_2 = 4.282 \times 10^{-3} \text{ m}^2$$

$$v_1 = \frac{Q}{A_1} = \frac{100 \text{ L/min}}{5.017 \times 10^{-4} \text{ m}^2} \times \frac{1 \text{ m}^3/\text{s}}{60\,000 \text{ L/min}} = 3.32 \text{ m/s}$$

$$\frac{v_1^2}{2g} = \frac{(3.32)^2}{(2)(9.81)} \text{ m} = 0.56 \text{ m}$$

**TABLE 10.1** Resistance coefficient—Sudden enlargement

| $D_2/D_1$ | Velocity, $v_1$ | | | | | | |
|---|---|---|---|---|---|---|---|
| | 0.6 m/s 2 ft/s | 1.2 m/s 4 ft/s | 3 m/s 10 ft/s | 4.5 m/s 15 ft/s | 6 m/s 20 ft/s | 9 m/s 30 ft/s | 12 m/s 40 ft/s |
| 1.0 | 0.0 | 0.0 | 0.0 | 0.0 | 0.0 | 0.0 | 0.0 |
| 1.2 | 0.11 | 0.10 | · 0.09 | 0.09 | 0.09 | 0.09 | 0.08 |
| 1.4 | 0.26 | 0.25 | 0.23 | 0.22 | 0.22 | 0.21 | 0.20 |
| 1.6 | 0.40 | 0.38 | 0.35 | 0.34 | 0.33 | 0.32 | 0.32 |
| 1.8 | 0.51 | 0.48 | 0.45 | 0.43 | 0.42 | 0.41 | 0.40 |
| 2.0 | 0.60 | 0.56 | 0.52 | 0.51 | 0.50 | 0.48 | 0.47 |
| 2.5 | 0.74 | 0.70 | 0.65 | 0.63 | 0.62 | 0.60 | 0.58 |
| 3.0 | 0.83 | 0.78 | 0.73 | 0.70 | 0.69 | 0.67 | 0.65 |
| 4.0 | 0.92 | 0.87 | 0.80 | 0.78 | 0.76 | 0.74 | 0.72 |
| 5.0 | 0.96 | 0.91 | 0.84 | 0.82 | 0.80 | 0.77 | 0.75 |
| 10.0 | 1.00 | 0.96 | 0.89 | 0.86 | 0.84 | 0.82 | 0.80 |
| $\infty$ | 1.00 | 0.98 | 0.91 | 0.88 | 0.86 | 0.83 | 0.81 |

**Source:** H. W. King and E. F. Brater. 1963. *Handbook of Hydraulics,* 5th ed. New York: McGraw-Hill. (Table 6–7. Velocities converted to SI units.)

To find a value for $K$, the diameter ratio is needed. We find that

$$D_2/D_1 = 73.8/25.3 = 2.92$$

From Fig. 10.2, $K = 0.72$. Then we have

$$h_L = K(v_1^2/2g) = (0.72)(0.56 \text{ m}) = 0.40 \text{ m}$$

This result indicates that 0.40 N·m of energy is dissipated from each newton of water that flows through the sudden enlargement. The following problem illustrates the calculation of the pressure difference between points 1 and 2.

□ **EXAMPLE PROBLEM 10.2**    Determine the difference between the pressure ahead of a sudden enlargement and the pressure downstream from the enlargement. Use the data from Example Problem 10.1.

*Solution*    First, the energy equation must be written:

$$\frac{p_1}{\gamma} + z_1 + \frac{v_1^2}{2g} - h_L = \frac{p_2}{\gamma} + z_2 + \frac{v_2^2}{2g}$$

Solving for $p_1 - p_2$ gives

$$p_1 - p_2 = \gamma[(z_2 - z_1) + (v_2^2 - v_1^2)/2g + h_L]$$

If the enlargement is horizontal, $z_2 - z_1 = 0$. Even if it were vertical, the distance between points 1 and 2 is so small that it is considered negligible. Now, calculating the velocity in the larger pipe, we get

$$v_2 = \frac{Q}{A_2} = \frac{100 \text{ L/min}}{4.282 \times 10^{-3} \text{ m}^2} \times \frac{1 \text{ m}^3/\text{s}}{60\,000 \text{ L/min}} = 0.39 \text{ m/s}$$

Using $\gamma = 9.81$ kN/m$^3$ for water and $h_L = 0.40$ m from Example Problem 10.1, we have

$$p_1 - p_2 = \frac{9.81 \text{ kN}}{\text{m}^3}\left[0 + \frac{(0.39)^2 - (3.32)^2}{(2)(9.81)} \text{ m} + 0.40 \text{ m}\right]$$

$$= -1.51 \text{ kN/m}^2 = -1.51 \text{ kPa}$$

Therefore, $p_2$ is 1.51 kPa greater than $p_1$.

■

---

**10.5**
**EXIT LOSS**
As a fluid flows from a pipe into a large reservoir or tank, as shown in Fig. 10.3, its velocity is decreased to very nearly zero. In the process, the kinetic energy that the fluid possessed in the pipe, indicated by the velocity head $v_1^2/2g$, is dissipated. Therefore, the energy loss for this condition is

$$h_L = 1.0(v_1^2/2g) \qquad (10\text{--}4)$$

This is called the *exit loss*. The value of $K = 1.0$ is used regardless of the form of the exit where the pipe connects to the tank wall.

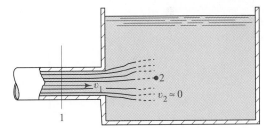

**FIGURE 10.3** Exit loss as fluid flows from a pipe into a static reservoir.

---

☐ **EXAMPLE PROBLEM 10.3**

Determine the energy loss that will occur as 100 L/min of water flows from a 1-in copper tube (Type K) into a large tank.

**Solution**

Using Eq. (10–4), we have

$$h_L = 1.0(v_1^2/2g)$$

From the calculations in Example Problem 10.1, we know that

$$v_1 = 3.32 \text{ m/s}$$
$$v_1^2/2g = 0.56 \text{ m}$$

Then the energy loss is

$$h_L = (1.0)(0.56 \text{ m}) = 0.56 \text{ m}$$

■

---

**10.6**
**GRADUAL**
**ENLARGEMENT**
If the transition from a smaller to a larger pipe can be made less abrupt than the square-edged sudden enlargement, the energy loss is reduced. This is normally done by placing a conical section between the two pipes as shown in Fig. 10.4. The sloping walls of the cone tend to guide the fluid during the deceleration and expansion

**FIGURE 10.4** Gradual enlargement.

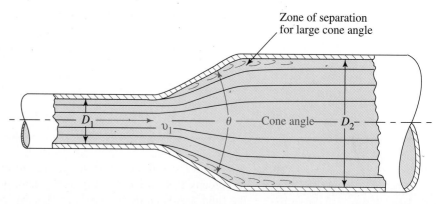

Zone of separation
for large cone angle

**FIGURE 10.5** Resistance coefficient—Gradual enlargement.

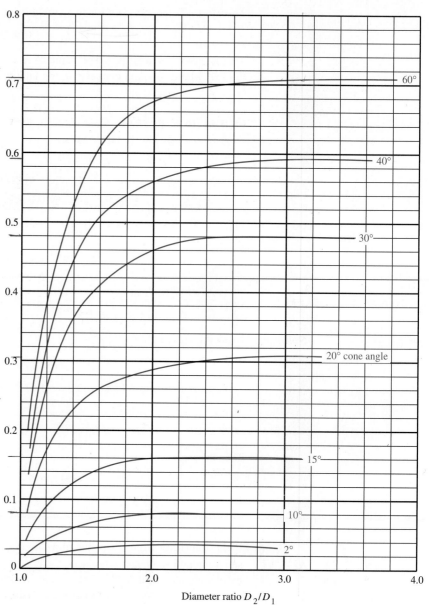

Diameter ratio $D_2/D_1$

of the flow stream. Therefore, the size of the zone of separation and the amount of turbulence is reduced as the cone angle is reduced.

The energy loss for a gradual enlargement is calculated from

$$h_L = K(v_1^2/2g) \tag{10–5}$$

where $v_1$ is the velocity in the smaller pipe ahead of the enlargement. The magnitude of $K$ is dependent on both the diameter ratio $D_2/D_1$ and the cone angle $\theta$. Data for various values of $\theta$ and $D_2/D_1$ are given in Fig. 10.5 and Table 10.2.

**TABLE 10.2**   Resistance coefficient—Gradual enlargement

| $D_2/D_1$ | Angle of Cone. $\theta$ | | | | | | | | | | | |
|---|---|---|---|---|---|---|---|---|---|---|---|---|
| | 2° | 6° | 10° | 15° | 20° | 25° | 30° | 35° | 40° | 45° | 50° | 60° |
| 1.1 | 0.01 | 0.01 | 0.03 | 0.05 | 0.10 | 0.13 | 0.16 | 0.18 | 0.19 | 0.20 | 0.21 | 0.23 |
| 1.2 | 0.02 | 0.02 | 0.04 | 0.09 | 0.16 | 0.21 | 0.25 | 0.29 | 0.31 | 0.33 | 0.35 | 0.37 |
| 1.4 | 0.02 | 0.03 | 0.06 | 0.12 | 0.23 | 0.30 | 0.36 | 0.41 | 0.44 | 0.47 | 0.50 | 0.53 |
| 1.6 | 0.03 | 0.04 | 0.07 | 0.14 | 0.26 | 0.35 | 0.42 | 0.47 | 0.51 | 0.54 | 0.57 | 0.61 |
| 1.8 | 0.03 | 0.04 | 0.07 | 0.15 | 0.28 | 0.37 | 0.44 | 0.50 | 0.54 | 0.58 | 0.61 | 0.65 |
| 2.0 | 0.03 | 0.04 | 0.07 | 0.16 | 0.29 | 0.38 | 0.46 | 0.52 | 0.56 | 0.60 | 0.63 | 0.68 |
| 2.5 | 0.03 | 0.04 | 0.08 | 0.16 | 0.30 | 0.39 | 0.48 | 0.54 | 0.58 | 0.62 | 0.65 | 0.70 |
| 3.0 | 0.03 | 0.04 | 0.08 | 0.16 | 0.31 | 0.40 | 0.48 | 0.55 | 0.59 | 0.63 | 0.66 | 0.71 |
| ∞ | 0.03 | 0.05 | 0.08 | 0.16 | 0.31 | 0.40 | 0.49 | 0.56 | 0.60 | 0.64 | 0.67 | 0.72 |

**Source:**   King, H. W., and E. F. Brater. 1963. *Handbook of Hydraulics,* 5th ed. New York: McGraw-Hill. (Table 6–8.)

The energy loss calculated from Eq. (10–5) does not include the loss due to friction at the walls of the transition. For relatively steep cone angles, the length of the transition is short and therefore the wall friction loss is negligible. However, as the cone angle decreases, the length of the transition increases and wall friction becomes significant. Taking both wall friction loss and the loss due to the enlargement into account, we can obtain the minimum energy loss with a cone angle of about 7°.

☐ **EXAMPLE PROBLEM 10.4**   Determine the energy loss that will occur as 100 L/min of water flows from a 1-in copper tube (Type K) into a 3-in copper tube (Type K) through a gradual enlargement having an included cone angle of 30°.

**Solution**   Using data from Appendix H and the results of some calculations in preceding example problems, we know that

$$v_1 = 3.32 \text{ m/s}$$
$$v_1^2/2g = 0.56 \text{ m}$$
$$D_2/D_1 = 73.8/25.3 = 2.92$$

From Fig. 10.5, we find that $K = 0.48$. Then we have

$$h_L = K(v_1^2/2g) = (0.48)(0.56 \text{ m}) = 0.27 \text{ m}$$

Compared with the sudden enlargement described in Example Problem 10.1, the energy loss decreases by 33 percent when the 30° gradual enlargement is used.

◼

### Diffuser

Another term for an enlargement is a *diffuser*. The function of a diffuser is to convert kinetic energy (represented by velocity head $v^2/2g$) to pressure energy (represented by the pressure head $p/\gamma$) by decelerating the fluid as it flows from the smaller to the larger pipe. The diffuser can be either sudden or gradual but the term is most often used to describe a gradual enlargement.

An ideal diffuser is one in which no energy is lost as the flow decelerates. Of course, no diffuser performs in the ideal fashion. But if it did, the theoretical maximum pressure after the expansion could be computed from Bernoulli's equation.

$$p_1/\gamma + z_1 + v_1^2/2g = p_2/\gamma + z_2 + v_2^2/2g$$

If the diffuser is in a horizontal plane, the elevation terms can be canceled out. Then the pressure increase across the ideal diffuser is,

▷ PRESSURE RECOVERY—
IDEAL DIFFUSER

$$\Delta p = p_2 - p_1 = \gamma(v_1^2 - v_2^2)/2g$$

This is often called *pressure recovery*.

In a *real diffuser*, energy losses do occur and the general energy equation must be used.

$$p_1/\gamma + z_1 + v_1^2/2g - h_L = p_2/\gamma + z_2 + v_2^2/2g$$

The pressure increase becomes,

▷ PRESSURE RECOVERY—
REAL DIFFUSER

$$\Delta p = p_2 - p_1 = \gamma[(v_1^2 - v_2^2)/2g - h_L]$$

The energy loss is computed using the data and procedures in this section. The ratio of the pressure recovery from the real diffuser to that of the ideal diffuser is a measure of the effectiveness of the diffuser.

### 10.7 SUDDEN CONTRACTION

The energy loss due to a sudden contraction, such as that sketched in Fig. 10.6, is calculated from

$$h_L = K(v_2^2/2g) \qquad (10\text{--}6)$$

where $v_2$ is the velocity in the small pipe downstream from the concentration. The resistance coefficient $K$ is dependent on the ratio of the sizes of the two pipes and on the velocity of flow, as Fig. 10.7 and Table 10.3 show.

**FIGURE 10.6**   Sudden contraction.

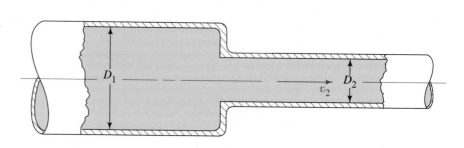

**FIGURE 10.7** Resistance coefficient—Sudden contraction.

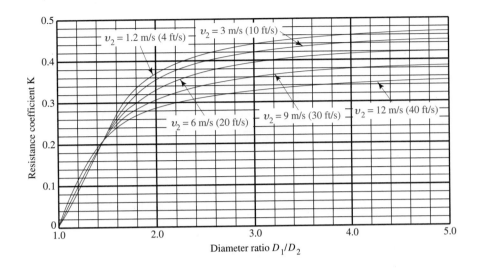

The mechanism by which energy is lost due to a sudden contraction is quite complex. Figure 10.8 illustrates what happens as the flow stream converges. The lines in the figure represent the paths of various parts of the flow stream called *streamlines*. As the streamlines approach the contraction, they assume a curved path and the total stream continues to neck down for some distance beyond the contraction. Thus, the effective minimum cross section of the flow is smaller than that of the smaller pipe. The section where this minimum flow area occurs is called the

**TABLE 10.3** Resistance coefficient—Sudden contraction

| | Velocity, $v_2$ | | | | | | | | |
|---|---|---|---|---|---|---|---|---|---|
| $D_1/D_2$ | 0.6 m/s 2 ft/s | 1.2 m/s 4 ft/s | 1.8 m/s 6 ft/s | 2.4 m/s 8 ft/s | 3 m/s 10 ft/s | 4.5 m/s 15 ft/s | 6 m/s 20 ft/s | 9 m/s 30 ft/s | 12 m/s 40 ft/s |
| 1.0 | 0.0 | 0.0 | 0.0 | 0.0 | 0.0 | 0.0 | 0.0 | 0.0 | 0.0 |
| 1.1 | 0.03 | 0.04 | 0.04 | 0.04 | 0.04 | 0.04 | 0.05 | 0.05 | 0.06 |
| 1.2 | 0.07 | 0.07 | 0.07 | 0.07 | 0.08 | 0.08 | 0.09 | 0.10 | 0.11 |
| 1.4 | 0.17 | 0.17 | 0.17 | 0.17 | 0.18 | 0.18 | 0.18 | 0.19 | 0.20 |
| 1.6 | 0.26 | 0.26 | 0.26 | 0.26 | 0.26 | 0.25 | 0.25 | 0.25 | 0.24 |
| 1.8 | 0.34 | 0.34 | 0.34 | 0.33 | 0.33 | 0.32 | 0.31 | 0.29 | 0.27 |
| 2.0 | 0.38 | 0.37 | 0.37 | 0.36 | 0.36 | 0.34 | 0.33 | 0.31 | 0.29 |
| 2.2 | 0.40 | 0.40 | 0.39 | 0.39 | 0.38 | 0.37 | 0.35 | 0.33 | 0.30 |
| 2.5 | 0.42 | 0.42 | 0.41 | 0.40 | 0.40 | 0.38 | 0.37 | 0.34 | 0.31 |
| 3.0 | 0.44 | 0.44 | 0.43 | 0.42 | 0.42 | 0.40 | 0.39 | 0.36 | 0.33 |
| 4.0 | 0.47 | 0.46 | 0.45 | 0.45 | 0.44 | 0.42 | 0.41 | 0.37 | 0.34 |
| 5.0 | 0.48 | 0.47 | 0.47 | 0.46 | 0.45 | 0.44 | 0.42 | 0.38 | 0.35 |
| 10.0 | 0.49 | 0.48 | 0.48 | 0.47 | 0.46 | 0.45 | 0.43 | 0.40 | 0.36 |
| ∞ | 0.49 | 0.48 | 0.48 | 0.47 | 0.47 | 0.45 | 0.44 | 0.41 | 0.38 |

**Source:** King, H. W., and E. F. Brater, 1963. *Handbook of Hydraulics*, 5th ed. New York: McGraw-Hill. (Table 6–9. Velocities converted to SI units.)

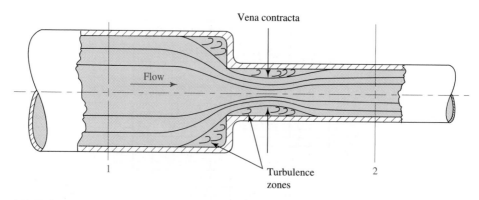

**FIGURE 10.8**   Vena contracta formed in a sudden contraction.

*vena contracta.* Beyond the vena contracta, the flow stream must decelerate and expand again to fill the pipe. The turbulence caused by the contraction and the subsequent expansion generates the energy loss.

☐ **EXAMPLE PROBLEM 10.5**   Determine the energy loss that will occur as 100 L/min of water from a 3-in copper tube (Type K) into a 1-in copper tube (Type K) through a sudden contraction.

*Solution*   From Eq. (10–6), we have

$$h_L = K(v_2^2/2g)$$

For the copper tube we know that $D_1 = 73.8$ mm, $D_2 = 25.3$ mm, and $A_2 = 5.017 \times 10^{-4}$ m². Then we can find the following values:

$$D_1/D_2 = 73.8/25.3 = 2.92$$

$$v_2 = \frac{Q}{A_2} = \frac{100 \text{ L/min}}{5.017 \times 10^{-4} \text{ m}^2} \times \frac{1 \text{ m}^3/\text{s}}{60\,000 \text{ L/min}} = 3.32 \text{ m/s}$$

$$v_2^2/2g = 0.56 \text{ m}$$

From Fig. 10.7 we can find $K = 0.42$. Then we have

$$h_L = K(v_2^2/2g) = (0.42)(0.56 \text{ m}) = 0.24 \text{ m}$$

■

**10.8**
**GRADUAL**
**CONTRACTION**

The energy loss in a contraction can be decreased substantially by making the contraction more gradual. Figure 10.9 shows such a gradual contraction, formed by a conical section between the two diameters with sharp breaks at the junctions. The angle $\theta$ is called the *cone angle.*

Figure 10.10 shows the data (from Reference 5) for the resistance coefficient versus the diameter ratio for several values of the cone angle. The energy loss is computed from Eq. (10–6), where the resistance coefficient is based on the velocity head in a smaller pipe after the contraction. These data are for Reynolds numbers greater than $1.0 \times 10^5$. Note that for angles over the wide range of 15° to 40°, $K = 0.05$ or less, a very low value. For angles as high as 60°, $K$ is less than 0.08.

**FIGURE 10.9** Gradual contraction.

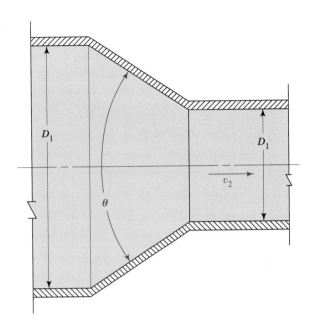

**FIGURE 10.10** Resistance coefficient—Gradual contraction.

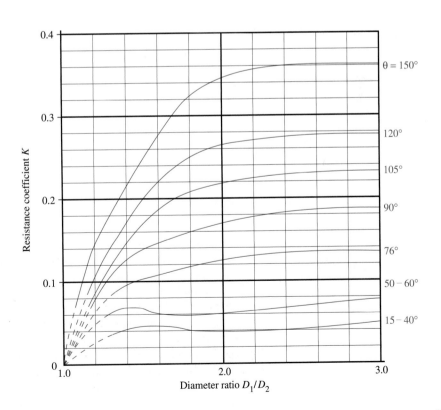

As the cone angle of the contraction decreases below 15 degrees, the resistance coefficient actually increases, as shown in Fig. 10.11. The reason is that the data include the effects of both the local turbulence caused by flow separation and pipe friction. For the smaller cone angles, the transition between the two diameters is very long, which increases the friction losses.

Rounding the end of the conical transition to blend it with the smaller pipe can decrease the resistance coefficient to below the values shown in Fig. 10.10. For example, in Fig. 10.12, which shows a contraction with a 120° included angle and

**FIGURE 10.11**   Resistance coefficient—Gradual contraction.

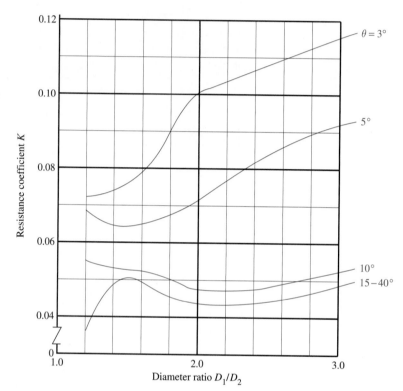

**FIGURE 10.12**   Gradual contraction with rounded end at small diameter.

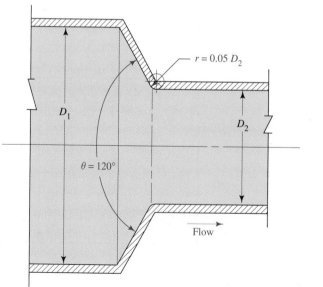

$D_1/D_2 = 2.0$, the value of $K$ decreases from approximately 0.27 to 0.10 with a radius of only $0.50(D_2)$, where $D_2$ is the inside diameter of the smaller pipe.

**10.9
ENTRANCE LOSS**

A special case of a contraction occurs when a fluid flows from a relatively large reservoir or tank into a pipe. The fluid must accelerate from a negligible velocity to the flow velocity in the pipe. The ease with which the acceleration is accomplished determines the amount of energy loss, and therefore, the value of the entrance resistance coefficient is dependent on the geometry of the entrance.

Figure 10.13 shows four different configurations and the suggested value of $K$ for each. The streamlines illustrate the flow of fluid into the pipe and show that the turbulence associated with the formation of a vena contracta in the tube is a major cause of the energy loss. This condition is most severe for the inward-projecting entrance, for which a conservative value of $K = 1.0$ is recommended for problems in this book. Reference 2 recommends $K = 0.78$. A more precise estimate of the resistance coefficient for an inward-projecting entrance is given in Reference 5. For a *well-rounded entrance with* $r/D_2 > 0.15$, no vena contracta is formed, the energy loss is quite small, and we use $K = 0.04$.

**FIGURE 10.13**   Entrance resistance coefficients.

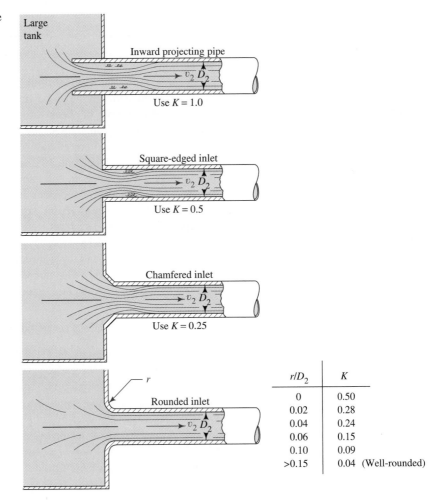

| $r/D_2$ | $K$ |
|---|---|
| 0 | 0.50 |
| 0.02 | 0.28 |
| 0.04 | 0.24 |
| 0.06 | 0.15 |
| 0.10 | 0.09 |
| >0.15 | 0.04  (Well-rounded) |

In summary, after selecting a value for the resistance coefficient from Fig. 10.13, we can calculate the energy loss at an entrance from

$$h_L = K(v_2^2/2g) \qquad (10\text{--}7)$$

where $v_2$ is the velocity of flow in the pipe.

□ **EXAMPLE PROBLEM 10.6**   Determine the energy loss that will occur as 100 L/min of water flows from a reservoir into a 1-in copper tube (Type K) (a) through an inward projecting tube and (b) through a well-rounded inlet.

**Solution**   Part (a): For the tube, $D_2 = 25.3$ mm and $A_2 = 5.017 \times 10^{-4}$ m$^2$. Then we get

$$v_2 = Q/A_2 = 3.32 \text{ m/s} \qquad \text{(from Example Problem 10.1)}$$
$$v_2^2/2g = 0.56 \text{ m}$$

For an inward projecting entrance, $K = 1.0$. Then we have

$$h_L = (1.0)(0.56 \text{ m}) = 0.56 \text{ m}$$

Part (b): For a well-rounded inlet, $K = 0.04$. Then we have

$$h_L = (0.04)(0.56 \text{ m}) = 0.02 \text{ m}$$

■

**10.10
RESISTANCE
COEFFICIENTS FOR
VALVES AND FITTINGS**

Many different kinds of valves and fittings are available from several manufacturers for specification and installation into fluid flow systems. Valves are used to control the amount of flow and may be globe valves, angle valves, gate valves, butterfly valves, any of several types of check valves, and many more. See Figs. 10.14 to 10.21 for some examples. Fittings direct the path of flow or cause a change in the size of the flow path. Included are elbows of several designs, tees, reducers, nozzles, and orifices. See Figs. 10.22 and 10.23.

It is important to determine the resistance data for the particular type and size chosen because the resistance is dependent on the geometry of the valve or fitting. Also, different manufacturers may report data in different forms.

Data reported here are taken from Reference 2, which includes a much more extensive list.

Energy loss incurred as fluid flows through a valve or fitting is computed from Eq. 10–1 as used for the minor losses already discussed. However, the method of determining the resistance coefficient $K$ is different. The value of $K$ is reported in the form

$$K = (L_e/D)f_T \qquad (10\text{--}8)$$

The value of $L_e/D$, called the equivalent length ratio, is reported in Table 10.4 and it is considered to be constant for a given type of valve or fitting. The value of $L_e$ itself is called the equivalent length and is the length of straight pipe of the same nominal diameter as the valve that would have the same resistance as the valve. The term $D$ is the actual inside diameter of the pipe.

The term $f_T$ is the friction factor in the pipe to which the valve or fitting is connected, *taken to be in the zone of complete turbulence*. Note in Fig. 9.2, the Moody diagram, that the zone of complete turbulence lies in the far right area where

**FIGURE 10.14** Globe value.
(Source: Crane Valves, Joliet, IL)

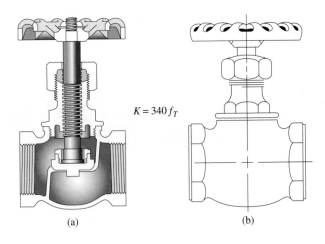

$K = 340\,f_T$

(a)                    (b)

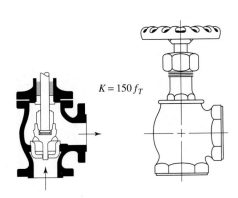

$K = 150\,f_T$

**FIGURE 10.15** Angle valve.
(Source: Crane Valves, Joliet, IL)

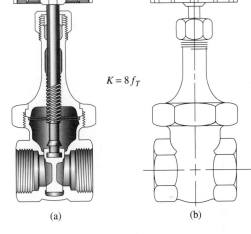

$K = 8\,f_T$

(a)                    (b)

**FIGURE 10.16** Gate valve.
(Source: Crane Valves, Joliet, IL)

**FIGURE 10.17** Check valve—
Swing type. (Source: Crane Valves,
Joliet, IL)

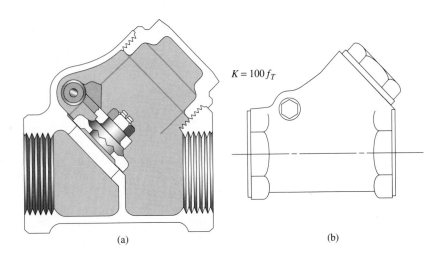

$K = 100\,f_T$

(a)                    (b)

**FIGURE 10.18**  Check valve—Ball type. (Source: Crane Valves, Joliet, IL)

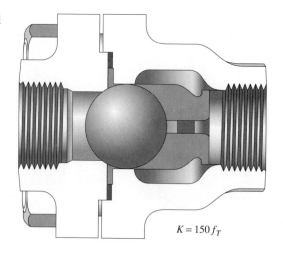

$K = 150\,f_T$

**FIGURE 10.19**  Butterfly valve. (Source: Crane Valves, Joliet, IL)

$K = 45\,f_T$

$K = 420\,f_T$

**FIGURE 10.20**  Foot valve with strainer—poppet disc type. (Source: Crane Valves, Joliet, IL)

**FIGURE 10.21**  Foot valve with strainer—hinged disc. (Source: Crane Valves, Joliet, IL)

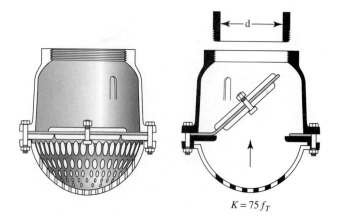

$K = 75\,f_T$

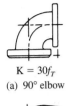

K = 30$f_T$
(a) 90° elbow

K = 20$f_T$
(b) 90° long radius elbow

K = 16$f_T$
(c) 45° elbow

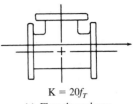

K = 20$f_T$
(a) Flow through run

K = 50$f_T$
(d) 90° street elbow

K = 26$f_T$
(e) 45° street elbow

K = 50$f_T$
(f) Return bend

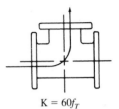

K = 60$f_T$
(b) Flow through branch

**FIGURE 10.22**  Pipe elbows.
(Source: Crane Valves, Joliet, IL)

**FIGURE 10.23**  Standard tees.
(Source: Crane Valves, Joliet, IL)

**TABLE 10.4**  Resistance in valves and fittings expressed as equivalent length in pipe diameters, $L_e/D$

| Type | Equivalent Length in Pipe Diameters, $L_e/D$ |
|---|---|
| Globe valve—fully open | 340 |
| Angle valve—fully open | 150 |
| Gate valve—fully open | 8 |
| —³/₄ open | 35 |
| —¹/₂ open | 160 |
| —¹/₄ open | 900 |
| Check valve—swing type | 100 |
| Check valve—ball type | 150 |
| Butterfly valve—fully open | 45 |
| Foot valve—poppet disc type | 420 |
| Foot valve—hinged disc type | 75 |
| 90° standard elbow | 30 |
| 90° long radius elbow | 20 |
| 90° street elbow | 50 |
| 45° standard elbow | 16 |
| 45° street elbow | 26 |
| Close return bend | 50 |
| Standard tee—with flow through run | 20 |
| —with flow through branch | 60 |

**Source:** Crane Valves, Joliet, IL

the friction factor is independent of Reynolds number. The dashed line running generally diagonally across the diagram divides the zone of complete turbulence from the transition zone to the left.

Values for $f_T$ vary with the size of the pipe and valve, causing the value of the resistance coefficient $K$ to also vary. Table 10.5 lists the values of $f_T$ for standard sizes of new, clean, commercial steel pipe.

**TABLE 10.5**   Friction factor in zone of complete turbulence for new, clean commercial steel pipe

| Nominal Pipe Size (in) | Friction Factor, $f_T$ | Nominal Pipe Size (in) | Friction Factor, $f_T$ |
|---|---|---|---|
| $^1/_2$ | 0.027 | $3^1/_2$, 4 | 0.017 |
| $^3/_4$ | 0.025 | 5 | 0.016 |
| 1 | 0.023 | 6 | 0.015 |
| $1^1/_4$ | 0.022 | 8–10 | 0.014 |
| $1^1/_2$ | 0.021 | 12–16 | 0.013 |
| 2 | 0.019 | 18–24 | 0.012 |
| $2^1/_2$, 3 | 0.018 | | |

Some system designers prefer to compute the equivalent length of pipe for a valve and combine that value with the actual length of pipe. Equation (10–8) can be solved for $L_e$.

$$L_e = K D/f_T \qquad (10\text{–}9)$$

Note, however, that this would be valid only if the flow in the pipe is in the zone of complete turbulence.

---

☐ **EXAMPLE PROBLEM 10.7**   Determine the resistance coefficient $K$ for a fully open globe valve placed in a 6-in Schedule 40 steel pipe.

*Solution*   From Table 10.4 we find that the equivalent length ratio $L_e/D$ for a fully open globe valve is 340. From Table 10.5 we find $f_T = 0.015$ for a 6-in pipe. Then,

$$K = (L_e/D)f_T = (340)(0.015) = 5.10$$

Using $D = 0.5054$ ft for the pipe, the equivalent length is

$$L_e = K D/f_T = (5.10)(0.5054 \text{ ft})/(0.015) = 172 \text{ ft}$$

---

☐ **EXAMPLE PROBLEM 10.8**   Calculate the pressure drop across a fully open globe valve placed in a 4-in Schedule 40 steel pipe carrying 400 gal/min of oil (sg = 0.87).

*Solution*   A sketch of the installation is shown in Fig. 10.24. In order to determine the pressure drop, the energy equation should be written for the flow between points 1 and 2:

$$\frac{p_1}{\gamma} + z_1 + \frac{v_1^2}{2g} + h_L = \frac{p_2}{\gamma} + z_2 + \frac{v_2^2}{2g}$$

**FIGURE 10.24**  Globe valve for
Example Problem 10.8.

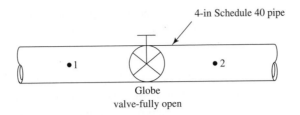

The energy loss $h_L$ is the minor loss due to the valve only. The pressure drop is the difference between $p_1$ and $p_2$. Solving the energy equation for this difference gives

$$p_1 - p_2 = \gamma\left[(z_2 - z_1) + \frac{v_2^2 - v_1^2}{2g} + h_L\right]$$

But $z_1 = z_2$ and $v_1 = v_2$. Then we have

$$p_1 - p_2 = \gamma h_L$$

Equation (10–1) is used to determine $h_L$:

$$h_L = K \times \frac{v^2}{2g} = f_T \times \frac{L_e}{D} \times \frac{v^2}{2g}$$

The velocity $v$ is the average velocity of flow in the 4-in pipe. For the pipe, $D = 0.3355$ ft
and $A = 0.0884$ ft². Then we have

$$v = \frac{Q}{A} = \frac{400 \text{ gal/min}}{0.0884 \text{ ft}^2} \times \frac{1 \text{ ft}^3/\text{s}}{449 \text{ gal/min}} = 10.08 \text{ ft/s}$$

From Table 10.5 we find $f_T = 0.017$ for a 4-in pipe. For the globe valve, $L_e/D = 340$. Then,

$$K = f_T \frac{L_e}{D} = (0.017)(340) = 5.78$$

$$h_L = K \times \frac{v^2}{2g} = (5.78)\frac{(10.08)^2}{(2)(32.2)} \text{ ft} = 9.12 \text{ ft}$$

For the oil, $\gamma = (0.870)(62.4 \text{ lb/ft}^3)$. Then we have

$$p_1 - p_2 = \gamma h_L = \frac{(0.870)(62.4) \text{ lb}}{\text{ft}^3} \times 9.12 \text{ ft} \times \frac{1 \text{ ft}^2}{144 \text{ in}^2}$$

$$p_1 - p_2 = 3.4 \text{ psi}$$

Therefore, the pressure in the oil drops by 3.4 psi as it flows through the valve. Also, an energy loss of 9.12 lb-ft is dissipated as heat from each pound of oil that flows through the
valve.

■

**10.11
APPLICATION OF
STANDARD VALVES**

The preceding section showed several types of valves typically used in fluid distribution systems. Figures 10.14 to 10.21 show drawings and cutaway photographs of
the configuration of these valves. The resistance is heavily dependent on the path
of the fluid as it travels into, through, and out from the valve. A valve with a more
constricted path will cause more energy losses. Therefore, you are advised to select
the valve type with care if you desire the system you are designing to be efficient

with relatively low energy losses. This section describes the general characteristics of the valves shown. You should seek similar data for other types of valves.

### Globe Valve

Figure 10.14 shows the internal construction and the external appearance of the globe valve. Turning the handle causes the sealing device to lift vertically off the seat. It is one of the most common valves and is relatively inexpensive. However, it is one of the poorest performing valves in terms of energy loss. Note that the resistance factor $K$ is:

$$K = f_T \left( L_e / D \right) = 340 f_T$$

This is among the highest of those listed in Table 10.4. It would be used where there is no real problem created by the energy loss. The energy loss occurs because the fluid must travel a complex path from input to output, first traveling upward, then down around the valve seat, then turning again to move to the outlet. Much turbulence is created.

Another use for the globe valve is to *throttle the flow* in a system. The term *throttle* refers to purposely adding resistance to the flow to control the amount of fluid delivered. An example is the simple faucet for a garden hose. You may choose to open the valve completely to get the maximum flow of water to your garden or lawn. But, by partially closing the valve, you could get a lesser flow rate for a more gentle spray or for washing the dog. Partially closing the valve provides more restriction and the pressure drop from the inlet to the outlet increases. The result is less flow.

If the globe valve is used in a commercial pipe line system where throttling is not needed, it would be very wasteful of energy. More efficient valves with lower $L_e / D$ values should be considered.

### Angle Valves

Figure 10.15 shows the external appearance of the angle valve and a sketch of its internal passages. The construction is very similar to the globe valve. However the path is somewhat simpler, as the fluid comes in through the lower port, moves around the valve seat, and turns to exit to the right. The resistance factor $K$ is:

$$K = f_T \left( L_e / D \right) = 150 f_T$$

### Gate Valves

The gate valve in Figure 10.16 is shown in the closed position. Turning the handle lifts the gate vertically out of the flow path. When fully open, there is very little obstruction in the flow path to cause turbulence in the fluid flow stream. Therefore, this is one of the best types of valve for limiting the energy loss. The resistance factor $K$ is:

$$K = f_T \left( L_e / D \right) = 8 f_T$$

In a given installation, the fully open gate valve would have only 2.4% ($8/340 \times 100\%$) of the amount of energy loss as a globe valve. The higher cost of the valve is usually justified by the continuous saving of energy during the lifetime of the system.

The gate valve could be used for throttling by partially closing the valve, bringing the gate back into the flow stream to some degree. Data are given in Table 10.4

for the partially closed positions. Note that it is highly nonlinear and care must be used to obtain the desired flow rate by throttling.

## Check Valves

The function of a check valve is to allow flow in one direction while stopping flow in the opposite direction. A typical use is shown in Figure 10.25 in which a sump pump is moving fluid from a sump below grade to the outside of a home or commercial building to maintain a dry basement area. The pump draws water from the sump and forces it up through the discharge pipe. When the water level in the sump drops to an acceptable level, the pump shuts off. At that time, you would not want the water that was in the pipe to flow back down through the pump and partially refill the sump. The use of a check valve just outside the discharge port of the pump precludes this from happening. The check valve closes immediately when the pressure on the output side exceeds that on the input side.

**FIGURE 10.25**   Sump pump system with check valve.

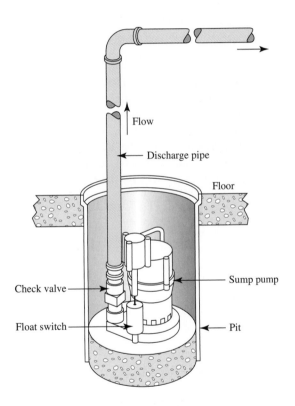

Two kinds of check valves are shown, the ball type and the swing type. There are several other designs available. When open, the swing check provides a modest restriction to the flow of fluid, resulting in the resistance factor of

$$K = f_T (L_e/D) = 100 f_T$$

The ball check causes more restriction as the fluid must flow completely around the ball. However, the ball check is typically smaller and simpler than the swing check. Its resistance is

$$K = f_T (L_e/D) = 150 f_T$$

An important application factor for check valves is that a certain minimum flow velocity is required to cause the valve to completely open. At lower flow rates, the partially open valve would provide more restriction and higher energy losses. Consult the manufacturer's data for the minimum required velocity for a particular type of valve.

### Butterfly Valve

Figure 10.19 shows a cutaway photograph of a typical butterfly valve in which a relatively thin smooth disc pivots about a vertical shaft. When fully open, only the thin dimension of the disc faces the flow, providing only a small obstruction. Closing the valve requires only one-quarter turn of the handle and this is often accomplished by a motorized operator for remote operation. The fully open butterfly valve has a resistance of

$$K = f_T \, (L_e/D) = 45 f_T$$

This value is for the smaller valves from 2 in to 8 in. From 10 in to 14 in, the factor would be $35 f_T$. Larger valves from 16 in to 24 in have a resistance factor of $25 f_T$.

### Foot Valves with Strainers

Foot valves perform a similar function to that of check valves. They are used at the inlet of suction pipes that deliver fluid from a source tank or reservoir to a pump as illustrated in Figure 10.26. They are typically equipped with an integral strainer to keep foreign objects out of the piping system. This is especially necessary when drawing water from an open pit or a natural lake or stream. There may be fish in the lake!

The resistances for the two kinds of foot valves shown are:

$$K = f_T \, (L_e/D) = 420 f_T \qquad \text{Poppet disc type}$$
$$K = f_T \, (L_e/D) = 75 f_T \qquad \text{Hinged disc type}$$

**FIGURE 10.26** Pumping system with foot valve in the suction line.

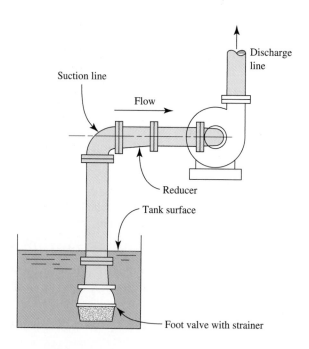

The poppet disc type is similar to the globe valve in internal construction but it is even more constricted. The hinge type is similar to the swing-type check valve. Some extra resistance should be planned for if the strainer could become clogged during service.

**10.12
PIPE BENDS**

It is frequently more convenient to bend a pipe or tube than to install a commercially made elbow. The resistance to flow of a bend is dependent on the ratio of the bend radius $r$ to the pipe inside diameter $D$. Figure 10.27 shows that the minimum resistance for a 90° bend occurs when the ratio $r/D$ is approximately three. The resistance is given in terms of the equivalent length ratio $L_e/D$, and therefore, Eq. (10–8) must be used to calculate the resistance coefficient. The resistance shown in Fig. 10.27 includes both the bend resistance and the resistance due to the length of the pipe in the bend.

When we compute the $r/D$ ratio, $r$ is defined as the radius to the *centerline* of the pipe or tube, called the *mean radius* (see Fig. 10.28). That is, if $R_o$ is the radius to the outside of the bend, $R_i$ is the radius to the inside of the bend, and $D_o$ is the *outside diameter* of the pipe or tube

$$r = R_i + D_o/2$$
$$r = R_o - D_o/2$$
$$r = (R_o + R_i)/2$$

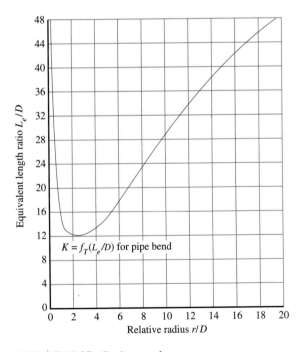

**FIGURE 10.27** Resistance due to 90° pipe bends. (Source: Beij, K. H. Pressure Losses for Fluid Flow in 90 Degree Pipe Bends. *Journal of Research of the National Bureau of Standards* 21 [July 1938]: 1–18.)

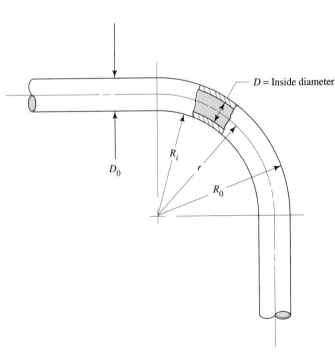

**FIGURE 10.28** 90° pipe bend.

☐ **EXAMPLE PROBLEM 10.9**   A distribution system for liquid propane is made from $1\frac{1}{4}$-in steel tubing with a wall thickness of 0.083 in. Several 90° bends are required to fit the tubes to the other equipment in the system. The specifications call for the radius to the inside of each bend to be 200 mm. When the system carries 160 L/min of propane at 25°C, compute the energy loss to each bend.

**Solution**   Darcy's equation should be used to compute the energy loss with the $L_e/D$ ratio for the bends found from Fig. 10.27. First, let's determine $r/D$, recalling that $D$ is the inside diameter of the tube and $r$ is the radius to the centerline of the tube. From Appendix G we find $D = 27.5$ mm $= 0.0275$ m. The radius $r$ must be computed from

$$r = R_i + D_o/2$$

where $D_o = 31.75$ mm, the outside diameter of the tube as found from Appendix G. Completion of the calculation gives

$$r = 200 \text{ mm} + (31.75 \text{ mm})/2 = 215.9 \text{ mm}$$

and

$$r/D = 215.9 \text{ mm}/27.5 \text{ mm} = 7.85$$

From Fig. 10.27 we find the equivalent length ratio to be 23.

   We now must compute the velocity to complete the evaluation of the energy loss from Darcy's equation:

$$v = \frac{Q}{A} = \frac{160 \text{ L/min}}{5.954 \times 10^{-4} \text{ m}^2} \frac{1.0 \text{ m}^3/\text{s}}{60\,000 \text{ L/min}} = 4.48 \text{ m/s}$$

The relative roughness is

$$D/\epsilon = (0.0275 \text{ m})/(4.6 \times 10^{-5} \text{ m}) = 598$$

Then, we can find $f_T = 0.022$ from Moody's diagram (Fig. 9.2) in the zone of complete turbulence. Then

$$K = f_T \left(\frac{L_e}{D}\right) = K = 0.022(23) = 0.506$$

Now the energy loss can be computed:

$$h_L = K \frac{v^2}{2g} = 5.06 \frac{(4.48)^2}{(2)(9.81)} = 0.517 \text{ m} = 0.517 \text{ N·m/N}$$

∎

## Bends at angles other than 90°

Reference 2 recommends the following formula for computing the resistance factor $K$ for bends at angles other than 90°:

$$K_B = (n - 1)[0.25 \pi f_T (r/D) + 0.5 K] + K \qquad (10\text{-}10)$$

where $K$ is the resistance for one 90° bend found from Figure 10.27. An example of the use of this equation is shown next.

☐ **EXAMPLE PROBLEM 10.10**   Evaluate the energy loss that would occur if the steel tubing described in Example Problem 10.9 is coiled for $4^1/2$ revolutions to make a heat exchanger. The inside radius of the bend is the same 200 mm used earlier and the other conditions are the same.

***Solution***   Let's start by bringing some data from Example Problem 10.9.

$$r/D = 7.85$$

$$f_T = 0.022$$

$$K = 5.06$$

$$v = 4.48 \text{ m/s}$$

Now we can compute the value of $K_B$ for the complete coil using Equation 10–10. Note that each revolution in the coil contains four 90° bends. Then,

$$n = 4.5 \text{ revolutions } (4.0 \text{ 90° bends/rev.}) = 18$$

The total bend resistance $K_B$ is,

$$K_B = (n - 1)[0.25\, \pi f_T\, (r/D) + 0.5\, K] + K$$

$$K_B = (18 - 1)[0.25\, \pi(0.022)(7.85) + 0.5(5.06)] + 5.06$$

$$K_B = 41.9$$

Then the energy loss is found from

$$h_L = K_B(v^2/2g) = 41.9(4.48)^2/[2(9.81)] = 42.8 \text{ N·m/N}$$

■

---

**10.13**
**PRESSURE DROP IN**
**FLUID POWER VALVES**

The field of fluid power encompasses both the flow of liquid hydraulic fluids and air flow systems called *pneumatic systems*. Liquid hydraulic fluids are generally some form of petroleum oil, although many types of blended and synthetic materials can be used. We will refer to the liquid hydraulic fluids simply as oil.

You may be familiar with fluid power systems that operate automation equipment in a production system. They move products through an assembly and packaging system. They actuate forming presses that can exert huge forces. They raise components or products to different elevations, similar to an elevator. They actuate processes to perform a variety of functions such as cutting metal, clamping, slitting, compressing bulk materials, and driving fasteners such as screws, bolts, nuts, nails, and staples.

Another large use is for agricultural and construction equipment. Consider the classic bulldozer that shapes the land for a construction project. The level of the bulldozer's blade is adjusted by the operator using fluid power controls to ensure that the grade of the land meets the design goals. When excess dirt must be removed, a front-end loader is often used to pick it up and dump it into a truck. Numerous hydraulic actuators drive the interesting linkage system that allow the bucket to pick up the dirt and maintain it in a safe position throughout the motion to the truck and then to dump it. The truck is then emptied at another site by actuating hydraulic cylinders to raise the truck bed. In farm work, most modern tractors and harvesting equipment employ hydraulic systems to raise and lower components, to drive rotary motors, and sometimes to even drive the units themselves.

Common elements for a liquid hydraulic system include:

- A pump to provide fluid to a system at an adequate pressure and at the appropriate volume flow rate to accomplish the desired task.
- A tank or reservoir of hydraulic fluid from which the pump draws fluid and to which the fluid is returned after accomplishing the task. Most fluid power systems are closed circuits in which the fluid is continuously circulated.

- One or more directional control valves to manage the flow as it moves through the system.
- Linear actuators, often called hydraulic cylinders, that provide the forces and motion needed to perform the actuation tasks.
- Rotary actuators, called fluid motors, to operate rotating cutting tools, agitators, wheels, linkages, and other devices needing rotary motion.
- Pressure control valves to ensure that an adequate and safe level of pressure exists at all parts of the system.
- Flow control devices that ensure that the correct volume flow rate is delivered to the actuators to provide the proper linear velocity or rotational angular velocity.

Fluid power systems consist of a very wide variety of components arranged in numerous ways to accomplish specific tasks. Also, the systems are inherently *not operating in steady flow* as was assumed in the examples in most of this book. Therefore, different methods of analysis are typically used for fluid power components than for the general-purpose fluid handling devices discussed earlier in the chapter.

However, the principles of energy loss that we discussed still apply. You should be concerned with the energy loss due to any change in direction, change in the size of the flow path, restrictions such as within the valves, and friction as the fluids flow through pipes and tubing.

### Example Fluid Power System

Consider the fluid power system shown in Figure 10.29. The basic purpose and operation of the system are described here.

**FORWARD ACTUATION OF THE LOAD TO THE RIGHT: PART (A) OF THE FIGURE**

- The function of the system is to exert a force of 20 000 lb on a load while providing linear actuation motion to the load. A large part of the force is required to accomplish a forming operation near the end of the stroke.
- An oil hydraulic linear actuator provides the force.
- Fluid is delivered to the actuator by a positive displacement pump that draws the fluid from a tank.
- The fluid leaves the pump and flows to the directional control valve. When it is desired to actuate the load, the flow passes through the valve from the $P$ port to the $A$ delivery port $(P - A)$.
- The flow control valve is placed between the directional control valve and the actuator to permit the system to be adjusted for optimum performance under load.
- The fluid flows into the piston end of the actuator.
- The fluid pressure acts on the face area of the piston exerting the force required to move the load and accomplish the forming operation.
- Simultaneously, the fluid in the rod end of the actuator flows out of the cylinder and proceeds to and through the directional control valve and back to the tank.
- A protection device called a pressure relief valve is placed in the line between the pump and the directional control valve to ensure that the pressure in the system never exceeds the level set by the relief valve. When the pressure rises above the set point, the valve opens and delivers part of the flow back to the tank. Flow can continue through the directional control valve but its pressure will be less than it would have been without the pressure relief valve in the system.

**RETURN ACTUATION OF THE PISTON ROD TO THE LEFT: PART (B) OF THE FIGURE**

The return action takes place with much less force required because the load is relatively light and no forming acting takes place. The sequence proceeds like this.

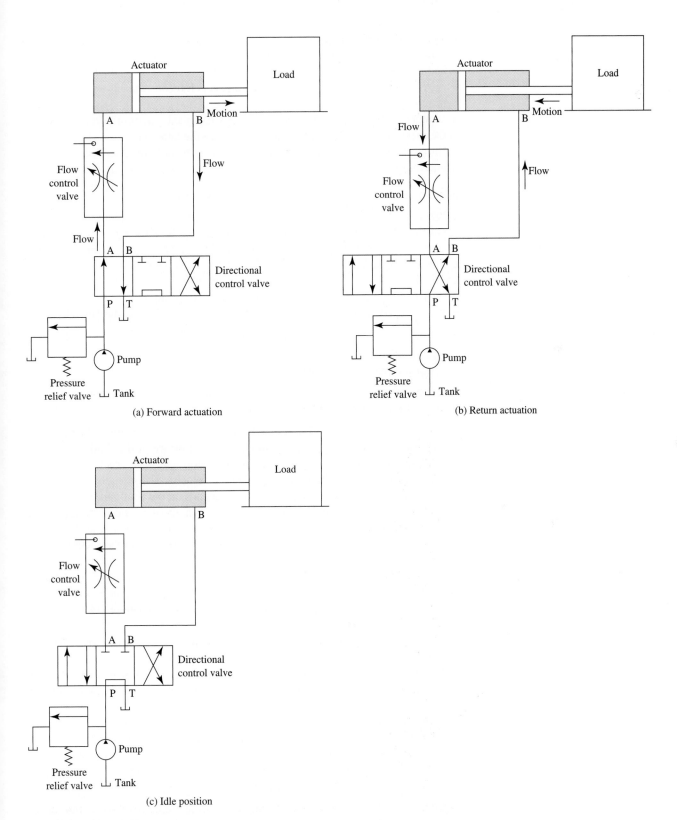

(a) Forward actuation

(b) Return actuation

(c) Idle position

**FIGURE 10.29**  Fluid power system.

- The directional control valve is shifted to the right changing the direction of the flow. The fluid that comes from the pump to the *P* port is directed to the *B* port and thus to the rod end of the actuator.
- As the fluid flows into the cylinder, the piston is forced to the left toward its home position.
- Simultaneously, the fluid in the piston end is forced out from port *A*, passes to the *A* port of the valve and is directed back to the tank.
- Because less pressure is required to accomplish this task, the pressure relief valve does not open.

### IDLE POSITION OF THE SYSTEM: PART (C) OF THE FIGURE

- When the load is returned to its home position, it may be required to idle at that position until some other action has been completed and a signal is received to start a new cycle. To accomplish that, the valve is placed in its center position.
- The flow from the pump is directed immediately to the tank.
- The *A* port and the *B* port are blocked in the valve and thus no flow can come back from the actuator. This holds the actuator in position.
- When the conditions are right for another forming stroke, the directional control valve is switched back to the left and the cycle begins again.

### PRESSURE LEVELS AND ENERGY LOSSES AND ADDITIONS IN THIS FLUID POWER SYSTEM

Let's now identify where energy additions and losses occur in this system and how the pressure levels will vary at critical points.

1. Let's start with the fluid in the tank. Assume that it is at rest and that the tank is vented with atmospheric pressure above the surface of the fluid.
2. As the pump draws fluid, we see that a suction line must accelerate the fluid from its resting condition in the tank to the velocity of flow in the suction line. Thus there will be *an entrance loss* that depends on the configuration of the inlet. The pipe may simply be submerged in the hydraulic fluid or it may have a strainer at the inlet to keep foreign particles out of the pump and the valves.
3. There will be *friction losses in the pipe* as the fluid flows to the suction port of the pump.
4. Along the way, there may be *energy losses in any elbows or bends* in the pipe.
5. We must be concerned with the pressure at the inlet to the pump to ensure that cavitation does not happen and that there is an adequate supply of fluid.
6. The *pump adds energy to the fluid* to cause the flow and to increase the fluid pressure to that required to operate the system. The energy comes from the prime mover, typically an electric motor or an engine. Some of the *input energy is lost because of the volumetric efficiency and mechanical efficiency* of the pump. (See Chapter 7.) Together these combine to produce the overall efficiency defined as follows:

   Overall efficiency, $e_o$ = (Volumetric efficiency, $e_V$)(Mechanical efficiency, $e_M$)

   Input power, $P_I$ = (Power delivered to the fluid)/$e_o$

7. As the fluid leaves the pump and travels to the directional control valve, *friction losses occur in the piping system, including any elbows, tees, or bends in the pipe.* These losses will cause the pressure that appears at the *P* port of the valve to be less than that at the outlet of the pump.

**FIGURE 10.30**  Pressure drop in a directional control valve.

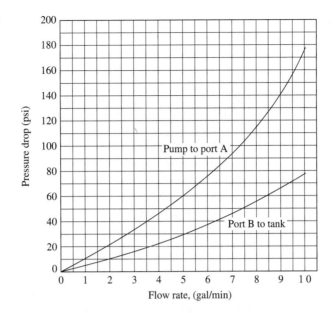

8. If the pressure relief valve has actuated because the pump discharge pressure exceeds the set point of the valve, there will be *pressure drop across this valve.* The pressure actually reduces from the discharge line pressure, $p_d$, to the atmospheric pressure in the tank, $p_T$. Much energy is lost during this process. If we apply the energy equation to the inlet and outlet of the pressure relief valve, we can show that,

$$h_L = (p_d - p_T)/\gamma$$

9. Back at the directional control valve, the fluid passes through the valve from the *P* port to the *A* port. *Energy losses occur in the valve* because the fluid must flow through several restrictions and changes in direction in the ports and around the movable spool in the valve that directs the fluid to the proper outlet port. These energy losses cause a pressure drop across the valve. The amount of the pressure drop is dependent on the design of the valve. Manufacturer's literature will typically include data from which you can estimate the magnitude of the pressure drop. Figure 10.30 shows a typical graph. These graphs are used rather than reporting resistance factors as is done for the standard fluid distribution valves discussed earlier in this chapter.

10. As the fluid flows from the *A* port to the flow control valve, *energy losses occur in the piping* as before.

11. The flow control valve ensures that the flow of fluid into the cylinder at the left end of the actuator is proper to cause the load to be moved at the desired speed. Control is effected by adjustable internal restrictions that can be set during system operation. *The restrictions cause energy losses* and therefore there is a pressure drop across the valve.

12. *Energy is lost at the actuator* as the fluid flows into the left end of the cylinder at *A* and out from the right end at *B*.

13. On the return path, *energy losses occur in the piping system.*

14. More *energy losses occur in the directional control valve* as the fluid passes back through the *B* port and on to the tank. The reasons for these losses are similar to those described in item 9.

This summary identifies fourteen ways in which energy is either added or lost from the hydraulic fluid in this relatively simple fluid power system. Each energy loss results in a pressure drop that could affect the performance of the system.

However, designers of fluid power systems do not always analyze each pressure drop. The transient nature of the operation makes it critical that there is sufficient pressure and flow at the actuator under all reasonable conditions. It is not uncommon for designers to provide extra capacity in the basic system design to overcome unforeseen circumstances. In the circuit just described, the critical pressure drops occur at the pressure relief valve, through the directional control valve, and through the flow control valve. These elements will be analyzed carefully. Other losses will often be only estimated in the initial design. In many cases, the actual configuration of the piping system is not defined during the design process, leaving it to skilled technicians to properly fit the components to the machine. Then when the system is in operation, some fine tuning will be done to ensure proper operation.

This scenario applies most to systems designed for a special purpose when one or only a few systems will be built. When a system is designed for a production application or for a very critical application, more time spent on analysis and optimization of system performance is justified. Examples are aircraft control systems and actuators for construction and agricultural equipment that are made in quantity.

## 10.14 FLOW COEFFICIENTS FOR VALVES USING $C_V$

A large number of manufacturers of valves used for controlling liquids, air, and other gases prefer to rate the performance of the valves using the flow coefficient, $C_V$. One basis for this flow coefficient is that a valve having a coefficient of 1.0 will pass 1.0 gal/min of water at 1.0 psi pressure drop across the valve. The test is convenient to run and gives a reliable means of comparing the overall performance characteristics of different valves.

When used with water, the basic flow equation is,

$$\text{Flow in gal/min} = C_V \sqrt{\Delta p}$$

where $\Delta p$ is in lb/in². $\Delta p$ is called the pressure drop, calculated from $p_U - p_D$, the difference in pressure between points upstream and downstream from the valve. *Be careful to note that $C_V$ is <u>not</u> a unitless factor.*

Data reported in the manufacturers catalog typically list the value of $C_V$ for the valve in the fully open condition. But the valve is often used to control flow rate by partially closing the valve manually or automatically. Therefore, many will report the effective $C_V$ as a function of the number of turns of the valve stem from full closed to full open. Such curves are highly dependent on the construction of the internal parts of the valve, particularly the closure device.

Some valves employ a pointed stem that is moved away from a seat as the valve is opened, progressively expanding the flow area around the stem. This type of valve is called *a needle valve.*

Users of such valves for controlling the flow of air or other gases must account for the compressibility of the gas and the effect of the overall pressure difference across the valve. As discussed in a later chapter on the flow of gases, when the ratio of the upstream to the downstream pressure in a gas reaches the *critical pressure ratio*, no further increase of flow occurs as the downstream pressure is lowered. At the critical pressure ratio, the velocity of flow through the nozzle or valve is equal to the speed of sound in the gas at the local conditions.

Practice Problems

10.1
10.8
10.16
10.28
10.33

301

# REFERENCES

1. Beij, K. H. 1938. Pressure Losses for Fluid Flow in 90 Degree Pipe Bends. *Journal of Research of the National Bureau of Standards* 21:1–18.

2. Crane Co. 1988. *Flow of Fluids Through Valves, Fittings, and Pipe*. Technical paper No. 410. Joliet, Illinois: Author.

3. Dodge L. 1968. How to Compute and Combine Fluid Flow Resistances in Components. *Hydraulics and Pneumatics* 21:118–21.

4. The Hydraulic Institute. 1994. *Engineering Data Book*. Parsippany, NJ: Author.

5. Idelchik, I. E. 1986. *Handbook of Hydraulic Resistance*. 2nd ed. New York: Harper & Row.

6. King, H. W., and E. F. Brater. 1976. *Handbook of Hydraulics*. 6th ed. New York: McGraw-Hill.

7. King, R. C., and Sabin Crocker. 1967. *Piping Handbook*. 5th ed. New York: McGraw-Hill.

8. Simpson, L. L. 1968. Sizing Piping for Process Plants. *Chemical Engineering* 75:193–214.

# PRACTICE PROBLEMS

**10.1M** *Determine the energy loss due to a sudden enlargement from a 50-mm pipe to a 100-mm pipe when the velocity of flow is 3 m/s in the smaller pipe.*

**10.2M** *Determine the energy loss due to a sudden enlargement from a standard 1-in Schedule 80 pipe to a 3$\frac{1}{2}$-in Schedule 80 pipe when the rate of flow is $3 \times 10^{-3}$ m$^3$/s.*

**10.3E** Determine the energy loss due to a sudden enlargement from a standard 1-in Schedule 80 pipe to a 3$\frac{1}{2}$-in Schedule 80 pipe when the rate of flow is 0.10 ft$^3$/s.

**10.4E** Determine the pressure difference between two points on either side of a sudden enlargement from a tube with a 2-in inside diameter to one with a 6-in inside diameter when the velocity of flow of water is 4 ft/s in the smaller tube.

**10.5E** Determine the pressure difference for the conditions in Problem 10.4 if the enlargement was gradual with a cone angle of 15°.

**10.6M** *Determine the energy loss due to a gradual enlargement from a 25-mm pipe to a 75-mm pipe when the velocity of flow is 3 m/s in the smaller pipe and the cone angle of the enlargement is 20°.*

**10.7M** *Determine the energy loss for the conditions in Problem 10.6 if the cone angle is increased to 60°.*

**10.8E** Compute the energy loss for gradual enlargements with cone angles from 2° to 60° in the increments shown in Fig. 10.5. For each case, water at 60°F is flowing at 85 gal/min in a 2-in Schedule 40 steel pipe that enlarges to a 6-in Schedule 40 pipe.

**10.9E** Plot a graph of energy loss versus cone angle for the results of Problem 10.8.

**10.10E** For the data of Problem 10.8, compute the length required to achieve the enlargement for each cone angle. Then compute the energy loss due to friction in that length using the velocity, diameter, and Reynolds number for the midpoint between the ends of the enlargement. Use water at 60°F.

**10.11E** Add the energy loss due to friction from Problem 10.10 to the energy loss for the enlargement from Problem 10.8 and plot the total versus the cone angle on the same graph used in Problem 10.9.

**10.12M** *Another term for an enlargement is a* diffuser. *A diffuser is used to convert kinetic energy ($v^2/2g$) to pressure energy ($p/\gamma$). An ideal diffuser is one in which no energy losses occur and Bernoulli's equation can be used to compute the pressure after the expansion. Compute the pressure after the expansion for an ideal diffuser for the flow of water at 20°C from a 1-in Type K copper tube to a 3-in Type K tube. The volume flow rate is 150 L/min, and the pressure before the expansion is 500 kPa.*

**10.13M** *Compute the resulting pressure after a "real" diffuser in which the energy loss due to the enlargement is considered for the data presented in Problem 10.12. The enlargement is sudden.*

**10.14M** *Compute the resulting pressure after a "real" diffuser in which the energy loss due to the enlargement is considered for the data presented in Problem 10.12. The enlargement is gradual with cone angles of (a) 60°, (b) 30°, and (c) 10°. Compare the results with those of Problems 10.12 and 10.13.*

**10.15M** *Determine the energy loss when 0.04 m$^3$/s of water flows from a 6-in standard Schedule 40 pipe into a large reservoir.*

**10.16E** Determine the energy loss when 1.50 ft$^3$/s of water flows from a 6-in standard Schedule 40 pipe into a large reservoir.

**10.17E** Determine the energy loss when oil with a specific gravity of 0.87 flows from a 4-in pipe to a 2-in pipe through a sudden contraction if the velocity of flow in the larger pipe is 4.0 ft/s.

**10.18E** For the conditions in Problem 10.17, if the pressure before the contraction were 80 psig, calculate the pressure in the smaller pipe.

**FIGURE 10.31**   Problem 10.37.

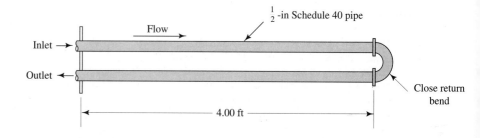

**10.19** True or False: For a sudden contraction with a diameter ratio of 3.0, the energy loss decreases as the velocity of flow increases.

**10.20M** *Determine the energy loss for a sudden contraction from a 5-in Schedule 80 steel pipe to a 2-in Schedule 80 pipe for a flow rate of 500 L/min.*

**10.21M** *Determine the energy loss for a gradual contraction from a 5-in Schedule 80 steel pipe to a 2-in Schedule 80 pipe for a flow rate of 500 L/min. The cone angle for the contraction is 105°.*

**10.22E** Determine the energy loss for a sudden contraction from a 4-in Schedule 80 steel pipe to a 1¹/₂-in Schedule 80 pipe for a flow rate of 250 gal/min.

**10.23E** Determine the energy loss for a gradual contraction from a 4-in Schedule 80 steel pipe to a 1¹/₂-in Schedule 80 pipe for a flow rate of 250 gal/min. The cone angle for the contraction is 76°.

**10.24E** For the data in Problem 10.22, compute the energy loss for gradual contractions with each of the cone angles listed in Figs. 10.10 and 10.11. Plot energy loss versus the cone angle.

**10.25E** For each contraction described in Problems 10.22 and 10.24, make a scale drawing of the device to observe its physical appearance.

**10.26E** Note in Figs. 10.10 and 10.11 that the minimum energy loss for a gradual contraction ($K = 0.04$ approximately) occurs when the cone angle is in the range of 15° to 40°. Make scale drawings of contractions at both of these extremes for a reduction from a 6-in to a 3-in ductile iron pipe.

**10.27E** If the contraction from a 6-in to a 3-in ductile iron pipe described in Problem 10.26 was made with a cone angle of 120°, what would the resulting resistance coefficient be? Make a scale drawing of this reducer.

**10.28E** Compute the energy loss that would occur as 50 gal/min flows from a tank into a steel tube with an outside diameter of 2.0 in and a wall thickness of 0.065 in. The tube is installed flush with the inside of the tank wall with a square edge.

**10.29M** *Determine the energy loss that will occur if water flows from a reservoir into a pipe with a velocity of 3 m/s if*

the configuration of the entrance is (a) an inward projecting pipe (using $K = 1.0$), (b) a square-edged inlet, (c) a chamfered inlet, or (d) a well-rounded inlet.

**10.30M** *Determine the equivalent length in meters of pipe of a fully open globe valve placed in a 10-in Schedule 40 pipe.*

**10.31M** *Repeat Problem 10.30 for a fully open gate valve.*

**10.32E** Calculate the resistance coefficient $K$ for a ball-type check valve placed in a 2-in Schedule 40 steel pipe if water at 100°F is flowing with a velocity of 10 ft/s.

**10.33E** Calculate the pressure difference across a fully open angle valve placed in a 5-in Schedule 40 steel pipe carrying 650 gal/min of oil (sg = 0.90).

**10.34M** *Determine the pressure drop across a 90° standard elbow in a 2¹/₂-in Schedule 40 steel pipe if water at 15°C is flowing at the rate of 750 L/min.*

**10.35M** *Repeat Problem 10.34 for a street elbow.*

**10.36M** *Repeat Problem 10.34 for a long radius elbow. Compare the results from Problems 10.34, 10.35, and 10.36.*

**10.37E** A simple heat exchanger is made by installing a close return bend on two ¹/₂-in Schedule 40 steel pipes as shown in Fig. 10.31. Compute the pressure difference between the inlet and outlet for a flow rate of 12.5 gal/min of ethylene glycol at 77°F.

**10.38E** A proposed alternate form for the heat exchanger described in Problem 10.37 is shown in Fig. 10.32. The entire flow conduit is a ³/₄-in steel tube with a wall thickness of 0.065 in. Note that the inside diameter for this tube is 0.620 in, slightly smaller than that of the ¹/₂-in Schedule 40 pipe ($D = 0.622$ in). The return bend is formed by two 90° bends with a short length of straight tube between them. Compute the pressure difference between the inlet and the outlet of this design and compare it with the system from Problem 10.37.

**10.39E** A piping system for a pump contains a tee, as shown in Fig. 10.33, to permit the pressure at the outlet of the pump to be measured. However, there is no flow into the line leading to the gage. Compute the energy loss as 0.40 ft³/s of water at 50°F flows through the tee.

**10.40M** *A piping system for supplying heavy fuel oil at 25°C is arranged as shown in Fig. 10.34. The bottom leg of the*

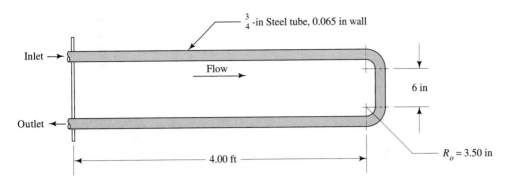

FIGURE 10.32 Problem 10.38.

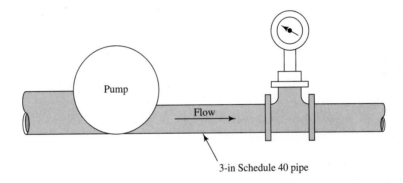

FIGURE 10.33 Problem 10.39.

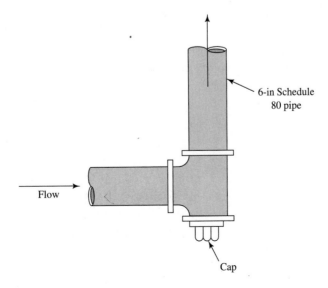

FIGURE 10.34 Problem 10.40.

*tee is normally capped, but the cap can be removed to clean the pipe. Compute the energy loss as 0.08 $m^3/s$ flows through the tee.*

**10.41M** *A 1-in Type K copper tube supplies hot water (80°C) to a washing system in a factory at a flow rate of 250 L/min. At several points in the system, a 90° bend is required. Compute the energy loss in each bend if the radius to the outside of the bend is 300 mm.*

**10.42M** *Specify the radius in mm to the centerline of a 90° bend in a 1-in Type K copper tube to achieve the minimum energy loss. For such a bend carrying 250 L/min of water at 80°C, compute the energy loss. Compare the results with those of Problem 10.41.*

**10.43M** *The inlet and the outlet shown in Fig. 10.35 (a) are to be connected with a 2-in Type K copper tube to carry 750 L/min of propyl alcohol at 25°C. Evaluate the two schemes shown in parts (b) and (c) of the figure with regard to the energy loss. Include the losses due to both the bend and the friction in the straight tube.*

**10.44M** *Compare the energy losses for the two proposals from Problem 10.43 with the energy loss for the proposal in Fig. 10.36.*

**10.45M** *Determine the energy loss that occurs as 40 L/min of water at 10°C flows around a 90° bend in a commercial steel tube having an outside diameter of 3/4 in and a wall thickness of 0.065 in. The radius of the bend to the centerline of the tube is 150 mm.*

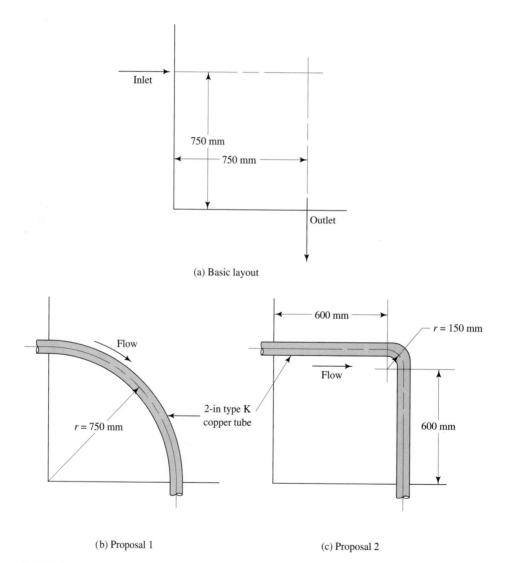

(a) Basic layout

(b) Proposal 1               (c) Proposal 2

**FIGURE 10.35**   Problem 10.43.

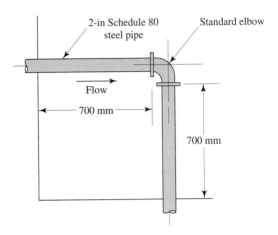

**FIGURE 10.36**   Problem 10.44.

**10.46M** *Figure 10.37 shows a test setup to determine the energy loss due to a heat exchanger. Water at 50°C is flowing vertically upward at $6.0 \times 10^{-3}$ m³/s. Calculate the energy loss between points 1 and 2. Determine the resistance coefficient for the heat exchanger based on the velocity in the inlet pipe.*

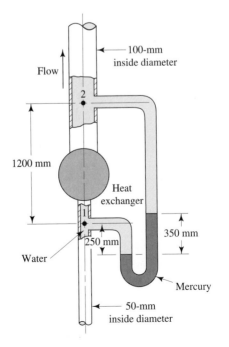

**FIGURE 10.37**   Problem 10.46.

**10.47E** Compute the energy loss in a 90° bend in a steel tube used for a fluid power system. The tube has a ¹/₂-in outside diameter and a wall thickness of 0.065 in. The mean bend radius is 2.00 in. The flow rate of hydraulic oil is 3.5 gal/min.

**10.48E** Compute the energy loss in a 90° bend in a steel tube used for a fluid power system. The tube has a 1¹/₄-in outside diameter and a wall thickness of 0.083 in. The mean bend radius is 3.25 in. The flow rate of hydraulic oil is 27.5 gal/min.

**10.49E** For the data of Problem 10.47, compute the resistance factor and the energy loss for a coil of the given tube that makes six complete revolutions. The mean bend radius is the same, 2.00 in.

**10.50E** For the data of Problem 10.48, compute the resistance factor and the energy loss for a coil of the given tube that makes 8.5 revolutions. The mean bend radius is the same, 3.50 in.

**10.51E** A tube similar to that in Problem 10.47 is being routed through a complex machine. At one point, the tube must be bent through an angle of 145 degrees. Compute the energy loss in the bend.

**10.52E** A tube similar to that in Problem 10.48 is being routed through a complex machine. At one point, the tube must be bent through an angle of 60 degrees. Compute the energy loss in the bend.

**10.53E** A fluid power system incorporates a directional control valve similar to that shown in Figure 10.29(a). Determine the pressure drop across the valve when 5.0 gal/min of hydraulic oil flows through the valve from the pump port to port A.

**10.54E** Repeat Problem 10.53 for flow rates of 7.5 and 10.0 gal/min.

**10.55E** For the data from Problem 10.53, compute the equivalent value of the resistance coefficient $K$ if the pressure drop is found from $\Delta p = \gamma h_L$ and $h_L = K(v^2/2g)$. The oil has a specific gravity of 0.90. The $K$ factor is based on the velocity head in a ⁵/₈ in OD steel tube with a wall thickness of 0.065 in.

**10.56E** Repeat Problem 10.55 for flow rates of 7.5 and 10.0 gal/min.

**10.57E** For the data from Problem 10.53, compute the flow coefficient $C_V$ as defined in Section 10.14.

**10.58E** Repeat Problem 10.57 for flow rates of 7.5 and 10.0 gal/min. (See Problem 10.54.)

## COMPUTER-AIDED ANALYSIS AND DESIGN ASSIGNMENTS

The purpose of the following assignments is to prepare aids that a designer of fluid power systems can use to specify appropriate sizes of steel tubing for a system being designed. Some also help to evaluate energy losses and to ensure that losses due to bends in the tubing are as low as practical.

1. Your company designs special-purpose fluid power systems for the industrial automation market. The normal technique used to fabricate the systems is to route steel tubing among the pumps, control valves, and actuators for the system using straight tubing and 90° bends. Many different sizes of tubing are used in the systems depending on the flow rate of hydraulic oil required for the application. You are asked to create a chart of the recommended bend radii for each nominal size of steel tubing listed in Appendix G. The wall thickness for each size will always be the largest listed in the table because of the high pressures used in the hydraulic systems. According to Figure 10.27, the minimum resistance will occur when the relative radius of the bend is approximately 3.0. Create the table of recommended bend radii, rounding the radii to the nearest $1/2$ in. But ensure that the relative radius of any bend is never less than 2.0. A spreadsheet approach is suggested.

2. Create a spreadsheet that will be a companion of that from Assignment 1 to include a recommendation of the size of discharge line tubing to be used as a function of the volume flow rate of hydraulic oil in the system. The two spreadsheets will be used by designers to select easily the size of tube for a given volume flow rate and to then specify a suitable bend radius to achieve a low energy loss in any bend in the system. Refer to Table 6.3 in Chapter 6 and use the recommended velocity of flow shown in the shaded parts of the table for discharge lines. Note that a range of velocities is given. Use your judgment to determine the appropriate velocity (or more than one) for any size tube within the given range. Round off the computed volume flow rates to convenient numbers in the units of gal/min and report the actual expected velocity in ft/s.

3. For each size of tubing used in Assignment 1, determine the value of $f_T$ for use in the energy loss equation for any minor loss calculation requiring that value for valves, fittings, and bends. See Example Problem 10.8 for an example. You will need to compute the ratio of $D/\epsilon$ for each tube size using the roughness for steel tubing. Then refer to the Moody diagram to determine the friction factor in the fully turbulent zone. List that value within the spreadsheet for Assignment 1 or make a separate spreadsheet for the list.

4. Combine Assignments 1–3 to include the computation of the energy loss for a given bend, using the following process:

   ■ Given a required volume flow rate for a fluid power system, determine an appropriate size for the discharge tubing to produce a velocity of flow in the recommended range.
   ■ For the selected tube size, recommend the bend radius for 90° bends.
   ■ For the selected tube size, determine the value of $f_T$, the friction factor in the fully turbulent range.
   ■ Compute the resistance factor $K$ for the bend from $K = f_T (L_e/D)$.
   ■ Compute the actual velocity of flow for the given volume flow rate in the selected tube size.
   ■ Compute the energy loss in the bend from $h_L = K(v^2/2g)$.

5. Repeat Assignment 1 for each tube size but use the smallest wall thickness rather than the largest. Such tubes could be used for the suction lines that draw the oil from the tank and deliver it to the inlet of the pump. The pressure in suction tubing is very low.

6. Repeat Assignment 2 for each volume flow rate used there. Recommend the size of suction line tubing to achieve the recommended velocity of flow in suction lines.

7. Repeat any of Assignments 1–6 using SI metric data. Volume flow rates are to be in appropriate units assigned by the instructor, such as m³/s, m³/hr, L/s, or L/min. Velocity calculations should be in m/s.

# 11 Series Pipe Line Systems

## 11.1
## The
## Big
## Picture

**Discussion Map**

☐ This chapter is the capstone for the preceding chapters 6 to 10 that considered specific aspects of the flow of fluids in pipes and tubes.

☐ A series pipe line system is one in which the fluid follows a single flow path throughout the system.

☐ You should develop the ability to identify three different classes of series pipe line systems and practice the techniques of analyzing them.

☐ Because most real systems include many different elements, the calculations can become highly involved. You should develop your ability to use computer assisted analysis of fluid flow systems to perform most of the calculations.

**Discover**

■ Review Chapters 6 to 10 to remind yourself of the analytical tools presented there: the continuity equation, the general energy equation, energy losses due to friction, and minor losses.

■ Study the various pipe line systems depicted in Chapter 7 and identify where energy losses occur.

■ Review your Big Picture discussions from Chapters 9 and 10 where you identified energy losses in many kinds of systems.

In this chapter you will learn how to analyze three different classes of series pipe line systems in which the fluid follows a single flow path throughout the system. You will also see some applications of computer assisted analysis of fluid flow systems using spreadsheet

This chapter is a capstone for the preceding chapters, which concerned the flow of fluids in pipes and tubes. We have developed the concepts of fluid flow rate, the continuity equation, Bernoulli's equation, and the general energy equation. Laminar and turbulent flow have been defined, and the Reynolds number has been used to determine the type of flow in a given system. The manner for computing energy losses due to friction has been presented. Also, we have discussed several types of minor losses for the flow of fluids through valves and fittings and for changes in the velocity or direction of the flow.

Of course, real fluid flow systems often contain several minor losses as well as the energy loss due to friction as the fluid is delivered from one point to another. More than one size of pipe may be used. This chapter presents the methods of analysis for real pipe line systems in which the fluid flows through a single continuous path. Such a system is called a *series pipe line system.*

Recall the discussion in the Big Picture section of Chapter 10. There you examined real systems to follow the path of the fluid flow and to identify the kinds of minor losses that occur in the systems. Each of these devices, such as valves, fittings, or changes in the size or direction of the flow path, cause energy loss from the system. The energy is lost in the form of heat dissipated from the fluid. The effect of the loss is to cause the fluid pressure to decrease. The lost energy was first delivered into the system by pumps or because the source was at a higher elevation. Therefore, the loss of energy is wasteful. Lower energy losses generally mean that a smaller pump and motor could be used or a given system could produce a greater output.

System analysis and design problems can be classified into three classes as follows:

*Class I.*    The system is completely defined in terms of the size of pipes, the types of minor losses that are present, and the volume flow rate of fluid in the system. The typical objective is to compute the pressure at some point of interest, to compute the total head on a pump, or to compute the elevation of a source of fluid to produce a desired flow rate or pressure at selected points in the system.

*Class II.*   The system is completely described in terms of its elevations, pipe sizes, valves and fittings, and allowable pressure drop at key points in the system. You desire to know the volume flow rate of the fluid that could be delivered by a given system.

*Class III.*  The general layout of the system is known along with the desired volume flow rate. The size of the pipe required to carry a given volume flow rate of a given fluid is to be determined.

As you study the methods of analyzing and designing these three classes of systems, you should also learn what are the desirable elements of a system. What are the better types of valves to use in given applications? Where are critical points in a system to evaluate pressures? Where should a pump be placed in a system relative to the source of the fluid? What are reasonable velocities of flow in different parts of the systems? Some of these issues were brought up in earlier chapters. Now you will be using them to evaluate the acceptability of a proposed system and to recommend improvements.

**11.2**
**OBJECTIVES**

After completing this chapter you should be able to:

1. Identify series pipe line systems.
2. Determine whether a given system is Class I, Class II, or Class III.
3. Compute the total energy loss, elevation differences, or pressure differences for Class I systems with any combination of pipes, minor losses, pumps, or reservoirs when the system carries a given flow rate.
4. Determine for Class II systems the velocity or volume flow rate through the system with known pressure differences and elevation heads.
5. Determine for Class III systems the size of pipe required to carry a given fluid flow rate with a specified limiting pressure drop or for a given elevation difference.

**11.3**
**CLASS I SYSTEMS**

This chapter deals only with series systems such as the one illustrated in Fig. 11.1. If the energy equation is written for this system, using the surface of each reservoir as the reference points, it would appear as

$$\frac{p_1}{\gamma} + z_1 + \frac{v_1^2}{2g} + h_A - h_L = \frac{p_2}{\gamma} + z_2 + \frac{v_2^2}{2g} \qquad (11\text{--}1)$$

The first three terms on the left side of this equation represent the energy possessed by the fluid at point 1 in the form of pressure head, elevation head, and velocity head. Likewise, the terms on the right side of the equation represent the energy possessed by the fluid at point 2. The term $h_A$ is the energy added to the fluid by a pump. A common name for this energy is *total head on the pump*, and it is used as one of the primary parameters in selecting a pump and in determining its performance. The term $h_L$ indicates the total energy lost from the system anywhere between

**FIGURE 11.1**   Series pipe line
system.

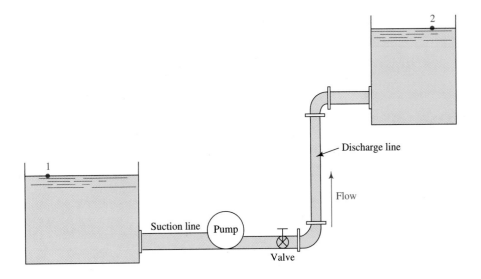

reference points 1 and 2. There are typically several factors that contribute to the
total energy loss. Six different factors apply in this problem.

$$h_L = h_1 + h_2 + h_3 + h_4 + h_5 + h_6 \qquad (11\text{--}2)$$

where        $h_L$ = total energy loss per unit weight of fluid flowing

$h_1$ = entrance loss

$h_2$ = friction loss in the suction line

$h_3$ = energy loss in the valve

$h_4$ = energy loss in the two 90° elbows

$h_5$ = friction loss in the discharge line

$h_6$ = exit loss

In a series pipe line the total energy loss is the sum of the individual minor
losses and all pipe friction losses. This statement is in agreement with the principle
that the energy equation is a means of accounting for all of the energy in the system between the two reference points.

Our approach to the analysis of Class I systems is identical to that used
throughout the previous chapters except that generally many types of energy losses
will exist. The following programmed example problem will illustrate the solution
of a Class I problem.

## PROGRAMMED EXAMPLE PROBLEM

☐ **EXAMPLE PROBLEM 11.1**        Calculate the power supplied to the pump shown in Fig. 11.2 if its efficiency is 76 percent.
Methyl alcohol at 25°C is flowing at the rate of 54.0 m³/hr. The suction line is a standard
4-in Schedule 40 steel pipe, 15 m long. The total length of 2-in Schedule 40 steel pipe in the
discharge line is 200 m. Assume that the entrance from reservoir 1 is through a square-edged
inlet and that the elbows are standard. The valve is a fully open globe valve.

**FIGURE 11.2**   System for Example
Problem 11.1.

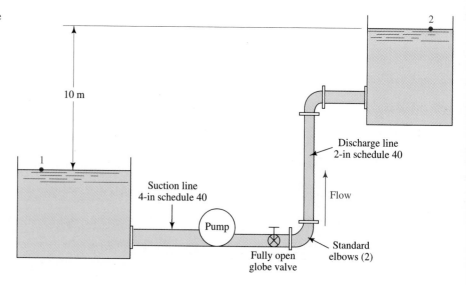

10 m

1

Suction line
4-in schedule 40

Pump

Fully open
globe valve

2

Discharge line
2-in schedule 40

Flow

Standard
elbows (2)

To begin the solution, write the energy equation for the system.

Using the surfaces of the reservoirs as the reference points, you should have

$$\frac{p_1}{\gamma} + z_1 + \frac{v_1^2}{2g} + h_A - h_L = \frac{p_2}{\gamma} + z_2 + \frac{v_2^2}{2g}$$

Since $p_1 = p_2 = 0$ and $v_1$ and $v_2$ are approximately zero, the equation can be simplified to

$$z_1 + h_A - h_L = z_2$$

Because the objective of the problem is to calculate the power supplied to the pump, solve
now for the total head on the pump, $h_A$.

The total head is

$$h_A = z_2 - z_1 + h_L$$

There are six components to the total energy loss. List them and write the formula for evaluating each one.

Your list should include the following items. The subscript $s$ indicates the suction line
and the subscript $d$ indicates the discharge line.

$h_1 = K(v_s^2/2g)$     (entrance loss)

$h_2 = f_s(L/D)(v_s^2/2g)$      (friction loss in suction line)

$h_3 = f_{dT}(L_e/D)(v_d^2/2g)$      (valve)

$h_4 = f_{dT}(L_e/D)(v_d^2/2g)$      (two 90° elbows)

$h_5 = f_d(L/D)(v_d^2/2g)$      (friction loss in discharge line)

$h_6 = 1.0(v_d^2/2g)$      (exit loss)

Since the velocity head in the suction or discharge line is required for each energy loss, calculate these values now.

You should have $v_s^2/2g = 0.17$ m and $v_d^2/2g = 2.44$ m.

$$Q = \frac{54.0 \text{ m}^3}{\text{hr}} \cdot \frac{1 \text{ hr}}{3600 \text{ s}} = 0.015 \text{ m}^3/\text{s}$$

$$v_s = \frac{Q}{A_s} = \frac{0.015 \text{ m}^3}{\text{s}} \times \frac{1}{8.213 \times 10^{-3} \text{ m}^2} = 1.83 \text{ m/s}$$

$$\frac{v_s^2}{2g} = \frac{(1.83)^2}{2(9.81)} \text{ m} = 0.17 \text{ m}$$

$$v_d = \frac{Q}{A_d} = \frac{0.015 \text{ m}^3}{\text{s}} \times \frac{1}{2.168 \times 10^{-3} \text{ m}^2} = 6.92 \text{ m/s}$$

$$\frac{v_d^2}{2g} = \frac{(6.92)^2}{2(9.81)} \text{ m} = 2.44 \text{ m}$$

To determine the friction losses in the suction line and the discharge line, and the minor losses in the discharge line, we need the Reynolds number, relative roughness, and friction factor for each pipe, and the friction factor in the zone of complete turbulence for the discharge line that contains a valve and pipe fittings. Find these values now.

---

For methyl alcohol at 25°C, $\rho = 789$ kg/m$^3$ and $\mu = 5.60 \times 10^{-4}$ Pa·s. Then, in the suction line, we have

$$N_R = \frac{vD\rho}{\mu} = \frac{(1.83)(0.1023)(789)}{5.60 \times 10^{-4}} = 2.64 \times 10^5$$

Because the flow is turbulent, the value of $f_s$ must be evaluated from the Moody diagram, Fig. 9.2. For steel pipe, $\epsilon = 4.6 \times 10^{-5}$ m.

$$D/\epsilon = 0.1023/(4.6 \times 10^{-5}) = 2224$$
$$N_R = 2.64 \times 10^5$$

$f = .0295$

Then $f_s = 0.018$.
In the discharge line, we have

$$N_R = \frac{vD\rho}{\mu} = \frac{(6.92)(0.0525)(789)}{5.60 \times 10^{-4}} = 5.12 \times 10^5$$

This flow is also turbulent. Evaluating the friction factor $f_d$ gives

$$D/\epsilon = 0.0525/(4.6 \times 10^{-5}) = 1141$$
$$N_R = 5.12 \times 10^5$$
$$f_d = 0.020$$

We can find from Table 10.5 that $f_{dT} = 0.019$ for the 2-in discharge pipe in the zone of complete turbulence.

Returning now to the energy loss calculations, evaluate $h_1$, the entrance loss, in N·m/N or m.

---

The result is $h_1 = 0.09$ m. For a square-edged inlet, $K = 0.5$ and

$$h_1 = 0.5(v_s^2/2g) = (0.5)(0.17 \text{ m}) = 0.09 \text{ m}$$

Now calculate $h_2$, the friction loss in the suction line.

---

The result is $h_2 = 0.45$ m.

$$h_2 = f_s \times \frac{L}{D} \times \frac{v_s^2}{2g} = f_s \left( \frac{15}{0.1023} \right)(0.17) \text{ m}$$

$$h_2 = (0.018)\left( \frac{15}{0.1023} \right)(0.17) \text{ m} = 0.45 \text{ m}$$

Now calculate $h_3$, the energy loss in the valve in the discharge line.

---

From the data in Chapter 10, the equivalent length ratio $L_e/D$ for a fully open globe valve is 340. The friction factor is $f_{dT} = 0.019$.
Then we have

$$h_3 = f_{dT} \times \frac{L_e}{D} \times \frac{v_d^2}{2g} = (0.019)(340)(2.44) \text{ m} = 15.76 \text{ m}$$

Now calculate $h_4$, the energy loss in the two 90° elbows.

---

For standard 90° elbows, $L_e/D = 30$. The value of $f_{dT}$ is 0.019, the same as that used in the preceding panel. Then we have

$$h_4 = 2f_{dT} \times \frac{L_e}{D} \times \frac{v_d^2}{2g} = (2)(0.019)(30)(2.44) \text{ m} = 2.78 \text{ m}$$

Now calculate $h_5$, the friction loss in the discharge line.

---

The discharge line friction loss is

$$h_5 = f_d \times \frac{L}{D} \times \frac{v_d^2}{2g} = (0.020)\left( \frac{200}{0.0525} \right)(2.44) \text{ m} = 185.9 \text{ m}$$

Now calculate $h_6$, the exit loss.

---

The exit loss is

$$h_6 = 1.0(v_d^2/2g) = 2.44 \text{ m}$$

This concludes the calculation of the individual energy losses. The total loss $h_L$ can now be determined.

---

$$h_L = h_1 + h_2 + h_3 + h_4 + h_5 + h_6$$

$$h_L = (0.09 + 0.45 + 15.76 + 2.78 + 185.9 + 2.44) \text{ m}$$

$$h_L = 207.4 \text{ m}$$

From the energy equation the expression for the total head on the pump was found to be

$$h_A = z_2 - z_1 + h_L$$

Then we have

$$h_A = 10 \text{ m} + 207.4 \text{ m} = 217.4 \text{ m}$$

Now calculate the power supplied to the pump, $P_A$.

---

$$\text{Power} = \frac{h_A \gamma Q}{e_M} = \frac{(217.4 \text{ m})(7.74 \times 10^3 \text{ N/m}^3)(0.015 \text{ m}^3/\text{s})}{0.76}$$

$$P_A = 33.2 \times 10^3 \text{ N·m/s} = 33.2 \text{ kW}$$

This concludes the programmed example problem.

■

## Critique of the System Shown in Fig. 11.2 and Analyzed in Example Problem 11.1

Problem solutions such as that just concluded can give you, the fluid flow system designer, much useful information on which you can evaluate the proposed design and make rational decisions about system improvement. The following are some observations:

1. The length of the suction line between the first reservoir and the pump, given to be 15 m, appears to be excessively long. We recommended that the pump be relocated closer to the reservoir so the suction line can be as short as practical. This is especially important when low pressure is observed at the inlet to the pump as will be discussed in Chapter 13 on pump application.

2. It may be desirable to place a valve in the suction line before the inlet to the pump to allow the pump to be removed or serviced without draining the reservoir. A gate valve should be used so the energy loss is small during normal operation with the valve completely open.

3. The size of the suction line is acceptable. Note that the velocity is 1.83 m/s. Table 6.3 recommends that the velocity at the inlet to a pump be approximately 1.0 m/s to 6.0 m/s.

4. The 2-in discharge line is much too small. This is the most serious problem with the design of the system in Figure 11.1. The velocity in the pipe is 6.92 m/s and the velocity head is 2.44 m. Referring to Table 6.3, we see that the velocity is outside the recommended range for the volume flow rate of 54.0 m³/hr. Furthermore, note that the energy loss in the discharge line almost 186 m and that it is the largest contributor to the total head on the pump. Because the length of the discharge line is very long at 200 m, it is desirable to reduce the velocity significantly. Let's specify a 3-in Schedule 40 steel pipe, which would reduce the velocity to 3.15 m/s and the velocity head is reduced to 0.504 m, *a reduction by almost five times!* The energy loss will be reduced proportionally.

5. The globe valve in the discharge line should be replaced by a type with less resistance. The equivalent length ratio $L_e/D$ of 340 is among the highest of any kind of valve. A fully open gate valve has $L_e/D = 8$, *a reduction of over 42 times!*

## Summary of Design Changes

The following changes are proposed:

1. Decrease the suction line length from 15 m to 1.5 m. Assuming that the two reservoirs must stay in the same position, the extra 13.5 m of length will be added to the discharge line, making it a total of 213.5 m long.

2. Add a fully open gate valve in the suction line.

3. Increase the discharge line size from 2-in to 3-in Schedule 40. Then, $v_d=$ 3.15 m/s and the velocity head is 0.504 m.

4. Replace the globe valve in the discharge line with a fully open gate valve.

Making all of these changes would result in the reduction of the energy to be added by the pump from 217.4 m to 37.9 m. The power supplied to the pump would decrease from 33.2 kW to 5.8 kW, *a reduction of almost a factor of 6!*

## 11.4 SPREADSHEET AID FOR CLASS I PROBLEMS

The solution procedure for Class I series pipe line problems is direct in that the system is completely defined and the analysis leads to the final solution with no iteration or estimates of values. But it is a cumbersome procedure requiring many calculations. If several systems are to be designed or if the designer wants to try several modifications to a given design, it can take much time.

The use of a spreadsheet can improve the procedure dramatically by doing most of the calculations for you after you enter the basic data.

Figure 11.3 shows one approach. It is designed to model a system similar to that shown in Figure 11.2, in which a pump draws fluid from some source and delivers it to a destination point. The data shown are from Example Problem 11.1, where the objective was to compute the power required to drive the pump. Compare the values in the spreadsheet with those found in the example problem. The minor differences are mostly due to rounding and the fact that the friction factors are computed by the spreadsheet whereas they were read manually from the Moody Diagram for the example problem.

The spreadsheet is somewhat more versatile, however. Its features are explained below.

### FEATURES OF THE SPREADSHEET TO COMPUTE THE POWER REQUIRED BY A PUMP IN A CLASS I SERIES PIPE LINE SYSTEM (SI METRIC UNITS VERSION)

1. Data that you must enter in appropriate cells are identified by the shaded areas.

2. At the top left of the sheet, you can enter the identification information for the system.

3. At the top right, you enter the description of the two reference points for use in the energy equation.

4. Then enter the system data. First the volume flow rate $Q$ is entered in the units of m$^3$/s. Then the pressures and elevations at both reference points are entered. In the example problem, the pressures are zero because both reference points are at the free surface of the reservoirs. The reference elevation is taken at the centerline of the suction line. Therefore the elevation of point 1 is 2.0 m and at point 2 it is 12.0 m.

5. Carefully study the required velocity data. In the example problem, the velocity at both reference points is zero because they are at the free, still surface of the reservoirs. The zero values were entered manually. But if either or both of the reference points are in a pipe instead of at the surface of a reservoir, actual pipe velocities are needed. The instruction to the right side of the spreadsheet calls for you to actually type a cell reference for the velocities. The cell reference "B20" refers to the cell where the velocity of flow in Pipe 1 is computed below. The cell reference "E20" is for the cell where the velocity of flow for Pipe 2 is computed. Then, after the proper data for the pipes are entered, the correct velocity and velocity head values will appear in the system data cells.

6. Enter the fluid properties data next. The specific weight, $\gamma$, and the kinematic viscosity, $\nu$, are needed to compute the Reynolds number and the power

| APPLIED FLUID MECHANICS | CLASS I SERIES SYSTEMS | |
|---|---|---|
| **Objective: *Pump power*** | *Reference points for the energy equation:* | |
| Example Problem 11.1 | Point 1: At surface of lower reservoir | |
| Figure 11.2 | Point 2: At surface of upper reservoir | |

| *System Data:* | **SI Metric Units** | | |
|---|---|---|---|
| Volume flow rate: $Q =$ | $0.015\ m^3/s$ | Elevation at point 1 = | $0\ m$ |
| Pressure at point 1 = | $0\ kPa$ | Elevation at point 2 = | $10\ m$ |
| Pressure at point 2 = | $0\ kPa$ | *If Ref. pt. is in pipe: Set v1 "= B20" OR Set v2 "= E20"* | |
| Velocity at point 1 = | $0\ m/s \rightarrow$ | Vel head at point 1 = | $0\ m$ |
| Velocity at point 2 = | $0\ m/s \rightarrow$ | Vel head at point 2 = | $0\ m$ |

| *Fluid Properties:* | | May need to compute $\nu = \mu/\rho$ | |
|---|---|---|---|
| Specific weight = | $7.74\ kN/m^3$ | Kinematic viscosity = $7.10E\text{-}07\ m^2/s$ | |

| **Pipe 1:** | | **Pipe 2:** | | |
|---|---|---|---|---|
| Diameter: $D =$ | $0.1023\ m$ | Diameter: $D =$ | $0.0525\ m$ | |
| Wall roughness: $\epsilon =$ 4.60E-05 m | | Wall roughness: $\epsilon =$ 4.60E-05 m | | [See Table 9.1] |
| Length: $L =$ | 15 m | Length: $L =$ | 200 m | |
| Area: $A =$ 8.22E-03 $m^2$ | | Area: $A =$ 2.16E-03 $m^2$ | | [$A = \pi D^2/4$] |
| $D/\epsilon =$ | 2224 | $D/\epsilon =$ | 1141 | Relative roughness |
| $L/D =$ | 147 | $L/D =$ | 3810 | |
| Flow velocity = | 1.82 m/s | Flow velocity = | 6.93 m/s | [$v = Q/A$] |
| Velocity head = | 0.170 m | Velocity head = | 2.447 m | [$v^2/2g$] |
| Reynolds No. = 2.63E+05 | | Reynolds No. = 5.13E+05 | | [$N_R = vD/\nu$] |
| Friction factor: $f =$ | 0.0182 | Friction factor: $f =$ | 0.0198 | Using Eq. 9-9 |

| **Energy losses in Pipe 1:** | | Qty. | | | |
|---|---|---|---|---|---|
| Pipe: $K_1 = f(L/D) =$ | 2.67 | 1 | Energy loss $h_{L1}$ = | 0.453 m | Friction |
| Entrance loss: $K_2 =$ | 0.50 | 1 | Energy loss $h_{L2}$ = | 0.085 m | |
| Element 3: $K_3 =$ | 0.00 | 1 | Energy loss $h_{L3}$ = | 0.000 m | |
| Element 4: $K_4 =$ | 0.00 | 1 | Energy loss $h_{L4}$ = | 0.000 m | |
| Element 5: $K_5 =$ | 0.00 | 1 | Energy loss $h_{L5}$ = | 0.000 m | |
| Element 6: $K_6 =$ | 0.00 | 1 | Energy loss $h_{L6}$ = | 0.000 m | |
| Element 7: $K_7 =$ | 0.00 | 1 | Energy loss $h_{L7}$ = | 0.000 m | |
| Element 8: $K_8 =$ | 0.00 | 1 | Energy loss $h_{L8}$ = | 0.000 m | |

| **Energy losses in Pipe 2:** | | Qty. | | | |
|---|---|---|---|---|---|
| Pipe: $K_1 = f(L/D) =$ | 75.35 | 1 | Energy loss $h_{L1}$ = | 184.40 m | Friction |
| Globe valve: $K_2 =$ | 6.46 | 1 | Energy loss $h_{L2}$ = | 15.81 m | |
| 2 std elbows: $K_3 =$ | 0.57 | 2 | Energy loss $h_{L3}$ = | 2.79 m | |
| Exit loss: $K_4 =$ | 1.00 | 1 | Energy loss $h_{L4}$ = | 2.45 m | |
| Element 5: $K_5 =$ | 0.00 | 1 | Energy loss $h_{L5}$ = | 0.00 m | |
| Element 6: $K_6 =$ | 0.00 | 1 | Energy loss $h_{L6}$ = | 0.00 m | |
| Element 7: $K_7 =$ | 0.00 | 1 | Energy loss $h_{L7}$ = | 0.00 m | |
| Element 8: $K_8 =$ | 0.00 | 1 | Energy loss $h_{L8}$ = | 0.00 m | |

| | | |
|---|---|---|
| | Total energy loss $h_{Ltot}$ = | 205.98 m |
| **Results:** | Total head on pump: $h_A$ = | 216.0 m |
| | Power added to fluid: $P_A$ = | 25.08 kW |
| | *Pump efficiency =* | *76.00 %* |
| | Power input to pump: $P_I$ = | 32.99 kW |

**FIGURE 11.3** Spreadsheet for Class I series pipe line systems. Data for Example Problem 11.1.

required by the pump. Note that you must compute kinematic viscosity from $\nu = \mu/\rho$ if you originally know only the dynamic viscosity, $\mu$, and the density of the fluid, $\rho$.

7. Pipe data are now entered. Provisions are made for systems with two different pipe sizes such as those in the example problem. It is typical for pumped systems to have a larger suction pipe and a smaller discharge pipe. For each, you must enter in the shaded areas the flow diameter, the wall roughness, and the total length of straight pipe. The system then computes the values in the unshaded areas. Note that the friction factors are computed using the Swamee-Jain equation from Chapter 9, Equation 9–5.

8. The energy losses are addressed next in the spreadsheet. The energy loss is computed using the appropriate resistance factor $K$ for each element. $K$ for pipe friction is computed automatically. For minor losses you will have to obtain values from charts or compute them as described below. These are entered in the shaded areas and brief descriptions of each element can be listed. Room for eight losses in each of two pipes is provided. Values for cells not used should be entered as zero. Recall from Chapters 9 and 10:

- For pipe friction: $K = f(L/D)$, where $f$ is the friction factor, $L$ is the length of straight pipe, and $D$ is the flow diameter of the pipe. These data values were computed in the pipe data section so this value is automatically computed by the spreadsheet.

- For minor losses due to changes in the size of the flow path, refer to Sections 10.4 through 10.9 for values for $K$. It is essential that these values be entered for the proper pipe. You must note which velocity is used as the reference velocity for the given type of minor loss. $K$ factors for enlargements and contractions are based on the velocity head in the smaller pipe.

- For minor losses due to valves, fittings, and bends: $K = f_T(L_e/D)$, where $f_T$ is the friction factor in the fully turbulent zone for the size and type of pipe to which the element is connected. Table 10.5 is the source of such data for steel pipe. For other types of pipe or tubing, the method shown in Example Problem 10.9 should be used. The relative roughness $D/\epsilon$ is used to find the value of $f$ in the zone of complete turbulence from the Moody diagram. The values for the equivalent length ratio, $L_e/D$, can be found in Table 10.4 or in Figure 10.27.

9. The results are computed automatically at the bottom of the sheet. The total energy loss is the sum of all pipe friction and minor losses in both pipes.

10. The total head on the pump, $h_A$, is found by solving the general energy equation for that value:

$$h_A = \frac{p_B - p_A}{\gamma} + (z_B - z_A) + \frac{v_B^2 - v_A^2}{2g} + h_L$$

The spreadsheet makes the necessary calculations using data from appropriate cells in the upper part of the sheet.

11. The power added to the fluid is computed from

$$P_A = h_A \gamma Q$$

12. The pump efficiency, $e_M$, must be entered as a percentage.

13. The power input to the pump is computed from

$$P_I = P_A/e_M$$

Other types of Class I series pipe line problems can be analyzed in a similar manner by adjusting this form. Different sheets for different unit systems should be created because certain unit-specific constants, such as $g = 9.81$ m/s², are used in this version.

For example, if the objective of the problem is to compute the pressure at a particular upstream point $A$ when the pressure is known at a downstream reference point $B$, the energy equation can be solved for the upstream pressure as:

$$p_A = p_B + \gamma \left[ (z_B - z_A) + \frac{v_B^2 - v_A^2}{2g} + h_L \right]$$

You must configure the spreadsheet to evaluate these terms as the final result. Note that it is assumed that no pump or fluid motor is in the system.

**11.5**
**CLASS II SYSTEMS**

A Class II series pipe line system is one for which you desire to know the volume flow rate of the fluid that could be delivered by a given system. The system is completely described in terms of its elevations, pipe sizes, valves and fittings, and allowable pressure drop at key points in the system.

You know that pressure drop is directly related to the energy loss in the system and that the energy losses are typically proportional to the velocity head of the fluid as it flows through the system. Because velocity head is $v^2/2g$, the energy losses are proportional to the square of the velocity. Your task as the designer is to determine how high the velocity can be and still meet the goal of a limited pressure drop.

We will suggest three different approaches to designing Class II systems. They vary in their complexity while also varying in the degree of precision of the final result. The following list gives the type of system for which each method is used and a brief overview of the method. More details for each method are presented within Example Problems 11.2, 11.3, and 11.4.

### Method II-A

Used for a series system in which only pipe friction losses are considered. This is a direct solution process using an equation that is based on the work of Swamee and Jain (Reference 4) that includes the direct computation of the friction factor. See Example Problem 11.2.

### Method II-B

Used for a series system in which relatively small minor losses exist along with a relatively large pipe friction loss. This method adds steps to the process of Method II-A. Minor losses are initially neglected and the same equation used in Method II-A is used to estimate the allowable velocity and volume flow rate. Then a modestly lower volume flow rate is decided upon, the minor losses are introduced, and the system is analyzed as a Class I system to determine the final performance at the specified flow rate. If the performance is satisfactory, the problem is finished. If not, different volume flow rates can be tried until satisfactory results are obtained. See the spreadsheet for Example Problem 11.3. This method requires some trial and error but the process goes quickly once the data are entered into the spreadsheet.

### Method II-C

Used for a series system in which minor losses are significant in comparison with the pipe friction losses and for which a high level of precision in the analysis is desired. This method is the most time-consuming, requiring an algebraic analysis of the behavior of the entire system and the expression of the velocity of flow in terms of the friction factor in the pipe. Both of these quantities are unknown because the friction factor also depends on velocity (Reynolds number). An *iteration* process is used to complete the analysis. Iteration is a controlled "trial-and-error" method in which each step of iteration yields a more accurate estimate of the limiting velocity of flow to meet the pressure drop limitation. The process typically converges in two to four iterations. See Example Problem 11.4.

☐ **EXAMPLE PROBLEM 11.2**

A lubricating oil must be delivered through a horizontal 6-in Schedule 40 steel pipe with a maximum pressure drop of 60 kPa per 100 m of pipe. The oil has a specific gravity of 0.88 and a dynamic viscosity of $9.5 \times 10^{-3}$ Pa s. Determine the maximum allowable volume flow rate of oil.

**Solution**

Figure 11.4 shows the system. This is a Class II series pipe line problem because the volume flow rate is unknown and, therefore, the velocity of flow is unknown. Method II-A is used here because only pipe friction losses exist in the system.

**FIGURE 11.4**  Reference points in pipe for Example Problem 11.2.

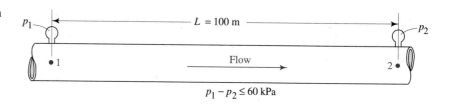

$$p_1 - p_2 \leq 60 \text{ kPa}$$

*Step 1.* Write the energy equation for the system.

*Step 2.* Solve for the limiting energy loss, $h_L$.

*Step 3.* Determine the following values for the system:

   Pipe flow diameter, $D$

   Relative roughness, $D/\epsilon$

   Length of pipe, $L$

   Kinematic viscosity of the fluid, $\nu$. May requiring using $\nu = \mu/\rho$.

*Step 4.* Use Equation 11–3 to compute the limiting volume flow rate, ensuring that all data are in the coherent units of the given system.

$$Q = -2.22 \, D^2 \sqrt{\frac{gDh_L}{L}} \log\left[\frac{1}{3.7D/\epsilon} + \frac{1.784 \, \nu}{D\sqrt{gDh_L/L}}\right] \qquad (11\text{--}3)$$

**Results**

We use points 1 and 2 shown in Figure 11.3 to write the energy equation.

$$\frac{p_1}{\gamma} + z_1 + \frac{v_1^2}{2g} - h_L = \frac{p_2}{\gamma} + z_2 + \frac{v_2^2}{2g}$$

We can cancel some terms because $z_1 = z_2$ and $v_1 = v_2$. The equation then becomes,

$$\frac{p_1}{\gamma} - h_L = \frac{p_2}{\gamma}$$

Then we solve algebraically for $h_L$ and evaluate the result.

$$h_L = \frac{p_1 - p_2}{\gamma} = \frac{60 \text{ kN}}{\text{m}^2} \times \frac{\text{m}^3}{(0.88)(9.81 \text{ kN})} = 6.95 \text{ m}$$

Other data needed are:

> Pipe flow diameter, $D = 0.1541$ m   [Appendix F]
>
> Pipe wall roughness, $\epsilon = 4.6 \times 10^{-5}$ m   [Table 9.1]
>
> Relative roughness, $D/\epsilon = (0.1541$ m$)/(4.6 \times 10^{-5}$ m$) = 3350$
>
> Length of pipe, $L = 100$ m
>
> Kinematic viscosity of the fluid. Use $\rho = (0.88)(1000$ kg/m$^3) = 880$ kg/m$^3$
>
> $$\nu = \mu/\rho = (9.5 \times 10^{-3} \text{ Pa·s})/(880 \text{ kg/m}^3) = 1.08 \times 10^{-5} \text{ m}^2/\text{s}$$

We place these values into Equation 11–3, ensuring that all data are in coherent SI units for this problem.

$$Q = -2.22(0.1541)^2 \sqrt{\frac{(9.81)(0.1541)(6.95)}{100}}$$

$$\log\left[\frac{1}{(3.7)(3350)} + \frac{(1.784)(1.08 \times 10^{-5})}{(0.1541)\sqrt{(9.81)(0.1541)(6.95)/100}}\right]$$

$$Q = 0.057 \text{ m}^3/\text{s}$$

**Comment**   Thus, if the volume flow rate of oil through this pipe is no greater than 0.057 m³/s, the pressure drop over a 100 m length of the pipe will be no greater than 60 kPa.

■

### Spreadsheet Solution for Method II-A Class II Series Pipe Line Problems

Figure 11.5 shows a simple spreadsheet to facilitate the calculations required for Method II-A. Its features are described here.

1. The heading identifies the nature of the spreadsheet and allows the problem number or other description of the problem to be entered in the shaded area.
2. The system data consist of the pressures and elevations at two reference points. If a given problem gives the allowable difference in pressure $\Delta p$, you may

| APPLIED FLUID MECHANICS | CLASS II SERIES SYSTEMS | |
|---|---|---|
| Objective: Volume flow rate | Method II-A: No minor losses | |
| Example Problem 11.2 Figure 11.4 | Uses Equation 11–3 to find maximum allowable volume flow rate to maintain desired pressure at point 2 for a given pressure at point 1 | |
| *System Data:*   SI Metric Units | | |
| Pressure at point 1 =   120 kPa | Elevation at point 1 =   0 m | |
| Pressure at point 2 =   60 kPa | Elevation at point 2 =   0 m | |
| Energy loss: $h_L$ =   6.95 m | | |
| *Fluid Properties:* | May need to compute $\nu = \mu/\rho$ | |
| Specific weight =   8.63 kN/m³ | Kinematic viscosity = 1.08E-05 m²/s | |
| Pipe data: | | |
| Diameter: D =   0.1541 m | | |
| Wall roughness: $\epsilon$ = 4.60E-05 m | | |
| Length: L =   100 m | **Results: Maximum values** | |
| Area: A =  0.01865 m² | Volume flow rate: Q =  0.0569 m³/s | |
| $D/\epsilon$ =   3350 | Velocity: v =   3.05 m/s | |

**FIGURE 11.5**   Spreadsheet for Method II-A Class II Series pipe line problems.

assign the value for pressure at one point and then compute the pressure at the second point from $p_2 = p_1 + \Delta p$.

3. The energy loss is calculated in the spreadsheet using,

$$h_L = (p_1 - p_2)/\gamma + z_1 - z_2$$

This is found from the energy equation noting that the velocities are equal at the two reference points.

4. The fluid properties of specific weight and kinematic viscosity are entered.

5. Pipe data for flow diameter, roughness, and length are entered.

6. The spreadsheet completes the remaining calculations for area and relative roughness that are needed in Equation 11–3.

7. The results are then computed using Equation 11–3 and the maximum allowable volume flow rate and the corresponding velocity are shown at the bottom right of the spreadsheet. These values compare favorably with those found in Example Problem 11.2.

### Spreadsheet for Solution Method II-B for Class II Series Pipe Line Problems

We use a new spreadsheet shown in Figure 11.6 for solution Method II-B that is an extension of that for Method II-A. In fact, the first part of the spreadsheet is identical to Figure 11.5 in which the allowable volume flow rate for a straight pipe with no minor losses is determined. Then a lower volume flow rate is assumed in the lower part of the spreadsheet that includes the effect of minor losses. Obviously, with minor losses added to the friction loss considered in Method II-A, a lower allowable volume flow rate will result. The method is inherently a two-step process and more than one trial for the second step may be required.

To illustrate the use of Method II-B, we create the new Example Problem 11.3 shown below. We take the same basic data from Example Problem 11.2 and add minor losses due to two standard elbows and a fully open butterfly valve.

---

☐ **EXAMPLE PROBLEM 11.3**     A lubricating oil must be delivered through the piping system shown in Figure 11.7 with a maximum pressure drop of 60 kPa between points 1 and 2. The oil has a specific gravity of 0.88 and a dynamic viscosity of $9.5 \times 10^{-3}$ Pa·s. Determine the maximum allowable volume flow rate of oil.

**Solution**     The system is similar to that in Example Problem 11.2. There are 100 m of 6-in Schedule 40 steel pipe in a horizontal plane. But the addition of the valve and the two elbows provide a moderate amount of energy loss.

Initially, we ignore the minor losses and use Equation 11–3 to compute a rough estimate of the allowable volume flow rate. This is accomplished in the upper part of the spreadsheet in Figure 11.6 and it is identical to the solution shown in Figure 11.5 for Example Problem 11.2. This is the starting point for Method II-B.

The features of the lower part of Figure 11.6 are described next.

1. A revised estimate of the allowable volume flow rate $Q$ is entered at the upper right, just under the computation of the initial estimate. The revised estimate must be lower than the initial estimate.

2. The spreadsheet then computes the "Additional Pipe Data" using the known pipe data from the upper part of the spreadsheet and the new estimated value for $Q$.

| APPLIED FLUID MECHANICS | CLASS II SERIES SYSTEMS |
|---|---|
| **Objective: Volume flow rate** | **Method II–A: No minor losses** |
| Example Problem 11.3<br>Figure. 11.7 | Uses Equation 11–3 to estimate the allowable volume flow rate<br>to maintain desired pressure at point 2 for a given pressure at point 1 |

| *System Data:* | SI Metric Units | | |
|---|---|---|---|
| *Pressure at point 1 =* | 120 kPa | *Elevation at point 1 =* | 0 m |
| *Pressure at point 2 =* | 60 kPa | *Elevation at point 2 =* | 0 m |
| Energy loss: $h_L$ = | 6.95 m | | |

| *Fluid Properties:* | | May need to compute $\nu = \mu/\rho$ |
|---|---|---|
| Specific weight = | 8.63 kN/m³ | Kinematic viscosity = 1.08E-05 m²/s |

| *Pipe data:* 6–is schedule 40 steel | | |
|---|---|---|
| Diameter: D = | 0.1541 m | |
| Wall roughness: $\epsilon$ = | 4.60E-05 m | **Results: Maximum values** |
| Length: L = | 100 m | Volume flow rate: Q = 0.0569 m³/s |
| Area: A = | 0.01865 m² | Velocity: v = 3.05 m/s |
| $D/\epsilon$ = | 3350 | |

| CLASS II SERIES SYSTEMS | Volume flow rate: Q = 0.0538 m³/s |
|---|---|
| **Method II-B:** Use results of Method IIA;<br>include minor losses;<br>then pressure at Point 2 is computed | Given: Pressure $p_1$ = 120 kPa<br>**Pressure $p_2$ = 60.18 kPa**<br>**NOTE: Should be >** 60 kPa |

| *Additional Pipe Data:* | | Adjust estimate for Q until $p_2$ |
|---|---|---|
| L/D = | 649 | **is greater than desired pressure.** |
| Flow velocity = | 2.88 m/s | Velocity at point 1 = 2.88 m/s  → If velocity is in pipe: |
| Velocity head = | 0.424 m | Velocity at point 2 = 2.88 m/s  → Enter "=B24" |
| Reynolds No. = 4.12E+04 | | Vel. head at point 1 = 0.424 m |
| Friction factor: f = | 0.0228 | Vel. head at point 2 = 0.424 m |

| *Energy losses in Pipe 1:* | | Qty. | | |
|---|---|---|---|---|
| Pipe: $K_1 = f(L/D)$ = | 14.76 | 1 | Energy loss $h_{L1}$ = | 6.26 m   Friction |
| 2 std elbows: $K_2$ = | 0.45 | 2 | Energy loss $h_{L2}$ = | 0.38 m |
| Butterfly valve: $K_3$ = | 0.68 | 1 | Energy loss $h_{L3}$ = | 0.29 m |
| Element 4: $K_4$ = | 0.00 | 1 | Energy loss $h_{L4}$ = | 0.00 m |
| Element 5: $K_5$ = | 0.00 | 1 | Energy loss $h_{L5}$ = | 0.00 m |
| Element 6: $K_6$ = | 0.00 | 1 | Energy loss $h_{L6}$ = | 0.00 m |
| Element 7: $K_7$ = | 0.00 | 1 | Energy loss $h_{L7}$ = | 0.00 m |
| Element 8: $K_8$ = | 0.00 | 1 | Energy loss $h_{L8}$ = | 0.00 m |
| | | | Total energy loss $h_{Ltot}$ = | 6.93 m |

**FIGURE 11.6**   Spreadsheet for Method II-B Class II series pipe line problems.

**FIGURE 11.7**   Piping system for
Example Problem 11.3

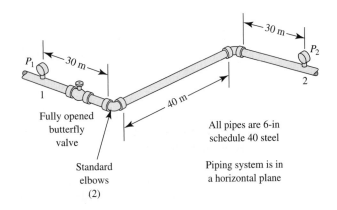

$P_1$ — 30 m —

Fully opened
butterfly
valve

Standard
elbows
(2)

— 40 m —

— 30 m —

$P_2$

2

All pipes are 6-in
schedule 40 steel

Piping system is in
a horizontal plane

3. Note at the middle right of the spreadsheet that the velocities at reference points 1 and 2 must be entered. If they are in the pipe, as they are in this problem, then the cell reference "B24" can be entered because that is where the velocity in the pipe is computed. Other problems may have the reference points elsewhere, such as the surface of a reservoir where the velocity is zero. The appropriate value should then be entered in the shaded area.

4. Now the data for minor losses must be added in the section called "Energy Losses in Pipe 1." The $K$ factor for the pipe friction loss is automatically computed from known data. The values for the other two $K$ factors must be determined and entered in the shaded area in a manner similar to that used in the Class I spreadsheet. In this problem they are both dependent on the value of $f_T$ for the 6-inch pipe. That value is 0.015 as found in Table 10.5.

- Elbow (Standard):  $K = f_T(L_e/D) = (0.015)(30) = 0.45$
- Butterfly Valve:  $K = f_T(L_e/D) = (0.015)(45) = 0.675$

5. The spreadsheet then computes the total energy loss and uses this value to compute the pressure at reference point 2. The equation is derived from the energy equation,

$$p_2 = p_1 + \gamma[z_1 - z_2 + v_1^2/2g - v_2^2/2g - h_L]$$

6. The computed value for $p_2$ must be larger than the desired value as entered in the upper part of the spreadsheet. This value is placed up close to the assumed volume flow rate to give you a visual cue as to the acceptability of your current estimate for the limiting volume flow rate. Adjustments in the value of $Q$ can then be quickly made until the pressure assumes an acceptable value.

**Result**    The spreadsheet in Figure 11.6 shows that a volume flow rate of 0.0538 m³/s through the system in Figure 11.7 will result in the pressure at point 2 being 60.18 kPa, slightly more that the minimum acceptable value.

■

## Method II-C: Iteration Approach for Class II Series Pipe Line Problems

Method II-C is presented here as a manual iteration process. It is used for Class II systems in which minor losses play a primary role in determining what the maximum volume flow rate can be while limiting the pressure drop in the system to a specified amount. As with all Class II systems except those for which pipe friction is the only significant loss, there are more unknowns than can be directly solved for. The process of iteration is used to guide you through the choices that need to be made to arrive at a satisfactory design or analysis.

Both the friction factor and the velocity of flow are unknown in a Class II system. And since they depend on each other, no direct solution is possible.

The iteration proceeds most efficiently if the problem is set up to facilitate the final cycle of estimating one unknown, the friction factor, to compute an approximate value of the other major unknown, the velocity of flow in the system. The procedure provides a means of checking the accuracy of the trial value of $f$ and also indicates the new trial value to be used if an additional cycle is required. This is what distinguishes iteration from "trial and error," in which there are no discrete guidelines for subsequent trials.

The complete iteration process is illustrated within Example Problem 11.4. The following step-by-step procedure is used.

**SOLUTION PROCEDURE FOR CLASS II SYSTEMS WITH ONE PIPE**

1. Write the energy equation for the system.
2. Evaluate known quantities such as pressure heads and elevation heads.
3. Express energy losses in terms of the unknown velocity $v$ and friction factor $f$.
4. Solve for the velocity in terms of $f$.
5. Express the Reynolds number in terms of the velocity.
6. Calculate the relative roughness $D/\epsilon$.
7. Select a trial value of $f$ based on the known $D/\epsilon$ and a Reynolds number in the turbulent range.
8. Calculate the velocity, using the equation from Step 4.
9. Calculate the Reynolds number from the equation in Step 5.
10. Evaluate the friction factor $f$ for the Reynolds number from Step 9 and the known value of $D/\epsilon$, using the Moody diagram, Fig. 9.2.
11. If the new value of $f$ is different from the value used in Step 8, repeat Steps 8 through 11 using the new value of $f$.
12. If there is no significant change in $f$ from the assumed value, then the velocity found in Step 8 is correct.

## PROGRAMMED EXAMPLE PROBLEM

☐ **EXAMPLE PROBLEM 11.4**     Water at 80°F is being supplied to an irrigation ditch from an elevated storage reservoir as shown in Fig. 11.8. Calculate the volume flow rate of water into the ditch.

Begin with Step 1 of the solution procedure by writing the energy equation. Use A and B as the reference points and simplify the equation as much as possible.

---

Compare this with your solution:

$$\frac{p_A}{\gamma} + z_A + \frac{v_A^2}{2g} - h_L = \frac{p_B}{\gamma} + z_B + \frac{v_B^2}{2g}$$

Since $p_A = p_B = 0$, and $v_A$ is approximately zero, then

$$z_A - h_L = z_B + (v_B^2/2g)$$

$$z_A - z_B = (v_B^2/2g) + h_L \tag{11-3}$$

**FIGURE 11.8**  Pipe line system for Example Problem 11.4.

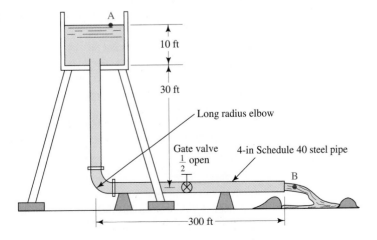

Notice that the stream of water at point B has the same velocity as that inside the pipe.

The elevation difference, $z_A - z_B$, is known to be 40 ft. However, the energy losses which make up $h_L$ all depend on the unknown velocity, $v_B$. Thus, iteration is required. Do Step 3 of the solution procedure now.

There are four components of the total energy loss $h_L$:

$$h_L = h_1 + h_2 + h_3 + h_4$$

where
$$h_1 = 1.0(v_B^2/2g) \qquad \text{(entrance loss)}$$
$$h_2 = f(L/D)(v_B^2/2g) \qquad \text{(pipe friction loss)}$$
$$= f(330/0.3355)(v_B^2/2g)$$
$$= 985f(v_B^2/2g)$$
$$h_3 = f_T(L_e/D)(v_B^2/2g) \qquad \text{(long radius elbow)}$$
$$= 20f_T(v_B^2/2g)$$
$$h_4 = f_T(L_e/D)(v_B^2/2g) \qquad \text{(half-open gate valve)}$$
$$= 160f_T(v_B^2/2g)$$

From Table 10.5 we find $f_T = 0.017$ for a 4-in steel pipe. Then we have

$$h_L = (1.0 + 985f + 20f_T + 160f_T)(v_B^2/2g)$$
$$= (4.06 + 985f)(v_B^2/2g) \qquad \text{(11–5)}$$

Now substitute this expression for $h_L$ into Eq. (11–4) and solve for $v_B$ in terms of $f$.

You should have $v_B = \sqrt{2580/(5.06 + 985f)}$.

$$z_A - z_B = (v_B^2/2g) + h_L$$
$$40 \text{ ft} = (v_B^2/2g) + (4.06 + 985f)(v_B^2/2g)$$
$$= (5.06 + 985f)(v_B^2/2g)$$

Solving for $v_B$, we get

$$v_B = \sqrt{\frac{2g(40)}{5.06 + 985f}} = \sqrt{\frac{2580}{5.06 + 985f}} \qquad \text{(11–6)}$$

Equation (11–6) represents the completion of Step 4 of the procedure. Now do Steps 5 and 6.

$$N_R = \frac{v_B D}{\nu} = \frac{v_B(0.3355)}{9.15 \times 10^{-6}} = (0.366 \times 10^5)v_B \qquad \text{(11–7)}$$

$$D/\epsilon = (0.3355/1.5 \times 10^{-4}) = 2235$$

Step 7 is the start of the iteration process. What is the possible range of values for the friction factor for this system?

Since $D/\epsilon = 2235$, the lowest possible value of $f$ is 0.0155 for very high Reynolds numbers and the highest possible value is 0.039 for a Reynolds number of 4000. The initial trial value of $f$ must be in this range. Use $f = 0.020$ and complete Steps 8 and 9.

We find the values for velocity and the Reynolds number by using Eqs. (11–6) and (11–7):

$$v_B = \sqrt{\frac{2580}{5.06 + (985)(0.02)}} = \sqrt{104} = 10.2 \text{ ft/s}$$

$$N_R = (0.366 \times 10^5)(10.2) = 3.73 \times 10^5$$

Now do Step 10.

---

You should have $f = 0.0175$. Since this is different from the initial trial value of $f$, Steps 8 through 11 must be repeated now.

---

Using $f = 0.0175$, we get

$$v_B = \sqrt{\frac{2580}{5.06 + (985)(0.0175)}} = \sqrt{116} = 10.8 \text{ ft/s}$$

$$N_R = (0.366 \times 10^5)(10.8) = 3.94 \times 10^5$$

The new value of $f$ is 0.0175, which is unchanged. Therefore, we have

$$v_B = 10.8 \text{ ft/s}$$

$$Q = A_B v_B = (0.0884 \text{ ft}^2)(10.8 \text{ ft/s}) = 0.955 \text{ ft}^3/\text{s}$$

This programmed example problem is concluded.

■

---

**11.6**
**CLASS III SYSTEMS**

A Class III series pipe line system is one for which you desire to know the size of pipe that will carry a given volume flow rate of a given fluid with a specified maximum pressure drop due to energy losses.

You can use a similar logic to that used to discuss Class II series pipe line systems to plan an approach to designing Class III systems. You know that pressure drop is directly related to the energy loss in the system and that the energy losses are typically proportional to the velocity head of the fluid as it flows through the system. Because velocity head is $v^2/2g$, the energy losses are proportional to the square of the velocity. But velocity is, in turn, inversely proportional to the flow area found from

$$A = \pi D^2/4$$

Therefore the energy loss is inversely proportional to the flow diameter *to the fourth power*. The size of the pipe is a major factor in how much energy loss occurs in a pipe line system. Your task as the designer is to determine how small the pipe can be and still meet the goal of a limited pressure drop. You don't want to use an unreasonably large pipe because the cost of piping increases with increasing size. But if the size of the pipe is too small, the energy wasted by excessive energy losses would generate a high operating cost for the life of the system. You should consider the total life-cycle cost.

We suggest two different approaches to designing Class III systems.

### Method III-A

This is the simplified approach that considers only energy loss due to friction in the pipe. We assume that the reference points for the energy equation are in the pipe to be designed and at a set distance apart. There may be an elevation difference between the two points. But, because the flow diameter is the same at the two reference points, there is no difference in the velocities or the velocity heads. We can write the energy equation and then solve for the energy loss.

$$\frac{p_1}{\gamma} + z_1 + \frac{v_1^2}{2g} - h_L = \frac{p_2}{\gamma} + z_2 + \frac{v_2^2}{2g}$$

But $v_1 = v_2$. Then we have

$$h_L = \frac{p_1 - p_2}{\gamma} + z_1 - z_2$$

This value, along with other system data, can be entered into the following design equation 11–8. See References 3 and 4.

$$D = 0.66 \left[ \epsilon^{1.25} \left( \frac{LQ^2}{gh_L} \right)^{4.75} + \nu Q^{9.4} \left( \frac{L}{gh_L} \right)^{5.2} \right]^{0.04} \tag{11–8}$$

The result is the smallest flow diameter that can be used for a pipe to limit the pressure drop to the desired value. Normally, you will specify a standard pipe or tube that has an inside diameter just larger than this limiting value.

□ **EXAMPLE PROBLEM 11.5**   Compute the required size of new clean Schedule 40 pipe that will carry 0.50 ft³/s of water at 60°F and limit the pressure drop to 2.00 psi over a length of 100 ft of horizontal pipe.

**Solution**   We first calculate the limiting energy loss. Note that the elevation difference is zero.

$$h_L = (p_1 - p_2)/\gamma + (z_1 - z_2) = (2.00 \text{ lb/in}^2)(144 \text{ in}^2/\text{ft}^2)/(62.4 \text{ lb/ft}^3) + 0 = 4.62 \text{ ft}$$

The following data are needed in Equation 11–8.

| | | |
|---|---|---|
| $Q = 0.50$ ft³/s | $L = 100$ ft | $g = 32.2$ ft/s² |
| $h_L = 4.62$ ft | $\epsilon = 1.5 \times 10^{-4}$ ft | $\nu = 1.21 \times 10^{-5}$ ft²/s |

Now we can enter these data into Equation 11–8:

$$D = 0.66 \left[ (1.5 \times 10^{-4})^{1.25} \left[ \frac{(100)(0.50)^2}{(32.2)(4.62)} \right]^{4.75} + (1.21 \times 10^{-5})(0.50)^{9.4} \left[ \frac{100}{(32.2)(4.62)} \right]^{5.2} \right]^{0.04}$$

$$D = 0.309 \text{ ft}$$

The result shows that the pipe should be larger than $D = 0.309$ ft. The next larger standard pipe size is a 4-in Schedule 40 steel pipe having an inside diameter of $D = 0.3355$ ft.

■

## Spreadsheet for Completing Method III-A for Class III Series Pipe Line Problems

Obviously, Equation 11–8 is cumbersome to evaluate and the opportunity for calculation error is great. The use of a spreadsheet to perform the calculation alleviates this problem.

Figure 11.9 shows an example of such a spreadsheet. Its features are described here.

■ The problem identification and given data are listed to the left side. Where the allowable pressure drop $\Delta p$ is given, as it is in Example Problem 11.5, we spec-

| APPLIED FLUID MECHANICS | | CLASS III SERIES SYSTEMS | | |
|---|---|---|---|---|
| **Objective: Minimum pipe diameter** | | **Method III-A:** *Uses Equation 11–8 to compute the* | | |
| Example Problem 11.5 | | *minimum size of pipe of a given length* | | |
| | | *that will flow a given volume flow rate of fluid* | | |
| *System Data:* | *SI Metric Units* | *with a limited pressure drop. (No minor losses)* | | |
| *Pressure at point 1 =* | 102 psig | **Fluid Properties:** | | |
| *Pressure at point 2 =* | 100 psig | *Specific weight =* | 62.4 lb/ft$^3$ | |
| *Elevation at point 1 =* | 0 ft | *Kinematic viscosity =* 1.21E-05 ft$^2$/s | | |
| *Elevation at point 2 =* | 0 ft | Intermediate Results in Eq. 11–8: | | |
| Allowable Energy Loss: $h_L$ = | 4.62 ft | $L/gh_L$ =  0.67  878 | | |
| *Volume flow rate: Q =* | 0.5 ft$^3$/s | Argument in bracket: 5.77E-09 | | |
| *Length of pipe: L =* | 100 ft | Final Minimum Diameter: | | |
| *Pipe wall roughness: $\epsilon$ =* | 1.50E-04 ft | Minimum diameter: $D$ =  0.3090 ft | | |

**FIGURE 11.9**  Spreadsheet for Method III-A for Class III series pipe line problems.

ify an arbitrary value for the pressure at point 2 and then set the pressure at point 2 as

$$p_2 = p_1 + \Delta p$$

- Note that the spreadsheet computes the allowable energy loss $h_L$ using the method shown in the solution of Example Problem 11.5.
- The fluid properties data are entered at the upper right side of the spreadsheet.
- The intermediate results are reported simply for reference. They represent factors from Equation 11–8 and can be used by those solving the equation manually as a check on their calculation procedure. If you prepare a spreadsheet yourself, you should carefully verify the form of the equation that solves Equation 11–8 because the programming is complex. Breaking it up into parts can simplify the final equation.
- The **Final Minimum Diameter** is the result of the calculation from Equation 11–8 and it represents the minimum acceptable size of pipe to carry the given volume flow rate with the stated limitation on pressure drop.

## Method III-B

When minor losses are to be considered, a modest extension of Method III-A can be used. The standard pipe size selected as a result of Method III-A is normally somewhat larger than the minimum allowable diameter. Therefore, modest additional energy losses due to a few minor losses will likely not produce a total pressure drop greater than that allowed. The selected pipe size will probably still be acceptable.

After making a tentative specification of pipe size, we can add the minor losses to the analysis and examine the resulting pressure at the end of the system to ensure that it is within the desired limits. If not, a simple adjustment to the next larger size pipe will almost surely produce an acceptable design. Implementing this procedure using a spreadsheet makes the calculations very rapid.

Figure 11.10 shows a spreadsheet that implements this design philosophy. It is actually a juncture of two spreadsheets already described in this chapter. The upper part is identical to Figure 11.9, which was used for solving Example Problem 11.5 using Method III-A. From that we gain an estimate of the size of pipe that will carry the desired amount of fluid without any minor losses.

The lower part of the spreadsheet uses a technique similar to that in Figure 11.3 for Class I series pipe line problems. It is simplified to include only one pipe

| APPLIED FLUID MECHANICS | | CLASS III SERIES PIPE LINE SYSTEMS | |
|---|---|---|---|
| **Objective: Minimum pipe diameter** | | **Method III-A:** *Uses Equation 11–8 to compute the minimum size of pipe of a given length that will flow a given volume flow rate of fluid with a limited pressure drop. (No minor losses)* | |
| Example Problem 11.6 | | | |
| *System Data:* | *SI Metric Units* | *Fluid Properties:* | |
| Pressure at point 1 = | 102 psig | Specific weight = $62.4 \text{ lb/ft}^3$ | |
| Pressure at point 2 = | 100 psig | Kinematic Viscosity = $1.21\text{E-}05 \text{ ft}^2/\text{s}$ | |
| Elevation at point 1 = | 0 ft | | |
| Elevation at point 2 = | 0 ft | **Intermediate Results in Eq. 11–8:** | |
| Allowable Energy Loss: $h_L$ = | 4.62 ft | $L/gh_L$ = 0.672878 | |
| Volume flow rate: Q = | 0.5 ft³/s | Argument in bracket: 5.77E-09 | |
| Length of pipe: L = | 100 ft | Final Minimum Diameter: | |
| Pipe wall roughness: $\epsilon$ = | 1.50E-04 ft | Minimum diameter: D = 0.3090 ft | |

| CLASS III SERIES SYSTEMS | | | Specified pipe diameter: D = 0.3355 ft | |
|---|---|---|---|---|
| **Method III-B:** *Use results of Method IIIA; Specify actual diameter; Include minor losses; then pressure at Point 2 is computed* | | | 4-inch schedule 40 steel pipe | |
| | | | *If velocity is in the pipe, enter "=B23" for value* | |
| *Additional Pipe Data:* | | | Velocity at point 1 = | 5.66 ft/s |
| Flow area: A = | 0.08840 ft² | | Velocity at point 2 = | 5.66 ft/s |
| Relative roughness: $D/\epsilon$ = | 2237 | | Vel. head at point 1 = | 0.497 ft |
| $L/D$ = | 298 | | Vel. head at point 2 = | 0.497 ft |
| Flow velocity = | 5.66 ft/s | | *Results:* | |
| Velocity head = | 0.497 ft | | Given pressure at point 1 = | 102 psig |
| Reynolds No. = | 1.57E+05 | | Desired pressure at point 2 = | 100 psig |
| Friction factor: f = | 0.0191 | | **Actual pressure at point 2 = 100.46 psig** | |
| | | | *(Compare actual with desired pressure at point 2)* | |
| *Energy losses in Pipe:* | | Qty. | | |
| Pipe friction: $K_1 = f(L/D)$ = | 5.70 | 1 | Energy loss $h_{L1}$ = | 2.83 ft |
| Two long rad. elbows: $K_2$ = | 0.34 | 2 | Energy loss $h_{L2}$ = | 0.34 ft |
| Butterfly valve: $K_3$ = | 0.77 | 1 | Energy loss $h_{L3}$ = | 0.38 ft |
| Element 4: $K_4$ = | 0.00 | 1 | Energy loss $h_{L4}$ = | 0.00 ft |
| Element 5: $K_5$ = | 0.00 | 1 | Energy loss $h_{L5}$ = | 0.00 ft |
| Element 6: $K_6$ = | 0.00 | 1 | Energy loss $h_{L6}$ = | 0.00 ft |
| Element 7: $K_7$ = | 0.00 | 1 | Energy loss $h_{L7}$ = | 0.00 ft |
| Element 8: $K_8$ = | 0.00 | 1 | Energy loss $h_{L8}$ = | 0.00 ft |
| | | | Total energy loss $h_{Ltot}$ = | 3.55 ft |

**FIGURE 11.10**   Spreadsheet for Method III-B for Class III series pipe line problems.

size. Its objective is to compute the pressure at point 2 in a system when the pressure at point 1 is given. Minor losses are included.

The following procedure illustrates the use of this spreadsheet.

### Spreadsheet for Method III-B Class III Series Pipeline Problems with Minor Losses

- Initially ignore the minor losses and use the upper part of the spreadsheet to estimate the size of pipe required to carry the given flow rate with less than the allowable pressure drop. This is identical to Method III-A described in the preceding example problem.
- Enter the next standard pipe size at the upper right part of the lower spreadsheet in the cell called "Specified pipe diameter: *D*."
- The spreadsheet automatically computes the values under *Additional Pipe Data*.
- The velocities listed in the right column are usually in the pipe being analyzed and are usually equal. The reference to cell *B23* will automatically enter the com-

puted velocity from the pipe data. However, if the system being analyzed has a reference point outside the pipe, the actual velocity there must be entered. Then the velocity heads at the reference points are calculated.

- The section headed as *Energy Losses in Pipe* require you to enter the resistance factors $K$ for each minor loss, as was done in earlier spreadsheet solution procedures. The $K$ factor for pipe friction loss is computed automatically from the pipe data.
- The *Results* section lists the given pressure at point 1 and the desired pressure at point 2 taken from the initial data at the top of the spreadsheet. The *Actual pressure at point 2* is computed from an equation derived from the energy equation

$$p_2 = p_1 - \gamma[z_1 - z_2 + v_1^2/2g - v_2^2/2g - h_L]$$

- You as the designer of the system must compare the actual pressure at point 2 with the listed desired pressure.
- If the actual pressure is greater that the desired pressure, you have a satisfactory result and the pipe size specified is acceptable.
- If the actual pressure is less that the desired pressure, simply pick the next larger standard pipe size and repeat the spreadsheet calculations. This step is virtually immediate because all calculations are automatic once the new pipe flow diameter is entered.
- Unless there are many high-loss minor losses this size pipe should be acceptable. If not, continue to specify larger pipes until a satisfactory solution is achieved. You are also advised to examine the magnitude of the energy losses contributed by the minor losses. You may be able to use a smaller pipe size if you change valves and fittings to more efficient, lower-loss designs.

The example problem that follows illustrates the use of this spreadsheet.

☐ **EXAMPLE PROBLEM 11.6**     Extend the situation described in Example Problem 11.5 by adding a fully open butterfly valve and two long-radius elbows to the 100 ft of straight pipe. Will the 4-inch Schedule 40 steel pipe size selected limit the pressure drop to 2.00 psi with these minor losses added?

*Solution*     To simulate the desired pressure drop of 2.00 psi, we have set the pressure at point 1 to be 102 psig. Then we examine the resulting value of the pressure at point 2 to see that it is at or greater than 100 psig.

The spreadsheet in Figure 11.10 shows the calculations. For each minor loss, a resistance factor, $K$, is computed as defined in Chapters 9 and 10. For the pipe friction loss,

$$K_1 = f(L/D)$$

and the friction factor, $f$, is computed by the spreadsheet using Equation 9–9.
For the elbows and the butterfly valve, the method of Chapter 10 is applied.

$$K = f_T(L_e/D)$$

The values of $(L_e/D)$ and $f_T$ are found from Tables 10.4 and 10.5 respectively.

*Result*     The result shows that the pressure at point 2 at the end of the system is 100.46 psig. Thus the design is satisfactory. Note that the energy loss due to pipe friction is 2.83 ft out of the total energy loss of 3.55 ft. The elbows and the valve contribute truly minor losses.

■

# REFERENCES

1. Crane Co. 1988. *Flow of Fluids Through Valves, Fittings, and Pipe*. Technical Paper No. 410. Joliet, Illinois: Author.
2. Heald, C. C., ed. 1998. *Cameron Hydraulic Data*, 18th ed. Liberty Corner, New Jersey: Ingersoll-Dresser Pumps.
3. Streeter, Victor L., and E. Benjamin Wylie. 1985. *Fluid*

*Mechanics*, 8th ed. New York: McGraw-Hill.
4. Swamee, P. K., and A. K. Jain. 1976. Explicit Equations for Pipe-flow Problems. *Journal of the Hydraulics Division*, 102 (HY5) 657–644. New York: American Society of Civil Engineers.

# PRACTICE PROBLEMS

## Class I Systems

**11.1M** *Water at 10°C flows from a large reservoir at the rate of 1.5 × 10⁻² m³/s through the system shown in Fig. 11.11. Calculate the pressure at B.*

*10 points*

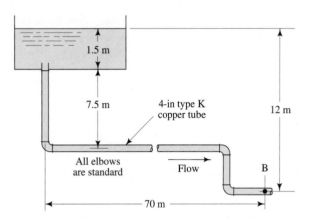

**FIGURE 11.11**   Problem 11.1.

**11.2M** *For the system shown in Fig. 11.12, kerosene (sg = 0.82) at 20°C is to be forced from tank A to reservoir B by increasing the pressure in the sealed tank A above the kerosene. The total length of 2-in Schedule 40 steel pipe*

*is 38 m. The elbow is standard. Calculate the required pressure in tank A to cause a flow rate of 435 L/min.*

**11.3E** Figure 11.13 shows a portion of a hydraulic circuit. The pressure at point B must be 200 psig when the volume flow rate is 60 gal/min. The hydraulic fluid has a specific gravity of 0.90 and a dynamic viscosity of

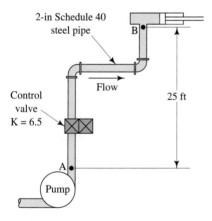

**FIGURE 11.13**   Problem 11.3.

**FIGURE 11.12**   Problem 11.2.

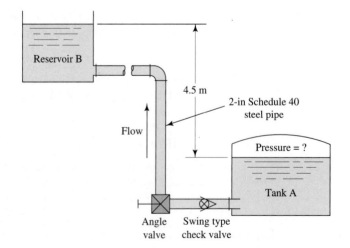

$6.0 \times 10^{-5}$ lb-s/ft$^2$. The total length of pipe between A and B is 50 ft. The elbows are standard. Calculate the pressure at the outlet of the pump at A.

**11.4E** Figure 11.14 shows part of a large hydraulic system in which the pressure at B must be 500 psig while the flow rate is 750 gal/min. The fluid is a medium machine tool hydraulic oil. The total length of 4-in pipe is 40 ft. The elbows are standard. Neglect the energy loss due to friction in the 6-in pipe. Calculate the required pressure at A if the oil is (a) at 104°F and (b) at 212°F.

**FIGURE 11.14** Problem 11.4.

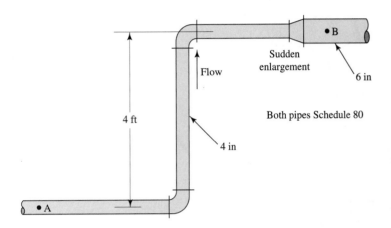

**11.5M** *Oil is flowing at the rate of 0.015 m³/s in the system shown in Fig. 11.15. Data for the system are*

*Oil specific weight = 8.80 kN/m³*
*Oil kinematic viscosity = 2.12 × 10⁻⁵ m²/s*
*Length of 6-in pipe = 180 m*
*Length of 2-in pipe = 8 m*
*Elbows are long radius type*
*Pressure at B = 12.5 MPa*

*Considering all pipe friction and minor losses, calculate the pressure at A.*

**FIGURE 11.15** Problem 11.5.

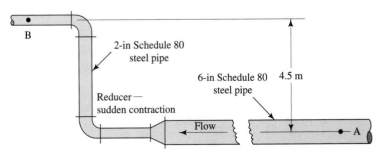

**11.6M** *For the system shown in Fig. 11.16, calculate the vertical distance between the surfaces of the two reservoirs when water at 10°C flows from A to B at the rate of 0.03 m³/s. The elbows are standard. The total length of the 3-in pipe is 100 m. For the 6-inch pipe it is 300 m.*

**FIGURE 11.16** Problem 11.6.

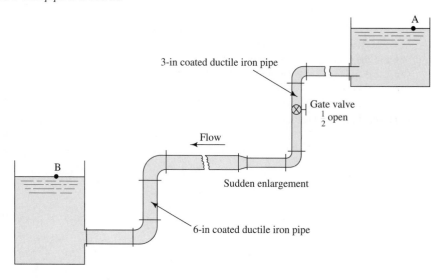

3-in coated ductile iron pipe

A

Gate valve
$\frac{1}{2}$ open

Flow

Sudden enlargement

B

6-in coated ductile iron pipe

**11.7M** *A liquid refrigerant flows through the system shown in Fig. 11.17 at the rate of 1.70 L/min. The refrigerant has a specific gravity of 1.25 and a dynamic viscosity of $3 \times 10^{-4}$ Pa·s. Calculate the pressure difference between points A and B. The tube is steel, with an outside diameter of $^1/_2$ in, a wall thickness of 0.049 in, and a total length of 30 m.*

**FIGURE 11.17** Problem 11.7.

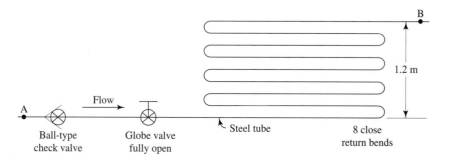

B

1.2 m

Flow

A

Ball-type
check valve

Globe valve
fully open

Steel tube

8 close
return bends

## Class II Systems

**11.8E** Water at 100°F is flowing in a 4-in Schedule 80 steel pipe which is 25 ft long. Calculate the maximum allowable volume flow rate if the energy loss due to pipe friction is to be limited to 30 ft-lb/lb.

**11.9M** *A hydraulic oil is flowing in a steel tube with an outside diameter of 2 in and a wall thickness of 0.083 in. A pressure drop of 68 kPa is observed between two points in the tube, 30 m apart. The oil has a specific* gravity of 0.90 and a dynamic viscosity of $3.0 \times 10^{-3}$ Pa·s. Assume the tube wall roughness to be $3 \times 10^{-5}$ m. Calculate the velocity of flow of oil.

**11.10E** In a processing plant, ethylene glycol at 77°F is flowing in 5000 ft of 6-in coated ductile iron pipe. Over this distance, the pipe falls 55 ft and the pressure drops from 250 psig to 180 psig. Calculate the velocity of flow in the pipe.

**11.11M** *Water at 15°C is flowing downward in a vertical tube 7.5 m long. The pressure is 550 kPa at the top and 585 kPa at the bottom. A ball-type check valve is installed near the bottom. The tube is steel, with a 1¹/₄-in outside diameter and a 0.083-in wall thickness. Compute the volume flow rate of the water.*

**11.12E** Turpentine at 77°F is flowing from A to B in a 3-in coated ductile iron pipe. Point B is 20 ft higher than A, and the total length of pipe is 60 ft. Two 90° long radius elbows are installed between A and B. Calculate the volume flow rate of turpentine if the pressure at A is 120 psig and the pressure at B is 105 psig.

**11.13E** A device designed to allow cleaning of walls and windows on the second floor of homes is similar to the system shown in Fig. 11.18. Determine the velocity of flow from the nozzle if the pressure at the bottom is (a) 20 psig and (b) 80 psig. The nozzle has a loss coefficient K of 0.15 based on the outlet velocity. The tube is smooth drawn aluminum and has an inside diameter of 0.5 in. The 90° bend has a radius of 6 in. The total length of straight tube is 20 ft. The fluid is water at 100°F.

**11.14M** *Kerosene at 25°C is flowing in the system shown in Fig. 11.19. The total length of 2-in Type K copper tubing is 30 m. The two 90° bends have a radius of 300 mm. Calculate the volume flow rate into tank B if a pressure of 150 kPa is maintained above the kerosene in tank A.*

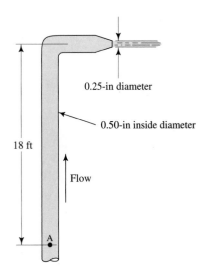

**FIGURE 11.18**   Problem 11.13.

**FIGURE 11.19**   Problem 11.14.

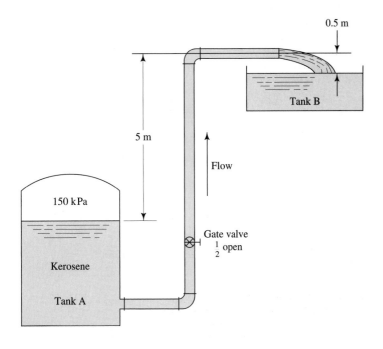

**11.15M** *Water at 40°C is flowing from A to B through the sys-*
*tem shown in Fig. 11.20. Determine the volume flow*
*rate of water if the vertical distance between the sur-*
*faces of the two reservoirs is 10 m. The elbows are*
*standard.*

**FIGURE 11.20**   Problem 11.15.

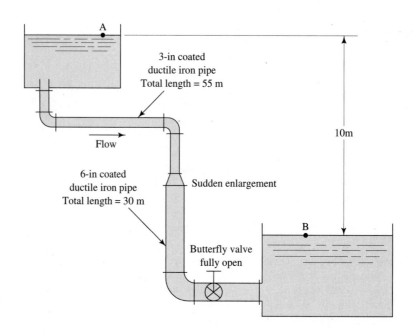

**11.16M** *Oil with a specific gravity of 0.93 and a dynamic vis-*
*cosity of $9.5 \times 10^{-3}$ Pa·s is flowing into the open tank*
*shown in Fig. 11.21. The total length of 2-in tubing is*
*30 m. For the 4-in tubing the total length is 100 m. The*
*elbows are standard. Determine the volume flow rate*
*into the tank if the pressure at point A is 175 kPa.*

**FIGURE 11.21**   Problem 11.16.

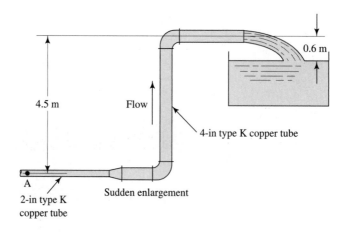

## Class III Systems

**11.17E** Determine the required size of new Schedule 80 steel pipe to carry water at 160°F with a maximum pressure drop of 10 psi per 1000 ft when the flow rate is 0.5 ft³/s.

**11.18M** *What size of standard type K copper tube is required to transfer 0.06 m³/s of water at 80°C from a heater where the pressure is 150 kPa to an open tank? The water flows from the end of the tube into the atmosphere. The tube is horizontal and 30 m long.*

**11.19E** Water at 60°F is to flow by gravity between two points, 2 mi apart, at the rate of 13 500 gal/min. The upper end is 130 ft higher than the lower end. What size concrete pipe is required? Assume that the pressure at both ends of the pipe is negligible.

**11.20E** The tank shown in Fig. 11.22 is to be drained to a sewer. Determine the size of new Schedule 40 steel pipe which will carry at least 400 gal/min of water at 80°F through the system shown. The total length of pipe is 75 ft.

**FIGURE 11.22** Problem 11.20.

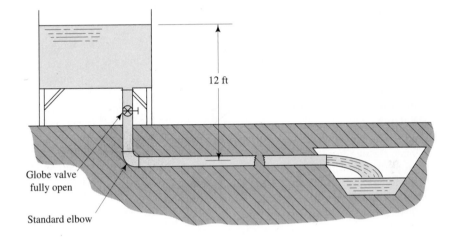

Globe valve
fully open

Standard elbow

## Practice Problems for Any Class

**11.21M** *Figure 11.23 depicts gasoline flowing from a storage tank into a truck for transport. The gasoline has a specific gravity of 0.68 and the temperature is 25°C.*

*Determine the required depth h in the tank to produce a flow of 1500 L/min into the truck. Since the pipes are short, neglect the energy losses due to pipe friction but do consider minor losses.*

**FIGURE 11.23** Problem 11.21.

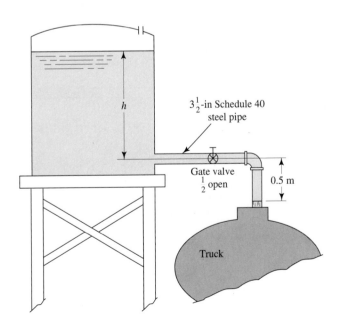

*Note:* Figure 11.24 shows a system used to pump coolant from a collector tank to an elevated tank, where it is cooled. The pump delivers 30 gal/min. The coolant then flows back to the machines as needed, by gravity. The coolant has a specific gravity of 0.92 and a dynamic viscosity of $3.6 \times 10^{-5}$ lb·s/ft². This system is used in Problems 11.22 through 11.24.

**11.22E** For the system in Fig. 11.24, compute the pressure at the inlet to the pump. The filter has a resistance coefficient of 1.85 based on the velocity head in the suction line.

**11.23E** For the system in Fig. 11.24, compute the total head on the pump and the power delivered by the pump to the coolant.

**11.24E** For the system in Fig. 11.24, specify the size of Schedule 40 steel pipe required to return the fluid to the machines. Machine 1 requires 20 gal/min and Machine 2 requires 10 gal/min. The fluid leaves the pipes at the machines at 0 psig.

**11.25E** A manufacturer of spray nozzles specifies that the maximum pressure drop in the pipe feeding a nozzle be 10.0 psi per 100 ft of pipe. Compute the maximum allowable velocity of flow through a 1-in Schedule 80 steel pipe feeding the nozzle. The pipe is horizontal and the fluid is water at 60°F.

**11.26E** Specify the size of new Schedule 40 steel pipe required to carry gasoline at 77°F through 120 ft of horizontal pipe with no more than 8.0 psi of pressure drop at a volume flow rate of 100 gal/min.

**11.27M** *Refer to Fig. 11.25. Water at 80°C is being pumped from the tank at the rate of 475 L/min. Compute the pressure at the inlet of the pump.*

**11.28M** *For the system shown in Fig. 11.25 and analyzed in Problem 11.27, it is desirable to change the system to increase the pressure at the inlet to the pump. The volume flow rate must stay at 475 L/min, but anything else can be changed. Redesign the system and recompute the pressure at the inlet to the pump for comparison with the result of Problem 11.27.*

**11.29E** In a water pollution control project, the polluted water is pumped vertically upward 80 ft and then sprayed into the air to increase the oxygen content in the water and to evaporate volatile materials. The system is sketched in Fig. 11.26. The polluted water has a specific weight of 64.0 lb/ft³ and a dynamic viscosity of $4.0 \times 10^{-5}$ lb·s/ft². The flow rate is 0.50 ft³/s. The pressure at the inlet to the pump is 3.50 psi below atmospheric pressure. The total length of discharge pipe is 82 ft. The nozzle has a resistance coefficient of 32.6 based on the velocity head in the discharge pipe.

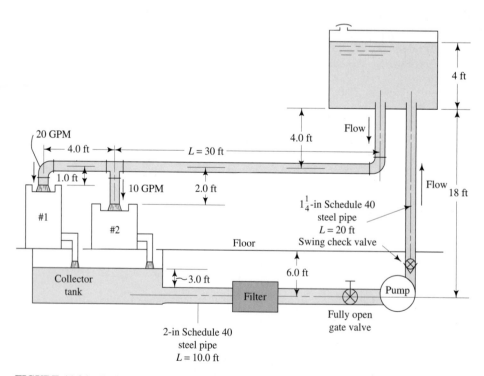

**FIGURE 11.24** Problems 11.22, 11.23, and 11.24.

**FIGURE 11.25**  Problems 11.27 and 11.28.

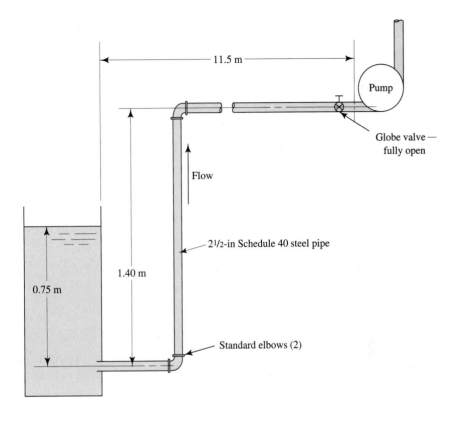

**FIGURE 11.26**  Problems 11.29 and 11.30.

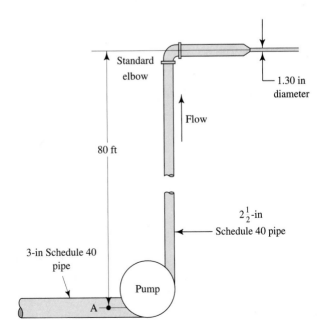

Compute the power delivered by the pump to the fluid. If the pump efficiency is 76%, compute the power input to the pump.

**11.30E**  Repeat Problem 11.29, but use a 3-in Schedule 40 steel pipe for the discharge line instead of the 2½-in pipe. Compare the power delivered by the pump for the two designs.

**11.31M**  *Water at 10°C is being delivered to a tank on the roof of a building, as shown in Fig. 11.27. The elbow is standard. What pressure must exist at Point A for 200 L/min to be delivered?*

**11.32M**  *If the pressure at point A in Fig. 11.27 is 300 kPa, compute the volume flow rate of water at 10°C delivered to the tank.*

**11.33M**  *Change the design of the system in Fig. 11.27 to replace the globe valve with a fully open gate valve. Then, if the pressure at point A is 300 kPa, compute the volume flow rate of water at 10°C delivered to the tank. Compare the result with that of Problem 11.32 to demonstrate the effect of the valve change.*

**11.34E**  It is desired to deliver 250 gal/min of ethyl alcohol at 77°F from the tank A to tank B in the system shown in Fig. 11.28. The total length of pipe is 110 ft. Compute the required pressure in tank A.

**11.35E**  For the system shown in Fig. 11.28, compute the volume flow rate of ethyl alcohol at 77°F that would occur if the pressure in tank A is 100 psig. The total length of pipe is 110 ft.

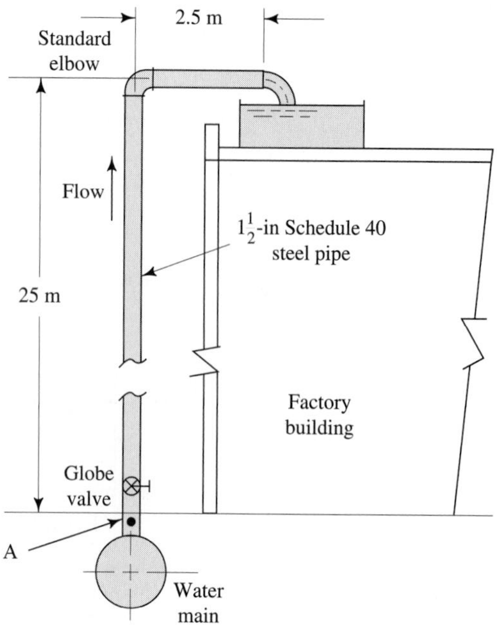

**FIGURE 11.27**   Problems 11.31, 11.32, and 11.33.

**FIGURE 11.28**   Problems 11.34, 11.35, 11.36, and 11.37.

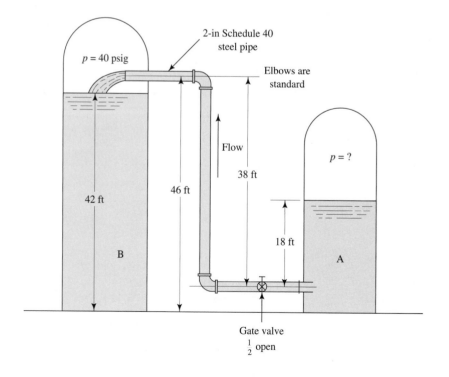

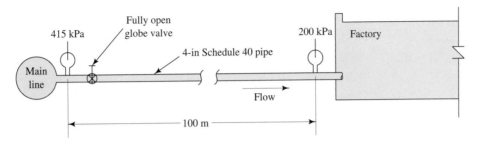

**FIGURE 11.29**   Problems 11.38, 11.39, 11.40, and 11.41.

**11.36E** Repeat Problem 11.35, but consider the valve to be fully open.

**11.37E** Repeat Problem 11.35, but consider the valve to be fully open and the elbows to be the long radius type instead of standard. Compare the results with those from Problems 11.35 and 11.36.

**11.38M** *Figure 11.29 depicts a pipe delivering water at 15°C from a main line to a factory. The pressure at the main is 415 kPa. Compute the maximum allowable flow rate if the pressure at the factory must be no less than 200 kPa.*

**11.39M** *Repeat Problem 11.38, but replace the globe valve with a fully open butterfly valve.*

**11.40M** *Repeat Problem 11.38, but use a 5-in Schedule 40 pipe.*

**11.41M** *Repeat Problem 11.38, but replace the globe valve with a butterfly valve and use a 5-in Schedule 40 steel pipe. Compare the results of Problems 11.38–11.41.*

**11.42E** It is desired to drive a small, positive displacement pump by chucking a household electric drill to the drive shaft of the pump. The pump delivers 1.0 in³ of water at 60°F per revolution, and the pump rotates at 2100 rpm. The outlet of the pump flows through a 100-ft smooth plastic hose with an inside diameter of 0.75 in. How far above the source can the outlet of the hose be if the maximum power available from the drill motor is 0.20 hp? The pump efficiency is 75 percent. Consider the friction loss in the hose, but neglect other losses.

**11.43E** Figure 11.30 shows a pipe delivering water to the putting green on a golf course. The pressure in the main is at 80 psig and it is necessary to maintain a minimum of 60 psig at point B to adequately supply a sprinkler system. Specify the required size schedule 40 steel pipe to supply 0.50 ft³/s of water at 60°F.

**11.44E** Repeat Problem 11.43 except consider that there will be the following elements added to the system:

- A fully open gate value near the water main
- A fully open butterfly valve near the green (but before point B)
- Three standard 90° elbows
- Two standard 45° elbows
- One swing type check valve

**11.45E** A sump pump in a commercial building sits in a sump at an elevation of 150.4 ft. The pump delivers 40 gal/min of water through a piping system that discharges the water at an elevation of 172.8 ft. The pressure at the pump discharge is 15.0 psig. The fluid is water at 60°F. Specify the required size of plastic pipe if the system contains the following elements.

- A ball-type check valve
- Eight standard elbows
- A total length of pipe of 55.3 ft.

The pipe is available in the same dimensions as Schedule 40 steel pipe.

**FIGURE 11.30**

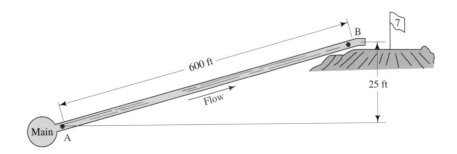

**FIGURE 11.31**

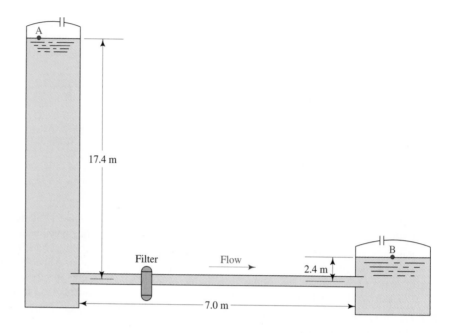

**11.46E** For the system designed in Problem 11.45, compute the total head on the pump.

**11.47M** *Figure 11.31 shows a part of a chemical processing system in which propyl alcohol at 25°C is taken from the bottom of a large tank and transferred by gravity to another part of the system. The length between the two tanks is 7.0 m. A filter is installed in the line and is known to have a resistance coefficient K of 8.5 based on the pipe velocity head. Drawn stainless steel tubing having a roughness ε of 3.0 × 10⁻⁵ m is to be used. Specify the standard size of tubing from Appendix G which would allow a volume flow rate of 150 L/min.*

**11.48M** *For the system described in Problem 11.47, and using the tube size found in that problem, compute the expected volume flow rate through the tube if the elevation in the large tank drops to 12.8 m.*

**11.49M** *For the system described in Problem 11.47, and using the tube size found in that problem, compute the expected volume flow rate through the tube if the pressure above the fluid in the large tank is −32.5 kPa gage.*

**11.50M** *For the system described in Problem 11.47, and using the tube size found in that problem, compute the expected volume flow rate through the tube if a ¹/₂ open gate valve is placed in the line ahead of the filter.*

## COMPUTER AIDED ANALYSIS AND DESIGN ASSIGNMENTS

**1.** Create a program or spreadsheet to analyze Class I pipe line systems, including energy losses due to friction and minor losses due to valves and fittings.

**2.** Create a program or spreadsheet to determine the velocity of flow and the volume flow rate in a given pipe with a limited pressure drop, considering energy loss due to friction only. Use the computational approach described in Section 11.5 and illustrated in Example Problem 11.2.

**3.** Create a program or spreadsheet to determine the size of pipe required to carry a specified flow rate with a limited pressure

drop, using the Class III solution procedure described in Example Problem 11.5.

**4.** Create a program or spreadsheet to determine the size of pipe required to carry a specified flow rate with a limited pressure drop. Consider both energy loss due to friction and minor losses. Use a method similar to that described in Example Problem 11.6.

# 12 Parallel Pipe Line Systems

| 12.1 The Big Picture |
|---|

## Discussion Map

☐ Parallel pipe line systems are those having more than one path for the fluid to take as it flows from a source to a destination point. See Figure 12.1.

☐ The principle of continuity for steady flow requires that the volume flow rate into the branching system is the same as that leaving the system.

☐ Continuity also requires that the sum of the flows in all of the branches must equal the total volume flow rate into the system.

☐ Each unit weight of fluid entering a parallel system experiences the same energy loss regardless of the path taken through the system.

☐ The fluid will tend to follow the path of lower resistance and, therefore, the incoming flow is divided among all of the branches with greater flow in the branches with the least resistance.

### Discover

■ Try to find examples of parallel flow systems around your home, in your car, or at your place of work.
■ Sketch any system you find, showing the main pipe, all of the branches, the sizes of pipe or tube used, and any valves or fittings.
■ Do the branches reconnect at some point or do they remain separate?

In this chapter you will learn analytical techniques to predict how the flow splits among all of the paths in a parallel system and how much the pressure drops across such a system.

Parallel pipe line systems are those having more than one path for the fluid to take as it flows from a source to a destination point. Look at Figure 12.1, for example. Imagine that you are a small part of the fluid stream entering the system from the left and you find yourself at point 1. The total volume flow rate here is called $Q_1$ and you are a part of it. As you then enter the junction point, you have a decision to make. Which way do you go as you proceed to the destination? All the other parts of the flow must make the same decision.

Of course, some of the flow goes into each of the three branches that lead away from the junction, called $a$, $b$, and $c$ in the figure. These flow rates are called $Q_a$, $Q_b$, and $Q_c$. You will learn in this chapter that the important thing for you to determine is how much fluid flows into each branch and how much pressure drop occurs as the fluid completes the circuit and ends up at the destination. In this case, the three paths rejoin at the right of the system

**FIGURE 12.1** Example of a parallel pipe line system with three branches.

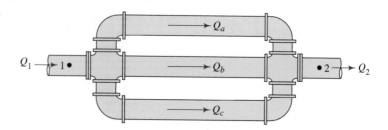

and flow on through an outlet pipe to point 2, the destination. Here the volume flow rate is called $Q_2$.

When we apply the principle of steady flow to a parallel system, we can reach the following conclusion,

**CONTINUITY EQUATION FOR
PARALLEL SYSTEMS**

$$Q_1 = Q_2 = Q_a + Q_b + Q_c \qquad (12\text{--}1)$$

The first part, $Q_1 = Q_2$, says just what we have said about previous steady flow systems; that the volume flow rate is the same at any particular cross section when the total flow is considered. No fluid has been added to or taken away from the system between points 1 and 2. The second part defines that the branch flows, $Q_a + Q_b + Q_c$, must sum to the total volume flow rate. This should seem logical because all the fluid that flows into the left junction must go somewhere and it splits into three parts. Finally, you should see that all of the flows from the branches must come together at the right junction in order for the total flow to continue as $Q_2$.

Now let's consider the pressure drop across the system. At point 1 there is a pressure $p_1$. At point 2 there is a different pressure $p_2$. The pressure drop is then, $p_1 - p_2$. To help analyze the pressures, use the energy equation between points 1 and 2.

$$\frac{p_1}{\gamma} + z_1 + \frac{v_1^2}{2g} - h_L = \frac{p_2}{\gamma} + z_2 + \frac{v_2^2}{2g}$$

Solving for the pressure drop, $p_1 - p_2$, gives,

$$p_1 - p_2 = \gamma[(z_2 - z_1) + (v_2^2 - v_1^2)/2g + h_L]$$

This form of the energy equation says that the difference in pressure between points 1 and 2 depends on the elevation difference, the difference in the velocity heads, and the energy loss per unit weight of fluid flowing in the system. When any element of fluid reaches point 2 in the system shown in Figure 12.1, each will have experienced the same elevation change and the same velocity change and the same energy loss per unit weight regardless of the path taken. All elements converging in the junction at the right side of the system have the same total energy per unit weight. That is, they all have the same total head. Therefore, each unit weight of fluid must have the same amount of energy. This can be stated mathematically as,

**HEAD LOSS EQUATION FOR
PARALLEL SYSTEMS**

$$h_{L\,1\text{--}2} = h_a = h_b = h_c \qquad (12\text{--}2)$$

Equations 12–1 and 12–2 are the governing relationships for parallel pipe line systems. The system automatically adjusts the flow in each branch until the total system flow satisfies these equations.

***Discover.***    Try to find examples of parallel flow systems. Look around your home, in your car, or at your place of work. Try to identify fluid flow systems in which the incoming flow splits into two or more branches. In some cases the branches will rejoin as they do in Figure 12.1. In other cases, the flow may continue along the branch lines until it gets to a point where it is delivered away from the system.

Consider the flow of water through your home. Follow its path as described next.

- There is a supply pipe coming into the house from the main source. That might come from a municipal water supply or from a well.
- The supply pipe then splits to take some water to a water heater while the rest continues on.
- The water line exiting from the water heater typically is routed back alongside the original pipe carrying cold water.
- When the two pipes get to their destination points, each can deliver a certain flow rate of water depending on the total resistance in the line that it followed. The resistances are made up from same elements as you learned in Chapters 9, 10, and 11. There will be friction in the pipes, energy losses due to elbows and bends, and energy losses in restrictions such as valves, faucets, and shower spray heads.

Let's say, for example, that the hot and cold water pipes end at separate faucets at a sink. And let's also assume that the hot water line offers more total resistance than the cold water line because it has to travel the extra path through the water heater. Then, if the two faucets were the same and if they were opened the same amount, more cold water would flow into the sink than hot water.

How can you obtain an equal flow from each faucet?

One way is to close the cold water faucet a bit more to provide a higher resistance. When the total resistance of the cold water line, including the faucet, is the same as the total resistance experienced by the water that traveled through the hot water line, the flow rate from each faucet will be equal.

But, now what will happen when someone else in the house opens another faucet? If they open a cold water faucet, some of the cold water supply will flow to that faucet and less will be available to the faucet at your sink. Conversely, if they open a hot water faucet, less hot water will flow into your sink.

Perhaps you experience this phenomenon while taking a shower. Did you freeze or did you get scalded from a change in the amount of hot or cold water flowing into your shower head?

The parallel flow system in your home looks different from that in Figure 12.1 because the flow does not rejoin to a single pipe at the end of the system. Instead, each branch terminates at an outlet such as a faucet, a shower head, a dishwasher, or some other appliance requiring water. But the principal is the same because, whenever any line discharges fluid, the total pressure head is zero because the water discharges to the atmosphere. The pressure at the supply line has been dissipated because of the myriad ways energy is lost from the water as it travels through the system.

What other parallel fluid flow systems did you discover?

In this chapter you will learn analytical techniques to predict how the flow splits among all of the paths in a parallel system and how much the pressure drops across such a system. You will see examples taken from commercial and industrial applications.

**12.2**
**OBJECTIVES**

After completing this chapter, you should be able to:

1. Discuss the difference between series pipe line systems and parallel pipe line systems.
2. State the general relationships for flow rates and head losses for parallel pipe line systems.
3. Compute the amount of flow that occurs in each branch of a two-branch parallel pipe line system and the head loss that occurs across the system when the total flow rate and the description of the system are known.
4. Determine the amount of flow that occurs in each branch of a two-branch parallel pipe line system and the total flow if the pressure drop across the system is known.

**5.** Use the Hardy Cross technique to compute the flow rates in all branches of a network having three or more branches.

**12.3**
**SYSTEMS WITH TWO**
**BRANCHES**

A common parallel piping system includes two branches arranged as shown in Fig. 12.2. The lower branch is added to allow some fluid to bypass the heat exchanger. The branch could also be used to isolate the heat exchanger, allowing continuous flow while the equipment is serviced. The analysis of this type of system is relatively simple and straightforward, although some iteration is typically required. Because velocities are unknown, friction factors are also unknown.

**FIGURE 12.2**   Parallel system with two branches.

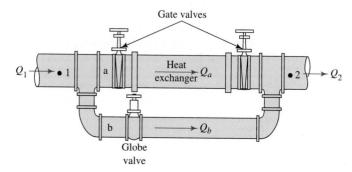

Parallel systems having more than two branches are more complex because there are many more unknown quantities than there are equations relating the unknowns. A solution procedure is described in Section 12.4.

We will use the system shown in Fig. 12.2 to illustrate the analysis of flow in two branches. The basic relationships which apply here are similar to Eqs. (12–1) and (12–2) except there are only two branches instead of three. These relationships are

$$Q_1 = Q_2 = Q_a + Q_b \qquad (12\text{--}3)$$
$$h_{L_{1-2}} = h_a = h_b \qquad (12\text{--}4)$$

The following example problems are presented in the programmed form. You should pay careful attention to the logic of the solution procedure as well as to the details performed.

**SOLUTION METHOD FOR SYSTEMS WITH TWO BRANCHES WHEN THE TOTAL**
**FLOW RATE AND THE DESCRIPTION OF THE BRANCHES ARE KNOWN**

Example Problem 12.1 is of this type. The method of solution is

**1.** Equate the total flow rate to the sum of the flow rates in the two branches, as stated in Eq. (12–3). Then express the branch flows as the product of the flow area and the average velocity; that is,

$$Q_a = A_a v_a \quad \text{and} \quad Q_b = A_b v_b$$

**2.** Express the head loss in each branch in terms of the velocity of flow in that branch and the friction factor. Include all significant losses due to friction and minor losses.

**3.** Compute the relative roughness $D/\epsilon$ for each branch, estimate the value of the friction factor for each branch, and complete the calculation of head loss in each branch in terms of the unknown velocities.

4. Equate the expression for the head losses in the two branches to each other as stated in Eq. (12–4).
5. Solve for one velocity in terms of the other from the equation in Step 4.
6. Substitute the result from Step 5 into the flow rate equation developed in Step 1, and solve for one of the unknown velocities.
7. Solve for the second unknown velocity from the relationship developed in Step 5.
8. If there is doubt about the accuracy of the value of the friction factor used in Step 2, compute the Reynolds number for each branch and reevaluate the friction factor from the Moody diagram or compute the values for the friction factors from Eq. (9–5) in Chapter 9.
9. If the values for the friction factor have changed significantly, repeat Steps 3–8, using the new values for friction factor.
10. When satisfactory precision has been achieved, use the now-known velocity in each branch to compute the volume flow rate for that branch. Check the sum of the volume flow rates to ensure that it is equal to the total flow in the system.
11. Use the velocity in either branch to compute the head loss across that branch, employing the appropriate relationship from Step 3. This head loss is also equal to the head loss across the entire branched system. You can compute the pressure drop across the system, if desired, by using the relationship $\Delta p = \gamma h_L$.

## PROGRAMMED EXAMPLE PROBLEM

☐ **EXAMPLE PROBLEM 12.1**   In Fig. 12.2, 100 gal/min of water at 60°F is flowing in a 2-in Schedule 40 steel pipe at Section 1. The heat exchanger in branch a has a loss coefficient of $K = 7.5$ based on the velocity head in the pipe. All three valves are wide open. Branch b is a bypass line composed of $1^1/_4$-in Schedule 40 steel pipe. The elbows are standard. The length of pipe between points 1 and 2 in branch b is 20 ft. Because of the size of the heat exchanger, the length of pipe in branch a is very short and friction losses can be neglected. For this arrangement, determine (a) the volume flow rate of water in each branch and (b) the pressure drop between points 1 and 2.

*Solution*   If we apply Step 1 of the solution method, Eq. (12–3) relates the two volume flow rates. How many quantities are unknown in this equation?

---

The two velocities $v_a$ and $v_b$ are unknown. Since $Q = Av$, Eq. (12–3) can be expressed as

$$Q_1 = A_a v_a + A_b v_b \qquad (12\text{–}5)$$

From the given data, $A_a = 0.02333$ ft², $A_b = 0.01039$ ft², $Q_1 = 100$ gal/min. Expressing $Q_1$ in the units of ft³/s gives

$$Q_1 = 100 \text{ gal/min} \times \frac{1 \text{ ft}^3/\text{s}}{449 \text{ gal/min}} = 0.223 \text{ ft}^3/\text{s}$$

Generate another equation which also relates $v_a$ and $v_b$ using Step 2.

---

Equation (12–4) states that the head loss in each branch is equal. Since the head losses $h_a$ and $h_b$ are dependent on the velocities $v_a$ and $v_b$, this equation can be used in conjunction with Eq. (12–5) to solve for the velocities. Now, express the head losses in terms of the velocities for each branch.

---

You should have something similar to this for branch a.

$$h_a = 2K_1(v_a^2/2g) + K_2(v_a^2/2g)$$

Where

$K_1 = f_{aT}(L_e/D) =$ resistance coefficient for each gate valve

$K_2 =$ resistance coefficient for the heat exchanger $= 7.5$ (Given in problem statement)

The following data are known:

$$f_{aT} = 0.019 \text{ for a 2-in Schedule 40 pipe (Table 10.5)}$$
$$L_e/D = 8 \text{ for a fully open gate valve (Table 10.4)}$$

Then,

$$K_1 = (0.019)(8) = 0.152$$

Then,

$$h_a = (2)(0.152)(v_a^2/2g) + 7.5(v_a^2/2g) = 7.80(v_a^2/2g) \qquad (12\text{--}6)$$

For branch b:

$$h_b = 2K_3(v_b^2/2g) + K_4(v_b^2/2g) + K_5(v_b^2/2g)$$

Where     $K_3 = f_{bT}(L_e/D) =$ resistance coefficient for each elbow

$K_4 = f_{bT}(L_e/D) =$ resistance coefficient for the glove valve

$K_5 = f_b(L_e/D) =$ friction loss in the pipe of branch b

The value of $f_b$ is not known and will be determined through iteration. The known data are

$$f_{bT} = 0.022 \text{ for a } 1^{1}/_4\text{-in Schedule 40 pipe (Table 10.5)}$$
$$L_e/D = 30 \text{ for each elbow (Table 10.4)}$$
$$L_e/D = 340 \text{ for a fully open globe valve (Table 10.4)}$$

Then,

$$K_3 = (0.022)(30) = 0.66$$
$$K_4 = (0.022)(340) = 7.48$$
$$K_5 = f_b(20/0.1150) = 173.9\,f_b$$

Then,

$$h_b = (2)(0.66)(v_b^2/2g) + (7.48)(v_b^2/2g) + f_b(173.9)(v_b^2/2g)$$
$$h_b = (8.80 + 173.9f_b)(v_b^2/2g)$$

This equation introduces the additional unknown, $f_b$. We can use an iteration procedure similar to that used for Class II series pipe line systems in Chapter 11. The relative roughness for branch b will aid in the estimation of the first trial value for $f_b$.

$$D/\epsilon = (0.1150/1.5 \times 10^{-4}) = 767$$

From the Moody diagram in Fig. 9.2, a logical estimate for the friction factor $f_b = 0.023$. Substituting this into the equation for $h_b$ gives

$$h_b = [8.80 + 173.9(0.023)](v_b^2/2g) = 12.80(v_b^2/2g) \qquad (12\text{--}7)$$

We now have completed Step 3 of the solution procedure. Steps 4 and 5 can be done now to obtain an expression for $v_a$ in terms of $v_b$.

---

You should have $v_a = 1.281v_b$. Here is how it is done.

$$h_a = h_b$$
$$7.80(v_a^2/2g) = 12.80(v_b^2/2g)$$

Solving for $v_a$ gives

$$v_a = 1.281v_b \qquad (12\text{--}8)$$

At this time, you can combine Eqs. (12–5) and (12–8) to calculate the velocities (Steps 6 and 7).

---

The solutions are $v_a = 5.54$ ft/s and $v_b = 7.09$ ft/s. Here are the details:

$$Q_1 = A_a v_a + A_b v_b \qquad (12\text{--}5)$$
$$v_a = 1.281v_b \qquad (12\text{--}8)$$

Then we have

$$Q_1 = A_a(1.281v_b) + A_b v_b = v_b(1.281A_a + A_b)$$

Solving for $v_b$, we get

$$v_b = \frac{Q_1}{1.281A_a + A_b} = \frac{0.223 \text{ ft}^3/\text{s}}{[(1.281)(0.02333) + 0.01039] \text{ ft}^2}$$
$$v_b = 5.54 \text{ ft/s}$$
$$v_a = (1.281)(5.54) \text{ ft/s} = 7.09 \text{ ft/s}$$

Since we made these calculations using an assumed value for $f_b$, we should check the accuracy of the assumption.

We can evaluate the Reynolds number for branch b.

$$N_{Rb} = v_b D_b/v$$

From Appendix Table A.2, we find $v = 1.21 \times 10^{-5}$ ft$^2$/s. Then,

$$N_{Rb} = (5.54)(0.1150)/(1.21 \times 10^{-5}) = 5.26 \times 10^4$$

Using this value and the relative roughness of 767 from before in the Moody diagram yields a new value for $f_b = 0.025$. Because this is significantly different from the assumed value of 0.023, we can repeat the calculations for Steps 3 to 8. The results are summarized below.

$$h_b = [8.80 + 173.9(0.025)](v_b^2/2g) = 13.15(v_b^2/2g) \qquad (12\text{--}7)$$
$$h_a = 7.80(v_a^2/2g) \qquad \text{(Same as for first trial)}$$

Equating the head losses in the two branches,

$$h_a = h_b$$
$$7.80(v_a^2/2g) = 13.15(v_b^2/2g)$$

Solving for the velocities,

$$v_a = 1.298v_b$$

Substituting this into the equation for $v_b$ used before gives

$$v_b = \frac{0.223 \text{ ft}^3/\text{s}}{[(1.298)(0.02333) + 0.01039] \text{ ft}^2} = 5.48 \text{ ft/s}$$

$$v_a = 1.298v_b = 1.298(5.48) = 7.12 \text{ ft/s}$$

Recomputing the Reynolds number for branch b gives

$$N_{Rb} = v_b D_b/\nu$$

$$N_{Rb} = (5.48)(0.1150)/(1.21 \times 10^{-5}) = 5.21 \times 10^4$$

There is no significant change in the value of $f_b$. Therefore the values of the two velocities computed above are correct. We can now complete Steps 10 and 11 of the procedure to find the volume flow rate in each branch and the head loss and pressure drop across the entire system.

Now calculate the volume flow rates $Q_a$ and $Q_b$ (Step 10.)

---

You should have

$$Q_a = A_a v_a = (0.02333 \text{ ft}^2)(7.12 \text{ ft/s}) = 0.166 \text{ ft}^3/\text{s}$$

$$Q_b = A_b v_b = (0.01039 \text{ ft}^2)(5.48 \text{ ft/s}) = 0.057 \text{ ft}^3/\text{s}$$

Converting these values to the units of gal/min gives $Q_a = 74.5$ gal/min and $Q_b = 25.5$ gal/min.

We are also asked to calculate the pressure drop. How can this be done?

---

We can write the energy equation using points 1 and 2 as reference points. Since the velocities and elevations are the same at these points, the energy equation is simply

$$\frac{p_1}{\gamma} - h_L = \frac{p_2}{\gamma}$$

Solving for the pressure drop, we get

$$p_1 - p_2 = \gamma h_L \tag{12-9}$$

What can be used to calculate $h_L$?

---

Since $h_{L_{1-2}} = h_a = h_b$, we can use either Eq. (12-6) or (12-7). Using Eq. (12-6), we get

$$h_a = 7.80(v_a^2/2g) = (7.80)(7.12)^2/64.4 \text{ ft} = 6.14 \text{ ft}$$

Note that this neglects the minor losses in the two tees. Then we have

$$p_1 - p_2 = \gamma h_L = \frac{62.4 \text{ lb}}{\text{ft}^3} \times 6.14 \text{ ft} \times \frac{1 \text{ ft}^2}{144 \text{ in}^2} = 2.66 \text{ psi}$$

This example problem is concluded.

■

**SOLUTION METHOD FOR SYSTEMS WITH TWO BRANCHES WHEN THE PRESSURE DROP ACROSS THE SYSTEM IS KNOWN AND THE VOLUME FLOW RATE IN EACH BRANCH AND THE TOTAL VOLUME FLOW RATE ARE TO BE COMPUTED**

Example Problem 12.2 is of this type. The method of solution is

1. Compute the total head loss across the system using the known pressure drop, $\Delta p$, in the relation $h_L = \Delta p / \gamma$
2. Write expressions for the head loss in each branch in terms of the velocity in that branch and the friction factor.
3. Compute the relative roughness $D/\epsilon$ for each branch, assume a reasonable estimate for the friction factor, and complete the calculation for the head loss in terms of the velocity in each branch.
4. Letting the magnitude of the head loss in each branch equal the total head loss as found in Step 1, solve for the velocity in each branch by using the expression found in Step 3.
5. If there is doubt about the accuracy of the value of the friction factor used in Step 3, compute the Reynolds number for each branch and reevaluate the friction factor from the Moody diagram in Fig. 9.2 or compute the value of the friction factor from Eq. (9–5).
6. If the values for the friction factor have changed significantly, repeat Steps 3 and 4, using the new values for friction factor.
7. When satisfactory precision has been achieved, use the now-known velocity in each branch to compute the volume flow rate for that branch. Then, compute the sum of the volume flow rates, which is equal to the total flow in the system.

## PROGRAMMED EXAMPLE PROBLEM

☐ **EXAMPLE PROBLEM 12.2**   The arrangement shown in Fig. 12.3 is used to supply lubricating oil to the bearings of a large machine. The bearings act as restrictions to the flow. The resistance coefficients are 11.0 and 4.0 for the two bearings. The lines in each branch are 1/2-in steel tubing with a wall thickness of 0.049 in. Each of the four bends in the tubing has a mean radius of 100 mm. Include the effect of these bends, but exclude the friction losses since the lines are short. Determine (a) the flow rate of oil in each bearing and (b) the total flow rate in L/min. The oil has a specific gravity of 0.881 and a kinematic viscosity of $2.50 \times 10^{-6}$ m²/s. The system lies in one plane so all elevations are equal.

*Solution*   Write the equation that relates the head loss $h_L$ across the parallel system to the head losses in each line $h_a$ and $h_b$.

**FIGURE 12.3**   Parallel system for Example Problem 12.2.

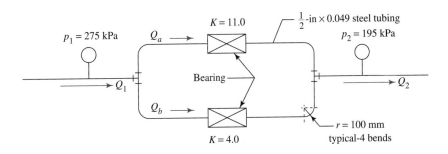

You should have

$$h_L = h_a = h_b \tag{12-10}$$

They are all equal. Determine the magnitude of these head losses by using Step 1.

---

We can find $h_L$ from the energy equation:

$$\frac{p_1}{\gamma} + z_1 + \frac{v_1^2}{2g} - h_L = \frac{p_2}{\gamma} + z_2 + \frac{v_2^2}{2g}$$

But $z_1 = z_2$ and $v_1 = v_2$. Then,

$$\frac{p_1}{\gamma} - h_L = \frac{p_2}{\gamma}$$

$$h_L = (p_1 - p_2)/\gamma \tag{12-11}$$

Using the given data, we get

$$h_L = \frac{(275 - 195)\text{ kN}}{\text{m}^2} \times \frac{\text{m}^3}{(0.881)(9.81)\text{ kN}}$$

$$h_L = 9.26 \text{ m}$$

Now write the expressions for $h_a$ and $h_b$, Step 2.

---

Considering the losses in the bends and in the bearings, you should have

$$h_a = 2K_1(v_a^2/2g) + K_2(v_a^2/2g) \tag{12-12}$$
$$h_b = 2K_1(v_b^2/2g) + K_3(v_b^2/2g) \tag{12-13}$$

Where

$K_1 = f_T(L_e/D)$ = resistance coefficient for each bend

$K_2$ = resistance coefficient for the bearing in branch a = 11.0 (Given in problem statement)

$K_3$ = resistance coefficient for the bearing in branch b = 4.0 (Given in problem statement)

$f_T$ = friction factor in the zone of complete turbulence in the steel tube

$(L_e/D)$ = equivalent length ratio for each bend. (Chapter 10, Figure 10.27)

We need the relative radius for the bends.

$$r/D = (100 \text{ mm})/(10.21 \text{ mm}) = 9.79$$

From Figure 10.27 we find $L_e/D = 29.5$

The friction factor in the zone of complete turbulence can be determined by using the relative roughness, $D/\epsilon$, and the Moody diagram, reading at the right end of the relative roughness curve where it approaches a horizontal line.

$$D/\epsilon = 0.010\ 21 \text{ m}/4.6 \times 10^{-5} \text{ m} = 222$$

We can read $f_T = 0.029$ from the Moody diagram. Now we can complete Step 3 by evaluating all of the resistance factors and expressing the energy loss in each branch in terms of the velocity head in the branch.

$$K_1 = f_T(L_e/D) = (0.029)(29.5) = 0.856$$

$$K_2 = 11.0$$

$$K_3 = 4.0$$

$$h_a = (2)(0.856)(v_a^2/2g) + 11.0(v_a^2/2g)$$

$$h_a = 12.71v_a^2/2g \qquad (12\text{--}14)$$

$$h_b = (2)(0.856)(v_b^2/2g) + 4.0(v_b^2/2g)$$

$$h_b = 5.71v_b^2/2g \qquad (12\text{--}15)$$

To complete Step 4, compute the velocities $v_a$ and $v_b$.

---

We found earlier that $h_L = 9.26$ m. Since $h_L = h_a = h_b$, Eqs. (12–14) and (12–15) can be solved directly for $v_a$ and $v_b$:

$$h_a = 12.71v_a^2/2g$$

$$v_a = \sqrt{\frac{2gh_a}{12.71}} = \sqrt{\frac{(2)(9.81)(9.26)}{12.71}} \ \text{m/s} = 3.78 \ \text{m/s}$$

$$h_b = 5.71v_b^2/2g$$

$$v_b = \sqrt{\frac{2gh_b}{5.71}} = \sqrt{\frac{(2)(9.81)(9.26)}{5.71}} \ \text{m/s} = 5.64 \ \text{m/s}$$

Now find the volume flow rates, as called for in Step 7.

---

You should have $Q_a = 18.6$ L/min, $Q_b = 27.7$ L/min, and the total volume flow rate $= 46.3$ L/min. The area of each tube is $8.189 \times 10^{-5}$ m². Then we have

$$Q_a = A_a v_a = 8.189 \times 10^{-5} \, \text{m}^2 \times 3.78 \ \text{m/s} \times \frac{60\ 000 \ \text{L/min}}{\text{m}^3/\text{s}}$$

$$Q_a = 18.6 \ \text{L/min}$$

Similarly,

$$Q_b = A_b v_b = 27.7 \ \text{L/min}$$

Then the total flow rate is

$$Q_1 = Q_a + Q_b = (18.6 + 27.7)\text{L/min} = 46.3 \ \text{L/min}$$

This example problem is concluded.

$\blacksquare$

---

**12.4**
**SYSTEMS WITH THREE
OR MORE BRANCHES—
NETWORKS**

When three or more branches occur in a pipe flow system, it is called a *network*. Networks are indeterminate because there are more unknown factors than there are independent equations relating the factors. For example, in Fig. 12.4 there are three unknown velocities, one in each pipe. The equations available to describe the system are

$$Q_1 = Q_2 = Q_a + Q_b + Q_c \qquad (12\text{--}16)$$

$$h_{L_{1-2}} = h_a = h_b = h_c \qquad (12\text{--}17)$$

**FIGURE 12.4**   Network with three branches.

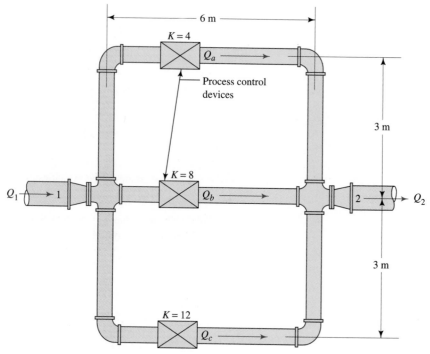

Note: Inlet and outlet pipes: 2-in Sch. 40
    Branch pipes a, b, and c: 1-in Sch. 40
    Elbows are standard

A third independent equation is required to solve explicitly for the three velocities, and none is available.

A rational approach to complete the analysis of a system such as that shown in Fig. 12.4 employing an iteration procedure has been developed by Hardy Cross.* This procedure converges on the correct flow rates quite rapidly. Many calculations are still required, but they can be set up in an orderly fashion for use on a calculator or digital computer.

The Cross technique requires that the head loss terms for each pipe in the system be expressed in the form

$$h = kQ^n \tag{12–18}$$

where $k$ is an equivalent resistance to flow for the entire pipe and $Q$ is the flow rate in the pipe. We will illustrate the creation of such an expression in the example problem to follow this general discussion of the Hardy Cross technique.

You should recall that both friction losses and minor losses are proportional to the velocity head, $v^2/2g$. Then, using the continuity equation we can express the velocity in terms of the volume flow rate. That is,

$$v = Q/A$$

* Hardy Cross, *Analysis of flow in networks of conduits or conductors,* University of Illinois Engineering Experiment Station Bulletin No. 286 (Urbana: University of Illinois, November 1936).

and

$$v^2 = Q^2/A^2$$

This will allow the development of an equation of the form shown in Eq. (12–18).

The Cross iteration technique requires that initial estimates for the volume flow rate in each branch of the system be made. Two factors that help in making these estimates are

1. At each junction in the network, the sum of the flow into the junction must equal the flow out.
2. The fluid tends to follow the path of least resistance through the network. Therefore, a pipe having a lower value of $k$ will carry a higher flow rate than those having higher values.

The network should be divided into a set of closed loop circuits prior to the beginning of the iteration process. Figure 12.5 shows a schematic representation of a 3-pipe system such as that shown in Fig. 12.4. The dashed arrows drawn in a clockwise direction assist in defining the signs for the flow rates $Q$ and the head losses $h$ in the various pipes of each loop according to the following convention:

*If the flow in a given pipe of a circuit is clockwise,* **Q** *and* **h** *are positive.*
*If the flow is counterclockwise,* **Q** *and* **h** *are negative.*

Then for circuit 1 in Fig. 12.5, $h_a$ and $Q_a$ are positive, while $h_b$ and $Q_b$ are negative. The signs are critical to the correct calculation of adjustments to the volume flow rates, indicated by $\Delta Q$, that are produced at the end of each iteration cycle. Notice that pipe b is common to both circuits. Therefore, the adjustments $\Delta Q$ for each circuit must be applied to the flow rate in this pipe.

**FIGURE 12.5**  Closed loop circuits used for the Hardy Cross technique for analysis of pipe networks.

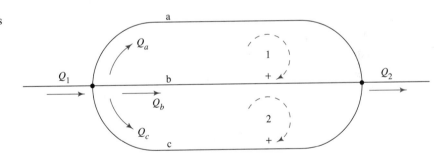

The Cross technique for analyzing the flow in pipe networks is presented in step-by-step form below. A programmed example problem follows to illustrate the application of the procedure.

### CROSS TECHNIQUE FOR ANALYSIS OF PIPE NETWORKS

1. Express the energy loss in each pipe in the form $h = kQ^2$.
2. Assume a value for the flow rate in each pipe such that the flow into each junction equals the flow out of the junction.
3. Divide the network into a series of closed loop circuits.
4. For each pipe, calculate the head loss $h = kQ^2$, using the assumed value of $Q$.

5. Proceeding around each circuit, algebraically sum all values for *h* using the following sign convention:
   If the flow is clockwise, *h* and *Q* are positive.
   If the flow is counterclockwise, *h* and *Q* are negative.
   The resulting summation is referred to as Σ*h*.
6. For each pipe, calculate $2kQ$.
7. Sum all values of $2kQ$ for each circuit, assuming all are positive. This summation is referred to as $\Sigma(2kQ)$.
8. For each circuit, calculate the value of $\Delta Q$ from

$$\Delta Q = \frac{\Sigma h}{\Sigma(2kQ)} \qquad (12\text{--}19)$$

9. For each pipe, calculate a new estimate for *Q* from

$$Q' = Q - \Delta Q$$

10. Repeat Steps 4–8 until $\Delta Q$ from Step 8 becomes negligibly small. The $Q'$ value is used for the next cycle of iteration.

## PROGRAMMED EXAMPLE PROBLEM

☐ **EXAMPLE PROBLEM 12.3**

For the system shown in Fig. 12.4, determine the volume flow rate of water at 15°C through each branch if 600 L/min (0.01 m³/s) is flowing into and out of the system through the 2-in pipes.

**Solution**

The head loss in each pipe should now be expressed in the form of $h = kQ^2$ as Step 1 of the procedure. Consider branch a first and write an expression for the head loss $h_a$.

---

The total head loss for the branch is due to the two elbows (each with $L_e/D = 30$), the restriction (with $K = 4.0$ based on the velocity head in the pipe), and friction in the pipe. Then,

$$h_a = 2(f_{aT})(30)(v_a^2/2g) + 4.0(v_a^2/2g) + f_a(L_a/D_a)(v_a^2/2g)$$
$$\text{(elbows)} \qquad \text{(restriction)} \qquad \text{(friction)}$$

The friction factor $f_a$ for flow in the pipe is dependent on the Reynolds number and therefore on the volume flow rate. Because that is the objective of the network analysis, we cannot determine that value explicitly at this time. Furthermore, the flow rate will, in general, be different in each segment of the flow system resulting in different values for the friction factor. We will take that into account in the present analysis by computing the value of the friction factor after assuming the magnitude of the volume flow rate in each pipe, a step that is inherent in the Hardy Cross technique. We will use the Swamee-Jain method to compute the friction factor from Eq. (9–5). Then we will recompute the values of the friction factors for each trial as the value of the volume flow rate is refined.

First, let's simplify the equation for $h_a$ by completing as many calculations as we can. What values can be determined?

---

The total length of pipe in branch a is 12 m and for the 1-in Schedule 40 pipe $D = 0.0266$ m and $A = 5.574 \times 10^{-4}$ m². From Table 10.5 we can find that the value of

$f_{aT} = 0.023$ for a 1-in Schedule 40 steel pipe with flow in the completely turbulent zone. The water at 15°C has a kinematic viscosity $\nu = 1.15 \times 10^{-6}$ m²/s. Also, we can introduce the volume flow rate $Q$ into the equation by noting that, as shown before,

$$v_a^2 = Q_a^2/A_a^2$$

Substitute these values into the equation for $h_a$ now and simplify it as much as possible.

---

You should have something like this.

$$h_a = [60(f_{aT}) + 4.0 + (f_a)(12/0.0266)](v_a^2/2g)$$
$$h_a = [60(f_{aT}) + 4.0 + 451(f_a)](Q_a^2/2gA^2)$$
$$h_a = [60(0.023) + 4.0 + 451(f_a)]\left[\frac{Q_a^2}{2(9.81)(5.574 \times 10^{-4})^2}\right]$$
$$h_a = [5.38 + 451(f_a)](1.64 \times 10^5)Q_a^2 \tag{12--20}$$

It is also convenient to express the Reynolds number in terms of the volume flow rate, $Q$, and to compute the value for the relative roughness, $D/\epsilon$. Do that now.

---

Because all three branches have the same size and type of pipe, these calculations apply to each branch. If different pipes are used throughout the network, these calculations must be redone for each pipe. For the 1-in steel pipe,

$$D/\epsilon = (0.0266 \text{ m})/(4.6 \times 10^{-5} \text{ m}) = 578$$

We should modify the Reynolds number formula as

$$N_{Ra} = \frac{v_a D_a}{\nu} = \frac{Q_a D_a}{A_a \nu} = \frac{Q_a(0.0266)}{(5.574 \times 10^{-4})(1.15 \times 10^{-6})}$$
$$N_{Ra} = (4.15 \times 10^7)Q_a \tag{12--21}$$

Now create expressions for the head losses in the other two pipes, $h_b$ and $h_c$, using similar procedures.

---

Compare your results with these. Note that the pipe size in branches b and c is the same as that in branch a. For branch b:

$$h_b = 8.0(v_b^2/2g) + f_b(L_b/D_b)(v_b^2/2g)$$
$$\text{(restriction)} \qquad \text{(friction)}$$
$$h_b = [8.0 + f_b(6/0.0266)](Q_b^2/2gA^2)$$
$$h_b = [8.0 + 225.6(f_b)](1.64 \times 10^5)Q_b^2 \tag{12--22}$$

For branch c:

$$h_c = 2(f_{cT})(30)(v_c^2/2g) + 12.0(v_c^2/2g) + f_c(L_c/D_c)(v_c^2/2g)$$
$$\text{(elbows)} \qquad \text{(restriction)} \qquad \text{(friction)}$$
$$h_c = [60(f_{cT}) + 12.0 + f_c(12/0.0266)](v_c^2/2g)$$
$$h_c = [60(0.023) + 12.0 + 451f_c](Q_c^2/2gA^2)$$
$$h_c = [13.38 + 451(f_c)](1.64 \times 10^5)Q_c^2 \tag{12--23}$$

Equations 12–20 to 12–23 will be used in the calculations of head losses as the Cross iteration process continues. When the values for the friction factors are known or assumed, the head loss equations can be reduced to the form of Eq. (12–18). Often it is satisfactory to assume reasonable values for the various friction factors, as minor changes have little effect on the flow distribution and the total head loss. However, we will demonstrate the more complete solution procedure in which new friction factors are calculated for each pipe for each trial.

Step 2 of the procedure calls for estimating the volume flow rate in each branch. Which pipe should have the greatest flow rate, and which should have the least?

---

Although the final values for the friction factors could affect the magnitudes of the resistances, it appears that pipe b has the least resistance and, therefore, it should carry the greatest flow. Pipe c has the most resistance and it should carry the least flow. Many different first estimates are possible for the flow rates, but we know that

$$Q_a + Q_b + Q_c = Q_1 = 0.01 \text{ m}^3/\text{s}$$

Let's use the initial assumptions,

$$Q_a = 0.0033 \text{ m}^3/\text{s} \qquad Q_b = 0.0036 \text{ m}^3/\text{s} \qquad Q_c = 0.0031 \text{ m}^3/\text{s}$$

Step 3 of the procedure is already shown in Fig. 12.5. To complete Step 4 we need values for the friction factor in each pipe. With the assumed values for the volume flow rates we can compute the Reynolds numbers and then the friction factors. Do that now.

---

You should have, using Equation 12–21 and $D/\epsilon = 578$

$$N_{Ra} = (4.15 \times 10^7)Q_a = (4.15 \times 10^7)(0.0033 \text{ m}^3/\text{s}) = 1.37 \times 10^5$$
$$N_{Rb} = (4.15 \times 10^7)Q_b = (4.15 \times 10^7)(0.0036 \text{ m}^3/\text{s}) = 1.49 \times 10^5$$
$$N_{Rc} = (4.15 \times 10^7)Q_c = (4.15 \times 10^7)(0.0031 \text{ m}^3/\text{s}) = 1.29 \times 10^5$$

We now use Eq. (9–5) to compute the friction factor for each pipe.

$$f_a = \frac{0.25}{\left[ \log_{10}\left( \dfrac{1}{3.7(D/\epsilon)} + \dfrac{5.74}{N_{Ra}^{0.9}} \right) \right]^2}$$

$$f_a = \frac{0.25}{\left[ \log_{10}\left( \dfrac{1}{3.7(578)} + \dfrac{5.74}{(1.37 \times 10^5)^{0.9}} \right) \right]^2} = 0.0241$$

In a similar manner we compute $f_b = 0.0240$ and $f_c = 0.0242$. These values are quite close in magnitude and such precision may not be justified. However, with greater disparity among the pipes in the network, more sizable differences would occur and the accuracy of the iteration technique would depend on the accuracy of evaluating the friction factors.

Insert the friction factors and the assumed values for $Q$ into Eqs. (12–20), (12–22), and (12–23) now to compute $k_a$, $k_b$, and $k_c$.

$$h_a = [5.38 + 451(f_a)](1.64 \times 10^5)Q_a^2 = k_aQ_a^2$$
$$h_a = [5.38 + 451(0.0241)](1.64 \times 10^5)Q_a^2 = 2.67 \times 10^6 Q_a^2$$

Then $k_a = 2.67 \times 10^6$. Completing the calculation gives,

$$h_a = 2.67 \times 10^6(0.0033)^2 = 29.05$$

Similarly, for branch b:

$$h_b = [8.0 + 225.6(f_b)](1.64 \times 10^5)Q_b^2 = k_bQ_b^2$$
$$h_b = [8.0 + 225.6(0.0240)](1.64 \times 10^5)Q_b^2 = 2.20 \times 10^6 Q_b^2$$
$$h_b = 2.20 \times 10^6(0.0036)^2 = 28.53$$

For branch c:

$$h_c = [13.38 + 451(f_c)](1.64 \times 10^5)Q_c^2 = k_cQ_c^2$$
$$h_c = [13.38 + 451(0.0242)](1.64 \times 10^5)Q_c^2 = 3.99 \times 10^6 Q_c^2$$
$$h_c = 3.99 \times 10^6(0.0031)^2 = 38.31$$

This completes Step 4. Now do Step 5.

---

For circuit 1,

$$\Sigma h_1 = h_a - h_b = 29.05 - 28.53 = 0.52.$$

For circuit 2,

$$\Sigma h_2 = h_b - h_c = 28.53 - 38.31 = -9.78$$

Now do Step 6.

---

Here are the correct values for the three pipes:

$$2k_aQ_a = (2)(2.67 \times 10^6)(0.0033) = 17\,609$$
$$2k_bQ_b = (2)(2.20 \times 10^6)(0.0036) = 15\,850$$
$$2k_cQ_c = (2)(3.99 \times 10^6)(0.0031) = 24\,717$$

Roundoff differences may occur. Now do Step 7.

---

For circuit 1,

$$\Sigma(2kQ)_1 = 17\,609 + 15\,850 = 33\,459$$

For circuit 2,

$$\Sigma(2kQ)_2 = 15\,850 + 24\,717 = 40\,567$$

Now you can calculate the adjustment for the flow rates, $\Delta Q$, for each circuit, using Step 8.

---

For circuit 1,

$$\Delta Q_1 = \frac{\Sigma h_1}{\Sigma(2kQ)_1} = \frac{0.52}{33\,459} = 1.56 \times 10^{-5}$$

For circuit 2,

$$\Delta Q_2 = \frac{\Sigma h_2}{\Sigma(2kQ)_2} = \frac{-9.78}{40\,567} = -2.41 \times 10^{-4}$$

The values for $\Delta Q$ are estimates of the error in the originally assumed values for $Q$. We recommended that the process be repeated until the magnitude of $\Delta Q$ is less than 1 percent of the assumed value of $Q$. Special circumstances may warrant using a different criterion for judging $\Delta Q$.

Step 9 can now be completed. Calculate the new value for $Q_a$ before looking at the next panel.

---

The calculation is as follows:

$$Q'_a = Q_a - \Delta Q_1 = 0.0033 - 1.56 \times 10^{-5}$$
$$= 0.003\ 28\ \text{m}^3/\text{s}$$

Calculate the new value for $Q_c$ before $Q_b$. Pay careful attention to algebraic signs.

---

You should have

$$Q'_c = Q_c - \Delta Q_2 = -0.0031 - (-2.41 \times 10^{-4})$$
$$= -0.002\ 86\ \text{m}^3/\text{s}$$

Notice that $Q_c$ is negative since it flows in a counterclockwise direction in circuit 2. We can interpret the calculation for $Q'_c$ as indicating that the magnitude of $Q_c$ must be decreased in absolute value.

Now calculate the new value for $Q_b$. Remember, pipe $b$ is in each circuit.

---

Both $\Delta Q_1$ and $\Delta Q_2$ must be applied to $Q_b$. For circuit 1,

$$Q'_b = Q_b - \Delta Q_1 = -0.0036 - 1.56 \times 10^{-5}$$

This would result in an increase in the absolute value of $Q_b$. For circuit 2,

$$Q'_b = Q_b - \Delta Q_2 = +0.0036 - (-2.41 \times 10^{-4})$$

This also results in increasing $Q_b$. Then $Q_b$ is actually increased in absolute value by the sum of $\Delta Q_1$ and $\Delta Q_2$. That is,

$$Q'_b = 0.0036 + 1.56 \times 10^{-5} + 2.41 \times 10^{-4}$$
$$= 0.003\ 86\ \text{m}^3/\text{s}$$

Remember that the sum of the absolute values of the flow rates in the three pipes must equal $0.01\ \text{m}^3/\text{s}$, the total $Q$.

We can continue the iteration by using $Q'_a$, $Q'_b$, and $Q'_c$ as the new estimates for the flow rates and repeating Steps 4–8. The results for four iteration cycles are summarized in Table 12.1. You should carry out the calculations yourself before looking at the table.

---

Notice that in Trial 4, the values of $\Delta Q$ are below 1 percent of the respective values of $Q$. This is an adequate degree of precision. The results show that $Q_a = 3.399 \times 10^{-3}\ \text{m}^3/\text{s}$, $Q_b = 3.789 \times 10^{-3}\ \text{m/s}$, and $Q_c = 2.812 \times 10^{-3}\ \text{m}^3/\text{s}$ in the directions shown in Fig. 12.5. The results, expressed more conveniently in L/min, are $Q_a = 204$ L/min, $Q_b = 227$ L/min, and $Q_c = 169$ L/min. The total $Q = 600$ L/min. Once again, observe that the pipes having the lower resistances carry the greater flow rates.

■

**TABLE 12.1**

| Trial | Circuit | Pipe | $Q$ | $N_R$ | $f$ | $k$ | $h = kQ^2$ | $2kQ$ | $\Delta Q$ | % Change |
|---|---|---|---|---|---|---|---|---|---|---|
| 1 | 1 | a | $3.300 \times 10^{-3}$ | $1.37 \times 10^5$ | 0.0241 | $2.67 \times 10^6$ | 29.054 | 17609 | | 0.48 |
| | | b | $-3.600 \times 10^{-3}$ | | | $2.20 \times 10^6$ | $-28.530$ | 15850 | | $-0.44$ |
| | | | | | Sum of $h$ and $2kQ =$ | | 0.524 | 33459 | $1.568 \times 10^{-5}$ | |
| | 2 | b | $3.600 \times 10^{-3}$ | $1.49 \times 10^5$ | 0.0240 | $2.20 \times 10^6$ | 28.530 | 15850 | | $-6.70$ |
| | | c | $-3.100 \times 10^{-3}$ | $1.29 \times 10^5$ | 0.0242 | $3.99 \times 10^6$ | $-38.312$ | 24717 | | 7.78 |
| | | | | | Sum of $h$ and $2kQ =$ | | $-9.782$ | 40567 | $-2.411 \times 10^{-4}$ | |
| 2 | 1 | a | $3.284 \times 10^{-3}$ | $1.36 \times 10^5$ | 0.0241 | $2.67 \times 10^6$ | 28.784 | 17528 | | $-3.46$ |
| | | b | $-3.857 \times 10^{-3}$ | | | $2.20 \times 10^6$ | $-32.700$ | 16957 | | 2.94 |
| | | | | | Sum of $h$ and $2kQ =$ | | $-3.916$ | 34485 | $-1.135 \times 10^{-4}$ | |
| | 2 | b | $3.857 \times 10^{-3}$ | $1.60 \times 10^5$ | 0.0239 | $2.20 \times 10^6$ | 32.700 | 16957 | | 0.03 |
| | | c | $-2.859 \times 10^{-3}$ | $1.19 \times 10^5$ | 0.0243 | $4.00 \times 10^6$ | $-32.654$ | 22844 | | $-0.04$ |
| | | | | | Sum of $h$ and $2kQ =$ | | 0.046 | 39801 | $1.151 \times 10^{-6}$ | |
| 3 | 1 | a | $3.398 \times 10^{-3}$ | $1.41 \times 10^5$ | 0.0241 | $2.67 \times 10^6$ | 30.770 | 18112 | | $-0.03$ |
| | | b | $-3.742 \times 10^{-3}$ | | | $2.20 \times 10^6$ | $-30.802$ | 16462 | | 0.02 |
| | | | | | Sum of $h$ and $2kQ =$ | | $-0.032$ | 34574 | $-9.176 \times 10^{-7}$ | |
| | 2 | b | $3.742 \times 10^{-3}$ | $1.55 \times 10^5$ | 0.0240 | $2.20 \times 10^6$ | 30.802 | 16462 | | $-1.28$ |
| | | c | $-2.860 \times 10^{-3}$ | $1.19 \times 10^5$ | 0.0243 | $4.00 \times 10^6$ | $-32.680$ | 22853 | | 1.67 |
| | | | | | Sum of $h$ and $2kQ =$ | | $-1.878$ | 39315 | $-4.776 \times 10^{-5}$ | |
| 4 | 1 | a | $3.399 \times 10^{-3}$ | $1.41 \times 10^5$ | 0.0241 | $2.67 \times 10^6$ | 30.787 | 18116 | | $-0.66$ |
| | | b | $-3.789 \times 10^{-3}$ | | | $2.20 \times 10^6$ | $-31.570$ | 16664 | | 0.59 |
| | | | | | Sum of $h$ and $2kQ =$ | | $-0.783$ | 34780 | $-2.252 \times 10^{-5}$ | |
| | 2 | b | $3.789 \times 10^{-3}$ | $1.57 \times 10^5$ | 0.0240 | $2.20 \times 10^6$ | 31.570 | 16664 | | $-0.03$ |
| | | c | $-2.812 \times 10^{-3}$ | $1.17 \times 10^5$ | 0.0244 | $4.00 \times 10^6$ | $-31.612$ | 22482 | | 0.04 |
| | | | | | Sum of $h$ and $2kQ =$ | | $-0.042$ | 39146 | $-1.073 \times 10^{-6}$ | |

The results of the iteration process for the Hardy Cross technique for the data of Example Problem 12.3 as shown in Table 12.1 were found using a spreadsheet on a computer. This facilitated the sequential, repetitive calculations typically required in such problems. A computer program written in BASIC, FORTRAN, or other technical language can also be used to advantage, especially if a large number of pipes and circuits exist in the network to be analyzed.

Many network analysis computer programs are commercially available.

## REFERENCES

1. Crane Valve Group. *Flow of Fluids, Version 4.0: Pipe line system analysis software*, Long Beach, CA.
2. Cross, Hardy. 1936 (November). *Analysis of Flow in Networks of Conduits or Conductors*. University of Illinois Engineering Experiment Station Bulletin No. 286. Urbana: University of Illinois.
3. Tahoe Design Software, *HYDROFLO II and HYDRONET 4: Pipe line system analysis software*, Truckee, CA.

## PRACTICE PROBLEMS

### Systems with Two Branches

**12.1M** *Figure 12.6 shows a branched system in which the pressure at A is 700 kPa and the pressure at B is 550 kPa. Each branch is 60 m long. Neglect losses at the junctions, but consider all elbows. If the system carries oil with a specific weight of 8.80 kN/m³, calculate the total volume flow rate. The oil has a kinematic viscosity of $4.8 \times 10^{-6}$ m²/s.*

**12.2E** Using the system shown in Fig. 12.2 and the data from Example Problem 12.1, determine (a) the volume flow rate of water in each branch and (b) the pressure drop between points 1 and 2 if the first gate valve is one-half closed and the other valves are wide open.

**12.3M** *In the branched pipe system shown in Fig. 12.7, 850 L/min of water at 10°C is flowing in a 4-in Schedule 40 pipe at A. The flow splits into two 2-in Schedule 40 pipes as shown and then rejoins at B. Calculate (a) the flow rate in each of the branches and (b) the pressure difference $p_A - p_B$. Include the effect of the minor losses in the lower branch of the system. The total length*

*of pipe in the lower branch is 60 m. The elbows are standard.*

**12.4E** In the branched pipe system shown in Fig. 12.8, 1350 gal/min of benzene (sg = 0.87) at 140°F is flowing in the 8-in pipe. Calculate the volume flow rate in the 6-in and the 2-in pipes. All pipes are standard Schedule 40 steel pipes.

**12.5M** *A 150-mm pipe branches into a 100-mm and a 50-mm pipe as shown in Fig. 12.9. Both pipes are copper and 30 m long. (The fluid is water at 10°C.) Determine what the resistance coefficient K of the valve must be in order to obtain equal volume flow rates of 500 L/min in each branch.*

**12.6E** For the system shown in Fig. 12.10, the pressure at A is maintained constant at 20 psig. The total volume flow rate exiting from the pipe at B depends on which valves are open or closed. Use K = 0.9 for each elbow, but neglect the energy losses in the tees. Also, since the length of each branch is short, neglect pipe friction losses. The pipe in branch 1 has a 2-in inside diameter,

**FIGURE 12.6**   Problem 12.1.

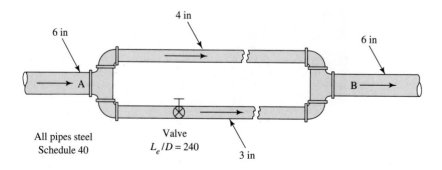

All pipes steel
Schedule 40

6 in

4 in

6 in

A

B

Valve
$L_e/D = 240$

3 in

**FIGURE 12.7** Problems 12.3 and 12.8.

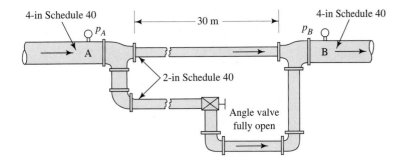

4-in Schedule 40    30 m    4-in Schedule 40

$p_A$    $p_B$

A    B

2-in Schedule 40

Angle valve fully open

**FIGURE 12.8** Problems 12.4 and 12.7.

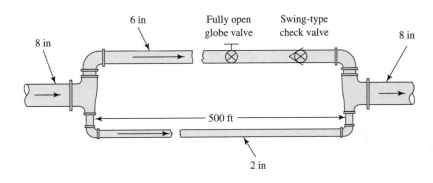

6 in    Fully open globe valve    Swing-type check valve

8 in    8 in

500 ft

2 in

**FIGURE 12.9** Problem 12.5.

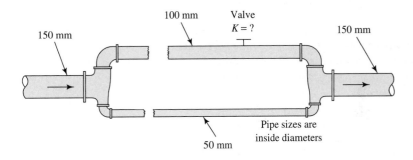

100 mm    Valve $K = ?$

150 mm    150 mm

50 mm    Pipe sizes are inside diameters

**FIGURE 12.10** Problem 12.6

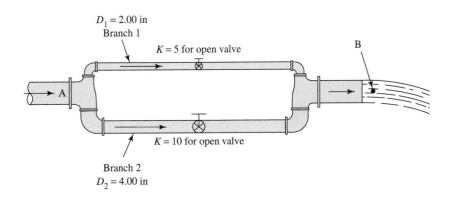

$D_1 = 2.00$ in
Branch 1

$K = 5$ for open valve

B

A

$K = 10$ for open valve

Branch 2
$D_2 = 4.00$ in

and branch 2 has a 4-in inside diameter. Calculate the volume flow rate of water for each of the following conditions:

**a.** Both valves open
**b.** Valve in branch 2 only open
**c.** Valve in branch 1 only open

**12.7E** Solve Problem 12.4, using the Hardy Cross technique.

**12.8M** *Solve Problem 12.3, using the Hardy Cross technique.*

## Networks

Note: Neglect minor losses.

**12.9E** Find the flow rate of water at 60°F in each pipe of Fig. 12.11.

**FIGURE 12.11**   Problem 12.9.

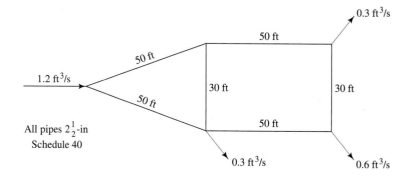

**12.10M** *Figure 12.12 represents a spray rinse system in which water at 15°C is flowing. All pipes are 3-in Type K copper tubing. Determine the flow rate in each pipe.*

**FIGURE 12.12**   Problem 12.10.

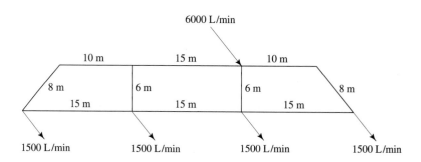

**12.11E** Figure 12.13 represents the water distribution network in a small industrial park. The supply of 15.5 ft³/s of water at 60°F enters the system at A. Manufacturing plants draw off the indicated flows at points C, E, F, G, H, and I. Determine the flow in each pipe in the system.

**FIGURE 12.13**    Problem 12.11.

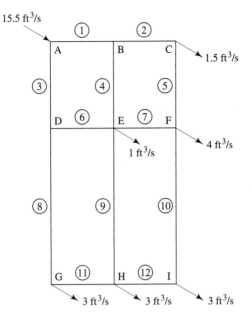

| Pipe no. | Length (ft) | Size (in) |
|---|---|---|
| 1 | 1500 | 16 |
| 2 | 1500 | 16 |
| 3 | 2000 | 18 |
| 4 | 2000 | 12 |
| 5 | 2000 | 16 |
| 6 | 1500 | 16 |
| 7 | 1500 | 12 |
| 8 | 4000 | 14 |
| 9 | 4000 | 12 |
| 10 | 4000 | 8 |
| 11 | 1500 | 12 |
| 12 | 1500 | 8 |

*Pipe Data*
All pipes Schedule 40

**12.12M** *Figure 12.14 represents the network for delivering coolant to five different machine tools in an automated machining system. The grid is a rectangle 7.5 m by 15 m. All pipes are commercial steel tubing with a 0.065-in wall thickness. Pipes 1 and 3 are 2-in diameter; pipe 2 is $1\frac{1}{2}$-in diameter; and all others are 1-in. The coolant has a specific gravity of 0.92 and a dynamic viscosity of $2.00 \times 10^{-3}$ Pa·s. Determine the flow in each pipe.*

**FIGURE 12.14**    Problem 12.12.

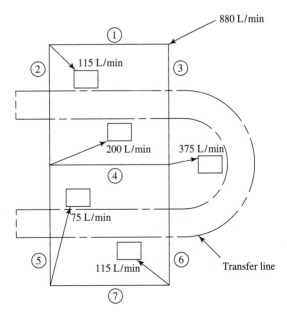

*Pipe data*
All pipes 7.5 m long
All pipes steel tubing
Wall thickness = 0.065 in

| Pipe no. | Outside diameter (in) |
|---|---|
| 1 | 2 |
| 2 | $1\frac{1}{2}$ |
| 3 | 2 |
| 4 | 1 |
| 5 | 1 |
| 6 | 1 |
| 7 | 1 |

## COMPUTER PROGRAMMING ASSIGNMENTS

1. Write a program or spreadsheet to analyze parallel pipe line systems with two branches of the type demonstrated in Example Problem 12.1. Part of the preliminary analysis, such as writing the expressions for head losses in the branches in terms of the velocities and the friction factors, may be done prior to entering data into the program.

2. Enhance the program from Assignment 1 so that it uses the Equation (9–5) from Chapter 9 to calculate the friction factor.

3. Write a program or spreadsheet to analyze parallel pipe line systems with two branches of the type demonstrated in Example Problem 12.2. Use an approach similar to that described for Program Assignment 1.

4. Enhance the program from Assignment 3 so that it uses the Equation (9–5) from Chapter 9 to calculate the friction factor.

5. Write a program or spreadsheet that uses the Hardy Cross technique, as described in Section 12.4 and illustrated in Example Problem 12.3, to perform the analysis of pipe flow networks. The following optional approaches may be taken:

   a. Consider single circuit networks with two branches as an alternative to the program from Assignment 1 or 2.

   b. Consider networks of two or more circuits similar to those described in Problems 12.9 through 12.12.

# 13 Pump Selection and Application

## 13.1
## The
## Big
## Picture

### Discussion Map

☐ Pumps are used to deliver liquids through piping systems.

☐ They must deliver the desired volume flow rate of fluid while developing the required total dynamic head, $h_a$, created by elevation changes, differences in the velocity heads, and all energy losses in the system.

☐ You need to develop the ability to specify suitable pumps to satisfy system requirements.

☐ You also need to learn how to design efficient piping systems for the inlet to a pump (the suction line) and for the discharge side of the pump.

☐ The pressure at the inlet to the pump must be analyzed to ensure proper operation of the pump.

### Discover

■ You probably encounter many different types of pumps performing many different jobs in the course of a given week. List some of them now.

■ For each pump, write down as much as you can about it and the system in which it operates.

■ Describe the function of the pump, the kind of fluid being pumped, the source of the fluid, the ultimate discharge point, and the piping system with its valves and fittings.

In this chapter you will learn how to analyze the performance of pumps and to select an appropriate pump for a given application. You will learn to design an efficient system that minimizes the amount of energy required to drive the pump.

Pumps are used to deliver liquids through piping systems. We have seen the general application of pumps in earlier chapters. In Chapter 7, when the general energy equation was introduced you learned how to determine the energy added by a pump to the fluid that we called $h_a$. Solving for $h_a$ from the general energy equation yields this form

 TOTAL HEAD ON A PUMP

$$h_a = \frac{p_2 - p_1}{\gamma} + z_2 - z_1 + \frac{v_2^2 - v_1^2}{2g} + h_L \qquad (13\text{–}1)$$

We will call this value of $h_a$ the *total head on the pump*. Some pump manufacturers refer to this as the *total dynamic head (TDH)*.

You should be able to interpret this equation as an expression for the total set of tasks the pump is asked to do in a given system.

■ It must, in general, increase the fluid pressure from the source $p_1$ to the destination point $p_2$.

■ It must raise the level of the fluid from the source $z_1$ to the destination $z_2$.

■ It must increase the velocity head from that at point 1 to that at point 2.

■ And it must overcome any energy losses that occur in the system due to friction in the pipes or energy losses in valves, fittings, process components, or changes in the flow area or direction of the flow.

It is your task to do the appropriate analysis to determine the value of $h_a$ using the techniques discussed in Chapters 11 and 12.

You also learned how to compute the power delivered to the fluid by the pump that we called $P_A$.

**POWER DELIVERED BY A PUMP TO THE FLUID**

$$P_A = h_a \gamma Q \tag{13–2}$$

There are inevitable energy losses in the pump itself because of mechanical friction and the turbulence created in the fluid as it passes through the pump. Therefore, there is more power required to drive the pump than the amount that eventually gets delivered to the fluid. You also learned in Chapter 7 to use the efficiency of the pump, $e_M$, to determine the power input to the pump, $P_I$.

**PUMP EFFICIENCY**

$$e_M = P_A/P_I \tag{13–3}$$

**POWER INPUT TO A PUMP**

$$P_I = P_A/e_M \tag{13–4}$$

For the list of pumps you developed earlier, ask yourself the following questions. Refer to Equation 13–1 as you do this.

■ Where does the fluid come from as it approaches the inlet of the pump?
■ What is the elevation, pressure and velocity of the fluid at the source?
■ What kind of fluid is in the system?
■ What is the temperature of the fluid?
■ Would you consider the fluid to have a low viscosity similar to water or a high viscosity like a heavy oil?
■ Can you name the type of pump?
■ How is the pump driven? By an electric motor? By a belt drive? Directly by an engine?
■ What elements make up the *suction line* that brings fluid to the pump inlet? Describe the pipe, valves, elbows or other elements.
■ To where is the fluid delivered? Consider its elevation, the pressure at the destination, and the velocity of flow there.
■ What elements make up the *discharge line* that takes fluid from the pump and delivers it to the destination? Describe the pipe, valves, elbows or other elements.

Compare your list with this example that describes the water pump on a household clothes washer. See Figure 7.23.

■ The task of the water pump is to bring wash water from the tub of the washer and deliver it through a hose to the laundry tub.
■ The fluid is a solution of water and a detergent or some other washing agent along with dirt from the clothes being washed.
■ The temperature of the fluid could range from cold at approximately 60°F (16°C) to hot at about 120°F (49°C).
■ The viscosity will be similar to that of water.
■ The pump is located near the bottom of the washer below the tub. Therefore, there is a positive pressure at the pump inlet due to the column of fluid above it.
■ The fluid exits from the tub through a port in its bottom and flows through a smooth rubber tube to the pump inlet. There is a 90° bend in the tube close to the pump inlet.

■ A short hose connects the discharge from the pump to a plastic fitting in the rear wall of the washer and the final discharge hose connects to the fitting.

■ The discharge hose has an inside diameter of approximately 1.0 inch (25 mm), a length of approximately 5 feet (1.5 m), and it carries the fluid to a height of approximately 36 inches.

■ The hose makes one 90° bend and one 180° bend.

■ The fluid discharges from the hose at a fairly high velocity into the tub at atmospheric pressure.

■ The pump appears to be a small centrifugal pump. (See Figure 13.12 later in this chapter.)

There are many other types of pumps described in this chapter. Look through the chapter now to get a feel of the scope of the topics covered here. Some have also been shown previously in Chapter 7 and you should review them now.

In this chapter you will learn how to analyze the performance of pumps and to select an appropriate pump for a given application. You will also see how the design of the fluid flow system affects the performance of the pump. This should help you to design an efficient system that minimizes the work required by the pump and, therefore, the amount of energy required to drive the pump.

## 13.2 OBJECTIVES

A wide variety of pumps is available to transport liquids in fluid flow systems. The proper selection and application of pumps require an understanding of their performance characteristics and typical uses.

After completing this chapter, you should be able to:

1. List the parameters involved in pump selection.
2. List the types of information that must be specified for a given pump.
3. Describe the basic pump classifications.
4. List four types of rotary positive displacement pumps.
5. List three types of reciprocating positive displacement pumps.
6. List three types of kinetic pumps.
7. Describe the main features of centrifugal pumps.
8. Describe *deep well jet pumps* and *shallow well jet pumps*.
9. Describe the typical performance curve for rotary positive displacement pumps.
10. Describe the typical performance curve for centrifugal pumps.
11. State the *affinity laws* for centrifugal pumps as they relate to the relationships among speed, impeller diameter, capacity, total head capability, and power required to drive the pump.
12. Describe how the operating point of a pump is related to the system resistance curve.
13. Define the *net positive suction head (NPSH)* for a pump, and discuss its significance in pump performance.
14. Describe the importance of the vapor pressure of the fluid in relation to the NPSH.
15. Compute the NPSH available for a given suction line design and a given fluid.
16. Define the *specific speed* for a centrifugal pump, and discuss its relationship to pump selection.
17. Describe the effect of increased viscosity on the performance of centrifugal pumps.
18. Describe the performance of parallel pumps and pumps connected in series.
19. Describe the features of a desirable suction line design.
20. Describe the features of a desirable discharge line design.

**13.3
PARAMETERS INVOLVED
IN PUMP SELECTION**

When selecting a pump for a particular application, the following factors must be considered:

1. The nature of the liquid to be pumped
2. The required capacity (volume flow rate)
3. The conditions on the suction (inlet) side of the pump
4. The conditions on the discharge (outlet) side of the pump
5. The total head on the pump (the term $h_a$ from the energy equation)
6. The type of system to which the pump is delivering the fluid
7. The type of power source (electric motor, diesel engine, steam turbine, etc.)
8. Space, weight, and position limitations
9. Environmental conditions
10. Cost of pump purchase and installation
11. Cost of pump operation
12. Governing codes and standards

The nature of the fluid is characterized by its temperature at the pumped condition, its specific gravity, its viscosity, its tendency to corrode or erode the pump parts, and its vapor pressure at the pumping temperature. The term *vapor pressure* is used to define the pressure at the free surface of a fluid due to the formation of a vapor. The vapor pressure gets higher as the temperature of the liquid increases, and it is essential that the pressure at the pump inlet stay above the vapor pressure of the fluid. We will learn more about vapor pressure in Section 13.12.

After pump selection, the following items must be specified:

1. Type of pump and manufacturer
2. Size of pump
3. Size of suction connection and type (flanged, screwed, etc.)
4. Size and type of discharge connection
5. Speed of operation
6. Specifications for driver (for example: for an electric motor—power required, speed, voltage, phase, frequency, frame size, enclosure type)
7. Coupling type, manufacturer, and model number
8. Mounting details
9. Special materials and accessories required, if any
10. Shaft seal design and seal materials

Pump catalogs and manufacturer's representatives will supply the necessary information to assist in the selection and specification of pumps and accessory equipment.

**13.4
TYPES OF PUMPS**

The types of pumps commonly used for fluid delivery can be classified as shown in Table 13.1.

**13.4.1
Positive Displacement
Pumps**

*Positive displacement pumps* ideally deliver a fixed quantity of fluid with each revolution of the pump rotor. Therefore, except for minor slippage because of clearances between the rotor and the casing, the delivery or capacity of the pump is unaffected by changes in the pressure it must develop. Most positive displacement pumps can also handle liquids with high viscosities.

In Chapter 7, Figs. 7.2 and 7.3 show two types of positive displacement pumps commonly used for fluid power applications. The gear pump (Fig. 7.2) is comprised of two counter-rotating, tightly meshing gears rotating within a housing. The outer

**TABLE 13.1** Classification of types of pumps

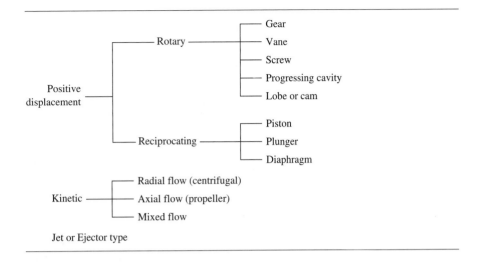

periphery of the gear teeth fit closely with the inside surface of the housing. Fluid is drawn in from the supply reservoir at the suction port and carried by the spaces between teeth to the discharge port where it is delivered at high pressure to the system. The delivery pressure is dependent on the resistance of the system. A cutaway of a commercially available gear pump is shown in part (a) of the figure. Gear pumps develop system pressures in the range of 1500 psi to 4000 psi (10.3 MPa to 27.6 MPa). Delivery varies with the size of the gears and the rotational speed which can be up to 4000 rpm. Deliveries from 1 to 50 gallons per minute (4 to 190 L/min) are possible with different size units.

Figure 7.3 shows an axial piston pump that uses a rotating swashplate that acts like a cam to reciprocate the pistons. The pistons alternately draw fluid into their cylinders through suction valves and then force it out the discharge valves against system pressure. Delivery can be varied from zero to maximum by changing the angle of the swashplate and thus changing the stroke of the pistons. Pressure capacity ranges up to 5000 psi (34.5 MPa).

Also used for fluid power, the vane pump (Fig. 13.1) consists of an eccentric rotor containing a set of sliding vanes that ride inside a housing. A cam ring in the housing controls the radial position of the vanes. Fluid enters the suction port at the left, is then captured in a space between two successive vanes, and is thus carried to the discharge port at the system pressure. The vanes then are retracted into their slots in the rotor as they travel back to the inlet, or suction, side of the pump. Variable displacement vane pumps can deliver from zero to the maximum flow rate by varying

**FIGURE 13.1** Vane pump. (Source of drawing: *Machine Design Magazine*)

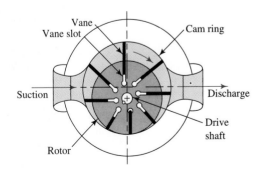

the position of the rotor with respect to the cam ring and the housing. The setting of the variable delivery can be manual, electric, hydraulic, or pneumatically actuated to tailor the performance of the fluid power unit to the needs of the system being driven. Typical pressure capacities are from 2000 to 4000 psi (13.8 to 27.6 MPa).

One disadvantage of the gear, piston, and vane pumps is that they deliver a pulsating flow to the output because each functional element moves a set, captured volume of fluid from suction to discharge. Screw pumps and progressing cavity pumps do not have this problem. Figure 13.2 shows a screw pump in which the central, thread-like power rotor meshes closely with the two idler rotors creating an enclosure inside the housing that moves axially from suction to discharge providing a continuous uniform flow. Screw pumps operate at nominally 3000 psi (20.7 MPa), can be run at high speeds, and run more quietly than most other types of hydraulic pumps.

**FIGURE 13.2**   Screw pump. (Source: IMO Industries Inc., IMO Pump Division, Monroe, NC)

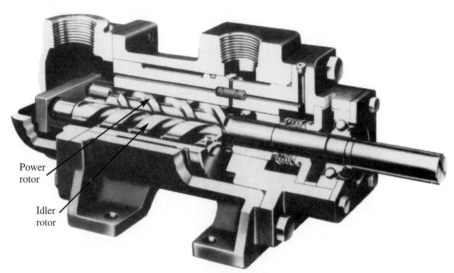

(a) Cutaway of pump assembly

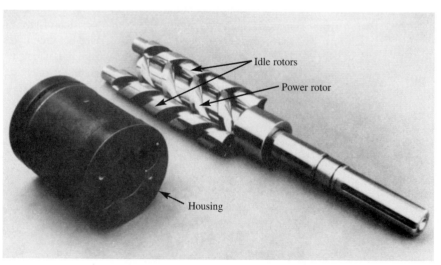

(b) Power rotor, idler rotors, and housing

The Moyno® progressing cavity pump shown in Fig. 13.3 also produces a smooth, nonpulsating flow and is used mostly for the delivery of process fluids rather than hydraulic applications. As the long central rotor turns within the stator, cavities are formed which progress toward the discharge end of the pump carrying the material being handled. The rotor is typically made from steel plated with heavy layers of hard chrome to increase resistance to abrasion. For most applications, stators are made from natural rubber or any of several types and formulations of synthetic rubbers. A compression fit exists between the metal rotor and the rubber stator to reduce slippage and improve efficiency. Delivery for a given pump is dependent on the dimensions of the rotor/stator combination and is proportional to the speed of rotation. Flow capacities range up to 1860 gal/min (7040 L/min) and pressure capability is up to 900 psi (6.2 MPa). This type of pump can handle a wide variety of fluids including clear water, slurries with heavy solids content, highly viscous liquids like adhesives and cement grout, abrasive fluids such as slurries of silicon carbide or ground limestone, pharmaceuticals such as shampoo and skin cream, corrosive chemicals such as cleaning solutions and fertilizers, and foods such as applesauce and even bread dough.

The lobe pump (Fig. 13.4), sometimes called a cam pump, operates in a similar fashion to the gear pump. The two counter-rotating rotors may have two, three, or more lobes that mesh with each other and fit closely with the housing. Fluid is conducted around by the cavity formed between successive lobes.

Piston pumps used for fluid transfer are classified as either single-acting simplex or double-acting duplex types as shown in Fig. 13.5. In principle these are similar to the fluid power piston pumps but they typically have a larger flow capacity and operate at lower pressures. Also, they are usually driven through a crank-type drive rather than the swashplate described before.

In the diaphragm pump shown in Fig. 13.6, a reciprocating rod moves a flexible diaphragm within a cavity, alternately discharging fluid as the rod moves to the left and drawing fluid in as it moves to the right. One advantage of this type of pump is that only the diaphragm contacts the fluid, eliminating contamination from the drive elements. The suction and discharge valves alternately open and close.

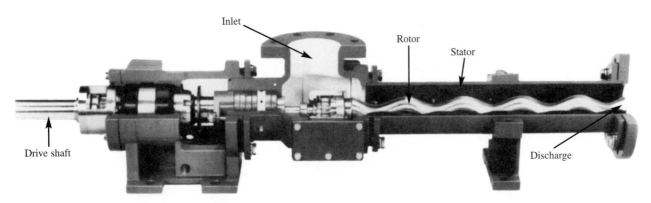

**FIGURE 13.3** Mayno® progressing cavity pump. (Source: Robbins & Myers, Inc., Fluids Handling Group, Springfield, OH)

**FIGURE 13.4** Lobe pump.

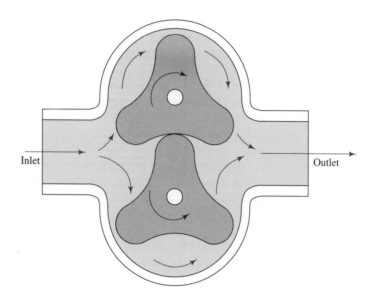

Inlet

Outlet

**FIGURE 13.5** Piston pumps for
fluid transfer.

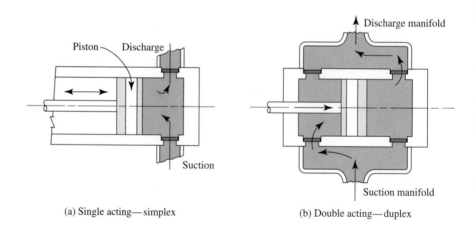

Piston   Discharge

Suction

(a) Single acting—simplex

Discharge manifold

Suction manifold

(b) Double acting—duplex

**FIGURE 13.6** Diaphragm-type
pump.

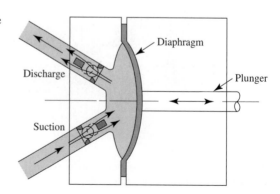

Diaphragm

Discharge

Plunger

Suction

**13.4.2**
**Kinetic Pumps**

*Kinetic pumps* add energy to the fluid by accelerating it through the action of a rotating *impeller*. Figure 13.7 shows the basic configuration of a radial flow centrifugal pump, the most common type of kinetic pump. The fluid is drawn into the center of the impeller and then thrown outward by the vanes. Leaving the impeller, the fluid passes through a spiral-shaped volute where it is gradually slowed, causing part of the kinetic energy to be converted to fluid pressure.

The propeller type of pump (axial flow) depends on the hydrodynamic action of the propeller blades to lift and accelerate the fluid axially, along a path parallel

(a) Pump and
motor

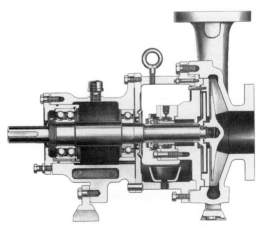

(b) Cutaway of pump

(c) Impeller

**FIGURE 13.7**   Centrifugal pump. (Source: Gould Pumps, Inc., Seneca Falls, NY)

to the axis of the propeller. The mixed flow pump incorporates some actions from both the centrifugal and propeller types.

Figure 13.8 shows the basic design of radial, axial, and mixed flow impellers.

**FIGURE 13.8**   Impellers for kinetic pumps.

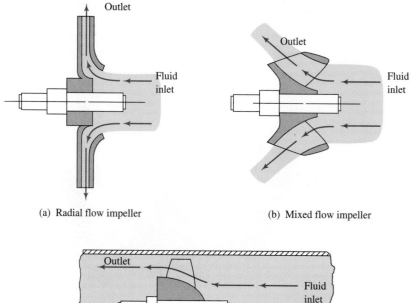

(a) Radial flow impeller          (b) Mixed flow impeller

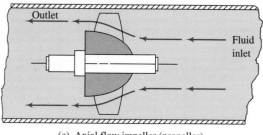

(c) Axial flow impeller (propeller)

*Jet pumps*, frequently used for household water systems, are composed of a centrifugal pump along with a jet or ejector assembly. Figure 13.9 shows a typical deep-well jet pump configuration where the main pump and motor are located above ground at the top of the well and the jet assembly is down near the water level. The pump delivers water under pressure down into the well through the pressure pipe to a nozzle. The jet issuing from the nozzle creates a vacuum behind it, which causes well water to be drawn up along with the jet. The combined stream passes through a diffuser where the flow is slowed, thus converting some of the kinetic energy of the water to pressure. Since the diffuser is inside the suction pipe, the water is carried to the inlet of the pump where it is acted on by the impeller. Part of the output is discharged to the system being supplied, while the remainder is recirculated to the jet to continue the operation.

If the well is shallow, with less than about 6.0 m (20 ft) from the pump to the water level, the jet assembly can be built into the pump body. Then the water is lifted through a single suction pipe, as shown in Fig. 13.10.

*Submersible pumps* are designed so the entire assembly of the centrifugal pump, the drive motor, and the suction and discharge apparatus can be submerged in the fluid to be pumped. Figure 13.11 shows one design that is portable and that can be installed in a pipe casing because of its small-diameter cylindrical housing. These pumps are useful for removing unwanted water from construction sites, mines, utility manholes, industrial tanks, and shipboard cargo holds. The suction for the

**FIGURE 13.9** Deep-well jet pump. (Source: Goulds Pumps, Inc. Seneca Falls, NY)

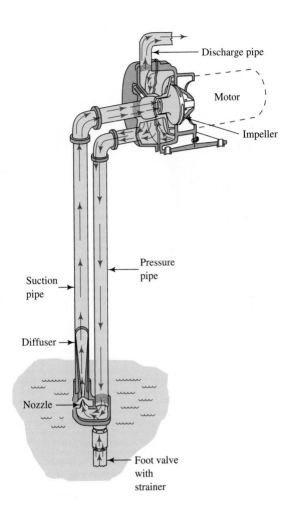

Discharge pipe

Motor

Impeller

Pressure pipe

Suction pipe

Diffuser

Nozzle

Foot valve with strainer

**FIGURE 13.10** Shallow-well jet pump. (Source: Goulds Pumps, Inc. Seneca Falls, NY)

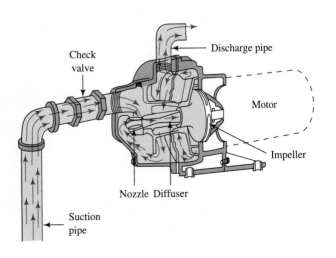

Check valve

Discharge pipe

Motor

Impeller

Nozzle  Diffuser

Suction pipe

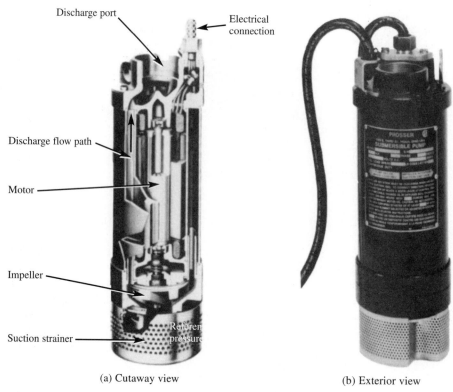

Discharge port

Electrical connection

Discharge flow path

Motor

Impeller

Suction strainer

Reference pressure

(a) Cutaway view                    (b) Exterior view

**FIGURE 13.11**   Portable submersible pump. (Source: Burks Pumps Inc., Piqua, OH)

pump is at the bottom where the water flows through a strainer and into the eye of the abrasion resistant impeller. The discharge flows upward through an annular passage between the shell and the motor housing. At the top of the unit the flow rejoins and flows into a centrally located discharge pipe or hose. The dry motor is sealed in the center of the pump.

Although most of the centrifugal pump styles discussed thus far have been fairly large and designed for commercial and industrial applications, small units are available for use in small appliances such as clothes washers and dishwashers and for small-scale products. Figure 13.12 shows one such design that is only about

**FIGURE 13.12**   Small centrifugal pump with integral motor. (Source: Gorman-Rupp Industries, Bellville, OH)

4 inches in diameter with a capacity up to 30 gal/min (114 L/min). Even smaller pumps are available.

**13.5**
**PERFORMANCE DATA**
**FOR POSITIVE**
**DISPLACEMENT PUMPS**

In this section we will discuss the general characteristics of direct-acting reciprocating pumps, and rotary pumps.

The operating characteristics of positive displacement pumps make them useful for handling such fluids as water, hydraulic oils in fluid power systems, chemicals, paint, gasoline, greases, adhesives, and some food products. Since delivery is proportional to the rotational speed of the rotor, these pumps can be used for metering. Some disadvantages of some designs include pulsating output, susceptibility to damage by solids and abrasives, and need for a relief valve. In general, they are used for high pressure applications requiring a relatively constant delivery.

**13.5.1**
**Reciprocating Pumps**

In its simplest form, the reciprocating pump (Figure 13.5) employs a piston that draws fluid into a cylinder through an intake valve as the piston draws away from the valve. Then, as the piston moves forward, the intake valve closes and the fluid is pushed out through the discharge valve. Such a pump is called *simplex*, and its curve of discharge versus time would look like that shown in Fig. 13.13(a). The resulting intermittent delivery is often undesirable. If the piston is *double acting* or *duplex*, one side of the piston delivers fluid while the other takes fluid in, resulting in the performance curve shown in Fig. 13.13(b). The delivery can be smoothed even more by having three or more pistons. Piston pumps for hydraulic systems often have five or six pistons.

**13.5.2**
**Rotary Pumps**

Figure 13.14 shows a typical set of performance curves for rotary pumps such as gear, vane, screw, and lobe pumps. It is a plot of capacity, efficiency, and power versus discharge pressure. As pressure is increased, a slight decrease in capacity

**FIGURE 13.13**   Simplex and duplex pump delivery.

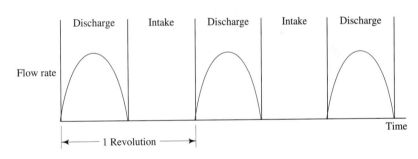

(a) Single acting pump—simplex

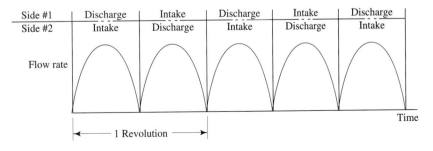

(b) Double acting pump—duplex

**FIGURE 13.14** Performance curves for a positive displacement rotary pump.

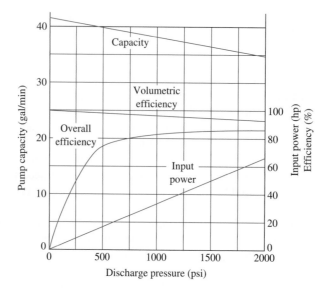

occurs, due to internal leakage from the high pressure side to the low pressure side. This is often insignificant. The power required to drive the pump varies almost linearly with pressure. Also, because of the positive displacement designs for rotary pumps, capacity varies almost linearly with the rotational speed, provided the suction conditions allow free flow into the pump.

Efficiency for positive displacement pumps is typically reported in two ways, as shown in Figure 13.14. *Volumetric efficiency* is a measure of the ratio of the volume flow rate delivered by the pump to the theoretical delivery based on the displacement per revolution of the pump times the speed of rotation. This efficiency is usually in the range from 90 to 100%, decreasing with increasing pressure in proportion to the decrease in capacity. *Overall efficiency* is a measure of the ratio of the power delivered to the fluid to the power input to the pump. Included in the overall efficiency is the volumetric efficiency, the mechanical friction from moving parts, and energy losses from the fluid as it passes through the pump. When operating at design conditions, rotary positive displacement pumps exhibit an overall efficiency ranging from 80 to 90%.

**13.6
PERFORMANCE DATA
FOR CENTRIFUGAL
PUMPS**

Because centrifugal pumps are not positive displacement types, there is a strong dependency between capacity and the pressure that must be developed by the pump. This makes their performance ratings somewhat more complex. The typical rating curve plots the total head on the pump $h_a$ versus the capacity or discharge $Q$, as shown in Fig. 13.15. The total head $h_a$ is calculated from the general energy equation, as described in Chapter 7. It represents the amount of energy added to a unit weight of the fluid as it passes through the pump. See also Equation 13–1.

As shown in Fig. 13.7, there are large clearances between the rotating impeller and the casing of the pump. This accounts for the decrease in capacity as the total head increases. Indeed, at a cut-off head, the flow is stopped completely when all of the energy input from the pump goes to maintain the head. Of course, the typical operating head is well below the cut-off head so that high capacity can be achieved.

Efficiency and power required are also important to the successful operation of a pump. Figure 13.16 shows a more complete performance rating of a pump, su-

**FIGURE 13.15**  Performance curve for a centrifugal pump—Total head versus capacity.

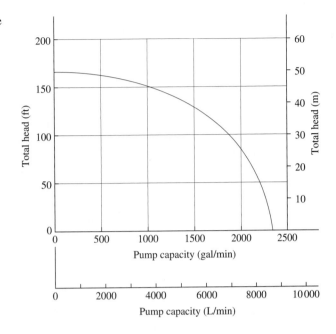

perimposing head, efficiency, and power curves and plotting all three versus capacity. Normal operation should be in the vicinity of the peak of the efficiency curve, with efficiencies in the range of 60 percent to 80 percent being typical for centrifugal pumps.

**13.7**
**AFFINITY LAWS FOR**
**CENTRIFUGAL PUMPS**

Most centrifugal pumps can be operated at different speeds to obtain varying capacities. Also, a given size pump casing can accommodate impellers of differing diameters. It is important to understand the manner in which capacity, head, and power vary when either speed or impeller diameter is varied. These relationships, called

**FIGURE 13.16**  Centrifugal pump performance curves.

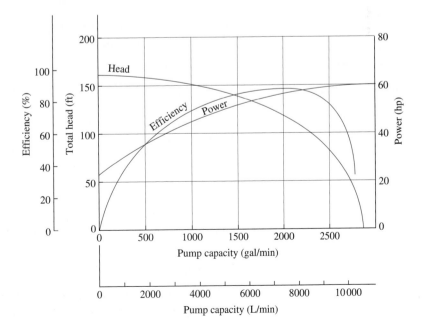

*affinity laws*, are listed here. The symbol $N$ refers to the rotational speed of the impeller, usually in revolutions per minute (r/min, or rpm).

When *speed varies*,

**a.** Capacity varies directly with speed:

$$\frac{Q_1}{Q_2} = \frac{N_1}{N_2} \tag{13-5}$$

**b.** The total head capability varies with the square of the speed:

$$\frac{h_{a_1}}{h_{a_2}} = \left(\frac{N_1}{N_2}\right)^2 \tag{13-6}$$

**c.** The power required by the pump varies with the cube of the speed:

$$\frac{P_1}{P_2} = \left(\frac{N_1}{N_2}\right)^3 \tag{13-7}$$

When *impeller diameter varies*,

**a.** Capacity varies directly with impeller diameter:

$$\frac{Q_1}{Q_2} = \frac{D_1}{D_2} \tag{13-8}$$

**b.** The total head varies with the square of the impeller diameter:

$$\frac{h_{a_1}}{h_{a_2}} = \left(\frac{D_1}{D_2}\right)^2 \tag{13-9}$$

**c.** The power required by the pump varies with the cube of the impeller diameter:

$$\frac{P_1}{P_2} = \left(\frac{D_1}{D_2}\right)^3 \tag{13-10}$$

Efficiency remains nearly constant for speed changes and for small changes in impeller diameter. (See Reference 2.)

---

☐ **EXAMPLE PROBLEM 13.1**     Assume that the pump for which the performance data are plotted in Fig. 13.16 was operating at a rotational speed of 1750 rpm and that the impeller diameter was 13 inches. First determine the head that would result in a capacity of 1500 gal/min and the power required to drive the pump. Then, compute the performance at a speed of 1250 rpm.

**Solution**     From Fig. 13.16, projecting upward from $Q_1 = 1500$ gal/min gives:

$$\text{Total head} = 130 \text{ ft} = h_{a_1}$$
$$\text{Power required} = 50 \text{ hp} = P_1$$

When the speed is changed to 1250 rpm, the new performance can be computed by using the affinity laws.

Capacity:    $Q_2 = Q_1(N_2/N_1) = 1500(1250/1750) = 1071$ gal/min

Head:    $h_{a_2} = h_{a_1}(N_2/N_1)^2 = 130(1250/1750)^2 = 66.3$ ft

Power:    $P_2 = P_1(N_2/N_1)^3 = 50(1250/1750)^3 = 18.2$ hp

Note the significant decrease in the power required to run the pump. If the capacity and the available head are adequate, large savings in energy costs can be obtained by varying the speed of operation of a pump.

■

**13.8
MANUFACTURER'S
DATA FOR
CENTRIFUGAL PUMPS**

Because they are able to use different impeller diameters and speeds, pump manufacturers can cover a wide range of requirements for capacity and head with a few basic pump sizes. Figure 13.17 shows a composite rating chart for one line of pumps which allows the quick determination of the pump size. Then, for each pump size, more complete performance charts are prepared as shown next.

**13.8.1
Effect of Impeller Size**

Figure 13.18 shows how the performance of a given pump varies as the size of the impeller varies. The 2 × 3 − 10 centrifugal pump is one with a 2-in discharge connection, a 3-in suction connection, and a casing that can accommodate an impeller with a diameter of 10 in or smaller. Shown are the capacity versus head curves for five different sizes of impellers in this same casing. The operating speed is 3500 rpm, which corresponds to the full load speed of a 2-pole electric motor.

**13.8.2
Effect of Speed**

Figure 13.19 is the performance of the same 2 × 3 − 10 pump operating at 1750 rpm (a standard 4-pole motor speed) instead of 3500 rpm. If we compare the maximum total heads for each impeller size, we illustrate the affinity law; that is, doubling the speed increases the total head capability by a factor of 4 (the square of the speed ratio). If the curves are extrapolated down to the zero total head point where the maximum capacity occurs, we see demonstrated that capacity doubles as the speed doubles.

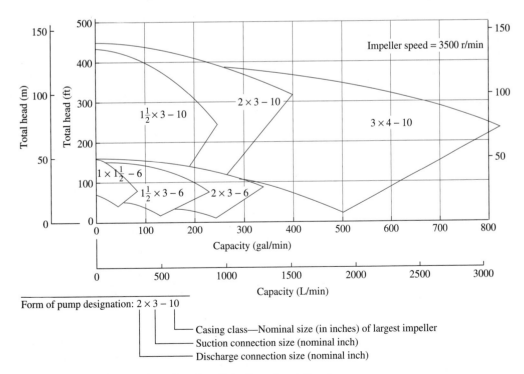

**FIGURE 13.17**   Composite rating chart for a line of centrifugal pumps.

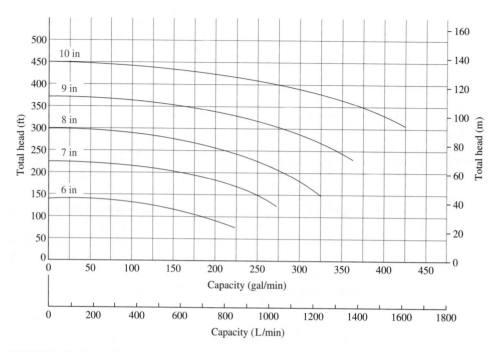

**FIGURE 13.18**   Illustration of pump performance for different impeller diameters.

**FIGURE 13.19**   Pump performance for 2 × 3 − 10 centrifugal pump operating at 1750 rpm.

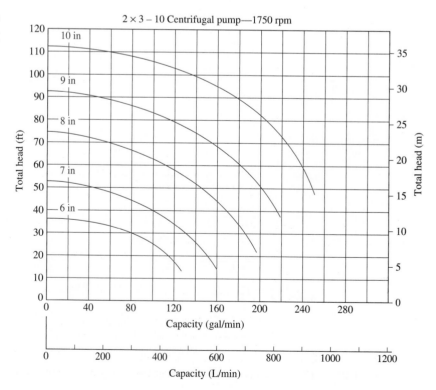

**13.8.3
Power Required**

Figure 13.20 is the same as Fig. 13.18, except that the curves showing the power required to drive the pump have been added. For example, the pump with an 8-in impeller would deliver 215 gal/min against a total head of 250 ft of fluid. Under these conditions, the pump would draw 23 hp. The same pump would deliver 280 gal/min at 200 ft of head and would draw 26 hp.

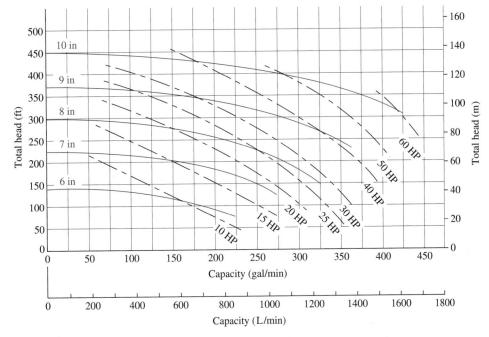

**FIGURE 13.20**   Illustration of pump performance for different impeller diameters with power required. Performance chart for $2 \times 3 - 10$ centrifugal pump at 3500 rpm.

**13.8.4
Efficiency**

Figure 13.21 is the same as Fig. 13.18, except that curves of constant efficiency have been added. The maximum efficiency for this pump is approximately 57 percent. Of course, it is desirable to operate a given pump near its best efficiency point.

**13.8.5
Net Positive Suction
Head Required**

Net positive suction head required ($NPSH_R$) is an important factor to consider in applying a pump, as will be discussed in Section 15.10. $NPSH_R$ is related to the pressure at the inlet to the pump. For this discussion, it is sufficient to say that a low $NPSH_R$ is desirable. For the pump in Fig. 13.22, the range is from about 4.5 ft of fluid at low capacities to over 12 ft of fluid at higher capacities.

**13.8.6
Composite Performance
Chart**

Figure 13.23 puts all these data together on one chart so the user can see all important parameters at the same time. The chart seems complicated at first, but, considering each individual part separately helps to interpret it correctly.

---

□ **EXAMPLE PROBLEM 13.2**

A centrifugal pump must deliver at least 250 gal/min of water at a total head of 300 ft of water. Specify a suitable pump. List its performance characteristics.

*Solution*

One possible solution can be found from Fig. 13.23. The $2 \times 3 - 10$ pump with a 9-in impeller will deliver approximately 275 gal/min at 300 ft of head. At this operating point, the

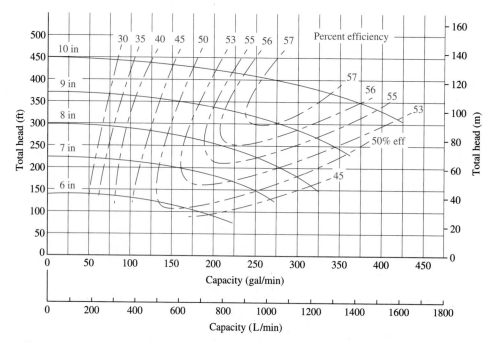

**FIGURE 13.21** Illustration of pump performance for different impeller diameters with efficiency. Performance chart for $2 \times 3 - 10$ centrifugal pump at 3500 rpm.

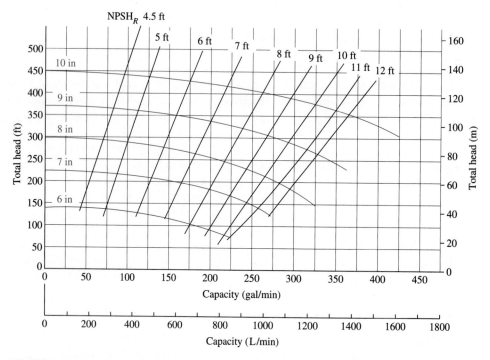

**FIGURE 13.22** Illustration of pump performance for different impeller diameters with net positive suction head required. Performance chart for $2 \times 3 - 10$ centrifugal pump at 3500 rpm.

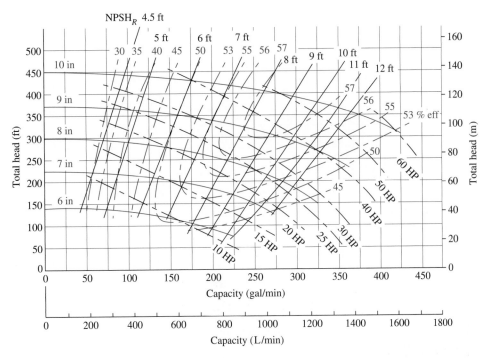

**FIGURE 13.23** Composite pump performance chart for $2 \times 3 - 10$ centrifugal pump at 3500 rpm.

efficiency would be 57 percent, near the maximum for this type of pump. Approximately 37 hp would be required. The $NPSH_R$ at the suction inlet to the pump is approximately 9.2 ft of water.

■

**13.8.7
Additional Performance
Charts**

Figures 13.24 to 13.29 show the composite performance charts for six other medium sized centrifugal pumps. They range in size from $1\frac{1}{2} \times 3 - 6$ to $6 \times 8 - 17$. Maximum capacities range from approximately 130 gal/min (492 L/min) up to nearly 4000 gal/min (15 140 L/min). A total head up to 700 ft (213 m) of fluid can be developed within the pumps in these figures. Note that Figs. 13.24 to 13.27 are for pumps operating at approximately 1750 rpm while Figs. 13.28 and 13.29 are for 3560 rpm.

Figures 13.30 and 13.31 show two additional performance curves for smaller centrifugal pumps. Because these pumps are generally offered with only one impeller size, the manner of displaying the performance parameters is somewhat different. Complete curves for total head, efficiency, input power required, and NPSH required are given versus the pump capacity. Each pump will deliver approximately 19 gal/min at the peak efficiency point. But the pump in Figure 13.30 has a smaller impeller diameter giving a total head capability of 32 ft at 19 gal/min while the larger pump in Figure 13.31 has a total head capability of 43 ft at the same capacity.

You need to develop the ability to interpret performance data from these charts so you can specify a suitable pump for a given application. Several design projects are described at the end of this chapter that require you to specify a pump

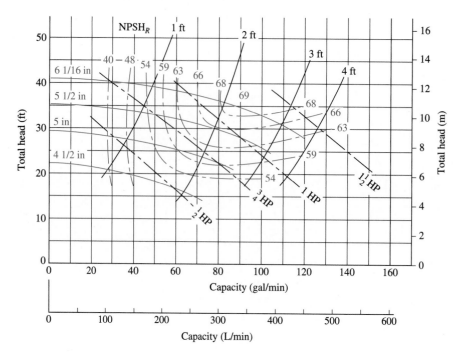

**FIGURE 13.24**  Performance for $1\frac{1}{2} \times 3 - 6$ centrifugal pump at 1750 rpm. (Source: Goulds Pumps, Inc., Seneca Falls, NY)

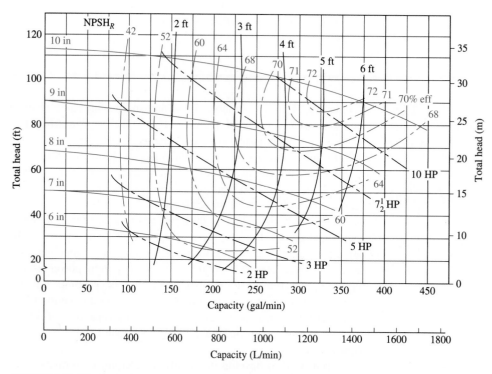

**FIGURE 13.25**  Performance for $3 \times 4 - 10$ centrifugal pump at 1750 rpm. (Source: Goulds Pumps, Inc., Seneca Falls, NY)

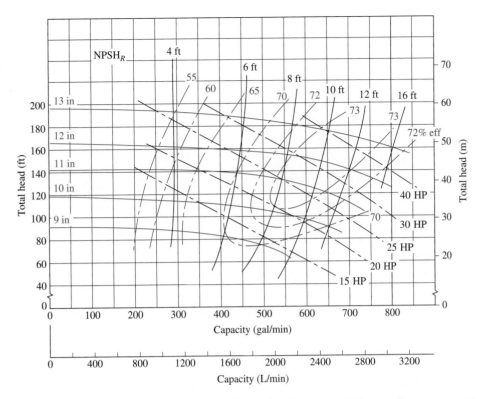

**FIGURE 13.26** Performance for 3 × 4 − 13 centrifugal pump at 1780 rpm. (Source: Goulds Pumps, Inc., Seneca Falls, NY)

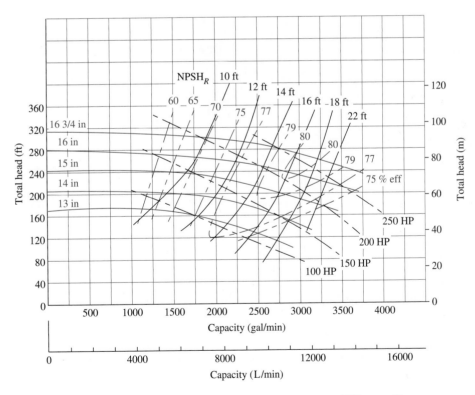

**FIGURE 13.27** Performance for 6 × 8 − 17 centrifugal pump at 1780 rpm. (Source: Goulds Pumps, Inc., Seneca Falls, NY)

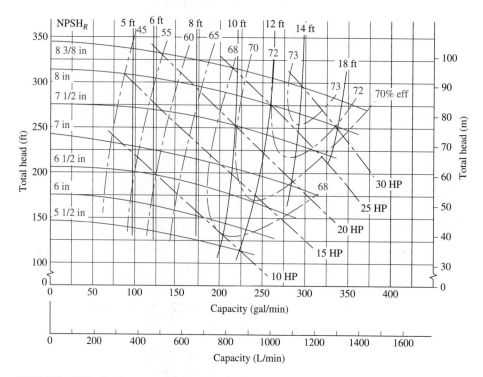

**FIGURE 13.28** Performance for 2 × 3 − 8 centrifugal pump at 3560 rpm. (Source: Goulds Pumps, Inc., Seneca Falls, NY)

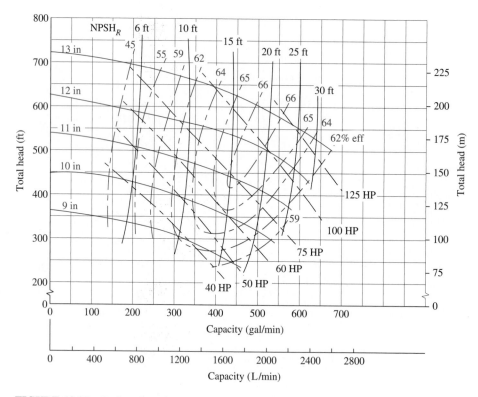

**FIGURE 13.29** Performance for 1¹/₂ × 3 − 13 centrifugal pump at 3560 rpm. (Source: Goulds Pumps, Inc., Seneca Falls, NY)

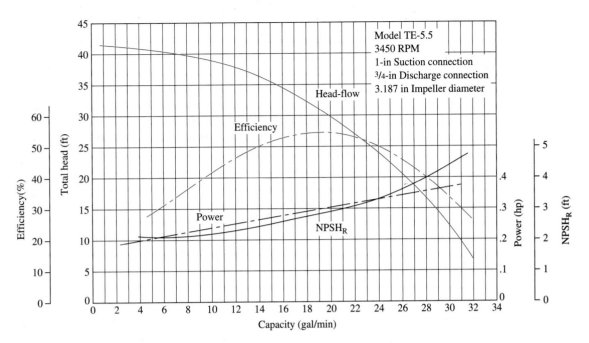

**FIGURE 13.30**  Model TE-5.5 Centrifugal pump. (Source: March Manufacturing, Inc., Glenview, IL)

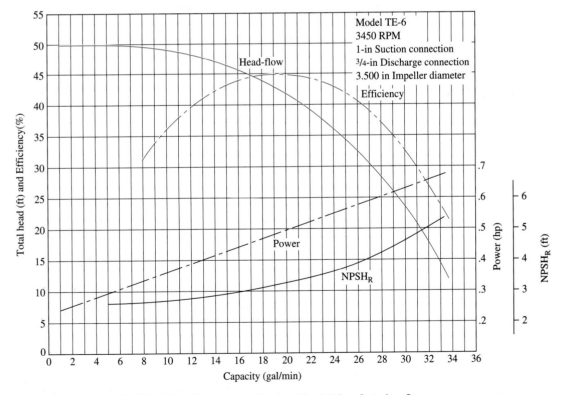

**FIGURE 13.31**  Model TE-6 Centrifugal pump. (Source: March Manufacturing, Inc., Glenview, IL)

to deliver a particular volume flow rate at a given head to meet the demands of a particular system.

**13.9**
**THE OPERATING POINT**
**OF A PUMP AND PUMP**
**SELECTION**

The operating point of a pump is defined as the volume flow rate it will deliver when installed in a given system. The total head developed by the pump is determined by the system resistance corresponding to the same volume flow rate. Figure 13.32 illustrates the concept. The pump rating curve is the plot of the volume flow rate that the pump delivers as a function of the total head it is subjected to by the system of which it is a part. Such curves are the basic elements of Figures 13.23 to 13.31.

**FIGURE 13.32**   Operating point of a pump.

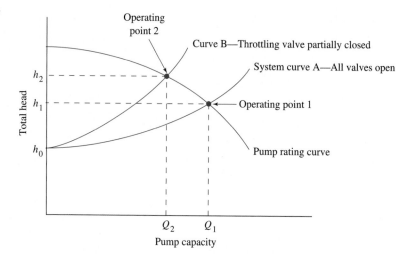

Now look at *System curve A* in Figure 13.32. This is a plot of the resistance exhibited by a given system with all of its valves fully open. Let's discuss the shape of this curve. At the left end, the curve starts at a particular value of total head corresponding to a volume flow rate of zero. This depicts the system resistance before any flow is developed. But the pump is holding the fluid at the elevation of the destination point in the system and maintaining the fluid pressure at that point. This point is called the *total static head*, $h_0$, where,

$$h_0 = (p_2 - p_1)/\gamma + (z_2 - z_1)$$

This equation, derived from the energy equation, says that the pump must develop a head equal to the pressure head difference between the two reference points plus the elevation head difference before any flow is delivered.

But the pump is capable of working against a higher head and, in fact, delivering fluid to the system. As soon as fluid starts to flow through the pipes, valves, and fittings of the system, more head is developed because of the energy losses that occur. Recall that the energy losses are proportional to the velocity head in the pipes and, therefore, they increase according to the square of the volume flow rate. This accounts for the curved (exponential) shape of the system curve.

As the flow increases with its corresponding increase in total head, the system curve eventually intersects the pump rating curve.

*Where the system curve and the pump rating curve intersect is the true operating point of the pump in this system.*

This determines how much flow is actually delivered into the system. The pump will automatically seek this operating point after it is energized. Thus when operating at this point, the pump delivers a flow rate $Q_1$ against a total head $h_1$.

But suppose that you really wanted to deliver a lower flow rate $Q_2$. One way to accomplish that with this particular pump is to increase the resistance (total head) on the pump which would cause the operating point to move back to the left along the pump rating curve. You could do this by partially closing a valve in the discharge line, a process called *throttling*. The increased resistance would shift the system curve to that labeled *Curve B* and the new operating point 2 would result in the delivery of the desired flow rate $Q_2$ at the new total head $h_2$.

It should be understood that throttling is generally not desirable because energy represented by the difference between the total heads at points 1 and 2 is virtually wasted. You should carefully specify a pump that has your desired operating point very close to the pump curve without throttling. If your system must be operated at different flow rates, it is more desirable to use a variable speed drive that is discussed later in this chapter.

### 13.9.1 Pump Selection Procedure

Example Problem 13.3 follows to demonstrate the general procedure for designing a system for a particular application. As a part of the process, the data for the system curve is generated, a pump is selected that will deliver at least the desired flow rate, and the resulting operating point for the selected pump is determined. Finally, we show the performance of the pump at the operating point, giving the actual expected flow rate, the total head on the pump, the power required to drive the pump, the efficiency of the pump, and the net positive suction head required $(NPSH_R)$.

More will be said in a later section of this chapter about the design details for suction lines and discharge lines for pumped systems. Also, the concept of *net positive suction head (NPSH)* and its relationship to the occurrence of *cavitation* will be discussed in more detail later.

---

☐ **EXAMPLE PROBLEM 13.3**

Figure 13.33 shows a system in which a pump is required to deliver at least 225 gal/min of water at 60°F from a lower reservoir to an elevated tank maintained at a pressure of 35.0 psig. Design the system and specify a suitable pump. Then determine the operating point for the pump in the system you have designed and give the performance parameters for the pump at the operating point.

**Solution**

*Step 1.* Propose the general layout of the system.

This is completed for this example problem as shown in Figure 13.33. The following are its primary features.

- The pump is placed near to the source and located below the level of fluid in the reservoir. This eases the ability of the pump to draw fluid without priming and it should maintain a relatively high pressure at the pump inlet to decrease the probability of creating cavitation in the pump.
- The suction line is a total of 8.0 feet long.
- A fully open gate valve is placed in the suction line before the pump to permit the inflow to be shut off when the pump is not in operation. It also may facilitate service or replacement of the pump.
- In the discharge line we have placed a check valve that prohibits the backflow of fluid from the upper tank when the pump is not operating.

**FIGURE 13.33**  System for
Example Problem 13.1.

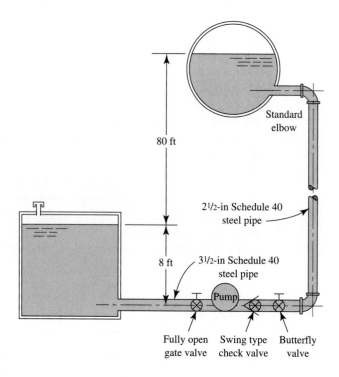

- We have also included a fully open butterfly valve beyond the check valve that can be shut off to isolate the pump for service or replacement. The butterfly valve could be used as well for a modest amount of throttling to fine tune the delivery of water.
- The discharge line contains two standard elbows and a total of 360 feet of pipe.

*Step 2.* Specify the sizes for the pipes. Use Table 6.3 in Chapter 6 as a guide.

- A 3½-inch Schedule 40 steel pipe is suitable for the suction line, giving a velocity of approximately 7.5 ft/sec. Its flow area is 0.06868 ft² and its inside diameter is 0.2957 ft. The actual flow velocity is

$$Q = (225 \text{ gal/min}) \frac{1 \text{ ft}^3/\text{s}}{449 \text{ gal/min}} = 0.5011 \text{ ft}^3/\text{s}$$

$$v_s = Q/A_s = (0.5011 \text{ ft}^3/\text{s})/(0.06868 \text{ ft}^2) = 7.296 \text{ ft/s}$$

- Using a 2½-inch Schedule 40 steel pipe for the discharge line will result in a velocity of approximately 15 ft/s. Its actual flow area is 0.03326 ft² and its inside diameter is 0.2058 ft. The actual flow velocity is

$$v_d = Q/A_d = (0.5011 \text{ ft}^3/\text{s})/(0.03326 \text{ ft}^2) = 15.07 \text{ ft/s}$$

*Step 3.* Using the energy equation, determine the equation for the total head on the pump at the desired operating conditions. First specify the reference points in the system that encompass all of the work the pump is required to do.

For this problem, let us pick point 1 to be at the surface of the lower reservoir where the pressure is 0 psig and the velocity is very nearly zero. Let's take point 2 at the surface of the upper tank where the pressure is 35.0 psig and the velocity is very nearly zero. The difference in elevation between these two points is 80 ft.

Then, using Equation 13.1 we can develop an expression for the total head on the pump.

$$h_a = \frac{p_2 - p_1}{\gamma} + z_2 - z_1 + \frac{v_2^2 - v_1^2}{2g} + h_L$$

The velocity head term is zero; $z_2 - z_1 = 80$ ft; and $p_1$ = zero. Also,

$$\frac{p_2}{\gamma} = \frac{35.0 \text{ lb}}{\text{in}^2} \frac{144 \text{ in}^2}{\text{ft}^2} \frac{\text{ft}^3}{62.4 \text{ lb}} = 80.77 \text{ ft}$$

Then,

$$h_a = p_2/\gamma + 80 \text{ ft} + h_L = 80.77 \text{ ft} + 80.0 \text{ ft} + h_L$$

$$h_a = 160.77 \text{ ft} + h_L$$

*Step 4.* Evaluate all energy losses at the desired flow rate.

You may choose to use a spreadsheet similar to that in Figure 11.3 in Chapter 11 for this part of the process. Figure 13.34 shows the results for this problem, giving the total energy loss of 139.0 ft and the total head on the pump of 299.8 ft when the flow rate is 225 gal/min (0.5011 ft$^3$/s). This is the *desired operating point* for the pump.

*Step 5.* Specify a suitable pump.

Note: Standards have been set jointly by the American National Standards Institute (ANSI) and the Hydraulic Institute (HI) calling for a *preferred operating region (POR)* for centrifugal pumps to be between 70% and 120% of the best efficiency point (BEP). See ANSI/HI 9.6.3-1997, *Standard for Centrifugal and Vertical Pumps for Allowable Operating Region.*

For this problem, let us inspect the pump curves shown in Figures 13.23 to 13.31. The 2 × 3 − 10 pump operating at 3500 rpm shown in Figure 13.23 is suitable. The desired operating point lies between the curves for the 8-inch and 9-inch diameter impellers. We must choose the 9-inch impeller to deliver at least the desired 225 gal/min flow rate.

*Step 6.* Create the system curve and determine the actual operating point for the pump.

The system curve is a plot of the total head $h_a$ versus the flow rate for the proposed system. We have used the spreadsheet shown in Figure 13.34 to complete the calculations by varying the flow rate from zero to 275 gal/min and recording the values for the resulting total head for each flow rate. Table 13.2 lists the results and Figure 13.35 is a graph of the system curve.

*Step 7.* Determine the operating point for the selected pump by superimposing the pump rating curve on the system curve and observing where they intersect.

Figure 13.36 is a graph showing the operating point.

*Step 8.* Determine the performance of the pump at the operating point.

The results for this problem are listed below, found by referring to the full chart of pump performance in Figure 13.23.

| 2 × 3 − 10 | Centrifugal pump with a 9-inch impeller operating at 3500 rpm |
|---|---|
| Flow rate (Capacity): | 240 gal/min |
| Total head: | 320 ft |
| Efficiency: | 57% |
| Input power: | 33 hp |
| NPSH required: | 8.0 ft |

**TABLE 13.2**   System curve

| Q (gpm) | Q (cfs) | h$_a$ (ft) |
|---|---|---|
| 0 | 0 | 160.8 |
| 25 | 0.056 | 162.9 |
| 50 | 0.111 | 168.6 |
| 75 | 0.167 | 177.6 |
| 100 | 0.223 | 189.9 |
| 125 | 0.278 | 205.4 |
| 150 | 0.334 | 224.1 |
| 175 | 0.390 | 246.1 |
| 200 | 0.445 | 271.3 |
| 225 | 0.501 | 299.8 |
| 250 | 0.557 | 331.4 |
| 275 | 0.612 | 366.3 |

| APPLIED FLUID MECHANICS | | CLASS I SERIES SYSTEMS | | |
|---|---|---|---|---|
| Objective: *System Curve* | | *Reference points for the energy equation:* | | |
| Example Problem 13.1 | | Point 1: Surface of lower reservoir | | |
| Figure 13.33 | | Point 2: Surface of upper reservoir | | |
| *System Data:*    **US Customary Units** | | | | |
| Volume flow rate: $Q =$    0.5011  $ft^3/s$ | | Elevation at point 1 =    0 ft | | |
| Pressure at point 1 =    0  psig | | Elevation at point 2 =    80 ft | | |
| Pressure at point 2 =    35  psig | | *If Ref. pt. is in pipe: Set $v_1" = B20"$ OR Set $v_2" = E20"$* | | |
| Velocity at point 1 =    0.00 ft/s $\rightarrow$ | | Vel. head at point 1 =    0.00 ft | | |
| Velocity at point 2 =    0.00 ft/s $\rightarrow$ | | Vel. head at point 2 =    0.00 ft | | |
| *Fluid Properties:* | | May need to compute $\nu = \mu/\rho$ | | |
| Specific weight =    62.40  $lb/ft^3$ | | Kinematic viscosity =    1.21E-05 $ft^2/s$ | | |
| *Pipe 1: 3½ in Schedule 40 steel pipe* | | *Pipe 2: 2½ in Schedule 40 steel pipe* | | |
| Diameter: $D =$    0.2957 ft | | Diameter: $D =$    0.2058 ft | | |
| Wall roughness: $\epsilon =$ 1.50E-04 ft | | Wall roughness: $\epsilon =$ 1.50E-04 ft | | |
| Length: $L =$    8 ft | | Length: $L =$    360 ft | | |
| Area: $A =$  0.06867 $ft^2$ | | Area: $A =$    0.03326 $ft^2$ | $[A = \pi D^2/4]$ | |
| $D/\epsilon =$    1971 | | $D/\epsilon =$    1372 | Rel. roughness | |
| $L/D =$    27 | | $L/D =$    1749 | | |
| Flow velocity =    7.30 ft/s | | Flow velocity =    15.06 ft/s | $[v = Q/A]$ | |
| Velocity head =    0.827 ft | | Velocity head =    3.524 ft | $[v^2/2g]$ | |
| Reynolds No. = 1.78E+05 | | Reynolds No. = 2.56E+05 | $[N_R = vD/\nu]$ | |
| Friction factor: $f =$    0.0192 | | Friction factor: $f =$    0.0197 | Using Eq. 9-9 | |
| *Energy losses in Pipe 1:* | Qty. | Total K | | |
| Pipe: $K_1 =$    0.519 | 1 | 0.519 | Energy loss $h_{L1} =$    0.43 ft | |
| Entrance: $K_2 =$    0.500 | 1 | 0.500 | Energy loss $h_{L2} =$    0.41 ft | |
| Gate valve: $K_3 =$    0.136 | 1 | 0.136 | Energy loss $h_{L3} =$    0.11 ft | |
| Element 4: $K_4 =$    0.000 | 1 | 0.000 | Energy loss $h_{L4} =$    0.00 ft | |
| Element 5: $K_5 =$    0.000 | 1 | 0.000 | Energy loss $h_{L5} =$    0.00 ft | |
| Element 6: $K_6 =$    0.000 | 1 | 0.000 | Energy loss $h_{L6} =$    0.00 ft | |
| Element 7: $K_7 =$    0.000 | 1 | 0.000 | Energy loss $h_{L7} =$    0.00 ft | |
| Element 8: $K_8 =$    0.000 | 1 | 0.000 | Energy loss $h_{L8} =$    0.00 ft | |
| *Energy losses in Pipe 2:* | Qty. | Total K | | |
| Pipe: $K_1 =$ 34.488 | 1 | 34.488 | Energy loss $h_{L1} =$    121.53 ft | |
| Check valve: $K_2 =$    1.800 | 1 | 1.800 | Energy loss $h_{L2} =$    6.34 ft | |
| Butterfly valve: $K_3 =$    0.810 | 1 | 0.810 | Energy loss $h_{L3} =$    2.85 ft | |
| Standard elbow: $K_4 =$    0.540 | 2 | 1.080 | Energy loss $h_{L4} =$    3.81 ft | |
| Exit loss: $K_5 =$    1.000 | 1 | 1.000 | Energy loss $h_{L5} =$    3.52 ft | |
| Element 6: $K_6 =$    0.000 | 1 | 0.000 | Energy loss $h_{L6} =$    0.00 ft | |
| Element 7: $K_7 =$    0.000 | 1 | 0.000 | Energy loss $h_{L7} =$    0.00 ft | |
| Element 8: $K_8 =$    0.000 | 1 | 0.000 | Energy loss $h_{L8} =$    0.00 ft | |
| | | | Total energy loss $h_{Ltot} =$ | 139.01 ft |
| | *Results:* | Total head on pump: $h_A =$ | 299.8 ft | |

**FIGURE 13.34**  Total head on the pump at the desired operating point for Example Problem 13.1.

*Step 9.* If necessary, provide a means of connecting the selected pipe sizes to the connections for the pump if they are different sizes. See Figure 7.1 for an example.

Connections are required in this design. We have selected a 3½-inch suction line and the pump has a 3-inch suction port. A reducer is required. Also, an enlargement is required to mate the 2-inch discharge port of the pump to the 2½-inch discharge pipe.

Gradual reducers and expanders are recommended to minimize the energy losses added to the system by these elements. See Section 10.8 for the K-factor for

**FIGURE 13.35** System curve for Example Problem 13.1.

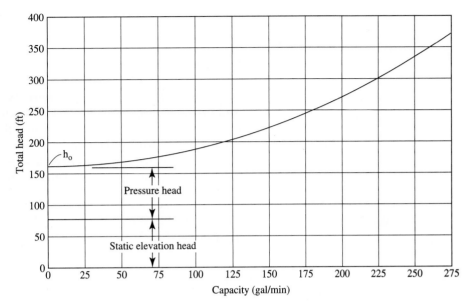

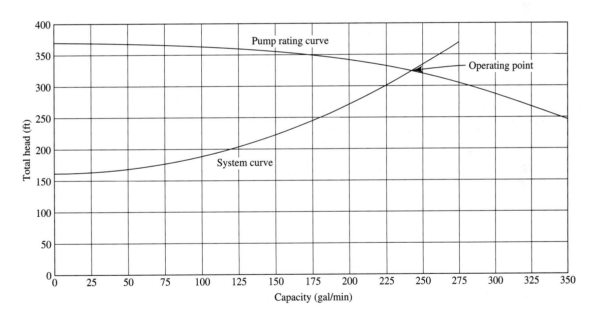

**FIGURE 13.36** Operating point for Example Problem 13.1.

a gradual reducer and Section 10.6 for a gradual enlargement. For each, the K-factor is typically much less than 1.0 and the added energy loss should have negligible effect on the operating point of the pump. Reducers in suction lines should be of the eccentric type and installed as shown in Figure 13.41 later in this chapter.

&#9632;

**13.10**
**ALTERNATE SYSTEM OPERATING MODES**

The discussion thus far has focused on the operation of a single pump at a single speed in a given system. Furthermore, the standard rating charts provided by manufacturers (such as those in Figures 13.23 to 13.31) are determined from test data with cool water as the fluid being pumped.

What happens if the speed of the pump varies? What if the fluid has a significantly higher or lower viscosity than water? What if we want to use two or more pumps in parallel to feed a system? What happens if we connect pumps in series where the output from one feeds to the input of the next?

This section discusses the basic principles involved in the answers to such questions. Consult References 3–7 or check with the pump manufacturer for more specific data and advice on applying pumps in these kinds of situations.

**13.10.1**
**Variable Speed Drives**

When a fluid transfer system must operate at a variety of capacities, a throttling valve is often used as depicted in Figure 13.32. The pump is sized for the largest capacity anticipated. If the delivery is lowered from $Q_1$ to $Q_2$, the energy represented by $h_2 - h_1$ is dissipated as the fluid passes through the valve. This energy is first delivered to the pump by the drive motor, then transferred to the fluid in the pump, and thus is wasted. High energy costs make it desirable to modify this manner of operation.

Variable speed drives offer an attractive alternative to throttling. Several types of mechanical variable speed drives and a variable frequency electronic control for a standard AC electric motor are available. The standard frequency for AC power in the United States and many other countries is 60 hertz (Hz), or 60 cycles per second. In Europe and some other countries, 50 Hz is standard. Because the speed of an AC motor is directly proportional to the frequency of the AC current, varying the frequency causes the motor speed to vary. Also, because of the affinity laws, as the motor speed decreases, its capacity decreases, which allows the pump to operate at the desired delivery without throttling. Further benefit is obtained because the power required by the pump decreases in proportion to the speed reduction ratio cubed. Of course, the variable speed drive is more expensive than a standard motor alone, and the overall economics of the system with time should be evaluated.

**13.10.2**
**Effect of Fluid Viscosity**

The performance rating curves for centrifugal pumps, such as those shown in Figs. 13.23 to 13.31, are generated from test data using water as the fluid. These curves are reasonably accurate for any fluid with a viscosity similar to that of water. However, pumping fluids with higher viscosities causes the capacity of the pump at a given head to decrease. Also the power required to drive the pump increases, and the efficiency decreases. Figure 13.37 illustrates the effect of increasing viscosity. Reference 4 gives data for correction factors that can be used to compute expected performance with fluids of different viscosities.

**13.10.3**
**Operating Pumps in Parallel**

Using two identical pumps in parallel to draw fluid from the same source and deliver it against the same system head doubles the flow rate delivered (see Fig. 13.38). This method is used when the desired capacity is beyond the range of any single suitable pump. This method also provides flexibility in the operation of the system, because one of the pumps can be shut down during low demand times or for service.

**13.10.4**
**Operating Pumps in Series**

Directing the output of one pump to the inlet of a second pump allows the same capacity to be obtained at a total head equal to the sum of the ratings of the two pumps. This method permits operation against unusually high heads.

**FIGURE 13.37**  Effect of increased viscosity on pump performance.

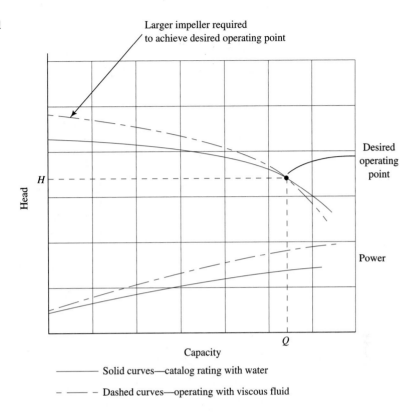

Larger impeller required to achieve desired operating point

Desired operating point

Power

Head

Capacity

———— Solid curves—catalog rating with water

– — – Dashed curves—operating with viscous fluid

**13.10.5**
**Multistage Pumps**

A performance similar to that achieved by using pumps in series can be obtained by using multistage pumps. Two or more impellers are arranged in the same housing in such a way that the fluid flows successively from one to the next. Each stage increases the fluid pressure, so that a high total head can be developed.

**13.11**
**PUMP SELECTION AND**
**SPECIFIC SPEED**

Figure 13.39 shows one method for deciding what type of pump is suitable for a given service. Some general conclusions can be drawn from such a chart, but it should be emphasized that boundaries between zones are approximate. Two or more

**FIGURE 13.38**  Performance of two pumps in parallel.

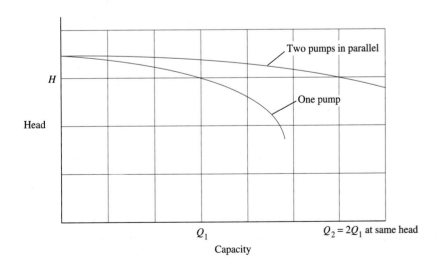

Two pumps in parallel

One pump

$H$

Head

$Q_1$

$Q_2 = 2Q_1$ at same head

Capacity

**FIGURE 13.39**   Pump selection chart.

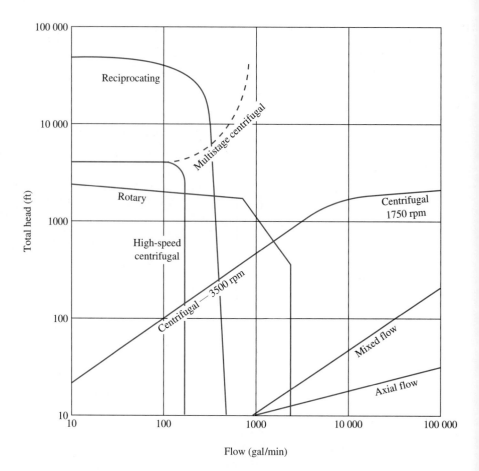

Flow (gal/min)

(Adapted from data in reference 6)

types of pumps may give satisfactory service under the same conditions. Such factors as cost, physical size, suction conditions, and the type of fluid may dictate a particular choice. In general:

1. Reciprocating pumps are used for flow rates up to about 500 gal/min and from very low heads to as high as 50 000 ft of head.
2. Centrifugal pumps are used over a wide range of conditions, mostly in high capacity, moderate head applications.
3. Single-stage centrifugal pumps operating at 3500 rpm are economical at lower flow rates and moderate heads.
4. Multistage pumps are desirable at high head conditions.
5. Rotary pumps (e.g., gear, vane, etc.) are used in applications requiring moderate capacities and high heads or for fluids with high viscosities.
6. Special high speed centrifugal pumps operating well above the 3500-rpm speed of standard electrical motors are desirable for high heads and moderate capacities. Such pumps are sometimes driven by steam turbines or gas turbines.
7. Mixed flow and axial flow pumps are used for very high flow rates and low heads.

Another parameter that is useful in selecting the type of pump for a given application is the *specific speed*, defined as

$$N_s = \frac{N\sqrt{Q}}{H^{3/4}} \tag{13-11}$$

where $\qquad$ $N$ = rotational speed of the impeller (rpm)

$Q$ = flow rate through the pump (gal/min)

$H$ = total head on the pump (ft)

Other units can be used.

The specific speed is often combined with the *specific diameter* to produce a chart like that shown in Fig. 13.40. The specific diameter is

$$D_s = \frac{DH^{1/4}}{\sqrt{Q}} \tag{13-12}$$

where $D$ is the impeller diameter in inches. The other terms are as defined before.

From Fig. 13.40 we can see that radial flow centrifugal pumps are recommended for specific speeds from about 400 to 4000. Mixed flow pumps are used from 4000 to about 7000. Axial flow pumps are used from 7000 to over 60 000.

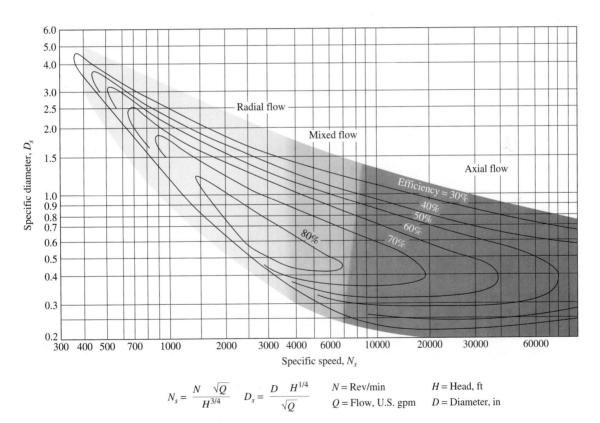

**FIGURE 13.40** Specific speed versus specific diameter for centrifugal pumps—An aid to pump selection. (Excerpted by special permission from *Chemical Engineering*, April 3, 1978. Copyright © 1978 by McGraw-Hill, Inc., New York, N.Y. 10020.)

**13.12**
**NET POSITIVE**
**SUCTION HEAD**

In addition to the total head, capacity, power, and efficiency requirements we have discussed, we must also consider that the condition at the inlet of a pump is critical. The inlet, or suction, system must allow a smooth flow of liquid to enter the pump at a sufficiently high pressure to avoid creating vapor bubbles in the fluid. As the pressure on a fluid decreases, the temperature at which vapor bubbles form (like boiling) also decreases. Therefore, it is essential that the suction pressure at the pump inlet be well above the pressure at which vaporization would occur for the operating temperature of the liquid. This is called providing a *net positive suction head*, sometimes abbreviated *NPSH*.

If the suction pressure is allowed to decrease to the point where vaporization occurs, *cavitation* is created inside the pump. Instead of a steady flow of liquid, the pump will draw a mixture of liquid and vapor, causing the delivery to decrease. Furthermore, as the vapor bubbles proceed through the pump, they encounter higher pressures which cause the bubbles to collapse rapidly. Excessive noise, vibration, and greatly increased wear of pump parts would result.

Pump manufacturers supply data about the *required* net positive suction head for satisfactory operation. The person selecting the pump must then ensure that there is a sufficiently high *NPSH available*. That is,

$$NPSH \text{ available} > NPSH \text{ required} \qquad (13\text{–}13)$$

Standards have been set jointly by the American National Standards Institute (ANSI) and the Hydraulic Institute (HI) calling for a minimum of a 10% margin for $NPSH_A$ over $NPSH_R$. Higher margins, up to 100% are expected for critical applications such as flood control, pipelines, and power generation service. See ANSI/HI 9.6.1-1998, *Standard for Centrifugal and Vertical Pumps for NPSH Margin*.

The value of *NPSH* available is dependent on the nature of the fluid being pumped, the suction piping, the location of the fluid reservoir, and the pressure applied to the fluid in the reservoir. This can be expressed as

$$NPSH_A = h_{sp} \pm h_s - h_f - h_{vp} \qquad (13\text{–}14)$$

Refer to Fig. 13.41 for an illustration of these terms:

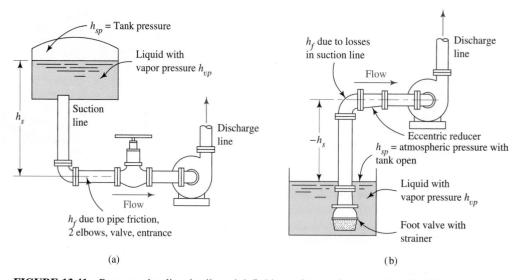

(a)                                    (b)

**FIGURE 13.41**   Pump suction line details and definitions of terms for computing *NPSH*.

$h_{sp}$ = static pressure head (absolute) applied to the fluid, expressed in *meters (or feet) of the liquid*

$h_s$ = elevation difference from the level of fluid in the reservoir to the pump inlet, expressed in *meters or feet*
If the pump is below the reservoir, $h_s$ is positive. (Preferred.)
If the pump is above the reservoir, $h_s$ is negative.

$h_f$ = friction loss in suction piping, expressed in *meters or feet*

$h_{vp}$ = vapor pressure of the liquid at the pumping temperature in *meters (or feet) of the liquid*

The pressure at which a liquid and its vapor can exist in equilibrium at any given temperature is called the vapor pressure at that temperature. If the pressure of the fluid entering the pump is at the vapor pressure of the fluid, vapor bubbles form throughout the fluid. The fluid actually boils, and this situation must be avoided to ensure that cavitation does not occur. Figure 13.42 and Table 13.3 give values of the vapor pressure of water as a function of temperature. You must use the values listed in *meters* or *feet* in the *NPSH* calculation.

Equation (13–14) can be developed by applying the energy equation to the suction piping system. Be careful to include pressure head in terms of *absolute* pressure. Strictly, Eq. (13–14) should have the velocity head for the fluid in the suction line added to the right side of the equation. However, most U.S. pump manufacturers already account for the velocity head in reporting the *NPSH required* by the pump. Therefore, it is not necessary to include it in the calculation of *NPSH available* (Reference 6, page 322). In critical installations, this matter should be checked with the pump supplier.

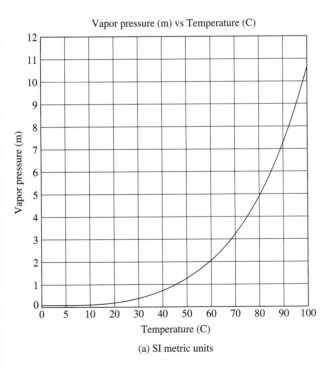

(a) SI metric units

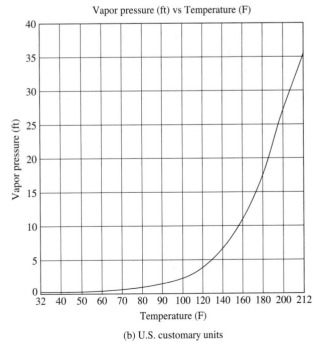

(b) U.S. customary units

**FIGURE 13.42**  Vapor pressure of water versus temperature.

**TABLE 13.3** Vapor pressure of water

| Temperature °C | Vapor Pressure kPa (abs) | Specific Weight (kN/m³) | Vapor Pressure (m) | Temperature °F | Vapor Pressure (psia) | Specific Weight (lb/ft³) | Vapor Pressure (ft) |
|---|---|---|---|---|---|---|---|
| 0 | 0.6105 | 9.806 | 0.06226 | 32 | 0.08854 | 62.42 | 0.2043 |
| 5 | 0.8722 | 9.807 | 0.08894 | 40 | 0.1217 | 62.43 | 0.2807 |
| 10 | 1.228 | 9.804 | 0.1253 | 50 | 0.1781 | 62.41 | 0.4109 |
| 20 | 2.338 | 9.789 | 0.2388 | 60 | 0.2563 | 62.37 | 0.5917 |
| 30 | 4.243 | 9.765 | 0.4345 | 70 | 0.3631 | 62.30 | 0.8393 |
| 40 | 7.376 | 9.731 | 0.7580 | 80 | 0.5069 | 62.22 | 1.173 |
| 50 | 12.33 | 9.690 | 1.272 | 90 | 0.6979 | 62.11 | 1.618 |
| 60 | 19.92 | 9.642 | 2.066 | 100 | 0.9493 | 62.00 | 2.205 |
| 70 | 31.16 | 9.589 | 3.250 | 120 | 1.692 | 61.71 | 3.948 |
| 80 | 47.34 | 9.530 | 4.967 | 140 | 2.888 | 61.38 | 6.775 |
| 90 | 70.10 | 9.467 | 7.405 | 160 | 4.736 | 61.00 | 11.18 |
| 100 | 101.3 | 9.399 | 10.78 | 180 | 7.507 | 61.58 | 17.55 |
| | | | | 200 | 11.52 | 60.12 | 27.59 |
| | | | | 212 | 14.69 | 59.83 | 35.36 |

**13.12.1**
**Effect of Pump Speed on NPSH**

The data given in pump catalogs for *NPSH* are for water and apply only to the listed operating speed. If the pump is operated at a different speed, the *NPSH* required at the new speed can be calculated from

$$(NPSH_R)_2 = \left(\frac{N_2}{N_1}\right)^2 (NPSH_R)_1$$

where the subscript 1 refers to catalog data and the subscript 2 refers to conditions at the new operating speed. The pump speed in rpm is $N$.

☐ **EXAMPLE PROBLEM 13.4**

Determine the available *NPSH* for the system shown in Fig. 13.41(a). The fluid reservoir is a closed tank with a pressure of −20 kPa above water at 70°C. The atmospheric pressure is 100.5 kPa. The water level in the tank is 2.5 m above the pump inlet. The pipe is a 1¹/₂-in Schedule 40 pipe with a total length of 12.0 m. The elbow is standard and the valve is a fully open globe valve. The flow rate is 95 L/min.

**Solution**

Using Eq. (13–14), first find $h_{sp}$:

$$\text{absolute pressure} = \text{atmospheric pressure} - \text{tank pressure}$$
$$p_{abs} = 100.5 \text{ kPa} - 20 \text{ kPa} = 80.5 \text{ kPa}$$

But we know that

$$h_{sp} = p_{abs}/\gamma$$
$$= \frac{80.5 \times 10^3 \text{ N/m}^2}{9.59 \times 10^3 \text{ N/m}^3} = 8.39 \text{ m}$$

Now, based on the elevation of the tank, we have

$$h_s = +2.5 \text{ m}$$

To find the friction loss $h_f$, we must find the velocity, Reynolds number, and friction factor:

$$v = \frac{Q}{A} = \frac{95 \text{ L/min}}{1.314 \times 10^{-3} \text{ m}^2} \times \frac{1.0 \text{ m}^3/\text{s}}{60\,000 \text{ L/min}} = 1.21 \text{ m/s}$$

$$N_R = \frac{vD}{\nu} = \frac{(1.21)(0.0409)}{4.11 \times 10^{-7}} = 1.20 \times 10^5 \quad \text{(turbulent)}$$

$$\frac{D}{\epsilon} = \frac{0.0409 \text{ m}}{4.6 \times 10^{-5} \text{ m}} = 889$$

Thus, from Fig. 9.2, $f = 0.0225$. From Table 10.5, $f_T = 0.021$. Now we have

$$h_f = \underset{\text{(pipe)}}{f(L/D)(v^2/2g)} + \underset{\text{(elbows)}}{2f_T(30)(v^2/2g)} + \underset{\text{(valve)}}{f_T(340)(v^2/2g)} + \underset{\text{(entrance)}}{1.0(v^2/2g)}$$

The velocity head is

$$\frac{v^2}{2g} = \frac{(1.21 \text{ m/s})^2}{2(9.81 \text{ m/s}^2)} = 0.0746 \text{ m}$$

Then, the friction loss is

$$\begin{aligned} h_f &= (0.0225)(12/0.0409)(0.0746) + (0.021)(60)(0.0746) \\ &\quad + (0.021)(340)(0.0746) + 0.0746 \\ &= (0.0746 \text{ m})[(0.0225)(12/0.0409) + (0.021)(60) + (0.021)(340) + 1.0] \\ &= 1.19 \text{ m} \end{aligned}$$

Finally, from Table 13.3 we get

$$h_{vp} = 3.25 \text{ m at } 70°\text{C}$$

Combining these terms gives

$$NPSH_A = 8.39 \text{ m} + 2.5 \text{ m} - 1.19 \text{ m} - 3.25 \text{ m} = 6.45 \text{ m}$$

Thus, a pump operating in this system must have a *required NPSH* less than 6.45 m.

■

---

**13.13**
**SUCTION LINE DETAILS**

The *suction line* refers to all parts of the flow system from the source of the fluid to the inlet of the pump. Great care should be exercised in designing the suction line to ensure an adequate net positive suction head, as we discussed in Section 13.12. Also, special conditions may require auxiliary devices.

Figure 13.41 shows two methods of providing fluid to a pump. In part (a), a positive head is created by placing the pump below the supply reservoir. This is an aid in ensuring a satisfactory *NPSH*. Also, the pump will always be primed with a column of liquid at start-up.

In part (b) of Fig. 13.41, a *suction lift* condition occurs because the pump must draw liquid from below. Most positive displacement pumps can lift fluids about 8 m and are called *self-priming*. For centrifugal pumps, however, the pump must be artificially primed by filling the suction line with fluid. This can be done by providing an auxiliary supply of liquid during start-up or by drawing a vacuum on the pump casing, causing the fluid to be sucked up from the source. Then, with the pump running, it will maintain the flow.

Unless the fluid is known to be very clean, a strainer should be installed either at the inlet or elsewhere in the suction piping to keep debris out of the pump and out of the process to which the fluid is to be delivered. A foot valve (Figures 10.20 and 10.21) at the inlet allows free flow to the pump but shuts when the pump stops. This maintains a column of liquid up to the pump and precludes the need to prime the pump each time it is started. If a valve is used near the pump, a gate valve, offering very little flow resistance when fully open, is preferred. The valve stem should be horizontal to avoid air pockets.

Although the pipe size for the suction line should never be smaller than the inlet connection on the pump, it can be somewhat larger to reduce flow velocity and friction losses. Pipe alignment should eliminate the possibility of forming air bubbles or air pockets in the suction line, as this will cause the pump to lose capacity and possibly to lose prime. Long pipes should slope upward toward the pump. Elbows in a horizontal plane should be avoided. If a reducer is required, it should be of the eccentric type, as shown in Fig. 13.41(b). Concentric reducers place part of the supply pipe above the pump inlet where an air pocket could form.

The discussion in Section 6.5 and Table 6.3 in Chapter 6 include recommendations for the ranges of desirable velocity of flow in suction lines and the corresponding pipe sizes to carry a given volume flow rate. In general, the lower velocities are recommended based on the ideal of minimizing the energy losses in the lines leading into pumps. But practical installation considerations and cost may lead to the selection of smaller pipes with the resulting higher velocities.

Some of these practical considerations include the cost of pipe, valves, and fittings; the physical space available to accommodate these elements; and the attachment of the suction pipe to the suction connection of the pump. References 3 to 8 include extensive discussion on the details of suction line design.

**13.14**
**DISCHARGE LINE**
**DETAILS**

In general, the discharge line should be as short and direct as possible to minimize the head on the pump. Elbows should be standard or long radius type if possible. Pipe size should be chosen according to velocity or allowable friction losses.

Table 6.3 in Chapter 6 includes recommendations for the ranges of desirable velocity of flow in discharge lines and the corresponding pipe sizes to carry a given volume flow rate. In general, the lower velocities are recommended based on the ideal of minimizing the energy losses. But practical installation considerations and cost may lead to the selection of smaller pipes with the resulting higher velocities.

Figure 13.43 shows the principle of considering the life cycle cost of a pumped fluid distribution system. The primary parameters are total cost on the vertical axis and pipe size on the horizontal axis. The system cost curve indicates that larger pipe sizes along with the larger valves and fittings are more expensive. The actual values for cost depend on the size and complexity of the system. Operating cost should include all forms of cost, most importantly the cost of energy to operate the pump. But installation cost, cost to house the pumping system, the cost of financing, and maintenance costs should also be considered. Time is an important element of the cost data and you should establish the expected life for the system as you make these calculations. References 3 and 6 discuss concepts of life cycle cost more completely.

The discharge line should contain a valve close to the pump to allow service or pump replacement. This valve acts with the valve in the suction line to isolate

**FIGURE 13.43** Life-cycle cost principle for pumped fluid distribution systems.

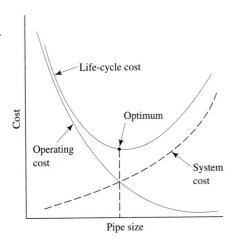

the pump. For low resistance, a gate or butterfly valve is preferred. If flow must be regulated during service, a globe valve is better because it allows a smooth throttling of the discharge. This, in effect, increases the system head and causes the pump delivery to decrease.

As shown in Fig. 13.44, other elements may be added to the discharge line as required. A pressure relief valve will protect the pump and other equipment in case of a blockage of the flow or accidental shut-off of a valve. A check valve prevents flow back through the pump when it is not running. A check valve should be placed between the shutoff valve and the pump. If an enlarger is used from the pump discharge port it should be placed between the check valve and the pump. A tap into the discharge line for a gage with its shut-off valve may be necessary. Also, a sample cock will allow a small flow of the fluid to be drawn off for testing without disrupting operation. Figure 7.1 in Chapter 7 shows a photograph of an actual installation.

**FIGURE 13.44** Discharge line details.

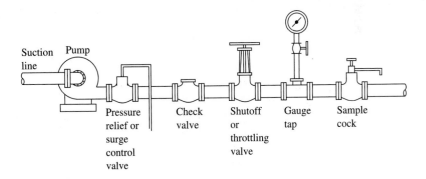

# REFERENCES

**1.** Cheremisinoff, N. P. 1984. *Fluid Flow Pocket Handbook.* Houston: Gulf Publishing Co.

**2.** Goulds Pumps, Inc. 1993. *Goulds Pump Manual.* Seneca Falls, New York: Goulds Pumps, Inc.

**3.** Hicks, T. G., and Edwards, T. W. 1971. *Pump Application Engineering.* New York: McGraw-Hill.

**4.** Hydraulic Institute. 1998. *1998 Hydraulic Institute Pump Standards S125, 5 Volumes.* Parsippany, NJ: Hydraulic Institute.

**5.** Hydraulic Institute. 1994. *1994 Hydraulic Institute Pump Standards S100, 13 Volumes.* Parsippany, NJ: Hydraulic Institute.

6. Mately, J., ed. 1979. *Fluid Movers: Pumps, Compressors, Fans and Blowers*. New York: Chemical Engineering McGraw-Hill Publications Co.

7. Pumps & Systems Magazine. 1998. *Pump Handbook Series, 5 Volumes*. Fort Collins, CO: Pumps & Systems Magazine.

8. Stewart, H. L. (Revised by T. Philbin). *Pumps*. 1986. New York: Macmillan Publishing Co.

## PRACTICE PROBLEMS

**13.1** List twelve factors that should be considered when selecting a pump.

**13.2** List ten items that must be specified for pumps.

**13.3** Describe a positive displacement pump.

**13.4** Name four examples of rotary positive displacement pumps.

**13.5** Name three types of reciprocating positive displacement pumps.

**13.6** Describe a kinetic pump.

**13.7** Name three classifications of kinetic pumps.

**13.8** Describe the action of the impellers and the general path of flow in the three types of kinetic pumps.

**13.9** Describe a jet pump.

**13.10** Distinguish between a shallow well jet pump and a deep well jet pump.

**13.11** Describe the difference between a simplex reciprocating pump and a duplex type.

**13.12** Describe the general shape of the plot of pump capacity versus discharge pressure for a positive displacement rotary pump.

**13.13** Describe the general shape of the plot of total head versus pump capacity for centrifugal pumps.

**13.14** To the head versus capacity plot of Practice Problem 13.13, add plots for efficiency and power required.

**13.15** To what do the *affinity laws* refer in regard to pumps?

**13.16** For a given centrifugal pump, if the speed of rotation of the impeller is cut in half, how does the capacity change?

**13.17** For a given centrifugal pump, if the speed of rotation of the impeller is cut in half, how does the total head capability change?

**13.18** For a given centrifugal pump, if the speed of rotation of the impeller is cut in half, how does the power required to drive the pump change?

**13.19** For a given size of centrifugal pump casing, if the diameter of the impeller is reduced by 25 percent, how much does the capacity change?

**13.20** For a given size of centrifugal pump casing, if the diameter of the impeller is reduced by 25 percent, how much does the total head capability change?

**13.21** For a given size of centrifugal pump casing, if the diameter of the impeller is reduced by 25 percent, how much does the power required to drive the pump change?

**13.22** Describe each part of this centrifugal pump designation: $1^{1}/_{2} \times 3 - 6$.

**13.23** For the line of pumps shown in Figure 13.17, specify a suitable size for delivering 100 gal/min of water at a total head of 300 ft.

**13.24** For the line of pumps shown in Fig. 13.17, specify a suitable size for delivering 600 L/min of water at a total head of 25 m.

**13.25** For the $2 \times 3 - 10$ centrifugal pump performance curve shown in Fig. 13.23, describe the performance that can be expected from a pump with an 8-in impeller operating against a system head of 200 ft. Give the expected capacity, the power required, the efficiency, and the required *NPSH*.

**13.26** For the $2 \times 3 - 10$ centrifugal pump performance curve shown in Fig. 13.23, at what head will the pump having an 8-in impeller operate at its highest efficiency? List the pump's capacity, power required, efficiency, and the required *NPSH* at that head.

**13.27** Using the result from Problem 13.26, describe how the performance of the pump changes if the system head increases by 15 percent.

**13.28** For the $2 \times 3 - 10$ centrifugal pump performance curve shown in Fig. 13.23, list the total head and capacity at which the pump will operate at maximum efficiency for each of the impeller sizes shown.

**13.29** For a given centrifugal pump and impeller size, describe how the *NPSH* required varies as the capacity increases.

**13.30** State some advantages of using a variable speed drive for a centrifugal pump that supplies fluid to a process requiring varying flow rates of a fluid, as compared with adjusting throttling valves.

**13.31** Describe how the capacity, efficiency, and power required for a centrifugal pump vary as the viscosity of the fluid pumped increases.

**13.32** If two identical centrifugal pumps are connected in parallel and operated against a certain head, how would the total capacity compare with that of a single pump operating against the same head?

**13.33** Describe the effect of operating two pumps in series.

**13.34** For each of the following sets of operating conditions, list at least one appropriate type of pump. See Figure 13.39.

   **a.** 500 gal/min of water at 80 ft of total head

   **b.** 500 gal/min of water at 800 ft of head

   **c.** 500 gal/min of a viscous adhesive at 80 ft of head

   **d.** 80 gal/min of water at 8000 ft of head

   **e.** 80 gal/min of water at 800 ft of head

   **f.** 8000 gal/min of water at 200 ft of head

   **g.** 8000 gal/min of water at 60 ft of head

   **h.** 8000 gal/min of water at 12 ft of head

**13.35** For the $1^{1}/_2 \times 3 - 13$ centrifugal pump performance curve shown in Fig. 13.29, determine the capacity that can be expected from a pump with a 12-in impeller operating against a system head of 550 ft. Then, compute the specific speed and specific diameter and locate the corresponding point on Fig. 13.40.

**13.36** For the $6 \times 8 - 17$ centrifugal pump performance curve shown in Fig. 13.27, determine the capacity that can be expected from a pump with a 15-in impeller operating against a system head of 200 ft. Then, compute the specific speed and specific diameter and locate the corresponding point on Fig. 13.40.

**13.37** Figure 13.39 shows that a mixed flow pump is recommended for delivering 10 000 gal/min of water at a head of 40 ft. If such a pump operates with a specific speed of 5000, compute the appropriate operating speed of the pump.

**13.38** Compute the specific speed for a pump operating at 1750 rpm delivering 5000 gal/min of water at a total head of 100 ft.

**13.39** Compute the specific speed for a pump operating at 1750 rpm delivering 12 000 gal/min of water at a total head of 300 ft.

**13.40** Compute the specific speed for a pump operating at 1750 rpm delivering 500 gal/min of water at a total head of 100 ft.

**13.41** Compute the specific speed for a pump operating at 3500 rpm delivering 500 gal/min of water at a total head of 100 ft. Compare the result with that of Problem 13.40 and with Fig. 13.40.

**13.42** It is desired to operate a pump at 1750 rpm by driving it with a 4-pole electric motor. For each of the following conditions, compute the specific speed using Eq. (13–11). Then, recommend whether you should use an axial pump, a mixed flow pump, a radial flow pump, or none of these, based on the discussion related to Fig. 13.40.

   **a.** 500 gal/min of water at 80 ft of total head

   **b.** 500 gal/min of water at 800 ft of head

   **c.** 3500 gal/min of water at 80 ft of head

   **d.** 80 gal/min of water at 8000 ft of head

   **e.** 80 gal/min of water at 800 ft of head

   **f.** 8000 gal/min of water at 200 ft of head

   **g.** 8000 gal/min of water at 60 ft of head

   **h.** 8000 gal/min of water at 12 ft of head

**13.43** Define *net positive suction head (NPSH)*.

**13.44** Distinguish between *NPSH available* and *NPSH required*.

**13.45** Describe what happens to the vapor pressure of water as the temperature increases.

**13.46** Describe why it is important to consider *NPSH* when designing and operating a pumping system.

**13.47** For what point in a pumping system is the *NPSH* computed? Why?

**13.48** Discuss why it is desirable to elevate the reservoir from which a pump draws liquid.

**13.49** Discuss why it is desirable to use relatively large pipe sizes for the suction lines in pumping systems.

**13.50** Discuss why an *eccentric reducer* should be used when it is necessary to decrease the size of a suction line as it approaches a pump.

**13.51** If we assume that a given pump requires 7.50 ft of *NPSH* when operating at 3500 rpm, what would be the *NPSH* required at 2850 rpm?

**13.52E** Determine the available *NPSH* for the pump in Practice Problem 7.14 if the water is at 80°F and the atmospheric pressure is 14.4 psia. Repeat the calculations for water at 180°F.

**13.53E** Find the available *NPSH* when a pump draws water at 140°F from a tank whose level is 4.8 ft below the pump inlet. The suction line losses are 2.2 lb-ft/lb and the atmospheric pressure is 14.7 psia.

**13.54M** *A pump draws benzene at 25°C from a tank whose level is 2.6 m above the pump inlet. The suction line has a head loss of 0.8 N·m/N. The atmospheric pressure is measured to be 98.5 kPa(abs). Find the available NPSH. The vapor pressure of benzene is 13.3 kPa.*

**13.55M** *Determine the available NPSH for the system shown in Fig. 13.41(b). The fluid is water at 80°C and the atmospheric pressure is 101.8 kPa. The water level in the tank is 2.0 m below the pump inlet. The vertical leg of the suction line is a 3-in Schedule 40 steel pipe, while the horizontal leg is a 2-in Schedule 40 pipe, 1.5 m long. The elbow is of the long radius type. Neglect the loss in the reducer. The foot valve and strainer are of the hinged-disk type. The flow rate is 300 L/min.*

## DESIGN PROBLEMS

Several situations are presented here in which a system is being designed to pump a fluid from some source to a given destination. In each case, the objective is to completely define the configuration of the system, including:

- Pipe sizes and types
- Location of the pump
- Length of pipe for all parts of the system
- Valves and fittings

Although the scope of the design problems can be adjusted, the following are suggested features of the exercises. See Example Problem 13.3 for the procedure.

1. Specify pipe types suitable to the application.
2. Use Table 6.3 to estimate the desirable sizes for pipes.
3. Select valves and fittings from the data given in Chapter 10.
4. If the head and flow characteristics of the system are appropriate, use the pump performance data from Figs. 13.23 through 13.31 to specify a suitable pump and evaluate its performance: efficiency, power required, and *NPSH* required.
5. Consider using other manufacturers' catalogs to select pumps for problems for which Figs. 13.23 to 13.31 are not suitable.

6. Analyze each system to determine the total head on the pump. This will require the calculation of energy losses throughout the system.
7. Compute the pressure at the inlet to the pump.
8. Compute the *NPSH* available at the pump inlet.
9. Where data are available, ensure that the *NPSH* available is greater than the *NPSH* required by the pump.
10. Make a list of materials required for the system.
11. Analyze the pressure at other points pertinent to the problem.

### Problem Statements

**1E** Design a system to pump water at 140°F from a sump below a heat exchanger to the top of a cooling tower, as sketched in Fig. 13.45. The desired minimum flow rate is 200 gal/min.

**2M** *Design a system to pump water at 80°C from a water heater to a washing system, as sketched in Fig. 13.46. The desired minimum flow rate is 750 L/min (198 gal/min).*

**FIGURE 13.45**   Design Problem 1.

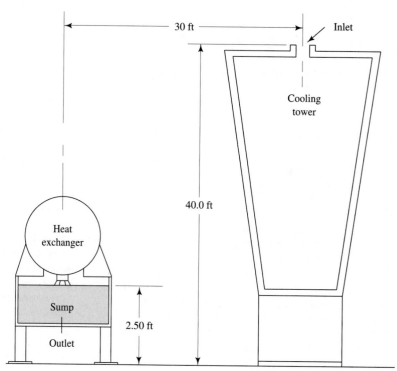

**FIGURE 13.46**   Design Problem 2.

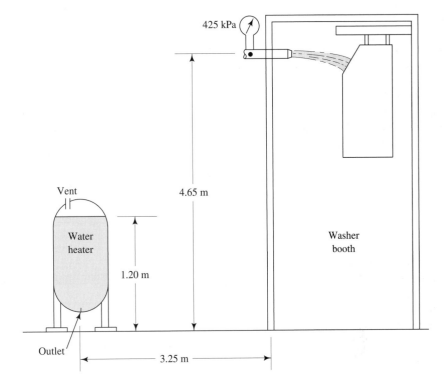

**3E** Design a system to pump water at 60°F from a river to a tank elevated 55 ft above the surface of the river. The desired minimum flow rate is 1500 gal/min. The tank is to be set back 125 ft from the river bank.

**4E** Design the water system for Professor Crocket's cabin, as described in Fig. 7.38. The desired minimum flow rate is 40 gal/min, and the distribution tank is to be maintained at a pressure of 30 psig above the water. The cabin sits 150 ft from the side of the stream from which the water is to be drawn. The slope of the hillside is approximately 30° from the vertical. The water is at 80°F.

**5M** *Design a system similar to that shown in Fig. 7.35, in which air pressure at 400 kPa above the kerosene at 25°C is used to cause the flow. The horizontal distance between the two tanks is 32 m. The desired minimum flow rate is 500 L/min.*

**6E** Design a system similar to that shown in Fig. 9.11, which must supply at least 1500 gal/min of water at 60°F for a fire protection system. The pressure at point B must be at least 85 psig. The depth of water in the tank is 5.0 ft. Ignore the specified sizes for the pipes and make your own decisions. Add appropriate valves and redesign the suction line.

**7E** Design a system similar to that shown in Fig. 9.17 and described in Problem 9.18. Ignore the given pipe sizes and the given pressure at the pump inlet. Add appropriate valves. The pump draws the polluted water from a still pond whose surface is 30 in below the centerline of the pump inlet. Use the vapor pressure for water at 100°F.

**8M** *Design a system similar to that shown in Fig. 7.22 to deliver 60 L/min of a water-based cutting fluid (sg = 0.95) to the cutter of a milling machine. Assume that the viscosity and vapor pressure are 10 percent greater than that for water at 15°C. Assume that the pump is submerged and that the depth above the suction inlet is 75 mm. The total length of the path required for the discharge line is 1.75 m.*

**9M** *Design a system similar to that shown in Fig. 7.21 to deliver 840 L/min of water at 100°F from an underground storage tank to a pressurized storage tank. Ignore original pipe sizes and make your own decision. Add appropriate valves. The upper tank pressure is 500 kPa.*

**10E** Specify a suitable pump for the system shown in Figure 13.47. It is a combination series/parallel system that operates as follows.

- Water at 160°F is drawn at the rate of 275 gal/min from a tank into the 4-in suction line of the pump. The suction line has a total length of 10 ft.
- The 3-in discharge line elevates the water 15 ft to the level of a large heat exchanger. The discharge line has a total length of 40 ft.
- The flow splits into two branches with the primary 3-in line feeding a large heat exchanger that has a K-factor of 12 based on the velocity head in the pipe. The total length of pipe in this branch is 8 ft.
- The 1-in line is a bypass around the heat exchanger with a total length of 30 ft.

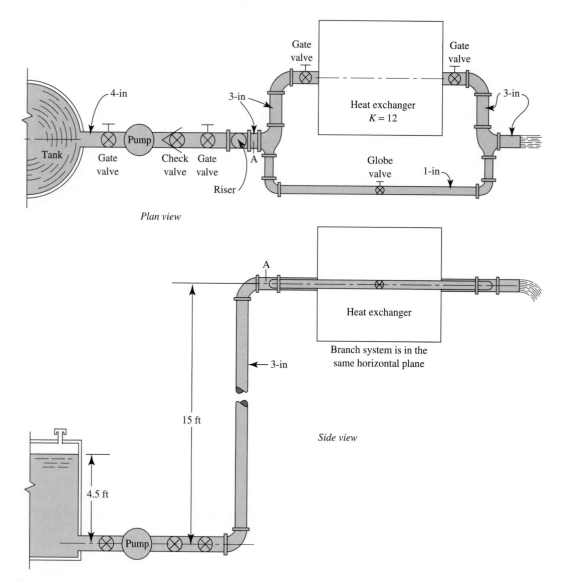

**FIGURE 13.47**    System for Design Problem 10

- The two lines join at the right and discharge to the atmosphere through a short 3-in pipe.
- All pipes are Schedule 40 steel.

For this system, operating at the desired operating conditions, determine the following:

**(a)** The pressure at the pump inlet

**(b)** The NPSH available at the pump inlet

**(c)** The pressure at point A before the branches

**(d)** The volume flow rate through the heat exchanger line

**(e)** The volume flow rate through the bypass line

**(f)** The total head on the pump

**(g)** The power delivered to the water by the pump.

Then specify a suitable pump for this system that will deliver at least the desired 275 gal/min of flow. For the selected pump, determine the following:

**(h)** The actual expected flow rate produced by the pump at the operating point

**(i)** The power input to the pump

**(j)** The NPSH required

**(k)** The efficiency at the operating point.

# COMPREHENSIVE DESIGN PROBLEM

Consider yourself to be a plant engineer for a company that is planning a new manufacturing facility. As a part of the new plant, there will be an automated machining line in which five machines will be supplied with coolant from the same reservoir. You are responsible for the design of the system to handle the coolant from the time it reaches the plant in railroad tank cars until the dirty coolant is removed from the premises by a contract firm for reclaim.

The layout of the planned facility is shown in Fig. 13.48. The following data, design requirements, and limitations apply.

1. New coolant is delivered to the plant by tank cars carrying 15 000 gal each. A holding tank for new coolant must be specified.

2. The reservoir for the automated machining system must have a capacity of 1000 gal.

3. The 1000-gal tank is normally emptied once per week. Emergency dumps are possible if the coolant becomes overly contaminated prior to the scheduled emptying.

4. The dirty fluid is picked up by truck only once per month.

5. A holding tank for the dirty fluid must be specified.

6. The plant is being designed to operate two shifts per day, 7 days per week.

7. Maintenance is normally performed during the third shift.

8. The building is one-story with a concrete floor.

9. The floor level is at the same elevation as the railroad track.

10. No storage tank can be inside the plant or under the floor except the 1000-gal reservoir that supplies the machining system.

11. The roof top is 32 ft from the floor level and the roof can be designed to support a storage tank.

12. The building is to be located in Dayton, Ohio, where the outside temperature may range from $-20°F$ to $+105°F$.

13. The frost line is 30 in below the surface.

14. The coolant is a solution of water and a soluble oil with a specific gravity of 0.94 and a freezing point of $0°F$. Its corrosiveness is approximately the same as that of water.

15. Assume that the viscosity and vapor pressure of the coolant are 1.50 times that of water at any temperature.

16. You are not asked to design the system to supply the machines.

17. The basic coolant storage and delivery system is to have the functional design sketched in the block diagram in Fig. 13.49.

The following tasks are to be completed by you, the system designer:

a. Specify the location and size of all storage tanks.

b. Specify the layout of the piping system, the types and sizes of all pipes, and the lengths required.

c. Specify the number, type, and size of all valves, elbows, and fittings.

d. Specify the number of pumps, their types, capacities, head requirements, and power required.

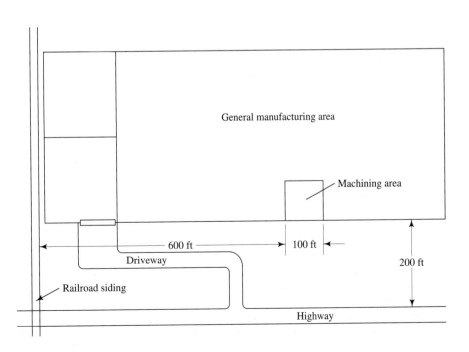

**FIGURE 13.48**  Plot plan for factory building for comprehensive design problem.

**FIGURE 13.49**    Block diagram of
coolant system.

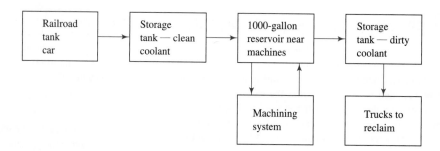

e. Specify the installation requirements for the pumps, including the complete suction line system. Evaluate the net positive suction head (*NPSH*) available for your design, and demonstrate that your pump has an acceptable *NPSH* required.

f. Determine the time required to fill and empty all tanks.

g. Sketch the layout of your design in both a plan view (top) and an elevation view (side). An isometric sketch also may be used.

h. Include the analysis of all parts of the system, including energy losses due to friction and minor losses.

i. Submit the results of your design in a neat and complete report, including a narrative description of the system, the sketches, a list of materials, and the analysis to show that your design meets the specifications.

# 14 Open Channel Flow

**14.1
The
Big
Picture**

### Discussion Map

- An open channel is a flow system in which the top surface of the fluid is exposed to the atmosphere.

- Examples are rain gutters on buildings, storm sewers, natural rivers and streams, and channels constructed to drain fluids in a controlled manner.

- The analysis of open channels requires special techniques somewhat different from those you have used to analyze flow in closed pipe and tubing.

**Discover**

- Take time to observe where open channels exist in areas with which you are familiar.
- Look for rain gutters, natural streams, and other drainage structures.
- What new ones can you find?

This chapter presents some basic methods of analyzing open channel flow.

In contrast to the closed conduits that have been discussed in preceding chapters, an *open channel* is a flow system in which the top surface of the fluid is exposed to the atmosphere.

Many examples of open channels occur in nature and in systems designed to supply water to communities or to carry storm drainage and sewage safely away. See Fig. 14.1. Rivers and streams are obvious examples of natural channels. Rain gutters on buildings and at the sides of streets carry rainwater. Storm sewers, usually beneath the streets, collect the runoff from the streets and conduct it to a stream or to a larger man-made ditch or canal. In industry, open channels are often used to convey cooling water away from heat exchangers or coolants away from machining systems.

Take time to observe where open channels exist in areas with which you are familiar. As you find actual open channels, try to describe them in as much detail as you can. Answer such questions as

- What is the channel used for?
- What fluid is flowing in the channel?
- Does the flow appear to be smooth and tranquil or chaotic and turbulent?
- What is the shape of the cross section of the channel and what are its dimensions?
- Is the cross section uniform along its length or does it vary?
- How deep was the fluid when you saw it? How deep can the fluid be under very heavy flow conditions before overflowing?
- How does the shape of the flow stream change, if at all, as the depth increases?
- Could you detect whether the channel is installed on a slope?

This chapter will present some of the methods of analyzing open channel flow. Complete coverage of the subject is an extensive undertaking and is treated in entire texts such as References 3, 4, 5, 6, and 7 (listed at the end of this chapter).

**FIGURE 14.1** Examples of cross sections of open channels.

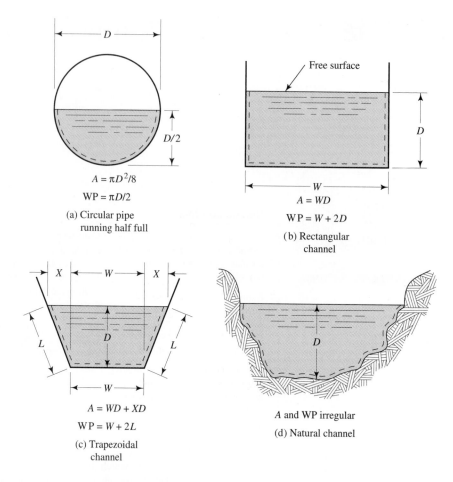

$$A = \pi D^2/8$$
$$WP = \pi D/2$$

(a) Circular pipe running half full

$$A = WD$$
$$WP = W + 2D$$

(b) Rectangular channel

$$A = WD + XD$$
$$WP = W + 2L$$

(c) Trapezoidal channel

$A$ and WP irregular

(d) Natural channel

**14.2
OBJECTIVES**

After completing this chapter, you should be able to:

1. Compute the hydraulic radius for open channels.
2. Describe *uniform flow* and *varied flow*.
3. Use Manning's equation to analyze uniform flow.
4. Define the slope of an open channel and compute its value.
5. Compute the normal discharge for an open channel.
6. Compute the normal depth of flow for an open channel.
7. Design an open channel to transmit a given discharge with uniform flow.
8. Define the *Froude number*.
9. Describe *critical flow*, *subcritical flow*, and *supercritical flow*.
10. Define the specific energy of the flow in open channels.
11. Define the terms *critical depth*, *alternate depth*, and *sequent depth*.
12. Describe the term *hydraulic jump*.

**14.3
CLASSIFICATION OF
OPEN CHANNEL FLOW**

Open channel flow can be classified into several types.

   *Uniform steady flow* occurs when the volume flow rate (typically called *discharge* in open channel flow analysis) remains constant in the section of interest and the depth of the fluid in the channel does not vary. To achieve uniform steady flow, the cross section of the channel may not change along its length. Such a channel is *prismatic*. Figure 14.2 shows uniform flow in a side view.

**FIGURE 14.2** Uniform steady open channel flow—side view.

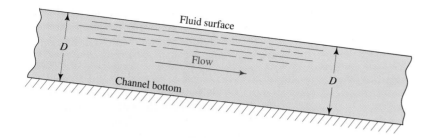

*Varied steady flow* occurs when the discharge remains constant but the depth of the fluid varies along the section of interest. This will occur if the channel is not prismatic.

*Unsteady varied flow* occurs when the discharge varies with time, resulting in changes in the depth of the fluid along the section of interest whether the channel is prismatic or not.

Varied flow can be further classified into *rapidly varying flow* or *gradually varying flow*. As the name implies, the difference refers to the rate of change in depth with position along the channel. Figure 14.3 illustrates a series of conditions in which varied flow occurs. The following discussion describes the flow in the various parts of this figure.

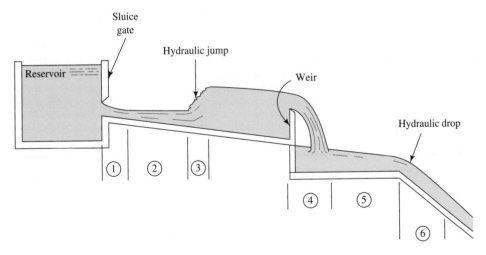

**FIGURE 14.3** Conditions causing varied flow.

- *Section 1* The flow starts from a reservoir in which the fluid is virtually at rest. The *sluice gate* is a device that allows the fluid to flow from the reservoir at a point beneath the surface. Rapidly varying flow occurs near the gate as the fluid accelerates, and the velocity of flow is likely to be quite large in this area.
- *Section 2* If the channel downstream from the sluice gate is relatively short, and if its cross section does not vary much, then gradually varying flow occurs. If the channel is prismatic and long enough, uniform flow could develop.
- *Section 3* The formation of a *hydraulic jump* is a curious open channel flow phenomenon. The flow before the jump is quite rapid and relatively shallow. In

the jump, the flow becomes very turbulent, and a large amount of energy is dissipated. Then, following the jump, the flow velocity is much lower, and the depth of the fluid is greater. More will be said later about the hydraulic jump.

- *Section 4*   A *weir* is an obstruction placed in the flow stream that causes an abrupt change in the cross section of the channel. Weirs can be used as control devices or to measure the volume flow rate. Typically the flow is rapidly varying as it travels over the weir, with a "waterfall" formed downstream.
- *Section 5*   As with Section 2, the flow downstream from a weir is usually gradually varying if the channel is prismatic.
- *Section 6*   The *hydraulic drop* occurs when the slope of the channel suddenly increases to a steep angle. The flow accelerates because of gravity, and rapidly varying flow occurs.

**14.4**
**HYDRAULIC RADIUS AND REYNOLDS NUMBER IN OPEN CHANNEL FLOW**

The characteristic dimension of open channels is the *hydraulic radius*, defined as the ratio of the net cross-sectional area of a flow stream to the wetted perimeter of the section. That is,

 HYDRAULIC RADIUS

$$R = \frac{A}{WP} = \frac{\text{area}}{\text{wetted perimeter}} \qquad (14\text{--}1)$$

The unit for $R$ is the meter in the SI unit system. In the English system, $R$ is expressed in feet.

In the calculation of the hydraulic radius, the net cross-sectional area should be evident from the geometry of the section. The *wetted perimeter* is defined as the sum of the length of the boundaries of the section actually in contact with (that is, wetted by) the fluid. Expressions for the area $A$ and the wetted perimeter $WP$ are given in Fig. 14.1 for those sections illustrated. In each case, the fluid flows in the shaded portion of the section. A dashed line is shown adjacent to the boundaries that make up the wetted perimeter. Notice that the length of the free surface of an open channel is *not* included in $WP$.

---

☐ **EXAMPLE PROBLEM 14.1**

Determine the hydraulic radius of the trapezoidal section shown in Fig. 14.1(c) if $W = 4$ ft, $X = 1$ ft, and $D = 2$ ft.

**Solution**

The net flow area is

$$A = WD + 2(XD/2) = WD + XD$$
$$= (4)(2) + (1)(2) = 10 \text{ ft}^2$$

To find the wetted perimeter, we must determine the value of $L$:

$$WP = W + 2L$$
$$L = \sqrt{X^2 + D^2} = \sqrt{(1)^2 + (2)^2} = 2.24 \text{ ft}$$
$$WP = 4 + 2(2.24) = 8.48 \text{ ft}$$

Then we have

$$R = A/WP = 10 \text{ ft}^2/8.48 \text{ ft} = 1.18 \text{ ft}$$

■

Recall that the Reynolds number for closed circular cross sections running full is

$$N_R = \frac{vD}{\nu} \qquad (14\text{--}2)$$

where $v$ = average velocity of flow, $D$ = pipe diameter, and $\nu$ = kinematic viscosity of the fluid. We have seen that laminar flow occurs when $N_R < 2000$ and turbulent flow occurs when $N_R > 4000$ for most practical pipe flow situations. The Reynolds number represents the effects of fluid viscosity relative to the inertia of the fluid.

In open channel flow, the characteristic dimension is the hydraulic radius, $R$. It was shown in Chapter 8 that, for a full circular cross section, $D = 4R$. For closed, noncircular cross sections, it was convenient to substitute $4R$ for $D$ so that the Reynolds number would have the same order of magnitude as that for circular pipes and tubes. However, this is not usually done in open channel flow analysis. The Reynolds number for open channel flow is then

⇨ REYNOLDS NUMBER FOR OPEN
CHANNELS

$$N_R = \frac{vR}{\nu} \qquad (14\text{--}3)$$

Experimental evidence (Reference 3) shows that, in open channels, laminar flow occurs when $N_R < 500$. The range from 500 to 2000 is the transition region. Turbulent flow normally occurs when $N_R > 2000$.

**14.5**
**KINDS OF OPEN CHANNEL FLOW**

The Reynolds number and the terms *laminar* and *turbulent* are not sufficient to characterize all kinds of open channel flow. In addition to the viscosity versus inertial effects, the ratio of inertial forces to gravity forces is also important. The *Froude number*, $N_F$, is defined as

⇨ FROUDE NUMBER

$$N_F = \frac{v}{\sqrt{gy_h}} \qquad (14\text{--}4)$$

where

⇨ HYDRAULIC DEPTH

$$y_h = A/T \qquad (14\text{--}5)$$

and $T$ is the width of the free surface of the fluid at the top of the channel.

When the Froude number is equal to 1.0, that is, when $v = \sqrt{gy_h}$, the flow is called *critical flow*. When $N_F < 1.0$, the flow is *subcritical*, and when $N_F > 1.0$, the flow is *supercritical*. See also Section 14.9.

Then, the following kinds of flow are possible:

**1.** Subcritical-laminar. $N_R < 500$, and $N_F < 1.0$.
**2.** Subcritical-turbulent. $N_R > 2000$, and $N_F < 1.0$.
**3.** Supercritical-turbulent. $N_R > 2000$, and $N_F > 1.0$.
**4.** Supercritical-laminar. $N_R < 500$, and $N_F > 1.0$.

In addition, flows can be in the transition region. However, such flows are unstable and very difficult to characterize.

In this discussion, the terms *laminar* and *turbulent* have the same significance as they did for pipe flow. In laminar flow, there is little or no mixing of the fluid, so that a stream of dye injected into the flow would remain virtually intact. In turbulent flow, however, chaotic intermixing occurs, and the dye stream rapidly dissipates throughout the fluid.

**14.6**
**UNIFORM STEADY**
**FLOW IN OPEN**
**CHANNELS**

Figure 14.2 is a schematic illustration of uniform steady flow in an open channel. The distinguishing feature of uniform flow is that the fluid surface is parallel to the slope of the channel bottom. We will use the symbol $S$ to indicate the slope of the channel bottom and $S_w$ for the slope of the water surface. Then for uniform flow, $S = S_w$. Theoretically, uniform flow can exist only if the channel is prismatic, that is, if its sides are parallel to an axis in the direction of flow. Examples of prismatic channels are rectangular, trapezoidal, triangular, and circular sections running partially full. Also, the channel slope $S$ must be constant. If the cross section or slope of the channel is changing, then the flow stream would be either converging or diverging and varied flow would occur.

In uniform flow, the driving force for the flow is provided by the component of the weight of the fluid which acts along the channel as shown in Fig. 14.4. This force is $w \sin \theta$, where $w$ is the weight of a certain element of fluid and $\theta$ is the angle of the slope of the channel bottom. If the flow is uniform, it cannot be accelerating. Therefore, there must be an equal opposing force acting along the channel surface. This is a friction force which depends on the roughness of the channel surfaces and on the cross-sectional size and shape.

**FIGURE 14.4**   Uniform open channel flow.

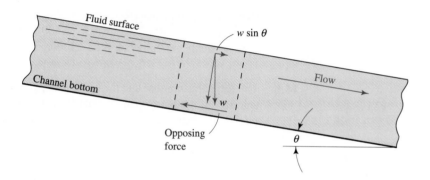

By equating the expressions for the driving force and the opposing force, an expression for the average velocity of uniform flow can be derived. A commonly used form of the resulting equation was developed by Robert Manning. In SI units, Manning's equation is written as

 MANNING'S EQUATION—
SI UNITS

$$v = \frac{1.00}{n} R^{2/3} S^{1/2} \tag{14–6}$$

Units must be consistent in Eq. (14–6). The average velocity of flow $v$ will be in m/s when the hydraulic radius $R$ is in m. The channel slope $S$, which we will define later, is dimensionless. The final term $n$ is a resistance factor sometimes called *Manning's n*. The value of $n$ depends on the condition of the channel surface and is therefore somewhat analogous to the pipe wall roughness $\epsilon$ used previously. The form of Manning's equation in English units is given later in this section.

Typical design values of $n$ are listed in Table 14.1 for materials commonly used for artificial channels and natural streams. A very extensive discussion of the determination of Manning's $n$ and a more complete table of values are given by V. T. Chow (Reference 3). The values listed in Table 14.1 are average values which will give good estimates for use in design or for rough analysis of existing channels. Variations from the average should be expected.

**TABLE 14.1**  Values for Manning's $n$

| Channel Description | $n$ |
| --- | --- |
| Glass, copper, plsatic, or other smooth surfaces | 0.010 |
| Smooth unpainted steel, planed wood | 0.012 |
| Painted steel or coated cast iron | 0.013 |
| Smooth asphalt, common clay drainage tile, trowel finished concrete, glazed brick | 0.013 |
| Uncoated cast iron, black wrought iron pipe, vitrified clay sewer tile | 0.014 |
| Brick in cement mortar, float finished concrete | 0.015 |
| Formed, unfinished concrete | 0.017 |
| Clean excavated earth | 0.022 |
| Corrugated metal storm drain | 0.024 |
| Earth with light brush | 0.050 |
| Earth with heavy brush | 0.100 |

The slope $S$ of a channel can be reported in several ways. It is ideally defined as the ratio of the vertical drop $h$ to the horizontal distance over which the drop occurs. For small slopes, which are typical in open channel flow, it is more practical to use $h/L$, where $L$ is the length of the channel as shown in Fig. 14.5. Normally, the magnitude of the slope for natural streams and drainage structures is very small, a typical value being 0.001. This number can also be expressed as a percentage, where $0.01 = 1$ percent. Then, $0.001 = 0.1$ percent. Since $\sin \theta = h/L$, the angle that the channel bottom makes with the horizontal could also be used. In summary, the slope of 0.001 could be reported as:

**1.** The channel falls 1 m per 1000 m of channel.
**2.** The slope is 0.1 percent.
**3.** $\sin \theta = 0.001$. Then $\theta = \sin^{-1}(0.001) = 0.057°$.

Because the angle is so small, it is rarely used as a measure of the slope.

We can calculate the volume flow rate in the channel from the continuity equation, which is the same as that used for pipe flow:

$$Q = Av \tag{14–7}$$

**FIGURE 14.5**  Slope of a channel.

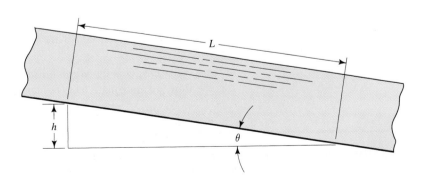

In open channel flow analysis, $Q$ is typically called the *discharge*. Substituting Eq. (14–6) into (14–7) gives an equation which directly relates the discharge to the physical parameters of the channel:

NORMAL DISCHARGE—SI UNITS

$$Q = \left(\frac{1.00}{n}\right)AR^{2/3}S^{1/2} \tag{14–8}$$

This is the only value of discharge for which uniform flow will occur for the given channel depth, and it is called the *normal discharge*. The units of $Q$ are m³/s when the area $A$ is expressed in square meters (m²) and the hydraulic radius in meters (m). Another useful form of this equation is

$$AR^{2/3} = \frac{nQ}{S^{1/2}} \tag{14–9}$$

The term on the left side of Eq. (14–9) is solely dependent on the geometry of the section. Therefore, for a given discharge, slope, and surface type, we can determine the geometrical features of a channel. Alternatively, for a given size and shape of channel, we can calculate the depth at which the normal discharge $Q$ would occur. This depth is called the *normal depth*.

In analyzing uniform flow, typical problems encountered are the calculations of the normal discharge, the normal depth, the geometry of the channel section, the slope, or the value of Manning's $n$. We can make these calculations by using Eqs. (14–6) through (14–9).

**14.6.1**
**Manning's Equation in U.S. Customary System**

Though not strictly true, it is conventional to take the values of Manning's $n$ to be dimensionless so that the same data can be used in either the SI form of the equation [Eq. (14–6)] or the U.S. Customary System form. A careful conversion of units (see Reference 3) allows the use of the same values of $n$ in the following equation:

MANNING'S EQUATION—
U.S. CUSTOMARY UNITS

$$v = \frac{1.49}{n}R^{2/3}S^{1/2} \tag{14–10}$$

The velocity will then be expressed in feet per second (ft/s) when $R$ is in ft. This is the form of Manning's equation for the U.S. Customary System.

We can also create forms of this equation parallel to Eqs. (14–8) and (14–9). That is,

NORMAL DISCHARGE—
U.S. CUSTOMARY UNITS

$$Q = AV = \left(\frac{1.49}{n}\right)AR^{2/3}S^{1/2} \tag{14–11}$$

and

$$AR^{2/3} = \frac{nQ}{1.49S^{1/2}} \tag{14–12}$$

In these equations, $Q$ is the *normal discharge* in cubic feet per second (ft³/s) when $A$ is the flow area in square feet (ft²) and $R$ is expressed in feet (ft).

□ **EXAMPLE PROBLEM 14.2**    Determine the normal discharge for a 200-mm inside diameter common clay drainage tile running half full if it is laid on a slope which drops 1 m over a run of 1000 m.

**Solution**   Equation (14–8) will be used:

$$Q = \left(\frac{1.00}{n}\right) AR^{2/3}S^{1/2}$$

The slope $S = 1/1000 = 0.001$. From Table 14.1 we find $n = 0.013$. Figure 14.6 shows the cross section of the tile half full.

$$A = \frac{1}{2}\left(\frac{\pi D^2}{4}\right) = \frac{\pi D^2}{8} = \frac{\pi(200)^2}{8}\ mm^2 = 5000\pi\ mm^2$$

$$= 15\ 708\ mm^2 = 0.0157\ m^2$$

$$WP = \pi D/2 = 100\pi\ mm$$

Then we have

$$R = A/WP = 5000\pi\ mm^2/100\pi\ mm = 50\ mm$$

or

$$R = 0.05\ m$$

Then in Eq. (14–8),

$$Q = \frac{(0.0157)(0.05)^{2/3}(0.001)^{1/2}}{0.013}$$

$$= 5.18 \times 10^{-3}\ m^3/s$$

**FIGURE 14.6**   Circular drain tile running half full for Example Problem 14.2.

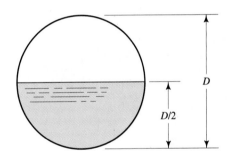

□ **EXAMPLE PROBLEM 14.3**   Calculate the minimum slope on which the channel shown in Fig. 14.7 must be laid if it is to carry 50 ft³/s of water with a depth of 2 ft. The sides and bottom of the channel are made of formed, unfinished concrete.

**FIGURE 14.7**   Trapezoidal channel for Example Problem 14.3.

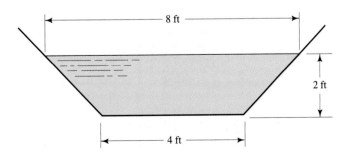

*Solution*    Equation (14–11) can be solved for the slope $S$:

$$Q = \left(\frac{1.49}{n}\right)AR^{2/3}S^{1/2}$$

$$S = \left(\frac{Qn}{1.49AR^{2/3}}\right)^2 \tag{14–13}$$

From Table 14.1 we find $n = 0.017$. The values of $A$ and $R$ can be calculated from the geometry of the section:

$$A = (4)(2) + (2)(2)(2)/2 = 12 \text{ ft}^2$$
$$WP = 4 + 2\sqrt{4 + 4} = 9.66 \text{ ft}$$
$$R = A/WP = 12/9.66 = 1.24 \text{ ft}$$

Then from Eq. (14–13) we have

$$S = \left[\frac{(50)(0.017)}{(1.49)(12)(1.24)^{2/3}}\right]^2 = 0.00169$$

Therefore, the channel must drop at least 1.69 ft per 1000 ft of length.

---

☐ **EXAMPLE PROBLEM 14.4**    Design a rectangular channel to be made of formed, unfinished concrete to carry 5.75 m³/s of water when laid on a 1.2 percent slope. The normal depth should be one-half the width of the channel bottom.

*Solution*    Since the geometry of the channel is to be determined, Eq. (14–9) is most convenient:

$$AR^{2/3} = \frac{nQ}{S^{1/2}} = \frac{(0.017)(5.75)}{(0.012)^{1/2}} = 0.892$$

Figure 14.8 shows the cross section. Since $y = b/2$, only $b$ must be determined. Both $A$ and $R$ can be expressed in terms of $b$:

$$A = by = \frac{b^2}{2}$$

$$WP = b + 2y = 2b$$

$$R = A/WP = \frac{b^2}{(2)(2b)} = \frac{b}{4}$$

Then we have

$$AR^{2/3} = 0.892$$

$$\frac{b^2}{2}\left(\frac{b}{4}\right)^{2/3} = 0.892$$

**FIGURE 14.8**   Rectangular channel for Example Problem 14.4.

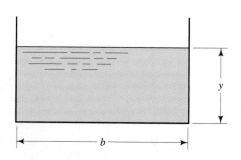

$$\frac{b^{8/3}}{5.04} = 0.892$$

$$b = (4.50)^{3/8} = 1.76 \text{ m}$$

The width of the channel must be 1.76 m.

---

☐ **EXAMPLE PROBLEM 14.5**   In the final design of the channel described in Example Problem 14.4, the width was made 2 m. The maximum expected discharge for the channel is 12 m³/s. Determine the normal depth for this discharge.

*Solution*   Equation (14–9) will be used again:

$$AR^{2/3} = \frac{nQ}{S^{1/2}} = \frac{(0.017)(12)}{(0.012)^{1/2}} = 1.86$$

Both $A$ and $R$ must be expressed in terms of the dimension $y$ in Fig. 14.8, with $b = 2.0$ m:

$$A = 2y$$

$$WP = 2 + 2y$$

$$R = A/WP = 2y/(2 + 2y)$$

Then we have

$$1.86 = AR^{2/3} = 2y \left( \frac{2y}{2 + 2y} \right)^{2/3}$$

Algebraic solution for $y$ is not simply done. A trial-and-error approach can be used. The results follow.

| $y$ | $A$ | $WP$ | $R$ | $R^{2/3}$ | $AR^{2/3}$ | Required Change in $y$ |
|-----|-----|------|-----|-----------|------------|------------------------|
| 2.0 m | 4.0 m² | 6.0 m | 0.667 m | 0.763 | 3.05 | $y$ too high |
| 1.5 | 3.0 | 5.0 | 0.600 | 0.711 | 2.13 | $y$ too high |
| 1.35 | 2.7 | 4.7 | 0.574 | 0.691 | 1.86 | OK |

Therefore, the channel depth would be 1.35 m when the discharge is 12 m³/s.

■

---

**14.7**
**THE GEOMETRY OF**
**TYPICAL OPEN**
**CHANNELS**

Frequently used shapes for open channels include circular, rectangular, trapezoidal, and triangular. Table 14.2 gives the formulas for computing the geometric features pertinent to open channel flow calculations.

The trapezoid is popular for several reasons. It is an efficient shape because it gives a large flow area relative to the wetted perimeter. The sloped sides are convenient for channels made in the earth, because the slopes can be set at an angle at which the construction materials are stable.

The slope of the sides can be defined by the angle with respect to the horizontal or by means of the *pitch*, the ratio of the horizontal distance to the vertical distance. The pitch in Table 14.2 is indicated by the value of $z$, which is the horizontal distance corresponding to one unit of vertical distance. Practical earth channels made in the trapezoidal shape use values of $z$ from 1.0 to 3.0.

**TABLE 14.2**  Geometry of open channel sections

| Section | Area $A$ | Wetted Perimeter $WP$ | Hydraulic Radius $R$ |
|---|---|---|---|
| **Rectangle** <br> $b = T$, $y$ | $by$ | $b + 2y$ | $\dfrac{by}{b + 2y}$ |
| **Triangle** <br> $T = 2zy$, $y$, $z$, $1$ | $zy^2$ | $2y\sqrt{1 + z^2}$ | $\dfrac{zy}{2\sqrt{1 + z^2}}$ |
| **Trapezoid** <br> $T = b + 2zy$, $y$, $z$, $1$, $b$ | $(b + zy)y$ | $b + 2y\sqrt{1 + z^2}$ | $\dfrac{(b + zy)y}{b + 2y\sqrt{1 + z^2}}$ |
| **Circle** <br> $T = 2\sqrt{y(D - y)}$, $D$, $y$, $\theta$ <br> $\theta$ is in radians | $\dfrac{(\theta - \sin\theta)\,D^2}{8}$ | $\theta D/2$ | $\left[\dfrac{(\theta - \sin\theta)}{\theta}\right]\dfrac{D}{4}$ |

Note: $\theta$ must be in radians.
For $y < D/2$, $\theta = \pi - 2\sin^{-1}[1 - (2y/D)]$
For $y > D/2$, $\theta = \pi + 2\sin^{-1}[(2y/D) - 1]$

The rectangle is a special case of the trapezoid with a side slope of 90° or $z = 0$. Formed concrete channels are often made in this shape. The triangular channel is also a special case of the trapezoid with a bottom width of zero. Simple ditches in earth are often made in this shape.

The computation of the data for circular sections at various depths can be facilitated by the graph in Fig. 14.9. At the left side of the figure is shown half of a circular section running partially full with the depth of the fluid called $y$. The vertical scale for the graph is the ratio, $y/D$. Curve A gives the ratio of $A/A_f$, in which $A$ is the actual fluid flow area and $A_f$ is the full cross-sectional area of the circle, easily calculated from $A_f = \pi D^2/4$. The use of Curve A is demonstrated by noting that the figure is drawn for the case $y/D = 0.65$. Follow the dashed horizontal line from this value on the vertical scale over to Curve A and then project down to the horizontal scale and read the value of 0.70. This means $A/A_f = 0.70$ for $y/D = 0.65$. As an example, assume $D = 2.00$ ft. Then,

$$A_f = \pi D^2/4 = \pi (2.00 \text{ ft})^2/4 = 3.14 \text{ ft}^2$$
$$A = (0.70)A_f = (0.70)(3.14 \text{ ft}^2) = 2.20 \text{ ft}^2$$

In a similar manner, you should be able to read the wetted perimeter ratio to be $WP/WP_f = 0.60$ and the hydraulic radius ratio to be $R/R_f = 1.16$. Then,

$$WP_f = \pi D \text{ for a full circle} = \pi (2.00 \text{ ft}) = 6.28 \text{ ft}$$
$$WP = (0.60)WP_f = (0.60)(6.28 \text{ ft}) = 3.77 \text{ ft}$$

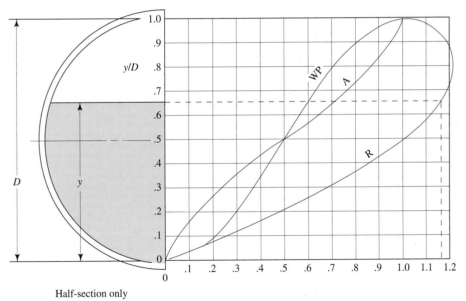

Half-section only shown

Curve A: Ratio of $A/A_f$; $A_f = \pi D^2/4$
Curve WP: Ratio of $WP/WP_f$; $WP_f = \pi D$
Curve R: Ratio of $R/R_f$; $R_f = D/4$

Example: $D = 2.0$ ft; $y = 1.30$ ft; $y/D = 0.65$

$A_f = 3.14 \text{ ft}^2$; $A/A_f = .7$; $A = 0.7(3.14) = 2.20 \text{ ft}^2$
$WP_f = 6.28$ ft; $WP/WP_f = 0.6$; $WP = 0.6(6.28) = 3.77$ ft
$R_f = 0.50$ ft; $R/R_f = 1.16$; $R = 1.16(0.50) = 0.580$ ft

**FIGURE 14.9** Geometry for partially full circular section.

and,

$$R_f = D/4 \text{ for a full circle} = (2.00 \text{ ft})/4 = 0.50 \text{ ft}$$
$$R = (1.16)R_f = (1.16)(0.50 \text{ ft}) = 0.580 \text{ ft}$$

Thus, the curves in Fig. 14.9 enable you to compute the values of $A$, $WP$, and $R$ for partially full circular sections with simple formulas using values of the three ratios read from the chart. Otherwise, the equations for direct calculation of $A$, $WP$, and $R$ are quite complex.

Figure 14.10 shows three other shapes used for open channels. Natural streams frequently can be approximated as shallow parabolas. The triangle with a rounded bottom is more practical to make in the earth than the sharp-V triangle. The round-cornered rectangle performs somewhat better than the square-cornered rectangle and is easier to maintain. However, it is more difficult to form. Reference 3 gives formulas for the geometric features of these types of cross sections.

**FIGURE 14.10**   Other shapes for open channels.

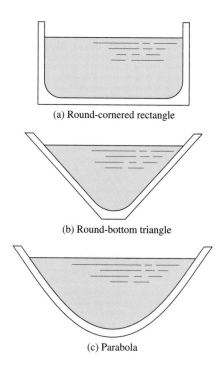

(a) Round-cornered rectangle

(b) Round-bottom triangle

(c) Parabola

**14.8**
**THE MOST EFFICIENT**
**SHAPES FOR OPEN**
**CHANNELS**

The term *conveyance* is used to indicate the carrying capacity of open channels. Its value can be deduced from Manning's equation. In SI metric units,

$$Q = \left(\frac{1.00}{n}\right) A R^{2/3} S^{1/2} \qquad (14\text{--}8)$$

Everything on the right side of this equation is dependent on the design of the channel except the slope. We can then define the conveyance $K$ to be

CONVEYANCE—SI UNITS

$$K = \left(\frac{1.00}{n}\right) A R^{2/3} \qquad (14\text{--}14)$$

In U.S. Customary units,

CONVEYANCE—

U.S. CUSTOMARY UNITS

$$K = \left(\frac{1.49}{n}\right)AR^{2/3} \tag{14–15}$$

Manning's equation is then

$$Q = KS^{1/2} \tag{14–16}$$

The conveyance of a channel would be maximum when the wetted perimeter is the least for a given area. Using this criterion, we find that the most efficient shape is the semicircle, that is, the circular section running half full. Table 14.3 on page 428 shows the most efficient designs of other shapes.

**14.9**
**CRITICAL FLOW AND**
**SPECIFIC ENERGY**

Consideration of energy in open channel flow usually involves a determination of the energy possessed by the fluid at a particular section of interest. The total energy is measured relative to the channel bottom and is composed of potential energy due to the depth of the fluid plus kinetic energy due to its velocity.

Letting $E$ denote the total energy, we get

$$E = y + v^2/2g \tag{14–17}$$

where $y$ is the depth and $v$ is the average velocity of flow. As with the energy equation used previously, the terms in Eq. (14–17) have the units of energy per unit weight of fluid flowing. In open channel flow analysis, $E$ is usually referred to as the *specific energy*. For a given discharge $Q$, the velocity is $Q/A$. Then

$$E = y + Q^2/2gA^2 \tag{14–18}$$

Since the area can be expressed in terms of the depth of flow, Eq. (14–18) relates the specific energy to the depth of flow. A graph of the depth $y$ versus the specific energy $E$ is useful in visualizing the possible regimes of flow in a channel. For a particular channel section and discharge, the specific energy curve appears as shown in Fig. 14.11.

**FIGURE 14.11** Variation of specific energy with depth.

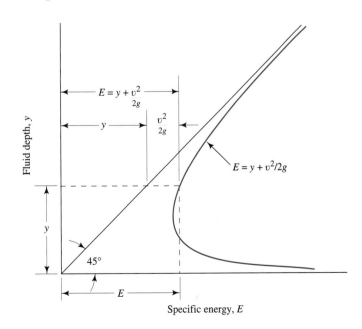

**TABLE 14.3**  Most efficient sections for open channels

| Section | Area $A$ | Wetted Perimeter $WP$ | Hydraulic Radius $R$ |
|---|---|---|---|
| **Rectangle** (half of a square) $b = 2y = T$, $y$ | $2.0y^2$ | $4y$ | $y/2$ |
| **Triangle** (half of a square) $T = 2y$, $z = 1$, $z = 1$, $45°$, $45°$, $y$ | $y^2$ | $2.83y$ | $0.354y$ |
| **Trapezoid** (half of a hexagon) $T = 2.309y$, $z = 0.577$, $z = 0.577$, $L = b$, $60°$, $60°$, $y$, $b = 1.155y$ | $1.73y^2$ | $3.46y$ | $y/2$ |
| **Semicircle** $T = 2y$, $D = 2y$, $y$ | $\frac{1}{2}\pi y^2$ | $\pi y$ | $y/2$ |

Several features of this curve are important. The 45° line on the graph represents the plot of $E = y$. Then for any point on the curve, the horizontal distance to this line from the $y$ axis represents the potential energy $y$. The remaining distance to the specific energy curve is the kinetic energy $v^2/2g$. A definite minimum value of $E$ appears, and we can show that this occurs when the flow is at the critical state, that is, when $N_F = 1$. See Section 14.5, Eq. (14–4) for the definition of the Froude number, $N_F$.

The depth corresponding to the minimum specific energy is therefore called the *critical depth* $y_c$. For any depth greater than $y_c$, the flow is subcritical. Conversely, for any depth lower than $y_c$, the flow is supercritical. Notice that for any energy level greater than the minimum, there can exist two different depths. In Fig. 14.12, both $y_1$ below the critical depth $y_c$, and $y_2$ above $y_c$, have the same energy. In the case of $y_1$, the flow is supercritical and much of the energy is kinetic energy due to the high velocity of flow. At the greater depth $y_2$, the flow is slower and only a small portion of the energy is kinetic energy. The two depths $y_1$ and $y_2$ are called the *alternate depths* for the specific energy $E$.

**FIGURE 14.12** Critical depth and alternate depths.

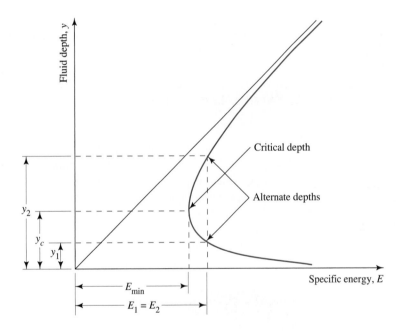

**14.10 HYDRAULIC JUMP**

To understand the significance of the phenomenon known as *hydraulic jump*, consider one of its most practical uses illustrated in Fig. 14.13. The water flowing over the spillway normally has a high velocity in the supercritical range when it reaches the bottom of the relatively steep slope at section 1. If this velocity were to be maintained into the natural stream bed beyond the paved spillway structure, the sides and bottom of the stream would be severely eroded. Instead, good design would cause a hydraulic jump to occur as shown, where the depth of flow abruptly changes from $y_1$ to $y_2$. Two beneficial effects result from the hydraulic jump. First, the velocity of flow is decreased substantially, decreasing the tendency for the flow to erode the stream bed. Second, much of the excess energy contained in the high velocity flow is dissipated in the jump. Energy dissipation occurs because the flow in the jump is extremely turbulent.

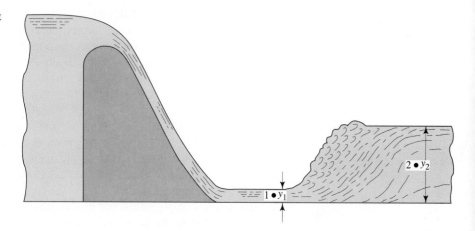

In order for a hydraulic jump to occur, the flow before the jump must be the supercritical range. That is, at section 1 in Fig. 14.13, $y_1$ is less than the critical depth for the channel and the Froude number $N_{F_1}$ is greater than 1.0. The depth at section 2 after the jump can be calculated from the equation

$$y_2 = (y_1/2)(\sqrt{1 + 8N_{F_1}^2} - 1) \qquad (14-19)$$

Also, the energy loss in the jump is dependent on the two depths $y_2$ and $y_1$:

$$E_1 - E_2 = \Delta E = (y_2 - y_1)^3/4y_1y_2 \qquad (14-20)$$

Figure 14.14 illustrates what happens in a hydraulic jump by using a specific energy curve. The flow enters the jump with an energy $E_1$ corresponding to a supercritical depth $y_1$. In the jump, the depth abruptly increases. If no energy were lost, the new depth would be $y'_2$, which is the alternate depth for $y_1$. However, since there was some energy dissipated, $\Delta E$, the actual new depth $y_2$ corresponds to the energy level $E_2$. Still, $y_2$ is in the subcritical range and tranquil flow would be maintained downstream from the jump. The name given to the actual depth $y_2$ after the jump is the *sequent depth*.

The following example problem illustrates another practical case in which hydraulic jump might occur.

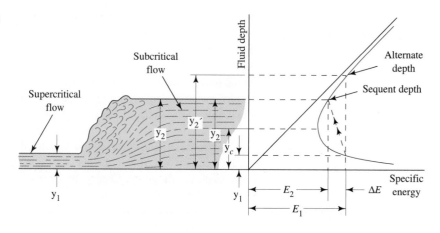

□ **EXAMPLE PROBLEM 14.6**   As shown in Fig. 14.15, water is being discharged from a reservoir under a sluice gate at the rate of 18 m³/s into a horizontal rectangular channel, 3 m wide, made of unfinished formed concrete. At a point where the depth is 1 m, a hydraulic jump is observed to occur. Determine the following:

a. The velocity before the jump
b. The depth after the jump
c. The velocity after the jump
d. The energy dissipated in the jump

**FIGURE 14.15** Hydraulic jump for Example Problem 13.6.

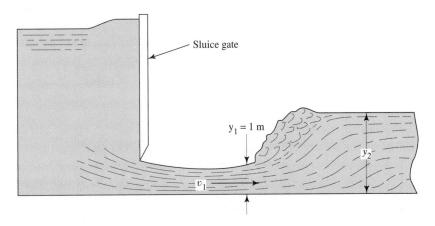

**Solution**   a. The velocity before the jump is

$$v_1 = Q/A_1$$
$$A_1 = (3)(1) = 3 \text{ m}^2$$
$$v_1 = (18 \text{ m}^3/\text{s})/3 \text{ m}^2 = 6.0 \text{ m/s}$$

b. Equation (13–19) can be used to determine the depth after the jump, $y_2$:

$$y_2 = (y_1/2)(\sqrt{1 + 8N_{F_1}^2} - 1)$$
$$N_{F_1} = v_1/\sqrt{gy_h}$$

The hydraulic depth is equal to $A/T$, where $T$ is the width of the free surface. Then for a rectangular channel $y_h = y$. Then we have

$$N_{F_1} = 6.0/\sqrt{(9.81)(1)} = 1.92$$

The flow is in the supercritical range.

$$y_2 = (1/2)(\sqrt{1 + (8)(1.92)^2} - 1) = 2.26 \text{ m}$$

c. Because of continuity,

$$v_2 = Q/A_2 = (18 \text{ m}^3/\text{s})/(3)(2.26) \text{ m}^2 = 2.65 \text{ m/s}$$

d. From Eq. (14–20), we get

$$\Delta E = (y_2 - y_1)^3/4y_1 y_2$$
$$= \frac{(2.26 - 1.0)^3}{(4)(1.0)(2.26)} \text{ m} = 0.221 \text{ m}$$

This means that 0.221 N·m of energy is dissipated from each newton of water as it flows through the jump.

■

## REFERENCES

1. Albertson, M. L., J. R. Barton, and D. B. Simons. 1960. *Fluid Mechanics for Engineers.* Englewood Cliffs, NJ: Prentice-Hall.

2. Binder, R. C. 1973. *Fluid Mechanics.* 5th ed. Englewood Cliffs, NJ: Prentice-Hall.

3. Chow, V. T. 1959. *Open Channel Hydraulics.* New York: McGraw-Hill.

4. Henderson, F. M. 1966. *Open Channel Flow.* New York: Macmillan.

5. Murdock, J. W. 1976. *Fluid Mechanics.* Boston: Houghton Mifflin.

6. Roberson, John A., John J. Cassidy, and M. Hanif Chaudhry. 1988. *Hydraulic Engineering.* Boston: Houghton Mifflin.

7. Simon, A. L. 1986. *Hydraulics.* 3rd ed. New York: John Wiley & Sons.

## PRACTICE PROBLEMS

**14.1M** *Compute the hydraulic radius for a circular drain pipe running half full if its inside diameter is 300 mm.*

**14.2M** *A rectangular channel has a bottom width of 2.75 m. Compute the hydraulic radius when the fluid depth is 0.50 m.*

**14.3E** A drainage structure for an industrial park has a trapezoidal cross section similar to that shown in Fig. 14.1(c). The bottom width is 3.50 ft, and the sides are inclined at an angle of 60° from the horizontal. Compute the hydraulic radius for this channel when the fluid depth is 1.50 ft.

**14.4E** Repeat Problem 14.3 if the side slope is 45°.

**14.5E** Compute the hydraulic radius for a trapezoidal channel with a bottom width of 150 mm and with sides that pitch 15 mm horizontally for a vertical change of 10 mm. That is, the ratio of *X/D* in Fig. 14.1(c) is 1.50. The depth of the fluid in the channel is 62 mm.

**14.6E** Compute the hydraulic radius for the section shown in Fig. 14.16 if water flows at a depth of 2.0 in. The section is that of a rain gutter for a house.

**14.7E** Repeat Problem 14.6 for a depth of 3.50 in.

**14.8M** *Compute the hydraulic radius for the channel shown in Fig. 14.17 if the water depth is 0.50 m.*

**14.9M** *Compute the hydraulic radius for the channel shown in Fig. 14.17 if the water depth is 2.50 m.*

**14.10M** *Water is flowing in a formed unfinished concrete rectangular channel 3.5 m wide. For a depth of 2.0 m, calculate the normal discharge and the Froude number of the flow. The channel slope is 0.1 percent.*

**14.11E** Determine the normal discharge for an aluminum rain spout with the shape shown in Fig. 14.16 that runs at the depth of 3.50 in. Use $n = 0.013$. The spout falls 4 in over a length of 60 ft.

**14.12E** A circular culvert under a highway is 6 ft in diameter and is made of corrugated metal. It drops 1 ft over a length of 500 ft. Calculate the normal discharge when the culvert runs half full.

**14.13M** *A wooden flume is being built to temporarily carry 5000 L/min of water until a permanent drain can be installed. The flume is rectangular, with a 205-mm bottom width and a maximum depth of 250 mm. Calculate the slope required to handle the expected discharge.*

**14.14M** *A storm drainage channel in a city where heavy sudden rains occur has the shape shown in Fig. 14.17. It is made of unfinished concrete and has a slope of 0.5 percent. During normal times, the water remains in the small rectangular section. The upper section allows large volumes to be carried by the channel. Determine the normal discharge for depths of 0.5 m and 2.5 m.*

**14.15E** Figure 14.18 represents the approximate shape of a natural stream channel with levees built on either side. The channel is earth with grass cover. Use $n = 0.04$. If the average slope is 0.000 15, determine the normal discharge for depths of 3 ft and 6 ft.

**14.16E** Calculate the depth of flow of water in a rectangular channel 10 ft wide, made of brick in cement mortar for a discharge of 150 ft³/s. The slope is 0.1 percent.

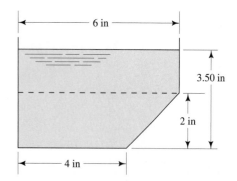

**FIGURE 14.16**   Problems 14.6, 14.7, and 14.11.

**14.17M** *Calculate the depth of flow in a trapezoidal channel with a bottom width of 3 m and whose walls slope 45° with the horizontal. The channel is made of unfinished concrete and is laid on a 0.1 percent slope. The discharge is 15 m³/s.*

**14.18M** *A rectangular channel must carry 2.0 m³/s of water from a water-cooled refrigeration condenser to a cooling pond. The available slope is 75 mm over a distance of 50 m. The maximum depth of flow is 0.40 m. Determine the width of the channel if its surface is trowel-finished concrete.*

**14.19M** *The channel shown in Fig. 14.19 has a surface of float-finished concrete and is laid on a slope which falls 0.1 m per 100 m of length. Calculate the normal discharge and the Froude number for a depth of 1.5 m. For that discharge, calculate the critical depth.*

**14.20E** A square storage room is equipped with automatic sprinklers for fire protection which spray 1000 gal/min of water. The floor is designed to drain this flow evenly to troughs near each outside wall. The troughs are shaped as shown in Fig. 14.20. Each trough carries 250 gal/min,

is laid on a 1 percent slope, and is formed of unfinished concrete. Determine the minimum depth *h*.

**14.21E** The flow from two of the troughs described in Problem 14.20 passes into a sump, from which a round common clay drainage tile carries it to a storm sewer. Determine the size of tile required to carry the flow (500 gal/min) when running half full. The slope is 0.1 percent.

**14.22M** *For a rectangular channel with a bottom width of 1.00 m, compute the flow area and hydraulic radius for depths ranging from 0.10 m to 2.0 m. Plot a graph of area and hydraulic radius versus depth.*

**14.23M** *It is desired to carry 2.00 m³/s of water at a velocity of 3.0 m/s in a rectangular open channel. The bottom width is 0.80 m. Compute the depth of the flow and the hydraulic radius.*

**14.24M** *For the channel designed in Problem 14.23, compute the required slope if the channel is float-finished concrete.*

**14.25M** *It is desired to carry 2.00 m³/s of water at a velocity of 3.0 m/s in a rectangular open channel. Compute the*

**FIGURE 14.17**   Problems 14.8, 14.9, and 14.14.

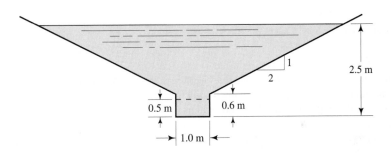

**FIGURE 14.18**   Problem 14.15.

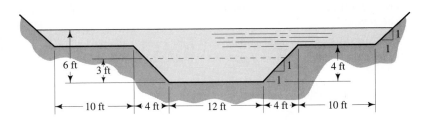

**FIGURE 14.19**   Problem 14.19.

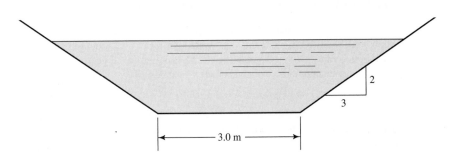

**FIGURE 14.20**   Problems 14.20
and 14.21.

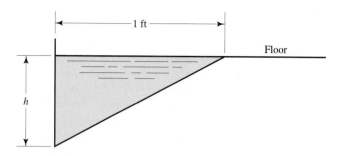

depth and hydraulic radius for a range of designs for
the channel, with bottom widths of 0.50 m to 2.00 m.
Plot depth and hydraulic radius versus bottom width.

**14.26M** *For each of the channels designed in Problem 14.25,
compute the required slope if the channel is float-
finished concrete. Plot slope versus bottom width.*

**14.27E** A trapezoidal channel has a bottom width of 2.00 ft and
a pitch of its sides of $z = 1.50$. Compute the flow area
and hydraulic radius for a depth of 20 in.

**14.28E** For the channel described in Problem 14.27, compute
the normal discharge that would be expected for a slope
of 0.005 if the channel is made from formed unfinished
concrete.

**14.29E** Repeat Problem 14.28, except that the channel is lined
with smooth plastic sheets.

**14.30E** A trapezoidal channel has a bottom width of 2.00 ft and
a pitch of its sides of $z = 1.50$. Compute the flow area
and hydraulic radius for depths ranging from 6.00 in to
24.00 in. Plot flow area and hydraulic radius versus
depth.

**14.31E** For each channel designed in Problem 14.30, compute
the normal discharge that would be expected for a slope
of 0.005 if the channel is made from formed unfinished
concrete.

**14.32M** *Compute the flow area and hydraulic radius for a circular drain pipe 375 mm in diameter for a depth of
225 mm.*

**14.33M** *Repeat Problem 14.32 for a depth of 135 mm.*

**14.34M** *For the channel designed in Problem 14.32, compute
the normal discharge that is expected for a slope of 0.12
percent if the channel is made from painted steel.*

**14.35M** *For the channel designed in Problem 14.33, compute
the normal discharge that is expected for a slope of 0.12
percent if the channel is made from painted steel.
Compare the result with that from Problem 14.34.*

**14.36E** It is desired to carry 1.25 ft³/s of water at a velocity of
2.75 ft/s. Design the channel cross section for each of
the shapes in Table 14.3, which lists the most efficient sections for open channels.

**14.37E** For each section designed in Problem 14.36, compute
the required slope if the channel is made of float-
finished concrete. Compare the results.

**14.38E** For each section designed in Problem 14.36, compute
the Froude number and tell whether the flow is subcritical or supercritical.

**14.39M** *A rectangular channel 2 m wide carries 5.5 m³/s of water and is made of unfinished concrete. Perform the following operations:*

   **a.** *Calculate the critical depth.*
   **b.** *Calculate the minimum specific energy.*
   **c.** *Plot the specific energy curve.*
   **d.** *Determine the specific energy when y = 0.5 m and
   the alternate depth for this energy.*
   **e.** *Determine the velocity of flow and the Froude numbers for each depth in (d).*
   **f.** *Calculate the required slopes of the channel if the
   depths from (d) are to be normal depths for 5.5 m³/s
   flow.*

## COMPUTER PROGRAMMING ASSIGNMENTS

1. Write a program to compute the geometry features for each
section shown in Table 14.2. Include the area, wetted perimeter, and hydraulic radius. These programs could be used as
subroutines in more comprehensive open channel flow analysis problems.

2. Write a program to compute the geometry features for each
section shown in Table 14.3. Include the area, wetted perimeter, and hydraulic radius.

3. Write a program to compute the normal discharge for a given
open channel and slope. Include the ability to compute the
geometry features of the channel. The program can be written for a given channel slope (e.g., rectangular, circular, etc.),
or it can include a set of subroutines for a variety of shapes.
The program can include the table of values for Manning's $n$
from which the user can select a design value. This program
will then perform the computations of the type shown in
Example Problem 14.2.

4. Write a program that computes the slope required for a channel of a given shape to carry a given discharge by using a method similar to that in Example Problem 14.3.

5. Write a program that computes the normal depth for a rectangular channel of a given width carrying a given normal discharge at a given slope. This program requires a trial-and-error solution method similar to that shown in Example Problem 14.5.

# ■ ■ ■ ■ ■  15 Flow Measurement

## 15.1 The Big Picture

### Discussion Map

■ *Flow Measurement* refers to the ability to measure the velocity, volume flow rate, or mass flow rate of any liquid or gas.

■ Accurate measurement of flow is essential to the control of industrial processes, the transfer of custody of fluids, the evaluation of the performance of engines, refrigeration systems, and other systems employing moving fluids.

■ There are many types of commercially available flow meters with which you should be familiar.

### Discover

Take a few minutes to think and to talk with colleagues about the ways in which your life has been affected by flow measurement recently.

List as many kinds of flow meters as you can think of at this time.

This chapter will increase your awareness of the many types of flow measurement equipment available, and you should develop your skill in making the appropriate calculations to interpret the results obtained from them.

*Flow measurement* is an important function within any organization that employs fluids to carry on its daily operations. It refers to the ability to measure the velocity, volume flow rate, or mass flow rate of any liquid or gas.

As you discuss flow measurement with your colleagues, compare the list of situations that you are aware of with the ones listed here.

■ You buy gasoline from a service station and the pump system includes a flow meter to indicate to you and to the station operator how many gallons or liters you pumped so you can pay for just the amount you put into your car.
■ The weather report indicates that showers are expected with winds of 30 miles per hour.
■ In the chemistry laboratory, you may monitor the heat input to a reaction by measuring the flow rate of fuel gas into a burner.

Now, how many can you add to this list? Consider the following general reasons for measuring the flow of fluids.

■ *Custody transfer and accounting.* Any time a person acquires a fluid product from a supplier, an accurate accounting is needed of the amount of the fluid transferred. Have you noticed that the gasoline pump meter is checked periodically by a public agency responsible for enforcing standards for accuracy of weights and measures in general commerce?

- *Performance evaluation.* An engine requires fuel that provides the basic energy needed to run it. One indication of the performance of the engine is to measure the power output (energy per unit time) in relation to the rate of fuel used by the engine (gallons per hour). This is directly related to the fuel efficiency measure you typically use for your car, miles per gallon or km/liter.
- *Process control.* Any industry that uses fluids in its processes must monitor the mass flow rate of key fluids into those processes. For example, beverages are blends of several constituents that must be precisely controlled to maintain the taste that the customer expects. Continuous monitoring and control of the volume flow rate of each constituent into the blending system is critical to producing consistently a quality product.
- *Research and development.* Numerous examples can be given. Consider the move from fluorocarbon refrigerants (freons) to more environmentally acceptable refrigerants. It is essential to test many candidate formulations to determine the refrigerating effect produced as a function of the mass flow rate of the refrigerant through the air conditioner or freezer.

This chapter will increaase your awareness of the many types of flow measurement equipment available and to develop your skill in making the appropriate calculations to interpret the results obtained from them. You should also be able to recommend suitable types of flow meters for a given application. It is most likely that you will use commercially available meters rather than designing and making your own. To do that efficiently and effectively, you must understand the physical principles on which the meters are based.

## 15.2 OBJECTIVES

After completing this chapter, you should be able to:

1. Describe six factors that should be considered when specifying a flow measurement system.
2. Describe four types of variable head meters: the *venturi tube*, the *flow nozzle*, the *orifice*, and the *flow tube*.
3. Compute the velocity of flow and the volume flow rate for variable head meters, including the determination of the discharge coefficient.
4. Describe the *rotameter variable area meter*, *turbine flowmeter*, *magnetic flowmeter*, *vortex flowmeter*, and *ultrasonic flowmeter*.
5. Describe two methods of measuring mass flow rate.
6. Describe the *pitot-static tube*, and compute the velocity of flow using data acquired from such a device.
7. Define the term *anemometer* and describe two kinds.
8. Describe *weirs* and *flumes* as they are used for measuring flow in open channels, and perform the necessary computations.

## 15.3 FLOWMETER SELECTION FACTORS

Many devices are available for measuring flow. Some measure volume flow rate directly, while others measure an average velocity of flow which can then be converted to volume flow rate by using $Q = Av$. Also, some provide direct primary measurements, while others require calibration or the application of a discharge coefficient to the observed output of the device. The form of the flow meter output also varies considerably from one type to another. The indication can be a pressure, a liquid level, a mechanical counter, the position of an indicator in the fluid stream, a continuous electrical signal, or a series of electrical pulses. The choice of the basic type of fluid meter and its indication system depends on several factors, some of which we will discuss here.

**15.3.1
Range**

Commercially available meters can measure flows from a few milliliters per second (mL/s) for precise laboratory experiments to several thousand cubic meters per second ($m^3$/s) for irrigation water or municipal water and sewage systems. Then, for a particular meter installation, the general order of magnitude of the flow rate must be known as well as the range of the expected variations.

**15.3.2
Accuracy Required**

Virtually any flow measuring device properly installed and operated can produce an accuracy within 5 percent of the actual flow. Most commercially made meters are capable of 2 percent accuracy, and several claim better than 0.5 percent. Cost usually becomes an important factor when great accuracy is desired.

**15.3.3
Pressure Loss**

Because the construction details of the various meters are quite different, they produce differing amounts of energy loss or pressure loss as the fluid flows through them. Except for a few types, fluid meters accomplish the measurement by placing a restriction or a mechanical device in the flow stream, thus causing the energy loss.

**15.3.4
Type of Indication**

Factors to consider when choosing the type of flow indication include whether remote sensing or recording is required, whether automatic control is to be actuated by the output, whether an operator needs to monitor the output, and whether severe environmental conditions exist.

**15.3.5
Type of Fluid**

The performance of some fluid meters is affected by the properties and condition of the fluid. A basic consideration is whether the fluid is a liquid or a gas. Other factors which may be important are viscosity, temperature, corrosiveness, electrical conductivity, optical clarity, lubricating properties, and homogeneity. Slurries and multiphase fluids require special meters.

**15.3.6
Calibration**

Calibration is required for some types of flowmeters. Some manufacturers provide a calibration in the form of a graph or chart of actual flow versus indicator reading. Some are equipped for direct reading, with scales calibrated in the desired units of flow. In the case of the more fundmantal types of meters, such as the variable head types, standard geometrical forms and dimensions have been determined for which empirical data are available. These data relate flow to an easily measured variable such as a pressure difference or fluid level. References at the end of this chapter give many of these calibration factors.

   If calibration is required by the user of the device, he or she may use another precision meter as a standard against which the reading of the test device can be compared. Alternatively, primary calibration can be performed by adjusting the flow to a constant rate through the meter and then collecting the output during a fixed time interval. The fluid thus collected can either be weighed for a weight per unit time calibration, or its volume can be measured for a volume flow rate calibration. Figure 15.1 shows a commercially available flow calibrator in which a precision piston moves at a controlled rate to move the test fluid through the flowmeter being calibrated. The meter output is compared with the known flow rate by a computer data acquisition and analysis system to prepare calibration charts and graphs.

**15.3.7
Other Factors**

In most cases the physical size of the meter, cost, the system pressure, and the operator's skill should also be considered.

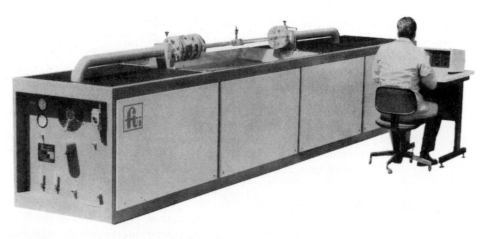

**FIGURE 15.1**   Flow calibration system. (Source: EG&G Flow Technology, Inc., Phoenix, AZ)

**15.4**
**VARIABLE HEAD**
**METERS**

The basic principle on which variable head meters are based is that when a fluid stream is restricted, its pressure decreases by an amount which is dependent on the rate of flow through the restriction. Therefore, the pressure difference between points before and after the restriction can be used to indicate flow rate. The most common types of variable head meters are the venturi tube, the flow nozzle, the orifice, and the flow tube. The derivation of the relationship between the pressure difference and the volume flow rate is the same regardless of which type of device is used. The venturi tube will be used as an example.

**15.4.1**
**Venturi Tube**

Figure 15.2 shows the basic appearance of a venturi tube. The flow from the main pipe at section 1 is caused to accelerate through a narrow section called the *throat*, where the fluid pressure is decreased. The flow then expands through the diverging portion to the same diameter as the main pipe. Pressure taps are located in the pipe wall at section 1 and in the wall of the throat, which we will call section 2. These pressure taps are attached to the two sides of a differential manometer so that the deflection $h$ is an indication of the pressure difference $p_1 - p_2$. Of course, other types of differential pressure gages could be used.

The energy equation and the continuity equation can be used to derive the relationship from which we can calculate the flow rate. Using sections 1 and 2 in Fig. 15.2 as the reference points, we can write the following equations:

$$\frac{p_1}{\gamma} + z_1 + \frac{v_1^2}{2g} - h_L = \frac{p_2}{\gamma} + z_2 + \frac{v_2^2}{2g} \qquad (15\text{--}1)$$

$$Q = A_1 v_1 = A_2 v_2 \qquad (15\text{--}2)$$

These equations are valid only for incompressible fluids, that is, liquids. For the flow of gases, we must give special consideration to the variation of the specific weight $\gamma$ with pressure. See Reference 5. The algebraic reduction of Eqs. (15–1) and (15–2) proceeds as follows:

$$\frac{v_2^2 - v_1^2}{2g} = \frac{p_1 - p_2}{\gamma} + (z_1 - z_2) - h_L$$

$$v_2^2 - v_1^2 = 2g[(p_1 - p_2)/\gamma + (z_1 - z_2) - h_L]$$

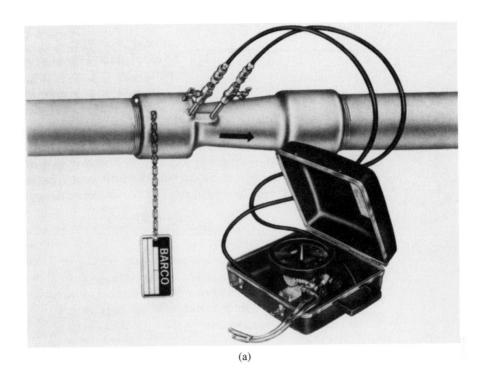

(a)

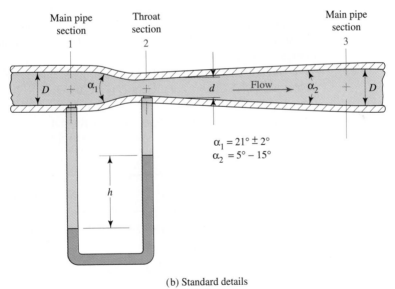

Main pipe section 1  Throat section 2  Main pipe section 3

$D$  $\alpha_1$  $d$  Flow  $\alpha_2$  $D$

$\alpha_1 = 21° \pm 2°$
$\alpha_2 = 5° - 15°$

$h$

(b) Standard details
of design

**FIGURE 15.2**  Venturi tube. (Source of photo: BARCO, Barrington, IL)

But $v_2^2 = v_1^2(A_1/A_2)^2$. Then we have

$$v_1^2[(A_1/A_2)^2 - 1] = 2g[(p_1 - p_2)/\gamma + (z_1 - z_2) - h_L]$$

$$v_1 = \sqrt{\frac{2g[(p_1 - p_2)/\gamma + (z_1 - z_2) - h_L]}{(A_1/A_2)^2 - 1}} \qquad (15\text{–}3)$$

We can make two simplifications at this time. First, the elevation difference $(z_1 - z_2)$ is very small, even if the meter is installed vertically. Therefore, this term is neglected. Second, the term $h_L$ is the energy loss from the fluid as it flows from section 1 to section 2. The value of $h_L$ must be determined experimentally. But it is more convenient to modify Eq. (15–3) by dropping $h_L$ and introducing a discharge coefficient $C$:

$$v_1 = C\sqrt{\frac{2g(p_1 - p_2)/\gamma}{(A_1/A_2)^2 - 1}} \qquad (15\text{–}4)$$

Equation (15–4) can be used to calculate the velocity of flow in the throat of the meter. Note that the velocity depends on the difference in the pressure head between points 1 and 2. This is the reason these meters are called variable head meters.

Normally we want to calculate the volume flow rate.

Since $Q = A_1 v_1$, we have

$$Q = CA_1\sqrt{\frac{2g(p_1 - p_2)/\gamma}{(A_1/A_2)^2 - 1}} \qquad (15\text{–}5)$$

The discharge coefficient $C$ represents the ratio of the actual velocity through the venturi to the ideal velocity for a venturi with no energy loss at all. Therefore, the value of $C$ will always be less than 1.0. The venturi of the Herschel type shown in Figure 15.2 is designed to minimize energy losses by employing a smooth gradual contraction to the throat and a smooth, long enlargement following the throat. Therefore, the discharge coefficient is typically close to 1.0.

Figure 15.3 indicates that the actual value of $C$ depends on the Reynolds number for the flow in the main pipe. Above a Reynolds number of $2 \times 10^5$ the value of $C$ is taken to be 0.984. This value applies to a venturi of the Herschel type that

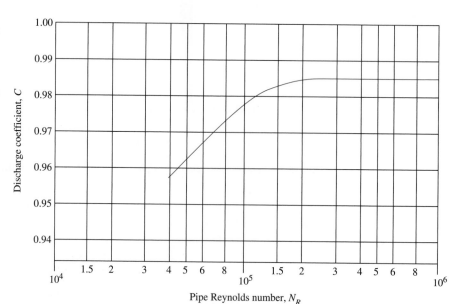

**FIGURE 15.3** Discharge coefficient for a rough cast venturi tube of the Herschel type. (Source: ASME Research Committee on Fluid Meters. 1959. *Fluid Meters: Their Theory and Application.* 5th ed. New York: The American Society of Mechanical Engineers, p. 125.)

is rough cast with a pipe diameter in the range from 4.0 in to 48.0 in (100 mm to 1200 mm). The throat diameter can vary over a fairly wide range, but the ratio of $d/D$, called the beta ratio or $\beta$, should be between 0.30 and 0.75.

For Reynolds numbers below $2 \times 10^5$ you must read the value of $C$ from Figure 15.3.

Smaller venturi meters, for pipe diameters in the range from 2-in to 10-in (50 mm to 250 mm), the venturi is typically machined resulting in a better surface finish than the rough casting. The value of $C$ is taken to be 0.995 for this type when $N_R > 2 \times 10^5$. Data are not available for $C$ for lower Reynolds numbers for the machined venturi.

References 3, 5, and 13 give more information about the application of venturi meters, including extensive discussions of the corrections that must be made when using them for measuring the flow of air and other gases. We will limit our use of Equations 15–4 and 15–5 to liquid flow.

---

☐ **EXAMPLE PROBLEM 15.1**   A venturi tube of the Herschel type shown in Figure 15.2 is being used to measure the flow rate of water at 140°F. The flow enters from the left in a 5-in Schedule 40 steel pipe. The throat diameter $d$ is 2.200 in. The venturi is rough cast. The manometer fluid is mercury (sg = 13.54) and the deflection $h$ is 7.40 in. Compute the velocity of flow in the pipe and the volume flow rate in gal/min.

**Solution**   We will use Equation 15–4 to compute the velocity of flow in the pipe, $v_1$. Then we will find the volume flow rate from $Q = A_1 v_1$.

First let's document pertinent data and compute some of the basic parameters in Equation 15–4.

Fluid flowing in the pipe: Water at 140°F. $\gamma_w = 61.4$ lb/ft³. $\nu = 5.03 \times 10^{-6}$ ft²/s. (From App. A)

Manometer fluid: Mercury (sg = 13.54). $\gamma_m = (13.54)(62.4 \text{ lb/ft}^3) = 844.9$ lb/ft³.

Pipe dimensions: $D = 0.4206$ ft. $A_1 = 0.1390$ ft². (From Appendix F)

Throat dimensions: $d = (2.20 \text{ in})(1.0 \text{ ft}/12 \text{ in}) = 0.1833$ ft. $A_2 = \pi d^2/4 = 0.02640$ ft²

Then,

$$A_1/A_2 = (0.1390 \text{ ft}^2/(0.0264 \text{ ft}^2) = 5.265$$

$$\beta = d/D = (0.1833 \text{ ft})/(0.4206 \text{ ft}) = 0.436. \text{ Note that } 0.30 < \beta < 0.75.$$

Figure 15.3 applies to give the value of the discharge coefficient $C$ for the rough-cast venturi. Let's assume that the Reynolds number for the flow of water in the pipe is greater than $2.0 \times 10^5$ and use the value of $C = 0.984$ as the first estimate. This must be checked later when the Reynolds number is known and adjusted according to Figure 15.3 if $N_R < 2.0 \times 10^5$.

The manometer equation must be used to determine the pressure head difference, $(p_1 - p_2)/\gamma_w$.

Starting at the centerline of the pipe at section 1, we write the equation for the pressure changes that occur down through the manometer and back up to the centerline of the throat of the venturi. We first note that the distance from the pipe centerline to the manometer is unknown. We saw in Chapter 3 that we can give that distance a name, say $y$, and proceed to write the equation. We will see that the value of $y$ cancels from the manometer equation.

$$p_1 + \gamma_w y + \gamma_w h - \gamma_m h - \gamma_w y = p_2$$

Here we see that the term $\gamma_w y$ appears with both a positive and a negative sign. Thus they can be canceled.

Let's now solve the equation for the pressure head difference we need in Equation 15–4.

$$p_1 - p_2 = -\gamma_w h + \gamma_m h = \gamma_m h - \gamma_w h = h(\gamma_m - \gamma_w)$$

Now divide both sides of the equation by $\gamma_w$.

$$(p_1 - p_2)/\gamma_w = h(\gamma_m - \gamma_w)/\gamma_w = h(\gamma_m/\gamma_w - 1)$$

We can now evaluate this term by inserting known values.

$$(p_1 - p_2)/\gamma_w = (7.40 \text{ in})(1.0 \text{ ft}/12 \text{ in})[(844.9 \text{ lb/ft}^3)/(61.4 \text{ lb/ft}^3) - 1] = 7.87 \text{ ft}$$

Finally, we place all these known data into Equation 15–4.

$$v_1 = C\sqrt{\frac{2g(p_1 - p_2/\gamma_w)}{(A_1/A_2)^2 - 1}} = 0.984\sqrt{\frac{2(32.2 \text{ ft/s}^2)(7.87 \text{ ft})}{(5.265)^2 - 1}} = 4.285 \text{ ft/s}$$

Now we must check the Reynolds number for the flow in the pipe using this value.

$$N_R = \frac{v_1 D}{\nu} = \frac{(4.285 \text{ ft/s})(0.4206 \text{ ft})}{5.03 \times 10^{-6}} = 3.58 \times 10^5$$

We note that this value is greater than $2 \times 10^5$ as we initially assumed. Then the value for the discharge coefficient, $C = 0.984$, is correct and the calculation for $v_1$ is also correct. If the Reynolds number was less than $2 \times 10^5$, read a new value of $C$ from Figure 15.3 and recompute the velocity.

**Result** Now we complete the problem by computing the volume flow rate $Q$.

$$Q = A_1 v_1 = (0.1390 \text{ ft}^2)(4.285 \text{ ft/s}) = 0.596 \text{ ft}^3/\text{s}$$

Converting this to gal/min,

$$Q = (0.596 \text{ ft}^3/\text{s})[(449 \text{ gal/min})/1.0 \text{ ft}^3/\text{s}] = 267 \text{ gal/min}$$

---

**15.4.2 Flow Nozzle**

The *flow nozzle* is a gradual contraction of the flow stream followed by a short, straight cylindrical section as illustrated in Fig. 15.4. Several standard geometries for flow nozzles have been presented and adopted by organizations such as the American Society of Mechanical Engineers (ASME) and the International Organization for Standardization.

Equation (15–5) is used for the flow nozzle and the orifice as well as for the venturi tube. Because of the smooth, gradual contraction, there is very little energy loss between points 1 and 2 for a flow nozzle. A typical curve of $C$ versus Reynolds number is shown in Fig. 15.5. At high Reynolds numbers $C$ is above 0.99. At lower Reynolds numbers the sudden expansion outside the nozzle throat causes greater energy loss and a lower value for $C$.

Reference 13 recommends Eq. (15–6) for $C$:

$$C = 0.9975 - 6.53\sqrt{\beta/N_R} \qquad (15\text{–}6)$$

where $\beta = d/D$. Figure 15.5 is a plot of Equation 15–6 for the value of $\beta = 0.50$.

References 3, 5, and 13 give extensive information on the proper selection and application of flow nozzles including corrections for gas flow.

**FIGURE 15.4**  Flow nozzle.

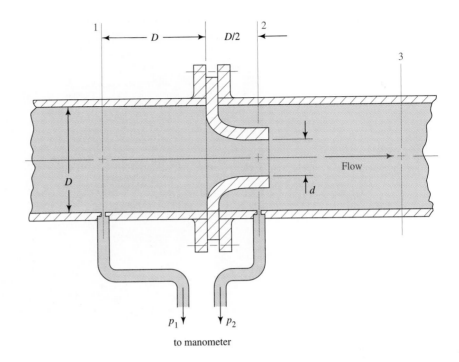

**FIGURE 15.5**  Flow nozzle discharge coefficient.

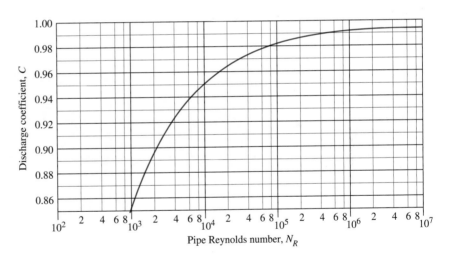

**15.4.3
Orifice**  A flat plate with an accurately machined, sharp-edged hole is referred to as an *orifice*. When placed concentrically inside a pipe as shown in Fig. 15.6(b) it causes the flow to suddenly contract as it approaches the orifice and then suddenly expand back to the full pipe diameter. The stream flowing through the orifice forms a vena contracta and the rapid flow velocity results in a decreased pressure downstream from the orifice. Pressure taps before and after the orifice (sections 1 and 2) allow the measurement of the differential pressure across the meter that is related to the volume flow rate by Eq. (15–5).

Figure 15.6(a) shows a commercially available unit that incorporates all of the major systems needed to measure flow. The orifice plate is part of an *integral flow orifice assembly* that also includes:

**FIGURE 15.6** Square-edged orifice with pressure taps at $D$ and $D/2$. (Source of photo: The Foxboro Company, Foxboro, MA)

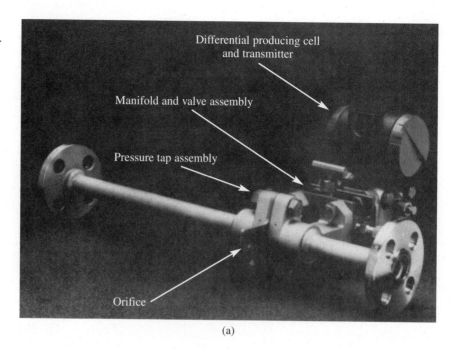

(a)

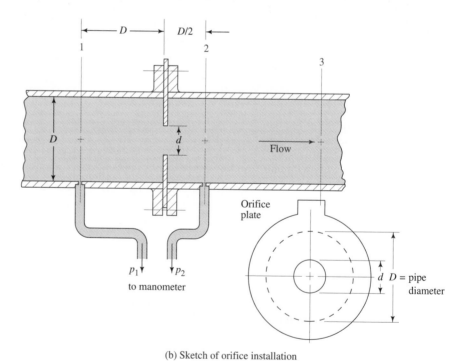

(b) Sketch of orifice installation

- pressure taps accurately located on both sides of the plate
- a manifold that facilitates the mounting of the differential producing (d/p) cell
- a d/p cell and transmitter to transmit the signal to a remote location
- a set of valves that allow fluid to bypass the d/p cell for service
- straight lengths of pipe into and out of the orifice to ensure predictable flow conditions at the orifice
- flanges to connect the unit to process piping
- a built-in microprocessor in the d/p cell that linearizes the output signal across the entire range of the meter giving a signal that is directly proportional to the flow. It performs the square root operation called for in Eq. (15–5).

The actual value of the discharge coefficient $C$ depends on the location of the pressure taps. Three possible locations are listed in Table 15.1.

**TABLE 15.1** Location of pressure taps for orifice meters

| | Inlet Pressure Tap, $p_1$ | Output Pressure Tap, $p_2$ |
|---|---|---|
| 1 | One pipe diameter upstream from plate | One-half pipe diameter downstream from inlet face of plate |
| 2 | One pipe diameter upstream from plate | At vena contracta (see Reference 5) |
| 3 | In flange, 1 inch upstream from plate | In flange, 1 inch downstream from outlet face of plate |

The value of $C$ also is affected by small variations in the geometry of the edge of the orifice. Typical curves for sharp-edged orifices are shown in Fig. 15.7, where $D$ is the pipe diameter and $d$ is the orifice diameter. The value of $C$ is much lower than that for the venturi tube or the flow nozzle since the fluid is forced to make a sudden contraction followed by a sudden expansion. Also, because measurements are based on the orifice diameter, the decrease in the diameter of the flow stream at the vena contracta tends to reduce the value of $C$.

References 3, 5, and 13 give extensive information on the proper selection and application of orifices including adjustments for gas flow.

**FIGURE 15.7** Orifice discharge coefficient. (Source: ASME Research Committee on Fluid Meters. 1959. *Fluid Meters: Their Theory and Application.* 5th ed. New York: The American Society of Mechanical Engineers, p. 148.)

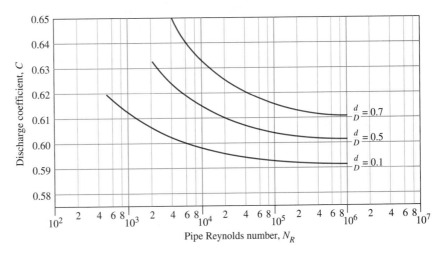

### 15.4.4
### Flow Tubes

Several proprietary designs for modified variable head flow meters called flow tubes are available. These can be used for applications similar to those for which the venturi, nozzle, or orifice meters are used, but flow tubes have somewhat lower pressure loss (higher pressure recovery). Figure 15.8 is a photograph of one manufacturer's flow tube.

**FIGURE 15.8** Flow tube. (Source: Fischer & Porter Company, Warminster, PA)

### 15.4.5
### Overall Pressure Loss

In each of the four types of variable head meters just described, the flow stream expands back to the main pipe diameter after passing the restriction. This is indicated as section 3 in Figs. 15.2, 15.4, and 15.6(b). Then the difference between the pressure $p_1$ and $p_3$ is due to the meter. The difference can be evaluated by considering the energy equation:

$$\frac{p_1}{\gamma} + z_1 + \frac{v_1^2}{2g} - h_L = \frac{p_3}{\gamma} + z_3 + \frac{v_3^2}{2g}$$

Because the pipe sizes are the same at both sections, $v_1 = v_3$. We may also assume $z_1 = z_3$. Then,

$$p_1 - p_3 = \gamma h_L$$

The pressure drop is proportional to the energy loss. The careful streamlining of the venturi tube and the long gradual expansion after the throat cause very little excess turbulence in the flow stream. Therefore, energy loss is low and the pressure loss is low. The lack of a gradual expansion causes the nozzle to have a higher pressure loss, while that for an orifice is still higher. The lowest pressure loss is obtained from the flow tube. Figure 15.9 shows the comparison among the several types of variable head meters with regard to pressure loss.

### 15.5
### VARIABLE AREA METERS

The *rotameter* is a common type of variable area meter. Figure 15.10 shows a typical geometry. The fluid flows upward through a clear tube which has an accurate taper on the inside. A float is suspended in the flowing fluid at a position proportional to the flow rate. The upward forces due to the fluid dynamic drag on the float and buoyancy just balance the weight of the float. A different flow rate causes the float to move to a new position, changing the clearance area between the float and the tube until equilibrium is achieved again. The position of the float is measured against a calibrated scale graduated in convenient units of volume flow rate or weight flow rate.

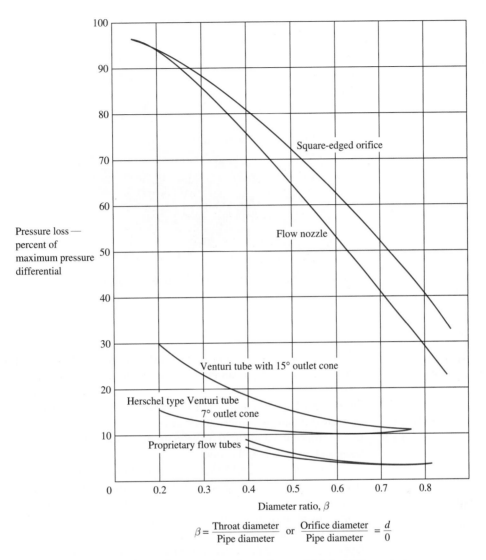

**FIGURE 15.9**  Comparison of pressure loss for various variable head flowmeters.
(Source: Bean, H.S., ed. 1971. *Fluid Meters: Their Theory and Application.* 6th ed. New
York: The American Society of Mechanical Engineers.)

Use of the type of rotameter shown in Figure 15.10 requires that the fluid be
transparent because the operator must visually see the position of the float. Also,
the transparent tube has a somewhat limited pressure capability. Some rotameters
are made from opaque tubing to withstand higher pressures. The position of the float
is sensed from outside the tube by an electromagnetic means and the flow rate is
indicated on a gage.

**15.6
TURBINE
FLOWMETER**

Figure 15.11 shows a turbine flowmeter in which the fluid causes the turbine rotor
to rotate at a speed dependent on the flow rate. As each blade of the rotor passes
the magnetic coil, a voltage pulse is generated which can be input to a frequency
meter, electronic counter, or other similar device whose readings can be converted

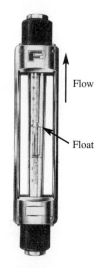

**FIGURE 15.10**    Rotameter.
(Source: Fischer & Porter Company,
Warminster, PA)

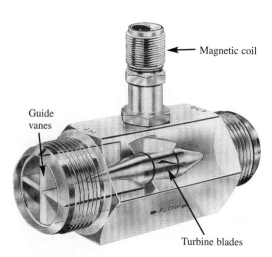

**FIGURE 15.11**    Turbine flowmeter.
(Source: EG&G Flow Technology,
Inc., Phoenix, AZ)

to flow rate. Flow rates from as low as 0.02 L/min (0.005 gal/min) to several thousand L/min or gal/min can be measured with turbine flowmeters of various sizes.

**15.7
VORTEX
FLOWMETER**

Figure 15.12 show a *vortex flowmeter*, in which a blunt obstruction placed in the flow stream causes vortices to be created and shed from the body at a frequency that is proportional to the flow velocity. A sensor in the flowmeter detects the vortexes and creates an indication for the meter readout device (see Reference 7).

Part (b) of Fig. 15.12 shows a sketch of the vortex shedding phenomenon. The shape of the blunt body, also called the *vortex shedding element*, may vary from manufacturer to manufacturer. As the flow approaches the front face of the shedding element, it divides into two streams. Fluid close to the body has a low velocity relative to that in the main streamlines. The difference in velocity causes shear layers to form which eventually break down into vortices alternately on the two sides of the shedding element. The frequency of the vortices created is directly proportional to the flow velocity and, therefore, to the volume flow rate. Sensors in the meter detect the pressure variations around the vortices and generate a voltage signal that alternates at the same frequency as the vortex shedding frequency. The output signal is either a stream of voltage pulses or a dc (direct current) analog signal. Standard instrumentation systems often use an analog signal that varies from 4 to 20 mA dc (milliamps dc). For the pulse output, the manufacturer supplies a flowmeter K-factor that indicates pulses per unit volume through the meter.

Vortex meters can be used for a wide range of fluids including clean and dirty liquids and gases and steam. The K-factor is the same for all these fluids.

**15.8
MAGNETIC
FLOWMETER**

Totally unobstructed flow is one of the advantages of a magnetic flowmeter like that shown in Fig. 15.13. The fluid must be slightly conducting because the meter operates on the principle that when a moving conductor cuts across a magnetic field, a voltage is induced. The primary components of the magnetic flowmeter include a tube lined with a nonconducting material, two electromagnetic coils, and two elec-

(a) Photograph of a vortex flowmeter

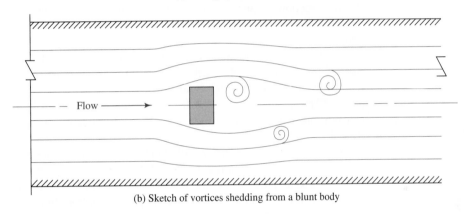

Flow →

(b) Sketch of vortices shedding from a blunt body

**FIGURE 15.12**   Vortex flowmeter. (Source of photo: Fischer & Porter Company, Warminster, PA)

**FIGURE 15.13**   Magnetic flowmeter. (Source: The Foxboro Company, Foxboro, MA)

One electrode

Electromagnetic coils

trodes mounted 180° apart in the tube wall. The electrodes detect the voltage generated in the fluid. Since the generated voltage is directly proportional to the fluid velocity, a greater flow rate generates a greater voltage. An important feature of this type of meter is that its output is completely independent of temperature, viscosity, specific gravity, or turbulence. Tube sizes from 2.5 mm to 2.4 m (0.1 in to 8.0 ft) in diameter are available.

**15.9**
**ULTRASONIC**
**FLOWMETERS**

A major advantage of an ultrasonic flowmeter is that it is not necessary to penetrate the pipe in any way. An ultrasonic generator is strapped to the outside of the pipe and a high frequency signal is transmitted through the wall of the pipe and across the flow stream, typically at an acute angle with respect to the axis of the pipe. The time for the signal to traverse the pipe depends on the velocity of flow of the fluid in the pipe. Some commercially available meters use detectors on the side opposite the transmitter while others employ reflectors to return the signal to a receiver built into the transmitter.

Another approach is to use two transmitter/receiver units aligned with the axis of the pipe. Each delivers a signal at an angle to the flow that is reflected from the opposite side of the pipe and received by the other. The signal that is directed in the same direction as the flow takes a different time to reach the receiver than the signal that opposes the flow. The difference between these two times is proportional to the velocity of flow.

A variety of orientations can be used for the transmitters, reflectors, and receivers of the signal. Most will use two sets to reduce the sensitivity of the meter to the velocity profile of the fluid flow stream.

Transit time meters work best wth clean fluids because entrained particles in dirty fluids can affect the time readings and the strength of the signal that reaches the detectors.

A second type of meter, called the *Doppler-type*, is preferred for dirty fluids, slurries, and other fluids that may inhibit the transmission of the ultrasonic signal. The ultrasonic pressure wave does not traverse completely to the opposite wall of the pipe. Rather, it is reflected from the particles in the fluid itself and back to the receiver.

Because the ultrasonic flowmeter is completely non-invasive, the pressure loss is due only to the friction in the pipe itself. The meter contributes no additional loss.

**15.10**
**POSITIVE**
**DISPLACEMENT**
**METERS**

Fluid entering a positive displacement meter fills up a chamber that is moved from the input to the output side of the meter. The meter records or indicates the cumulative volume of fluid that has passed through the meter. The chambers can take many forms and are often proprietary to a given manufacturer. Rotating disks or meshing rotors are often used for liquids. Gas meters like those used in homes employ flexible diaphragms that continuously capture and then deliver known volumes of the low pressure natural gas.

Typical uses for positive displacement meters are water delivered from the municipal system to a home or business, natural gas delivered to a customer, and gasoline delivered at a service station. They are also used in certain industrial applications in which blended materials are required to have a set volume of different constituents.

**15.11**
**MASS FLOW**
**MEASUREMENT**

The flowmeters discussed thus far in this chapter are designed to produce an output signal that is proportional to the average velocity of flow or the volume flow rate. This is satisfactory when only the *volume* delivered through the meter is needed.

However, some processes require a measurement of the *mass* of fluid delivered. For example, in food processing plants the production is often indicated as the amount delivered in kilograms, pounds-mass, or slugs. Some chemical processes are sensitive to the mass of the various constituents that are blended or that are introduced into a reaction. Two-phase fluids, such as steam, may be difficult to measure accurately if the temperature and pressure vary enough to cause significant changes in the amount of liquid and vapor in the steam.

One way to obtain mass flow rate measurements is to use a flowmeter of the type just discussed that indicates volume flow rate and then simultaneously measure the density of the fluid. Then the mass flow rate would be

$$M = \rho Q$$

That is, mass flow rate equals density times volume flow rate as discussed in Chapter 6 of this book. If the density of the fluid is known or can be conveniently measured, this is a simple calculation. For some fluids, the density can be calculated if the temperature of the fluid is known. Sometimes, particularly with gases, the pressure is also needed. Temperature probes and pressure transducers are readily available to provide the necessary data. Specific gravity can be measured with a device called a *gravitometer*. Density can be measured directly for some fluids with a *densitometer*. The signals related to volume flow rate, temperature, pressure, specific gravity, or density can all be input to special electronic devices that effectively perform the calculation of $M = \rho Q$. This is shown schematically in Fig. 15.14. This process, while straightforward, requires several separate measurements to be made, each of which is subject to small errors. Then the errors are compounded in the final calculation.

True mass flowmeters avoid the problems discussed above by generating a signal proportional to the mass flow rate directly. One such mass flowmeter is called the *Coriolis mass flowtube*, shown in Fig. 15.15. The fluid enters the flowmeter from

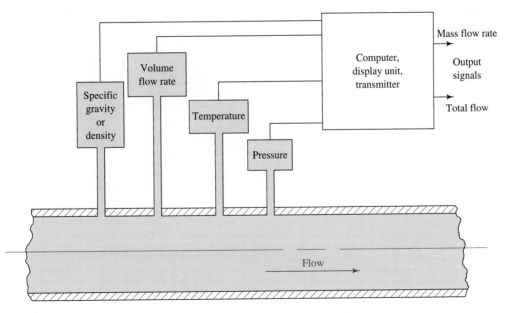

**FIGURE 15.14**  Schematic representation of mass flow measurement using multiple sensors.

**FIGURE 15.15** Coriolis mass flowtube. (Source: The Foxboro Company, Foxboro, MA)

(a) External view with programmer and indicator

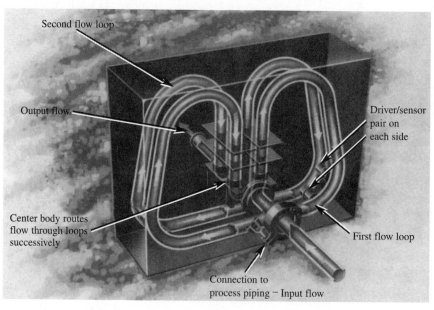

Second flow loop

Output flow

Driver/sensor pair on each side

Center body routes flow through loops successively

First flow loop

Connection to process piping − Input flow

(b) Internal view

the process piping and is directed through a continuous path of the same size that routes the fluid first through one loop, through a center body, then through the second loop, and then out to the exit pipe. Two electromagnetic drivers bridge both loops at opposite extremities, equidistant from the center. The vibratory motion created moves the two parallel loops alternately toward each other and then apart. Fluid in the tubes is simultaneously following the path of the loops and moving perpendicular to that path because of the action of the drivers. A Coriolis acceleration (and a corresponding Coriolis force) is produced that is proportional to the mass of fluid flowing through the tubes. Sensors mounted near the drivers detect the Coriolis force and transmit a signal that can be related to the true mass flow rate through the meter. Accuracy is reported to be 0.2 percent of the indicated rate or 0.02 percent of full-scale capacity, whichever is greater.

Density of the fluid can also be measured with the Coriolis mass flowtube because the driving frequency of the tubes is dependent on the density of the fluid flowing through the tubes. A temperature probe is also included in the system, completing a comprehensive set of fluid properties and mass flow rate data.

## 15.12 VELOCITY PROBES

Several devices are available which measure velocity of flow at a specific location rather than an average velocity. These are referred to as *velocity probes*. Some of the more common types will be described in this section.

### 15.12.1 Pitot Tube

When a moving fluid is caused to stop because it encounters a stationary object, a pressure is created which is greater than the pressure of the fluid stream. The magnitude of this increased pressure is related to the velocity of the moving fluid. The *pitot tube* uses this principle to indicate velocity, as illustrated in Fig. 15.16. The pitot tube is a hollow tube positioned so that the open end points directly into the fluid stream. The pressure at the tip causes a column of fluid to be supported. The fluid at or just inside the tip is then stationary or stagnant, and this point is referred to as the *stagnation point*. We can use the energy equation to relate the pressure at the stagnation point with the fluid velocity. If point 1 is in the undisturbed stream ahead of the tube and point $s$ is at the stagnation point, then

$$\frac{p_1}{\gamma} + z_1 + \frac{v_1^2}{2g} - h_L = \frac{p_s}{\gamma} + z_s + \frac{v_s^2}{2g} \qquad (15\text{--}7)$$

**FIGURE 15.16**   Pitot tube.

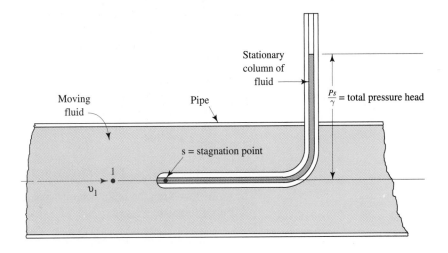

Observe that $v_s = 0$, $z_1 = z_2$ or very nearly so, and $h_L = 0$ or very nearly so. Then we have

$$\frac{p_1}{\gamma} + \frac{v_1^2}{2g} = \frac{p_s}{\gamma} \qquad (15-8)$$

The names given to the terms in Eq. (15–8) are as follows:

$$p_1 = \text{static pressure in the main fluid stream}$$

$$p_1/\gamma = \text{static pressure head}$$

$$p_s = \text{stagnation pressure or total pressure}$$

$$p_s/\gamma = \text{total pressure head}$$

$$v_1^2/2g = \text{velocity pressure head}$$

The total pressure head is equal to the sum of the static pressure head and the velocity pressure head. Solving Eq. (15–8) for the velocity gives

$$v_1 = \sqrt{2g(p_s - p_1)/\gamma} \qquad (15-9)$$

Notice that only the difference between $p_s$ and $p_1$ is required to calculate the velocity. For this reason, most pitot tubes are made as shown in Fig. 15.17, providing for the measurement of both pressures with the same device.

The device shown in Fig. 15.17 facilitates the measurement of both the static pressure and the stagnation pressure simultaneously and so it is sometimes called a *pitot-static* tube. Its construction shown in part (b) is actually a tube within a tube. The small central tube is open at the end and functions in the same manner as the single pitot tube shown in Fig. 15.16. Thus the stagnation pressure, also called the *total pressure*, is sensed through this tube. The *total pressure tap* at the end of this tube allows connection to a pressure-measuring device.

The larger outer tube is sealed around the central tube at its end, thus creating a closed annular cavity between the central and the outer tube. Section A-A shows that a series of small radial holes are drilled through the outer tube but not through the central tube. When the tube is aligned with the direction of flow, these radial holes are perpendicular to the flow and thus they sense the local static pressure which we have called $p_1$. Notice that a static pressure tap is affixed at the end of the tube to allow connection to a measuring instrument.

The measuring instrument need not measure either $p_s$ or $p_1$ because it is the *difference* $(p_s - p_1)$ that is needed in Eq. (15–9). Several manufacturers make differential pressure-measuring devices for such applications.

If a differential manometer is used as shown in Fig. 15.18, the manometer deflection $h$ can be related directly to the velocity. We can write the equation describing the difference between $p_s$ and $p_1$ by starting at the static pressure holes in the side of the tube, proceeding through the manometer, and ending at the open tip of the tube at point $s$:

$$p_1 - \gamma x + \gamma y + \gamma_g h - \gamma h - \gamma y + \gamma x = p_s$$

The terms involving the unknown distances $x$ and $y$ drop out. Then, solving for the pressure difference, we get

$$p_s - p_1 = \gamma_g h - \gamma h = h(\gamma_g - \gamma) \qquad (15-10)$$

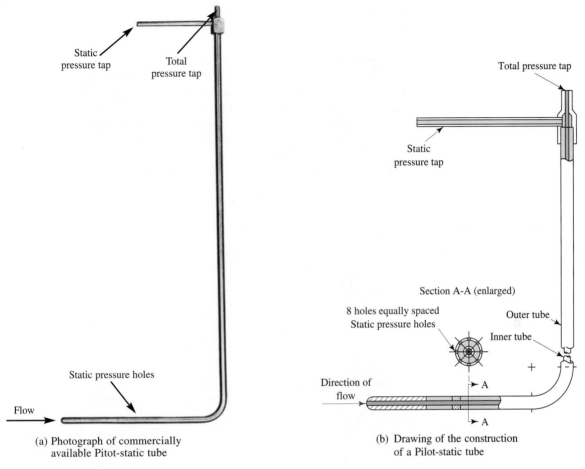

Static
pressure tap

Total
pressure tap

Total pressure tap

Static
pressure tap

Section A-A (enlarged)

8 holes equally spaced
Static pressure holes

Outer tube

Inner tube

Static pressure holes

+

Direction of
flow

A

Flow

A

(a) Photograph of commercially
available Pitot-static tube

(b) Drawing of the construction
of a Pilot-static tube

**FIGURE 15.17**   Pitot-static tube. (Source: Dwyer Instruments, Inc., Michigan City, IN)

**FIGURE 15.18**   Differential
manometer used with a pitot-static
tube.

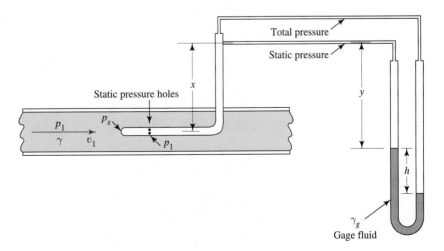

Total pressure

Static pressure

Static pressure holes

$x$

$y$

$p_1$

$p_s$

$\gamma$

$v_1$

$p_1$

$h$

$\gamma_g$
Gage fluid

Substituting this into Eq. (15–9) gives

$$v_1 = \sqrt{2gh(\gamma_g - \gamma)/\gamma} \qquad (15\text{–}11)$$

The velocity calculated by either Eq. (15–9) or (15–11) is the local velocity at the particular location of the tip of the tube. In Chapters 8 and 9 we found that the velocity of flow varies from point to point across a pipe. Therefore, if the average velocity of flow is desired, a traverse of the pipe should be made with the tip of the tube placed at the specific ten points indicated in Fig. 15.19. The dashed circles define concentric annular rings which have equal areas. The velocity at each point can be calculated, using Eq. (15–11). Then the average velocity of flow is the average of these ten values. We can find the volume flow rate from $Q = Av$, using the average velocity.

**FIGURE 15.19** Velocity measurement points within a pipe for computing average velocity.

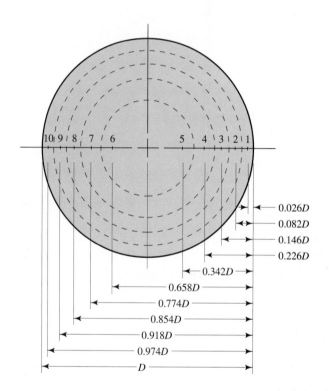

□ **EXAMPLE PROBLEM 15.2**    For the apparatus shown in Fig. 15.18, the fluid in the pipe is water at 60°C and the manometer fluid is mercury with a specific gravity of 13.54. If the manometer deflection $h$ is 264 mm, calculate the velocity of the water.

*Solution*    Equation (15–11) will be used:

$$v_1 = \sqrt{2gh(\gamma_g - \gamma)/\gamma}$$

$$\gamma = 9.65 \text{ kN/m}^3 \qquad \text{(water at 60°C)}$$

$$\gamma_g = (13.54)(9.81 \text{ kN/m}^3) = 132.8 \text{ kN/m}^3 \qquad \text{(mercury)}$$

$$h = 264 \text{ mm} = 0.264 \text{ m}$$

Since all terms are in SI units, the velocity is in m/s:

$$v_1 = \sqrt{\frac{(2)(9.81)(0.264)(132.8 - 9.65)}{9.65}}$$

$$= 8.13 \text{ m/s}$$

■

The pressure differential created by a pitot tube can also be read by an electronic device such as that shown in Fig. 15.20. The individual readings taken during a traverse of a pipe or duct can be recorded on the portable printer. Then the average is automatically computed and printed in either SI units or U.S. Customary System units.

### 15.12.2 Cup Anemometer

Air velocity is often measured with a *cup anemometer* such as that shown in Fig. 15.21. The moving air strikes the open cups, causing rotation of the shaft on which they are mounted. The rotational speed of the shaft is proportional to the air velocity, which is indicated on a meter or transmitted electrically.

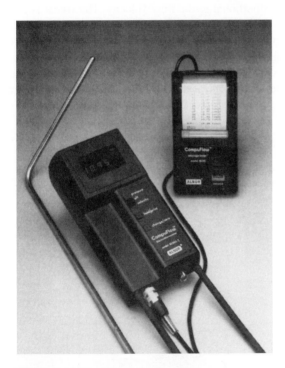

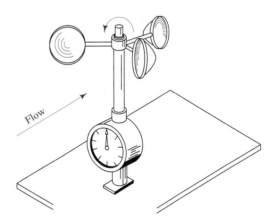

**FIGURE 15.20** Electronic readout device for pitot tubes. (Source: Alnor Instruments Co., Skokie, IL)

**FIGURE 15.21** Rotating cup anemometer.

### 15.12.3 Hot Wire Anemometer

This type of velocity probe employs a very thin wire, about 12 $\mu$m in diameter, through which an electric current is passed. The wire is suspended on two supports as shown in Fig. 15.22 and inserted into the fluid stream. The wire tends to heat because of the current flowing in it, but it is cooled by convection heat transfer to the moving fluid stream. The amount of cooling depends on the velocity of the fluid.

**FIGURE 15.22**  Hot wire anemometer tip.

Hot wire

In one type of hot wire anemometer, a constant current is applied to the wire. A variation in the flow velocity causes a change in the wire temperature, and therefore its resistance changes. The electronic measurement of the resistance change can be related to flow velocity. Another type senses a change in the resistance of the wire, but then varies the current flow to maintain a set wire temperature regardless of fluid velocity. The magnitude of the current flow is then related to fluid velocity.

**15.13
COMPUTER-BASED
DATA ACQUISITION
AND PROCESSING**

Microcomputers, programmable controllers, and other microprocessor-based electronic instrumentation greatly simplify the acquisition, processing, and recording of flow measurement data. As shown in this chapter, many of the flowmeters produce an electrical signal that is proportional to the flow velocity. The signal is either an analog voltage that varies with the velocity or a pulse frequency that can be counted electronically. Analog signals can be converted to digital signals by analog-to-digital converters, often called *A-D converters*, for input to digital computers.

The computers can total the fluid flow rate over time to determine the total quantity of fluid transferred to a given location. Figure 15.23 shows an indicating totalizer that is battery powered for remote operation.

**FIGURE 15.23**  Flow totalizer.
(Source: Fischer & Porter Company,
Warminster, PA)

**FIGURE 15.24** Electronic interface device for flowmeter data acquisition system. (Source: Fischer & Porter Company, Warminster, PA)

A comprehensive measurement and control system can consist of pressure, temperature, and flow measurement devices; automatic process controllers; interface units; operator control stations; and large host computers. Figure 15.24 shows an interface device that can communicate with up to 32 controllers, combining and reformatting data for transmission to an operator's station like the one shown in Fig. 15.25. The video terminal can display the status of several measurements

**FIGURE 15.25** Operator's station for computer-based process control system. (Source: Fischer & Porter Company, Warminster, PA)

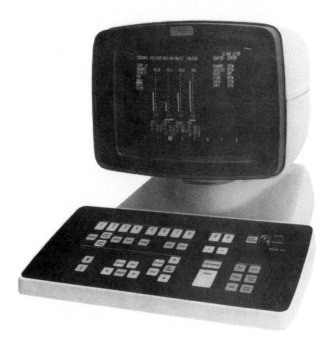

simultaneously for the operator while monitoring the data for values that fall outside prescribed levels. The host computer can acquire data from several places within
the plant and maintain the central data base for quality control, production data, and
inventory control.

**15.14**
**OPEN CHANNEL FLOW**
**MEASUREMENT**

An open channel is one that has its top surface open to the prevailing atmosphere.
Familiar examples are natural streams, sewers running partially full, waste water
management systems, and storm drainage structures. Industries often use open channels to conduct coolants away from machinery and to collect excess process fluids
and return them to holding tanks. See Chapter 14.

Two widely used devices for open channel flow measurement are *weirs* and
*flumes*. Each causes the area of the stream to change, which in turn changes the
level of the fluid surface. The resulting level of the surface relative to some feature
of the device is related to the quantity of flow. Large volume flow rates of liquids
can be measured with weirs and flumes.

**15.14.1**
**Weirs**

A *weir* is a barrier or dam placed in the channel so that the fluid backs up behind
it and then falls through a notch cut into the face of the weir. Two common notch
geometries are the rectangular and the triangular. Figure 15.26 shows a side view
of a weir in operation. Front views showing different notch geometries are shown
in Fig. 15.27. The discharge over the weir is dependent on the dimensions of the
notch and on the head $H$ of the fluid. Figure 15.26 shows that the fluid surface is
somewhat curved as it passes over the crest of the weir. In order to ensure consistent measurements, the value of $H$ should be the difference between the height to
the crest $H_c$ and the total height to the liquid surface $H_t$, with $H_t$ measured upstream
from the weir plate where the surface profile is undisturbed. Normally, this upstream
distance is approximately six times the maximum expected head $H$.

**FIGURE 15.26**    Flow over a weir.

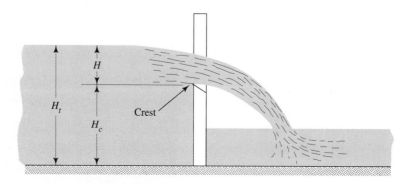

THEORETICAL DISCHARGE OVER
A WEIR

The theoretical equation for discharge over a rectangular notched weir is

$$Q = \tfrac{2}{3}L\sqrt{2gH^3} \qquad\qquad (15\text{--}12)$$

where $L$ is the length of the crest between the sides of the notch and $H$ is the head
above the crest. If $L$ and $H$ are measured in ft, then $Q$ is in cubic feet per second
(ft³/s).

The actual discharge is different from the theoretical for a variety of reasons;
therefore, relationships that are more accurate and easier to use have been

**FIGURE 15.27**  Notch geometry for weirs.

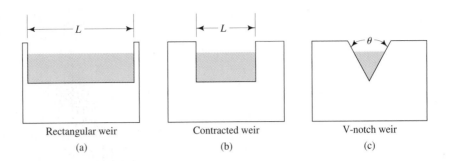

| Rectangular weir | Contracted weir | V-notch weir |
| (a) | (b) | (c) |

developed. Most formulas take the form

$$Q = CLH^{3/2} \qquad (15\text{--}13)$$

where $C$ is a discharge coefficient, $L$ is the effective length of the crest, and $H$ is the head above the weir crest.

For the full width rectangular weir in Fig. 15.27(a), the following equation can be used:

▷ **RECTANGULAR WEIR**

$$Q = (3.27 + 0.40H/H_c)LH^{3/2} \qquad (15\text{--}14)$$

The two end contractions in the contracted weir in Fig. 15.27(b) cause the stream to be curved in from the sides, decreasing the effective length of the crest. The discharge for this type of weir can be calculated from

▷ **CONTRACTED WEIR**

$$Q = (3.27 + 0.40H/H_c)(L - 0.2H)H^{3/2} \qquad (15\text{--}15)$$

The triangular weir is used primarily for low flow rates since the V-notch produces a larger head $H$ than can be obtained with a rectangular notch. The angle of the V-notch is a factor in the discharge equation. Angles from 35° to 120° are satisfactory, but angles of 60° and 90° are quite commonly used. The theoretical equation for a triangular weir is

▷ **GENERAL EQUATION FOR TRIANGULAR WEIR**

$$Q = {}^{8}/_{15}C\sqrt{2g}\,\tan(\theta/2)H^{5/2} \qquad (15\text{--}16)$$

where $\theta$ is the total included angle between the sides of the notch. An additional reduction of this equation gives

$$Q = 4.28C\tan(\theta/2)H^{5/2} \qquad (15\text{--}17)$$

The value of $C$ is somewhat dependent on the head $H$, but a nominal value is 0.58. Using this and the common values of 60° and 90° for $\theta$, we get

▷ **60° V-NOTCH WEIR**

$$Q = 1.43H^{5/2} \qquad (60° \text{ notch}) \qquad (15\text{--}18)$$

▷ **90° V-NOTCH WEIR**

$$Q = 2.48H^{5/2} \qquad (90° \text{ notch}) \qquad (15\text{--}19)$$

**15.14.2 Flumes**

*Critical flow flumes* are contractions in the stream which cause the flow to achieve its critical depth within the structure. There is a definite relationship between depth and discharge when critical flow exists. A widely used type of critical flow flume is the *Parshall flume*, the geometry of which is shown in Fig. 15.28. The discharge is dependent on the width of the throat section $L$ and the head $H$, where $H$ is measured at a specific location along the converging section of the flume.

The discharge equations for the Parshall flume were developed empirically for flumes designed and constructed in dimensions in the U.S. customary system. Table 15.2 lists the discharge equations for several sizes of flumes. The resulting

**FIGURE 15.28**    Parshall flume.

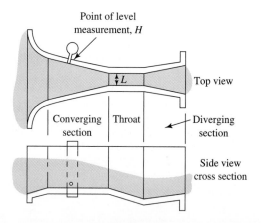

**TABLE 15.2**    Discharge equations for Parshall flumes

| Throat Width $L$ | Flow Range (ft³/s) | | Equation ($H$ and $L$ in ft, $Q$ in ft³/s) |
|---|---|---|---|
| | Min. | Max. | |
| 3 in. | 0.03 | 1.9 | $Q = 0.992H^{1.547}$ |
| 6 in | 0.05 | 3.9 | $Q = 2.06H^{1.58}$ |
| 9 in | 0.09 | 8.9 | $Q = 3.07H^{1.53}$ |
| 1 ft | 0.11 | 16.1 | |
| 2 ft | 0.42 | 33.1 | |
| 4 ft | 1.3 | 67.9 | $Q = 4LH^{1.522L^{0.26}}$ |
| 6 ft | 2.6 | 103.5 | |
| 8 ft | 3.5 | 139.5 | |
| 10 ft | 6 | 200 | |
| 20 ft | 10 | 1000 | |
| 30 ft | 15 | 1500 | $Q = (3.6875L + 2.5)H^{1.6}$ |
| 40 ft | 20 | 2000 | |
| 50 ft | 25 | 3000 | |

value of $Q$ can be converted to SI units by use of the factor

$$1.0 \text{ ft}^3/\text{s} = 0.028\ 32 \text{ m}^3/\text{s}$$

Chow (Reference 11) gives additional data about flumes and weirs.

---

## REFERENCES

1. American Society of Mechanical Engineers. 1989. *Introductory Guide to Flow Measurement*. New York: The American Society of Mechanical Engineers.
2. ———. 1996. *Introductory Guide to Industrial Flow Measurement*. Ed. Roger C. Baker. New York: The American Society of Mechanical Engineers.
3. ———. 1971. *Fluid Meters: Their Theory and Application*, *6th ed.* Ed. Howard S. Bean. New York: The American Society of Mechanical Engineers.
4. ———. 1997. *Glossary of Terms Used in the Measurement of Fluid Flow in Pipes. (Standard MFC-1M)*. New York: The American Society of Mechanical Engineers.
5. ———. 1995. *Measurement of Fluid Flow in Pipes Using Orifice, Nozzle, & Venturi. (Standard MFC-3M)*. New York: The American Society of Mechanical Engineers.
6. ———. 1994. *Measurement of Liquid Flow in Closed Conduits Using Transit-Time Ultrasonic Flowmeters (Standard MFC-5M)*. New York: The American Society of Mechanical Engineers.

7. ———. 1998. *Measurement of Fluid Flow in Pipes Using Vortex Flowmeters (Standard MFC-6M)*. New York: The American Society of Mechanical Engineers.

8. ———. 1998. *Measurement of Liquid Flow in Closed Conduits by Weighing Method (Standard MFC-9M)*. New York: The American Society of Mechanical Engineers.

9. ———. 1994. *Measurement of Fluid Flow in Pipes by Means of Coriolis Mass Flowmeters (Standard MFC-11M)*. New York: The American Society of Mechanical Engineers.

10. ———. 1995. *Measurement of Fluid Flow in Closed Conduits by Means of Electromagnetic Flowmeters (Standard MFC-16M)*. New York: The American Society of Mechanical Engineers.

11. Chow, V. T. 1959. *Open-Channel Hydraulics*. New York: McGraw-Hill.

12. DeCarlo, J. P. 1984. *Fundamentals of Flow Measurement*. Research Triangle Park, North Carolina: The Instrument Society of America.

13. Miller, Richard W. 1996. *Flow Measurement Engineering Handbook, 3rd ed*. New York: McGraw-Hill.

14. Spitzer, D. W. 1991. *Flow Measurement*. New York: The American Society of Mechanical Engineers.

15. Spitzer, D. W. 1990. *Industrial Flow Measurement, 2nd ed*. New York: The American Society of Mechanical Engineers.

## REVIEW QUESTIONS

1. List six factors that affect the selection and use of flowmeters.
2. Define *range* as it relates to flowmeters.
3. Describe three methods for calibrating flowmeters.
4. Name four types of variable head meters.
5. Describe the venturi tube.
6. What is meant by the *throat* of a venturi tube?
7. What is the nominal included angle of the convergent section for a venturi tube?
8. What is the nominal included angle of the divergent section for a venturi tube?
9. Why is there such a difference between the angles of the convergent and the divergent sections for a venturi tube?
10. Describe the term *discharge coefficient* as it relates to variable head meters.
11. Describe a flow nozzle and how it is used.
12. Describe an orifice meter and how it is used.
13. Describe a flow tube and how it is used.
14. Of the venturi, the flow nozzle, the flow tube, and the orifice, which has the lowest discharge coefficient? Why?
15. Describe *pressure loss* as it relates to flowmeters.
16. Rank the venturi, the flow nozzle, the orifice, and the flow tube on the basis of pressure loss.
17. Describe a rotameter variable area meter.

18. Describe a turbine flowmeter and how it is used.
19. Describe a vortex flowmeter and how it is used.
20. Describe a magnetic flowmeter and how it is used.
21. Describe how mass flow rate can be measured.
22. Describe a pitot tube and how it is used.
23. Define *stagnation pressure* and show how it can be derived from Bernoulli's equation.
24. Define *static pressure head*.
25. Define *velocity pressure head*.
26. Why is a differential manometer a convenient device for use with a pitot tube?
27. Describe the method used to measure the average velocity of flow in a pipe using a pitot tube.
28. Describe a cup anemometer.
29. Describe a hot wire anemometer and how it is used.
30. Describe the use of an analog-digital converter in the practice of flow measurement.
31. Name two devices used to measure the flow rate in open channels.
32. Describe a weir and three different geometries used for weirs.
33. What is the parameter that must be measured to determine flow using a weir?
34. Describe a Parshall flume and how it is used.

## PRACTICE PROBLEMS

**15.1M** *A venturi meter similar to the one in Fig. 15.2 has a pipe diameter of 100 mm and a throat diameter of 50 mm. While it is carrying water at 80°C, a pressure difference of 55 kPa is observed between sections 1 and 2. Calculate the volume flow rate of water.*

**15.2M** *Air with a specific weight of 12.7 N/m³ and a kinematic viscosity of $1.3 \times 10^{-5}$ m²/s is flowing through a flow nozzle similar to that shown in Fig. 15.4. A manometer using water as the gage fluid reads 81 mm of deflection. Calculate the volume flow rate if the nozzle diameter is 50 mm. Pipe inside diameter = 100 mm.*

**15.3E** The flow of kerosene is being measured with an orifice meter similar to that shown in Fig. 15.6. The pipe is a 2-in Schedule 40 pipe and the orifice diameter is 1.00 in. The kerosene is at 77°F. For a pressure difference of 0.53 psi across the orifice, calculate the volume flow rate of kerosene.

**15.4E** A sharp-edged orifice is placed in a 10-in diameter pipe carrying ammonia. If the volume flow rate is 25 gal/min, calculate the deflection of a water manometer (a) if the orifice diameter is 1.0 in and (b) if the orifice diameter is 7.0 in. The ammonia has a specific gravity of 0.83 and a dynamic viscosity of $2.5 \times 10^{-6}$ lb-s/ft².

**15.5M** *A pitot-static tube is inserted into a pipe carrying methyl alcohol at 25°C. A differential manometer using mercury as the gage fluid is connected to the tube and shows a deflection of 225 mm. Calculate the velocity of flow of the alcohol.*

**15.6M** *A pitot-static tube is connected to a differential manometer using water at 40°C as the gage fluid. The velocity of air at 40°C and atmospheric pressure is to be measured, and it is expected that the maximum velocity will be 25 m/s. Calculate the expected manometer deflection.*

**15.7M** *A pitot-static tube is inserted in a pipe carrying water at 10°C. A differential manometer using mercury as the gage fluid shows a deflection of 106 mm. Calculate the velocity of flow.*

**15.8E** Determine the maximum possible flow rate over a 60° V-notch weir if the width of the notch at the top is 12 in.

**15.9E** Determine the required length of a contracted weir similar to that shown in Fig. 15.27(b) to pass 15 ft³/s of water. The height of the crest is to be 3 ft from the channel bottom, and the maximum head above the crest is to be 18 in.

**15.10E** Plot a graph of $Q$ versus $H$ for a full width weir with a crest length of 6 ft and whose crest is 2 ft from the channel bottom. Consider values of the head $H$ from zero to 12 inches in 2-in steps.

**15.11E** Repeat the calculations of $Q$ versus $H$ for a weir with the same dimensions as used in Problem 15.10 except that it is placed in a channel wider than 6 ft. It thus becomes a contracted weir.

**15.12E** Compare the discharges over the following weirs when the head $H$ is 18 in:
  **a.** Full width rectangular: $L = 3$ ft, $H_c = 4$ ft
  **b.** Contracted rectangular: $L = 3$ ft, $H_c = 4$ ft
  **c.** 90° V-notch (top width also 3 ft)

**15.13E** Plot a graph of $Q$ versus $H$ for a 90° V-notch weir for values of the head from zero to 12 inches in 2-in steps.

**15.14E** For a Parshall flume with a throat width of 9 in, calculate the head $H$ corresponding to the minimum and maximum flows.

**15.15E** For a Parshall flume with a throat width of 8 ft, calculate the head $H$ corresponding to the minimum and maximum flows. Plot a graph of $Q$ versus $H$, using five values of $H$ spaced approximately equally between the minimum and maximum.

**15.16E** A flow rate of 50 ft³/s falls within the range of both the 4-ft and the 10-ft wide Parshall flume. Compare the head $H$ for this flow rate in each size.

## COMPUTER PROGRAMMING ASSIGNMENTS

1. Write a program using Eq. (15–5) to compute the volume flow rate for any variable head meter. Include the computation of the area in the main pipe, the area in the throat, the diameter ratio $\beta$, and the Reynolds number. Ask the user to input a value for $C$. Use Eq. (15–6) to compute the discharge coefficient for a nozzle. For the orifice, prompt the user to find the value of $C$ from Fig. 15.7 when given the Reynolds number and the diameter ratio. Permit the user to input the differential pressure in SI units (Pascals), U.S. Customary System units (psi), or in terms of the deflection of a differential manometer with a known gage fluid.

2. Write a program to accept data for the ten measurements required to complete a traverse of a circular pipe using a pitot tube as shown in Fig. 15.19. Compute the velocity of flow for each point by using Eq. (15–11). Then compute the average of ten values to determine the average velocity. Finally, compute the volume flow rate from $Q = Av$.

3. Write a program to compute the discharge through a rectangular weir by using Eq. (15–14); through a contracted weir by using Eq. (15–15); and through a V-notch weir by using Eqs. (15–18) and (15–19).

4. Write a program to compute the discharge through any of the standard Parshall flumes listed in Table 15.2.

5. Use the program from Assignment 3 to solve Practice Problems 15.10, 15.11, 15.12, and 15.13.

6. Use the program from Assignment 4 to solve Practice Problems 15.14, 15.15, and 15.16.

# 16 Forces Due to Fluids in Motion

**16.1
The
Big
Picture**

## Discussion Map

- Whenever a fluid stream is deflected from its initial direction or if its velocity is changed, a force is required to accomplish the change.

- You must be able to determine the magnitude and direction of such forces in order to design the structure to contain the fluid flow safely.

- Sometimes the force of the fluid causes a desired motion, such as when a jet of water strikes the blades of a turbine. The rotation of the turbine generates useful power.

### Discover

- How have you experienced forces due to fluids in motion?
- Consider situations in your home, your car, in a factory, or in some public utilities.
- Try to describe the effect of the forces caused by the fluids in motion as they are deflected from their initial direction or as the velocity of flow is changed.

In this chapter you will learn the fundamental principles that govern the generation of forces due to fluids in motion.

Whenever a fluid stream is deflected from its initial direction or if its velocity is changed, a force is required to accomplish the change. Sometimes the force is desired, sometimes it is destructive.

List as many situations as you can in which you have observed the effects of forces created when a fluid stream has been deflected or when its velocity has been changed. Consider these examples.

- Have you ever stuck your head outside an open window of a car traveling at highway speed?
- Have you ever been buffeted by the wind as you try to walk in a storm?
- Have you ever used the spray from a garden hose to knock some dirt loose from the sidewalk?
- Have you ever watched fire fighters struggle to control the nozzle of a fire hose that shoots a large stream of water at high velocity? They must exert large forces to hold it steady and if they lose control, the nozzle thrashes wildly and is very dangerous.
- The winds acting on the sail of a sailboat cause large forces to propel the boat through the water. This can be exhilarating. At the same time, the boat hull experiences drag forces that tend to slow it down because of the relative motion between the hull and the water.
- But the winds can also be very damaging. Thunderstorms with winds of 60–100 miles per hour (96 to 160 km/hr) can damage roofs, topple signs, or blow over a truck or mobile home. Tornadoes and hurricanes can generate winds up to 300 miles per hour (482 km/hr) and can cause great devastation. Have you ever experienced such a storm?
- Drag forces on automobiles, trucks, boats, and aircraft retard their motion. Additional expensive power must be generated by their engines to overcome the drag.

- Useful energy can be derived from the forces due to fluids in motion. High velocity jets of water impacting on the blades or buckets of a turbine wheel cause it to rotate and enable it to drive a generator to produce electric power.
- Hot combustion gases in a gas turbine engine expand through the turbine wheels to develop very high levels of power to propel an airplane, a helicopter, or a ship.
- The flow of compressed air from a nozzle is often used to move products in a production system or to remove metal chips and other debris.
- Very high-velocity narrow streams of water are used to cut fibrous material such as carpet and fabric in water-jet cutting systems.
- Piping systems that carry large volumes of fluids under pressure exert high forces as the fluid passes around elbows or is restricted by a contraction in the flow stream. Thus any part of the system where the flow direction is changed or where the magnitude of the velocity is changed must be anchored securely.

In this chapter, you will learn the fundamental principles that govern the generation of forces due to fluids in motion. Several types of practical problems will be demonstrated. Then in Chapter 17, you will extend this topic to include drag forces on many shapes of objects and lift on aerodynamic devices.

---

**16.2**
**OBJECTIVES**

After completing this chapter, you should be able to:

1. Use Newton's second law of motion, $F = ma$, to develop the *force equation*, which is used to compute the force exerted by a fluid as its direction of motion or its velocity is changed.
2. Relate the force equation to *impulse-momentum*.
3. Use the force equation to compute the force exerted on a stationary object that causes the change in direction of a fluid flow stream.
4. Use the force equation to compute the force exerted on bends in pipe lines.
5. Use the force equation to compute the force on moving objects, such as the vanes of a pump impeller.

**16.3**
**FORCE EQUATION**

Whenever the magnitude or direction of the velocity of a body is changed, a force is required to accomplish the change. Newton's second law of motion is often used to express this concept in mathematical form; the most common form is

$$F = ma \qquad (16–1)$$

Force equals mass times acceleration. Acceleration is the time rate of change of velocity. However, since velocity is a vector quantity having both magnitude and direction, changing either the magnitude or the direction will result in an acceleration. According to Eq. (16–1), an external force is required to cause the change.

Equation (16–1) is convenient for use with solid bodies since the mass remains constant and the acceleration of the entire body can be determined. In fluid flow problems, a continuous flow is caused to undergo the acceleration, and a different form of Newton's equation is desirable. Because acceleration is the time rate of change of velocity, Eq. (16–1) can be written as

$$F = ma = m\frac{\Delta v}{\Delta t} \qquad (16–2)$$

The term $m/\Delta t$ can be interpreted as the mass flow rate, that is, the amount of mass flowing in a given amount of time. In the discussion of fluid flow in Chapter 6, mass flow rate was indicated by the symbol $M$. Also, $M$ is related to the volume flow rate $Q$ by the relationship

$$M = \rho Q \tag{16-3}$$

where $\rho$ is the density of the fluid. Then Eq. (16–2) becomes

⟹ GENERAL FORM OF FORCE
EQUATION

$$F = (m/\Delta t)\Delta v = M\,\Delta v = \rho Q\,\Delta v \tag{16-4}$$

This is the general form of the force equation for use in fluid flow problems because it involves the velocity and volume flow rate, items generally known in a fluid flow system.

**16.4**
**IMPULSE-MOMENTUM**
**EQUATION**

The force equation, Eq. (16–4), is related to another principle of fluid dynamics, the *impulse-momentum equation*. Impulse is defined as a force acting on a body for a period of time and it is indicated by

$$\text{Impulse} = F(\Delta t)$$

This form, relying on the total change in time $\Delta t$, is suitable for dealing with steady flow conditions. When conditions vary, the instantaneous form of the equation is used.

$$\text{Impulse} = F(dt)$$

where $dt$ is the differential amount of change in time.

*Momentum* is defined as the product of the mass of a body and its velocity. The *change* in momentum is

$$\text{Change in momentum} = m(\Delta v)$$

In an instantaneous sense,

$$\text{Change in momentum} = m(dv)$$

Now Eq. (16–2) can be rearranged to the form

$$F(\Delta t) = m(\Delta v)$$

Here we have shown the impulse-momentum equation for steady flow conditions. In an instantaneous sense,

$$F(dt) = m(dv)$$

**16.5**
**PROBLEM-SOLVING**
**METHOD USING THE**
**FORCE EQUATIONS**

We emphasize that problems involving forces must account for the directions in which the forces act. In Eq. (16–4), force and velocity are both vector quantities. The equation is valid only when all terms have the same direction. For this reason, different equations are written for each direction of concern in a particular case. In general, if three perpendicular directions are called $x$, $y$, and $z$, a separate equation can be written for each direction:

⟹ FORCE EQUATIONS IN $x$, $y$, AND
$z$ DIRECTIONS

$$F_x = \rho Q\,\Delta v_x = \rho Q(v_{2_x} - v_{1_x}) \tag{16-5}$$

$$F_y = \rho Q\,\Delta v_y = \rho Q(v_{2_y} - v_{1_y}) \tag{16-6}$$

$$F_z = \rho Q\,\Delta v_z = \rho Q(v_{2_z} - v_{1_z}) \tag{16-7}$$

This is the form of the force equation which will be used in this book, with the directions chosen according to the physical situation. In a particular direction, say $x$, the term $F_x$ refers to the net external force which acts on the fluid in that direction.

Therefore, it is the algebraic sum of *all* external forces, including that exerted by a solid surface and forces due to fluid pressure. The term $\Delta v_x$ refers to the change in velocity in the *x* direction. Also, $v_1$ is the velocity as the fluid enters the device and $v_2$ is the velocity as it leaves. Then $v_{1_x}$ is the component of $v_1$ in the *x* direction and $v_{2_x}$ is the component of $v_2$ in the *x* direction.

The specific approach to problems using the force equations depends somewhat on the nature of the data given. A general procedure follows:

**PROCEDURE FOR USING THE FORCE EQUATIONS**

1. Identify a portion of the fluid stream to be considered a free body. This will be the part where the fluid is changing direction or where the geometry of the flow stream is changing.
2. Establish reference axes for directions of forces. Usually one axis is chosen to be parallel to one part of the flow stream. In the example problems to follow, the positive *x* and *y* directions are chosen to be in the same direction as the reaction forces.
3. Identify and show on the free body diagram all external forces acting on the fluid. All solid surfaces which affect the direction of the flow stream exert forces. Also, the fluid pressure acting on the cross-sectional area of the stream exerts a force in a direction parallel to the stream at the boundary of the free body.
4. Show the direction of the velocity of flow as it enters the free body and as it leaves the free body.
5. Using the data thus shown for the free body, write the force equations in the pertinent directions. Use Eqs. (16–5), (16–6), or (16–7).
6. Substitute data and solve for the desired quantity.

The example problems presented in the following sections illustrate this procedure.

**16.6**
**FORCES ON STATIONARY OBJECTS**

When free streams of fluid are deflected by stationary objects, external forces must be exerted to maintain the object in equilibrium. Some examples follow.

☐ **EXAMPLE PROBLEM 16.1**    A 1-in diameter jet of water having a velocity of 20 ft/s is deflected by a curved vane 90°, as shown in Figure 16.1. The jet flows freely in the atmosphere in a horizontal plane. Calculate the *x* and *y* forces exerted on the water by the vane.

**FIGURE 16.1**    Water jet deflected by a curved vane.

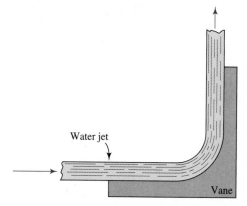

Water jet

Vane

**FIGURE 16.2**   Force diagram for the fluid deflected by the vane.

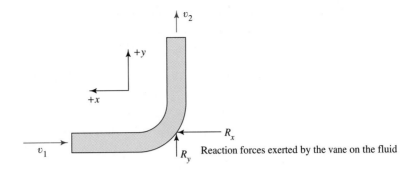

Reaction forces exerted by the vane on the fluid

***Solution***   Using the force diagram of Fig. 16.2, the force equation for the $x$ direction is

$$F_x = \rho Q(v_{2_x} - v_{1_x})$$
$$R_x = \rho Q[0 - (-v_1)] = \rho Q v_1$$

But, we know that

$$Q = Av = (0.00545 \text{ ft}^2)(20 \text{ ft/s}) = 0.109 \text{ ft}^3/\text{s}$$

Then, assuming $\rho = 1.94$ slugs/ft$^3$ = 1.94 lb·s$^2$/ft$^4$,

$$R_x = \rho Q v_1 = \frac{1.94 \text{ lb·s}^2}{\text{ft}^4} \times \frac{0.109 \text{ ft}^3}{\text{s}} \times \frac{20 \text{ ft}}{\text{s}} = 4.23 \text{ lb}$$

For the $y$ direction, assuming $v_2 = v_1$, the force is

$$F_y = \rho Q(v_{2_y} - v_{1_y})$$
$$R_y = \rho Q(v_2 - 0) = (1.94)(0.109)(20) \text{ lb} = 4.23 \text{ lb}$$

---

☐ **EXAMPLE PROBLEM 16.2**   In a decorative fountain, 0.05 m$^3$/s of water having a velocity of 8 m/s is being deflected by the angled chute shown in Fig. 16.3. Determine the reactions on the chute in the $x$ and $y$ directions shown. Also, calculate the total resultant force and the direction in which it acts. Neglect elevation changes.

***Solution***   Figure 16.4 shows the $x$ and $y$ components of the velocity vectors and the assumed directions for $R_x$ and $R_y$. The force equation in the $x$ direction is:

$$F_x = \rho Q(v_{2_x} - v_{1_x})$$

But, we know that

$$v_{2_x} = -v_2 \sin 15° \qquad \text{(toward the right)}$$
$$v_{1_x} = -v_1 \cos 45° \qquad \text{(toward the right)}$$

Neglecting friction in the chute, we can assume that $v_2 = v_1$. Also, the only external force is $R_x$. Then we have

$$R_x = \rho Q[-v_2\sin 15° - (-v_1\cos 45°)]$$
$$= \rho Q v(-\sin 15° + \cos 45°) = 0.448 \, \rho Q v$$

Using $\rho = 1000$ kg/m$^3$ for water, we get

$$R_x = \frac{(0.448)(1000 \text{ kg})}{\text{m}^3} \times \frac{0.05 \text{ m}^3}{\text{s}} \times \frac{8 \text{ m}}{\text{s}} = \frac{179 \text{ kg·m}}{\text{s}^2} = 179 \text{ N}$$

**FIGURE 16.3** Decorative fountain deflecting a water jet.

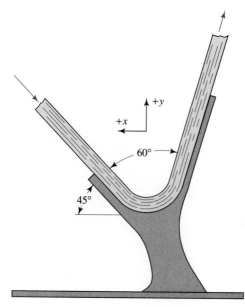

In the $y$ direction, the force equation is

$$F_y = \rho Q(v_{2_y} - v_{1_y})$$

But, we know that

$$v_{2_y} = v_2\cos 15° \qquad \text{(upward)}$$
$$v_{1_y} = -v_1\sin 45° \qquad \text{(downward)}$$

**FIGURE 16.4** Force diagram for the fluid deflected by the vane.

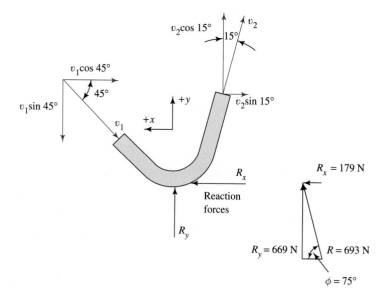

Then we have

$$R_y = \rho Q[v_2 \cos 15° - (-v_1 \sin 45°)]$$
$$= \rho Q v(\cos 15° + \sin 45°)$$
$$= (1000)(0.05)(8)(0.966 + 0.707) \text{ N}$$
$$R_y = 699 \text{ N}$$

The resultant force $R$ is

$$R = \sqrt{R_x^2 + R_y^2} = \sqrt{179^2 + 669^2} = 693 \text{ N}$$

For the direction of $R$, we get

$$\tan \phi = R_y/R_x = 669/179 = 3.74$$
$$\phi = 75.0°$$

Therefore, the resultant force that the chute must exert on the water is 693 N acting 75° from the horizontal, as shown in Fig. 16.4.

■

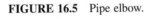

**16.7**
**FORCES ON BENDS**
**IN PIPE LINES**

Figure 16.5 shows a typical 90° elbow in a pipe carrying a steady volume flow rate $Q$. In order to ensure proper installation, it is important to know how much force is required to hold it in equilibrium. The following problem demonstrates an approach to this type of situation.

**FIGURE 16.5**  Pipe elbow.

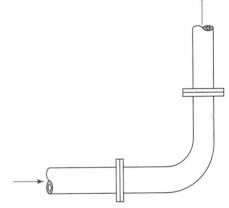

☐ **EXAMPLE PROBLEM 16.3**

Calculate the force that must be exerted on the pipe shown in Fig. 16.5 to hold it in equilibrium. The elbow is in a horizontal plane and is connected to two 4-in Schedule 40 pipes carrying 3000 L/min of water at 15°C. The inlet pressure is 550 kPa.

*Solution*

The problem may be visualized by considering the fluid within the elbow to be a free body, as shown in Fig. 16.6. Forces are shown in black vectors, while the direction of the velocity of flow is shown by blue vectors. A convention must be set for the directions of all vectors. Here we assume that the positive $x$ direction is to the left and the positive $y$ direction is up. The forces $R_x$ and $R_y$ are the external reactions required to maintain equilibrium. The forces

FIGURE 16.6   Force diagram on the fluid
in the elbow.

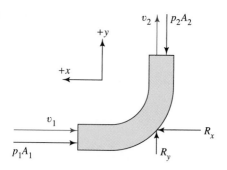

$p_1A_1$ and $p_2A_2$ are the forces due to the fluid pressure. The two directions will be analyzed
separately.

We find the net external force in the $x$ direction by using the equation

$$F_x = \rho Q(v_{2_x} - v_{1_x})$$

But, we know that

$$F_x = R_x - p_1A_1$$
$$v_{2_x} = 0$$
$$v_{1_x} = -v_1$$

Then we have

$$R_x - p_1A_1 = \rho Q[0 - (-v_1)]$$
$$R_x = \rho Q v_1 + p_1A_1 \qquad (16\text{–}8)$$

From the given data, $p_1 = 550 \text{ kPa}$, $\rho = 1000 \text{ kg/m}^3$, and $A_1 = 8.213 \times 10^{-3} \text{ m}^2$.
Then,

$$Q = 3000 \text{ L/min} \times \frac{1 \text{ m}^3/\text{s}}{60\,000 \text{ L/min}} = 0.05 \text{ m}^3/\text{s}$$

$$v_1 = \frac{Q}{A_1} = \frac{0.05 \text{ m}^3/\text{s}}{8.213 \times 10^{-3} \text{ m}^2} = 6.09 \text{ m/s}$$

$$\rho Q v_1 = \frac{1000 \text{ kg}}{\text{m}^3} \times \frac{0.05 \text{ m}^3}{\text{s}} \times \frac{6.09 \text{ m}}{\text{s}} = 305 \text{ kg·m/s}^2 = 305 \text{ N}$$

$$p_1A_1 = \frac{550 \times 10^3 \text{ N}}{\text{m}^2} \times (8.213 \times 10^{-3} \text{ m}^2) = 4517 \text{ N}$$

Substituting these values into Eq. (16–8) gives

$$R_x = (305 + 4517)\,\text{N} = 4822 \text{ N}$$

In the $y$ direction, the equation for the net external force is

$$F_y = \rho Q(v_{2_y} - v_{1_y})$$

But, we know that

$$F_y = R_y - p_2A_2$$
$$v_{2_y} = +v_2$$
$$v_{1_y} = 0$$

Then we have

$$R_y - p_2A_2 = \rho Q v_2$$
$$R_y = \rho Q v_2 + p_2A_2$$

If energy losses in the elbow are neglected, $v_2 = v_1$ and $p_2 = p_1$ since the sizes of the inlet and outlet are equal. Then,

$$\rho Q v_2 = 305 \text{ N}$$
$$p_2A_2 = 4517 \text{ N}$$
$$R_y = (305 + 4517) \text{ N} = 4822 \text{ N}$$

The forces $R_x$ and $R_y$ are the reactions caused at the elbow as the fluid turns 90°. They may be supplied by anchors on the elbow or taken up through the flanges into the main pipes.

---

☐ **EXAMPLE PROBLEM 16.4**   Linseed oil with a specific gravity of 0.93 enters the reducing bend shown in Fig. 16.7 with a velocity of 3 m/s and a pressure of 275 kPa. The bend is in a horizontal plane. Calculate the $x$ and $y$ forces required to hold the bend in place. Neglect energy losses in the bend.

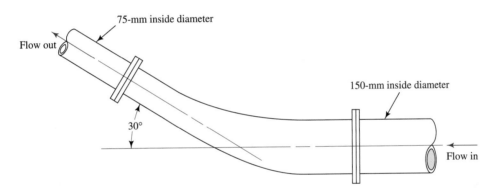

**FIGURE 16.7**   Reducing bend.

*Solution*   The fluid in the bend is shown as a free body in Fig. 16.8. We must first develop the force equations for the $x$ and $y$ directions shown.

The force equation for the $x$ direction is

$$F_x = \rho Q(v_{2_x} - v_{1_x})$$
$$R_x - p_1A_1 + p_2A_2\cos 30° = \rho Q[-v_2\cos 30° - (-v_1)] \qquad \text{(16–9)}$$
$$R_x = p_1A_1 - p_2A_2\cos 30° - \rho Q v_2\cos 30° + \rho Q v_1$$

Algebraic signs must be carefully included according to the sign convention shown in Fig. 16.8. Notice that all forces and velocity terms are the components *in the x direction*.

In the $y$ direction, the force equation is

$$F_y = \rho Q(v_{2_y} - v_{1_y})$$
$$R_y - p_2A_2\sin 30° = \rho Q(v_2\sin 30° - 0)$$
$$R_y = p_2A_2\sin 30° + \rho Q v_2\sin 30° \qquad \text{(16–10)}$$

**FIGURE 16.8**   Force diagram for fluid in the reducing bend.

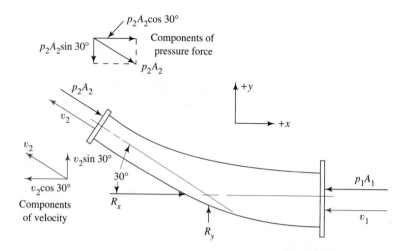

The numerical values of several items must now be calculated. For the entering and leaving pipes, $A_1 = 1.767 \times 10^{-2}$ m$^2$ and $A_2 = 4.418 \times 10^{-3}$ m$^2$.

$$\rho = (sg)(\rho_w) = (0.93)(1000 \text{ kg/m}^3) = 930 \text{ kg/m}^3$$

$$\gamma = (sg)(\gamma_w) = (0.93)(9.81 \text{ kN/m}^3) = 9.12 \text{ kN/m}^3$$

$$Q = A_1 v_1 = (1.767 \times 10^{-2} \text{ m}^2)(3 \text{ m/s}) = 0.053 \text{ m}^3/\text{s}$$

Because of continuity, $A_1 v_1 = A_2 v_2$. Then we have

$$v_2 = v_1(A_1/A_2) = (3 \text{ m/s})(1.767 \times 10^{-2}/4.418 \times 10^{-3}) = 12 \text{ m/s}$$

Bernoulli's equation can be used to find $p_2$:

$$\frac{p_1}{\gamma} + z_1 + \frac{v_1^2}{2g} = \frac{p_2}{\gamma} + z_2 + \frac{v_2^2}{2g}$$

But $z_1 = z_2$. Then we have

$$p_2 = p_1 + \gamma(v_1^2 - v_2^2)/2g$$

$$= 275 \text{ kPa} + \left[ \frac{(9.12)(3^2 - 12^2)}{(2)(9.81)} \times \frac{\text{kN}}{\text{m}^3} \times \frac{\text{m}^2}{\text{s}^2} \times \frac{\text{s}^2}{\text{m}} \right]$$

$$= 275 \text{ kPa} - 62.8 \text{ kPa}$$

$$p_2 = 212.2 \text{ kPa}$$

The quantities needed for Eqs. (16–9) and (16–10) are

$$p_1 A_1 = (275 \text{ kN/m}^2)(1.767 \times 10^{-2} \text{ m}^2) = 4859 \text{ N}$$

$$p_2 A_2 = (212.2 \text{ kN/m}^2)(4.418 \times 10^{-3} \text{ m}^2) = 938 \text{ N}$$

$$\rho Q v_1 = (930 \text{ kg/m}^3)(0.053 \text{ m}^3/\text{s})(3 \text{ m/s}) = 148 \text{ N}$$

$$\rho Q v_2 = (930 \text{ kg/m}^3)(0.053 \text{ m}^3/\text{s})(12 \text{ m/s}) = 591 \text{ N}$$

From Eq. (16–9), we get

$$R_x = (4859 - 938 \cos 30° - 591 \cos 30° + 148) \text{ N} = 3683 \text{ N}$$

From Eq. (16–10), we get

$$R_y = (938 \sin 30° + 591 \sin 30°) \text{ N} = 765 \text{ N}$$

<table>
<tr><td>

**16.8**
**FORCES ON MOVING OBJECTS**

</td><td>

The vanes of turbines and other rotating machinery are familiar examples of moving objects that are acted upon by high velocity fluids. A jet of fluid with a velocity greater than that of the blades of the turbine exerts a force on the blades, causing them to accelerate or to generate useful mechanical energy. When dealing with forces on moving bodies, the *relative motion* of the fluid with respect to the body must be considered.

</td></tr>
</table>

☐ **EXAMPLE PROBLEM 16.5**

Figure 16.9(a) shows a jet of water with a velocity $v_1$ striking a vane that is moving with a velocity $v_0$. Determine the forces exerted by the vane on the water if $v_1 = 20$ m/s and $v_0 = 8$ m/s. The jet is 50 mm in diameter.

**FIGURE 16.9**  Flow deflected by a moving vane.

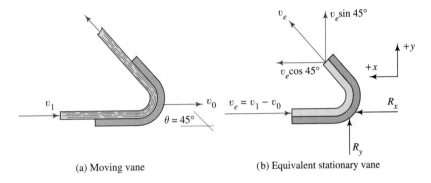

(a) Moving vane                    (b) Equivalent stationary vane

*Solution*

The system with a moving vane can be converted into an equivalent stationary system as shown in Fig. 16.9(b) by defining an effective velocity $v_e$ and an effective volume flow rate $Q_e$. We then have

EFFECTIVE VELOCITY AND
VOLUME FLOW RATE

$$v_e = v_1 - v_0 \tag{16–11}$$

$$Q_e = A_1 v_e \tag{16–12}$$

where $A_1$ is the area of the jet as it enters the vane. It is only the difference between the jet velocity and the vane velocity that is effective in creating a force on the vane. The force equations can be written in terms of $v_e$ and $Q_e$. In the $x$ direction,

$$R_x = \rho Q_e v_e \cos\theta - (-\rho Q_e v_e)$$
$$= \rho Q_e v_e (1 + \cos\theta) \tag{16–13}$$

In the $y$ direction,

$$R_y = \rho Q_e v_e \sin\theta - 0 \tag{16–14}$$

But, we know that

$$v_e = v_1 - v_0 = (20 - 8) \, \text{m/s} = 12 \, \text{m/s}$$
$$Q_e = A_1 v_e = (1.964 \times 10^{-3} \, \text{m}^2)(12 \, \text{m/s}) = 0.0236 \, \text{m}^3/\text{s}$$

Then the reactions are calculated from Eqs. (16–13) and (16–14):

$$R_x = (1000)(0.0236)(12)(1 + \cos 45°) = 483 \, \text{N}$$
$$R_y = (1000)(0.0236)(12)(\sin 45°) = 200 \, \text{N}$$

■

## PRACTICE PROBLEMS

**16.1M** *Calculate the force required to hold a flat plate in equilibrium perpendicular to the flow of water at 25 m/s issuing from a 75-mm diameter nozzle.*

**16.2E** What must be the velocity of flow of water from a 2-in diameter nozzle to exert a force of 300 lb on a flat wall?

**16.3E** Calculate the force exerted on a stationary curved vane that deflects a 1-in diameter stream of water through a 90° angle. The volume flow rate is 150 gal/min.

**16.4M** *A highway sign is being designed to withstand winds of 125 km/h. Calculate the total force on a sign 4 m by 3 m if the wind is flowing perpendicular to the face of the sign. Calculate the equivalent pressure on the sign in Pa. The air is at −10°C. (See Chapter 17 and Problem 17.9 for a more complete analysis of this problem.)*

**16.5E** Compute the forces in the vertical and horizontal directions on the block shown in Fig. 16.10. The fluid stream is a 1.75-in diameter jet of water at 60°F with a velocity of 25 ft/s. The velocity leaving the block is also 25 ft/s.

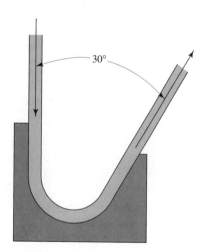

**FIGURE 16.10**　Problem 16.5.

**FIGURE 16.12**　Problem 16.7.

**16.6E** Figure 16.11 shows a free stream of water at 180°F being deflected by a stationary vane through a 130° angle. The entering stream has a velocity of 22.0 ft/s. The cross sectional area of the stream is constant at 2.95 in² throughout the system. Compute the forces in the horizontal and vertical directions exerted on the water by the vane.

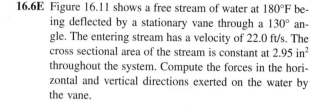
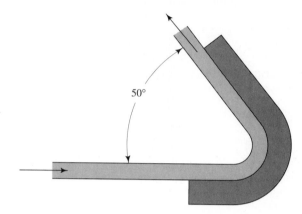

**FIGURE 16.11**　Problem 16.6.

**16.7M** *Compute the horizontal and vertical forces exerted on the vane shown in Fig. 16.12 due to a flow of water at 50°C. The velocity is constant at 15 m/s.*

**16.8E** In a plant where hemispherical cup-shaped parts are made, an automatic washer is being designed to clean the parts prior to shipment. One scheme being evaluated uses a stream of water at 180°F shooting vertically upward into the cup. The stream has a velocity of 30 ft/s and a diameter of 1.00 in. As shown in Fig. 16.13, the water leaves the cup vertically downward in the form of an annular ring having an outside diameter of 4.00 in and an inside diameter of 3.80 in. Compute the external force required to hold the cup down.

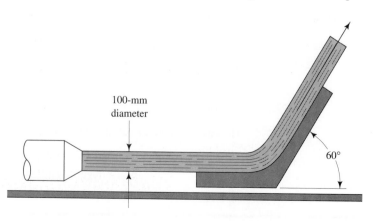

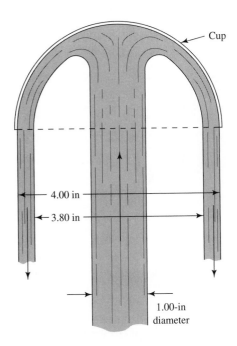

Cup

4.00 in

3.80 in

1.00-in
diameter

**FIGURE 16.13**   Problem 16.8.

**16.9M** *A stream of oil (sg = 0.90) is directed onto the center of the underside of a flat metal plate to keep it cool during a welding operation. The plate weighs 550 N. If the stream is 35 mm in diameter, calculate the velocity of the stream that will lift the plate. The stream strikes the plate perpendicularly.*

**16.10E** A 2-in diameter stream of water having a velocity of 40 ft/s strikes the edge of a flat plate such that half the stream is deflected downward as shown in Fig. 16.14. Calculate the force on the plate and the moment due to the force at point A.

**16.11E** Figure 16.15 represents a type of flowmeter in which the flat vane is rotated on a pivot as it deflects the fluid

stream. The fluid force is counterbalanced by a spring. Calculate the spring force required to hold the vane in a vertical position when water at 100 gal/min flows from the 1-in Schedule 40 pipe to which the meter is attached.

**16.12E** Water is piped vertically from below a boat and discharged horizontally in a 4-in diameter jet with a velocity of 60 ft/s. Calculate the force on the boat.

**16.13E** A 2-in nozzle is attached to a hose with an inside diameter of 4 in. The resistance coefficient $K$ of the nozzle is 0.12 based on the outlet velocity head. If the jet issuing from the nozzle has a velocity of 80 ft/s, calculate the force exerted by the water on the nozzle.

**16.14M** *Seawater (sg = 1.03) enters a heat exchanger through a reducing bend connecting a 4-in Type K copper tube with a 2-in Type K tube. The pressure upstream from the bend is 825 kPa. Calculate the force required to hold the bend in equilibrium. Consider the energy loss in the bend, assuming it has a resistance coefficient $K$ of 3.5 based on the inlet velocity. The flow rate is 0.025 $m^3/s$.*

**16.15E** A reducer connects a standard 6-in Schedule 40 pipe to a 3-in Schedule 40 pipe. The walls of the conical reducer are tapered at an included angle of 40°. The flow rate of water is 500 gal/min and the pressure ahead of the reducer is 125 psig. Considering the energy loss in the reducer, calculate the force exerted on the reducer by the water.

**16.16E** Calculate the force on a 45° elbow attached to an 8-in Schedule 80 steel pipe carrying water at 80°F at 6.5 $ft^3/s$. The outlet of the elbow discharges into the atmosphere. Consider the energy loss in the elbow.

**16.17M** *Calculate the force required to hold a 90° elbow in place when attached to 6-in Schedule 40 pipes carrying water at 125 $m^3/s$ and 1050 kPa. Neglect energy lost in the elbow.*

**16.18M** *Calculate the force required to hold a 180° close return bend in equilibrium. The bend is in a horizontal plane and is attached to a 4-in Schedule 80 steel pipe carry-*

**FIGURE 16.14**   Problem 16.10.

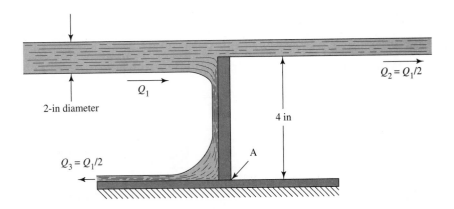

2-in diameter

$Q_1$

$Q_2 = Q_1/2$

4 in

A

$Q_3 = Q_1/2$

**FIGURE 16.15**   Problem 16.11.

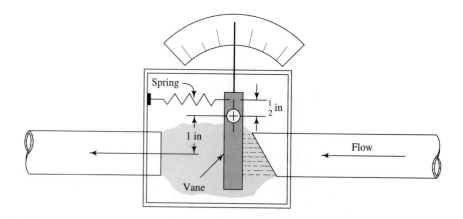

ing 2000 L/min of a hydraulic fluid at 2.0 MPa. The fluid has a specific gravity of 0.89. Neglect energy losses.

**16.19M** *A bend in a tube causes the flow to turn through an angle of 135°. The pressure ahead of the bend is 275 kPa. If the 6-in Type K copper tube carries 0.12 $m^3/s$ of carbon tetrachloride at 25°C, determine the force on the bend. Neglect energy losses.*

**16.20M** *A vehicle is to be propelled by a jet of water impinging on a vane as shown in Fig. 16.16. The jet has a velocity of 30 m/s and issues from a nozzle with a diameter of 200 mm. Calculate the force on the vehicle (a) if it is stationary and (b) if it is moving at 12 m/s.*

**16.21M** *A part of an inspection system in a packaging operation uses a jet of air to remove imperfect cartons from a conveyor line, as shown in Fig. 16.17. The jet is initiated by a sensor and timed so that the product to be rejected is in front of the jet at the right moment. The product is to be tipped over a ledge on the side of the conveyor as shown in the figure. Compute the required velocity of the air jet to tip the carton off the conveyor. The density of the air is 1.20 $kg/m^3$. The carton has a mass of 0.10 kg. The jet has a diameter of 10.0 mm.*

**16.22M** *Shown in Fig. 16.18 is a small decorative wheel fitted with flat paddles so the wheel turns about its axis when*

acted on by a blown stream of air. Assuming that all the air in a 15-mm diameter stream moving at 0.35 m/s strikes one paddle and is deflected by it at right angles, compute the force exerted on the wheel initially when it is stationary. The air has a density of 1.20 $kg/m^3$.

**16.23M** *For the wheel described in Problem 16.22, compute the force exerted on the paddle when the wheel rotates at 40 rpm.*

**16.24E** A set of louvers deflects a stream of warm air onto painted parts, as illustrated in Fig. 16.19. The louvers are rotated slowly to distribute the air evenly over the parts. Compute the torque required to rotate the louvers toward the stream of air when it is flowing at a velocity of 10 ft/s. Assume that all the air that approaches a given louver is deflected to the angle of the louver. The air has a density of 2.06 × $10^{-3}$ slugs/$ft^3$. Use $\theta = 45°$.

**16.25E** For the louvers shown in Fig. 16.19 and described in Problem 16.24, compute the torque required to rotate the louvers when the angle $\theta = 20°$.

**16.26E** For the louvers shown in Fig. 16.19 and described in Problem 16.24, compute the torque required to rotate the louvers for several settings of the angle $\theta$ from 10° to 90°. Plot a graph of torque versus angle.

**FIGURE 16.16**   Problem 16.20.

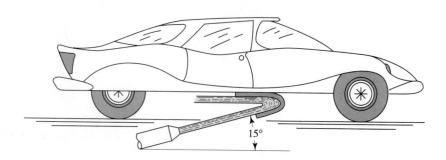

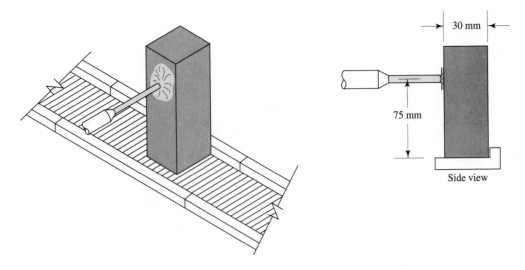

**FIGURE 16.17**   Problem 16.21.

**FIGURE 16.18**   Problems 16.22
and 16.23.

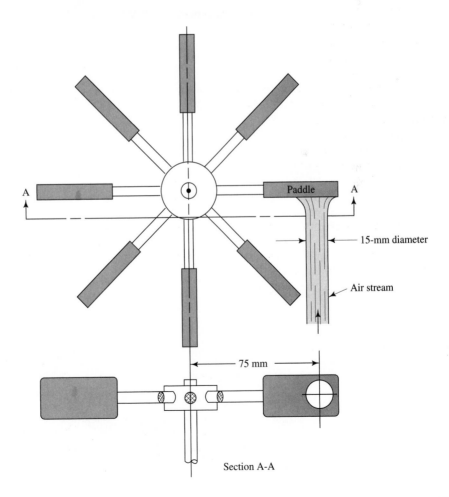

481

**FIGURE 16.19**  Problems 16.24, 16.25, and 16.26.

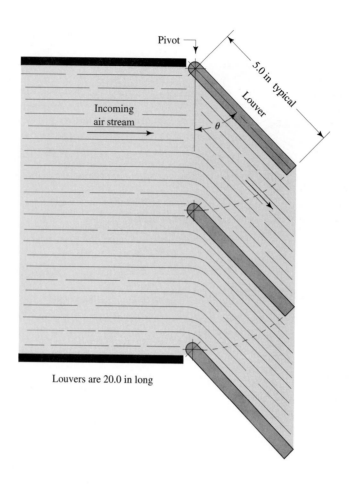

Pivot

5.0 in typical

Incoming air stream

Louver

$\theta$

Louvers are 20.0 in long

**FIGURE 16.20**  Problems 16.27 and 16.28.

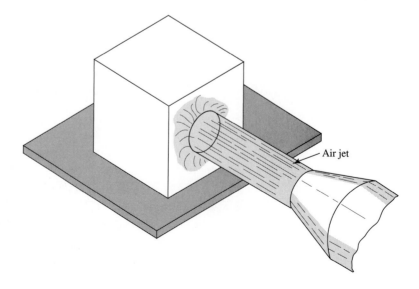

Air jet

**16.27E** Figure 16.20 shows a device for clearing debris using a 1¹/₂-in diameter jet of air issuing from a blower nozzle. As shown, the jet is striking a rectangular box-shaped object sitting on a floor. If the air velocity is 25 ft/s and the entire jet is deflected by the box, what is the heaviest object that could be moved? Assume that the box slides rather than tumbling over and that the coefficient of friction is 0.60. The air has a density of $2.40 \times 10^{-3}$ slugs/ft³.

**16.28E** Repeat Problem 16.27, except change the jet to water at 50°F and the diameter to 0.75 in.

**16.29M** *Figure 16.21 is a sketch of a turbine in which the incoming stream of water at 15°C has a diameter of*

7.50 mm and is moving with a velocity of 25 m/s. Compute the force on one blade of the turbine if the stream is deflected through the angle shown and the blade is stationary.*

**16.30M** *Repeat Problem 16.29 with the blade rotating as a part of the wheel at a radius of 200 mm and with a linear tangential velocity of 10 m/s. Also compute the rotational speed of the wheel in rpm.*

**16.31M** *Repeat Problem 16.29, except with the blade rotating as a part of the wheel at a radius of 200 mm and with a linear tangential velocity ranging from 0 to 25 m/s in 5 m/s steps.*

**FIGURE 16.21** Problems 16.29, 16.30, and 16.31.

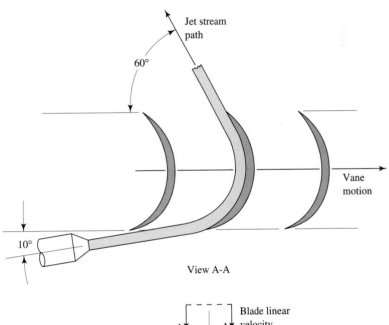

View A-A

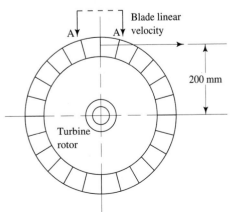

# 17 Drag and Lift

## 17.1 The Big Picture

### Discussion Map

- ☐ A moving body immersed in a fluid experiences forces caused by the action of the fluid. The force is called *drag*.

- ☐ Similarly, a moving fluid impinging on a stationary object or on one that is moving more slowly than the fluid exerts a force on the object.

- ☐ When a special shaped body, called an *airfoil*, moves through air, the flow of air around it causes a net upward force called *lift*. This is the fundamental reason that aircraft can fly.

- ☐ Similarly shaped bodies moving through water are called *hydrofoils*.

- ☐ You should develop your ability to analyze drag and lift forces.

### Discover

- ■ Seek examples of products and equipment where the drag or lift forces have an effect on its behavior or its performance.
- ■ Consider some of the examples mentioned in the Big Picture section of Chapter 16.
- ■ As you find examples, document the shape and size with as much detail as you can.

This chapter presents the principles of analysis of drag and lift forces. Data for drag coefficients for many shapes are presented.

A moving body immersed in a fluid experiences forces caused by the action of the fluid. The total effect of these forces is quite complex. However, for the purposes of design or for the analysis of the behavior of a body in a fluid, two resultant forces—drag and lift—are the most important. Lift and drag forces are the same regardless of whether the body is moving in the fluid or the fluid is moving over the body.

*Drag* is the force on a body caused by the fluid which resists motion in the direction of travel of the body. The most familiar applications requiring the study of drag are in the transportation fields. *Wind resistance* is the term often used to describe the effects of drag on aircraft, automobiles, trucks, and trains. The drag force must be opposed by a propulsive force in the opposite direction to maintain or increase the velocity of the vehicle. Since the production of the propulsive force requires added power, it is desirable to minimize drag.

*Lift* is a force caused by the fluid in a direction perpendicular to the direction of travel of the body. Its most important application is the design and analysis of aircraft wings called *airfoils*. The geometry of an airfoil is such that a lift force is produced as air passes over and under it. Of course, the magnitude of the lift must at least equal the weight of the aircraft in order for it to fly.

The study of the performance of bodies in moving air streams is called *aerodynamics*. Gases other than air could be considered in this field, but due to the obvious importance of the applications in aircraft design, the majority of work has been done with air as the fluid.

*Hydrodynamics* is the name given to the study of moving bodies immersed in liquids, particularly water. Many concepts concerning lift and drag are similar regardless of whether the fluid is a liquid or a gas. This is not true, however, at high velocities where the effects of the compressibility of the fluid must be taken into account. Liquids can be considered incompressible in the study or lift and drag. Conversely, a gas such as air is readily compressible.

What kinds of examples have you found of products or equipment where drag or lift forces have an effect on its behavior or performance? Consider the following questions and observations as you describe your examples.

- Are the edges sharp or smooth and well rounded?
- Is the shape flat or does it have a rounded surface?
- If the object is cup-shaped, is the open side of the cup facing into the wind (or other fluid) or away from it?
- What attempts have been made to streamline the shape?
- Try to find two automobiles that have radically different shapes; one that is highly streamlined and another that is more boxy. Perhaps the boxy shape will be an older car, even an antique. How do you think the shape affects drag?
- Describe the shape of fast trains such as the *TGV* that operates in France or the *Bullet Trains* that run in Japan. What approaches have been used to decrease drag? How do their shapes compare with conventional freight locomotives? See if you can find data for the drag characteristics of the fast trains on the Internet or some other information source.
- Compare race cars from varying periods of time. Consider the Indy-type, sports cars, and stock car racers. How are they similar in their approach to reducing drag? How do they differ?
- Compare aircraft from varying periods of time. What attempts were made in the early days of flight from the Wright brothers through the decade of the 1930s to reduce drag. How did military aircraft change from the advent of World War II through the Korean War and on through more recent conflicts? How do jet aircraft compare with propeller-driven planes with regard to their aerodynamics? Compare the supersonic Concorde with the subsonic Boeing 777 commercial aircraft.

Much of the practical data concerning lift and drag has been generated experimentally. We will report some of these data here to illustrate the concepts. The references listed at the end of this chapter include more comprehensive treatments of the subject.

---

**17.2 OBJECTIVES**

After completing this chapter, you should be able to:

1. Define *drag*.
2. Define *lift*.
3. Write the expression for computing the drag force on a body moving relative to a fluid.
4. Define the *drag coefficient*.
5. Define the term *dynamic pressure*.
6. Describe the stagnation point for a body moving relative to a fluid.
7. Distinguish between *pressure drag* and *friction drag*.
8. Discuss the importance of flow separation on pressure drag.
9. Determine the value of the pressure drag coefficient for cylinders, spheres, and other shapes.
10. Discuss the effect of Reynolds number and surface geometry on the drag coefficient.

11. Compute the magnitude of the pressure drag force on bodies moving relative to a fluid.
12. Compute the magnitude of the friction drag force on smooth spheres.
13. Discuss the importance of drag on the performance of ground vehicles.
14. Discuss the effects of compressibility and cavitation on drag and the performance of bodies immersed in fluids.
15. Define the lift coefficient for a body immersed in a fluid.
16. Compute the lift force on a body moving relative to a fluid.
17. Describe the effects of friction drag, pressure drag, and induced drag on airfoils.

**17.3**
**DRAG FORCE EQUATION**

 **DRAG FORCE**

Drag forces are usually expressed in the form

$$F_D = \text{drag} = C_D(\rho v^2/2)A \tag{17–1}$$

The terms of this equation are explained below.

- $C_D$ is the *drag coefficient*. It is a dimensionless number that depends on the shape of the body and its orientation relative to the fluid stream.
- $\rho$ is the density of the fluid. Because the density of liquids is much greater than that of a gas, the general order of magnitude of the drag forces on objects moving through water is far greater than when objects move through air. The compressibility of the air affects its density somewhat.
- $v$ is the velocity of the free stream of the fluid relative to the body. In general it does not matter if the body is moving or if the fluid is moving. However, the location of other surfaces near the body of interest can affect the drag. For example, when a truck or car travels on a highway, the interaction of the underside of the vehicle with the roadway affects the drag.
- $A$ is some characteristic area of the body. Be careful to note in later sections, just what area is to be used in a given situation. Most often the area of interest is the maximum cross sectional area of the body, sometimes called the *projected area*. Think of what the largest two-dimensional shape would be if you looked straight into the front of your car. That is the area you would use to compute the drag on a car, called the *form drag* or the *pressure drag*. But for very long, smooth shapes such as a passenger train car or a blimp, the surface area may be used. Here we are concerned with the *friction drag* as the air flows along the surface of the vehicle.
- The combined term $\rho v^2/2$, is called the *dynamic pressure* that we define next. But notice that the drag force is proportional to the dynamic pressure and therefore it is proportional to the velocity *squared*. This means, for example, that doubling the velocity for a given object will increase the drag force by a factor of four.

You can visualize the influence of the dynamic pressure on drag by referring to Fig. 17.1, which shows a sphere in a fluid stream. The streamlines depict the path of the fluid as it approaches and flows around the sphere. At point $s$ on the surface of the sphere, the fluid stream is at rest or "stagnant." The term *stagnation point* is used to describe this point. The relationship between the pressure $p_s$ and that in the undisturbed stream at point 1 can be found using Bernoulli's equation along a streamline.

$$\frac{p_1}{\gamma} + \frac{v_1^2}{2g} = \frac{p_s}{\gamma} \tag{17–2}$$

**FIGURE 17.1** Sphere in a fluid stream showing the stagnation point on the front surface and the turbulent wake behind.

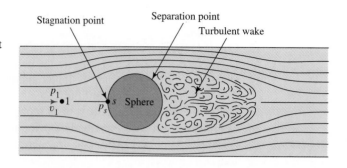

Solving for $p_s$, we get

$$p_s = p_1 + \gamma v_1^2/2g$$

But, since $\rho = \gamma/g$, we have

$$p_s = p_1 + \rho v_1^2/2 \qquad (17\text{--}3)$$

The stagnation pressure is greater than the static pressure in the free stream by the magnitude of the dynamic pressure $\rho v_1^2/2$. The kinetic energy of the moving stream is transformed into a kind of potential energy in the form of pressure.

The increase in pressure at the stagnation point can be expected to produce a force on the body opposing its motion, that is, a drag force. However, the magnitude of the force is dependent not only on the stagnation pressure but also on the pressure at the back side of the body. Since it is difficult to predict the actual variation in pressure on the back side, the drag coefficient is typically used.

The total drag on a body is due to two components. (For a lifting body such as an airfoil, a third component exists as described in Section 17.8.) *Pressure drag* (also called *form drag*) is due to the disturbance of the flow stream as it passes the body, creating a turbulent wake. The characteristics of the disturbance are dependent on the form of the body and sometimes on the Reynolds number of flow and the roughness of the surface. *Friction drag* is due to shearing stresses in the thin layer of fluid near the surface of the body called the *boundary layer*. These two types of drag are described in the following sections.

**17.4**
**PRESSURE DRAG**

As a fluid stream flows around a body, it tends to adhere to the surface for a portion of the length of the body. Then at a certain point, the thin boundary layer separates from the surface, causing a turbulent wake to be formed (see Fig. 17.1). The pressure in the wake is significantly lower than that at the stagnation point at the front of the body. A net force is thus created which acts in a direction opposite to that of the motion. This force is the pressure drag.

If the point of separation can be caused to occur farther back on the body, the size of the wake can be decreased and the pressure drag will be lower. This is the reasoning for streamlining. Figure 17.2 illustrates the change in the wake caused by the elongation and tapering of the tail of the body. Thus, the amount of pressure drag is dependent on the form of the body, and the term *form drag* is often used.

The pressure drag force is calculated from Eq. (17–1) in which *A is taken to be the maximum cross-sectional area of the body perpendicular to the flow.* The coefficient $C_D$ is the pressure drag coefficient.

As an illustration of the importance of streamlining, the value of $C_D$ for the drag on a smooth sphere moving through air with a Reynolds number of approxi-

**FIGURE 17.2** Effect of streamlining on the wake.

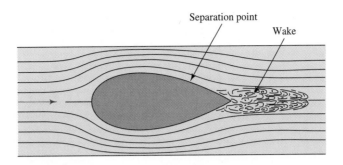

mately $10^5$ is 0.5. A highly streamlined shape like that used in most airships (blimps) has a $C_D$ of approximately 0.04, a reduction by more than a factor of 10!

**17.4.1**
**Properties of Air**

Drag on bodies moving in air is often the goal for drag analysis. In order to use Eq. (17–1) to calculate the drag forces, we need to know the density of the air. As with all gases, the properties of air change drastically with temperature. Also, as altitude above sea level increases, the density decreases. Appendix E presents the properties of air at various temperatures and altitudes.

**17.5**
**DRAG COEFFICIENT**

The magnitude of the drag coefficient for pressure drag depends on many factors, most notably the shape of the body, the Reynolds number of the flow, the surface roughness, and the influence of other bodies or surfaces in the vicinity. Two of the simpler shapes, the sphere and the cylinder, are discussed first.

**17.5.1**
**Drag Coefficient for**
**Spheres and Cylinders**

Data plotted in Fig. 17.3 give the value of the drag coefficient versus Reynolds number for *smooth* spheres and cylinders. For spheres and cylinders, the Reynolds number is computed from the familiar *looking* relation

$$N_R = \frac{\rho v D}{\mu} = \frac{v D}{\nu} \tag{17–4}$$

However, the diameter, $D$, is the diameter of the body itself, rather than the diameter of a flow conduit, which $D$ represented earlier.

Note the very high values of $C_D$ for low Reynolds numbers, over 100 for a smooth sphere at $N_R = 0.10$. This corresponds to motion through very viscous fluids. It drops rapidly to a value of about 4 for $N_R = 10$ and then to 1.0 for $N_R = 100$. The value of $C_D$ ranges from about 0.38 to 0.46 for the higher Reynolds numbers from 1000 to $10^5$.

For cylinders, $C_D \approx 60$ for the very low Reynolds number of 0.10. It drops to a value of 10 for $N_R = 1.0$ and to a value of 1.0 for $N_R = 1000$. In the higher ranges of Reynolds number, $C_D$ ranges from about 0.90 to 1.30 for $N_R$ from 1000 to $10^5$.

For very small Reynolds numbers ($N_R < 1.0$ approximately), the drag is almost entirely due to friction and will be discussed later. At higher Reynolds numbers, the importance of flow separation and the turbulent wake behind the body make pressure drag predominant. The following discussion relates only to pressure drag.

At a value of the Reynolds number of about $2 \times 10^5$, the drag coefficient for spheres drops sharply from approximately 0.42 to 0.17. This is caused by the abrupt change in the nature of the boundary layer from laminar to turbulent. Concurrently,

**FIGURE 17.3**   Drag coefficients for spheres and cylinders.

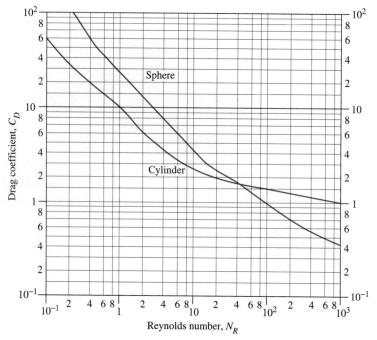

(a) $C_D$ vs. $N_R$ for lower values of $N_R$

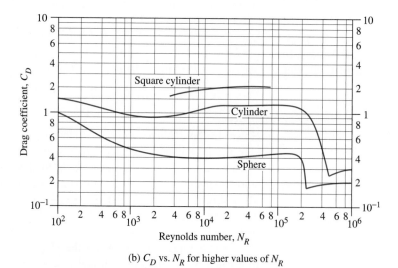

(b) $C_D$ vs. $N_R$ for higher values of $N_R$

the point on the sphere where separation occurs moves farther back, decreasing the size of the wake. For cylinders, a similar phenomenon occurs at approximately $N_R = 4 \times 10^5$ where $C_D$ changes from about 1.2 to 0.30.

Either roughening the surface or increasing the turbulence in the flow stream can decrease the value of the Reynolds number at which the transition from a laminar to a turbulent boundary layer occurs, as illustrated in Fig. 17.4. This graph is meant to show typical curve shapes only and should not be used for numerical values.

Golf balls are dimpled to optimize the turbulence of air as it flows around the ball and to cause the abrupt decrease in the drag coefficient to occur at a low ve-

**FIGURE 17.4**  Effect of turbulence and roughness on $C_D$ for spheres.

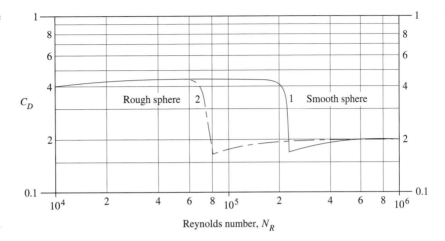

locity (low Reynolds number), resulting in longer flights. A perfectly smooth golf ball could be driven only about 100 yd by even the best golfers, whereas the familiar dimpled design allows the average golfer to far exceed this distance. Highly skilled professional golfers can make 300-yd drives (Reference 3).

### 17.5.2
### Drag Coefficients for Other Shapes

Also shown in Fig. 17.3 is the drag coefficient for a square cylinder with a flat side perpendicular to the flow for Reynolds numbers from $3.5 \times 10^3$ to $8 \times 10^4$. The values range from approximately 1.60 to 2.05, somewhat higher than for the circular cylinder. Significant reductions can be obtained by small to moderate corner radii bringing values of $C_D$ down to as low as 0.55 at high Reynolds numbers. However, the values tend to be highly affected by changes in Reynolds numbers for such designs. Testing is advised.

Figure 17.5 gives data for $C_D$ for three versions of elliptical cylinders for Reynolds numbers from $3.0 \times 10^4$ to $2 \times 10^5$. These shapes have an ellipse for a

**FIGURE 17.5**  Drag coefficients for elliptical cylinders and struts.

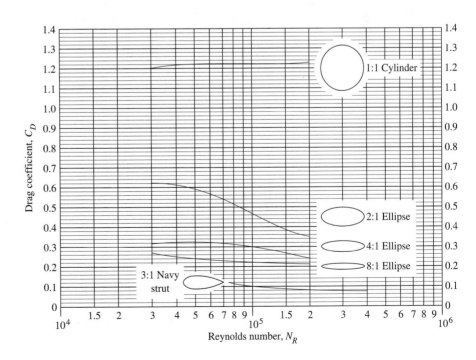

**FIGURE 17.6**   Geometry of the
Navy strut.

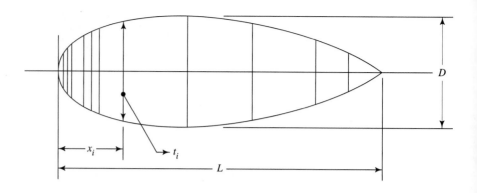

| $x/L$ | 0.00 | .0125 | .025 | .040 | .075 | .100 | .125 | .200 |
|---|---|---|---|---|---|---|---|---|
| $t/D$ | 0.00 | .260 | .371 | .525 | .630 | .720 | .785 | .911 |

| $x/L$ | .400 | .600 | .800 | .900 | 1.00 |
|---|---|---|---|---|---|
| $t/D$ | .995 | .861 | .562 | .338 | 0.00 |

cross section with different ratios of cross-sectional length to maximum thickness, sometimes called *fineness ratio*. Also shown for comparison is the circular cylinder that can be considered as a special case of the elliptical cylinder with a fineness ratio of 1:1. Note the dramatic reduction of drag coefficient to about 0.21 for the elliptical cylinders of high fineness ratio.

But even more reduction in drag coefficient can be made with the familiar "teardrop" shape, also shown in Fig. 17.5. This is a standard shape called a *Navy strut* that has values for $C_D$ in the range of 0.07 to 0.11. Figure 17.6 shows the strut geometry. (See Reference 1.)

Table 17.1 lists values of the drag coefficients for several simple shapes. Note the orientation of the shape relative to the direction of the oncoming flow. The $C_D$ values for such shapes are nearly independent of Reynolds numbers because they have sharp edges that cause the boundary layer to separate at the same place. Most of the testing for these shapes was done in the range of Reynolds numbers from $10^4$ to $10^5$.

The computation of the Reynolds number for the shapes shown in Table 17.1 uses the *length of the body parallel to the flow* as the characteristic dimension for the body. The formula then becomes

$$N_R = \frac{\rho v L}{\mu} = \frac{v L}{\nu} \tag{17–5}$$

For the square cylinder, semitubular cylinders, and triangular cylinders, the data are for models that are long relative to the major thickness dimension. For short cylinders of all shapes, the modified flow around the ends will tend to decrease the values for $C_D$ below those listed in Table 17.1.

**TABLE 17.1** Typical drag coefficients

| Shape of Body | Orientation | | $C_D$ |
|---|---|---|---|
| Square cylinder | Flow → | ◇ | 1.60 |
| Semitubular cylinders | → | ( | 1.12 |
| | → | ) | 2.30 |
| Triangular cylinders | → | ◁ 30° | 1.05 |
| | → | 30° ▷ | 1.85 |
| | → | ◁ | 1.39 |
| | → | 60° ▷ | 2.20 |
| | → | ◁ | 1.60 |
| | → | 90° ▷ | 2.15 |
| | → | ◁ | 1.75 |
| | → | 120° ▷ | 2.05 |
| Rectangular plate | Flow → $b$ ⟷ $a$ | *a/b* 1 | 1.16 |
| | | 4 | 1.17 |
| | | 8 | 1.23 |
| | | 12.5 | 1.34 |
| | | 25 | 1.57 |
| | | 50 | 1.76 |
| | | ∞ | 2.00 |
| Tandem disks $L$ = spacing $d$ = diameter | → ⟵$L$→ $d$ | *L/d* 1 | 0.93 |
| | | 1.5 | 0.78 |
| | | 2 | 1.04 |
| | | 3 | 1.52 |
| One circular disk | → $d$ ○ | | 1.11 |
| Cylinder $L$ = length $d$ = diameter | → ⟵$L$→ $d$ | *L/d* 1 | 0.91 |
| | | 2 | 0.85 |
| | | 4 | 0.87 |
| | | 7 | 0.99 |

**TABLE 17.1**    *Continued*

| Shape of Body | Orientation | $C_D$ |
|---|---|---|
| Hemispherical cup, open back | 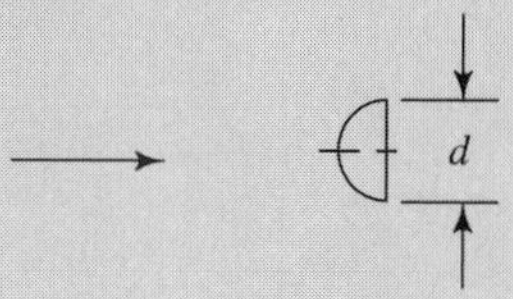 | 0.41 |
| Hemispherical cup, open front | 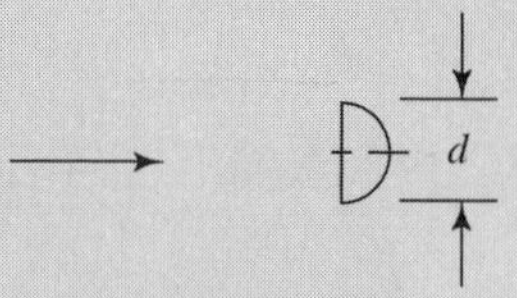 | 1.35 |
| Cone, closed base | 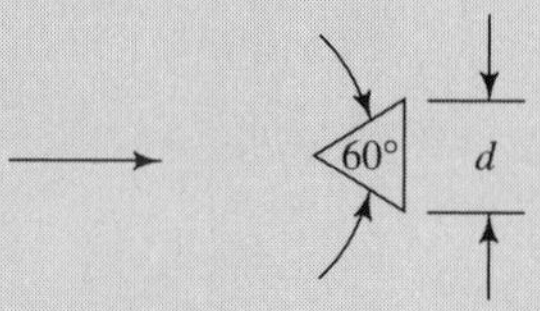 60° | 0.51 |
| | 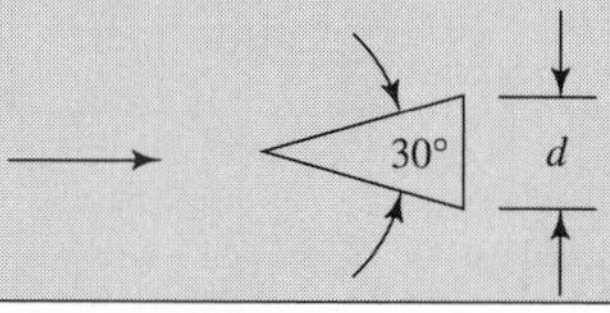 30° | 0.34 |

*Note:* Reynolds numbers are typically from $10^4$ to $10^5$ and are based on the length of the body parallel to the flow direction except for the semitubular cylinders, for which the characteristic length is the diameter. Data adapted from Eugene A. Avallone and Theodore Baumeister III, eds., *Marks' standard handbook for mechanical engineers*, 9th ed. (New York: McGraw-Hill, 1987), Table 4, and W. F. Lindsey, *Drag of cylinders of simple shapes*, Report No. 619 (National Advisory Committee for Aeronautics, 1938).

☐ **EXAMPLE PROBLEM 17.1**

Compute the drag force on a 6.00-ft square bar with a cross section of 4.00 in × 4.00 in when the bar is moving at 4.00 ft/s through water at 40°F. The long axis of the bar and a flat face are placed perpendicular to the flow.

**Solution**

We can use Eq. (17–1) to compute the drag force:

$$F_D = C_D(\rho v^2/2)A$$

Figure 17.3 shows that the drag coefficient depends on the Reynolds number found from Eq. (17–5).

$$N_R = \frac{vL}{\nu}$$

where $L$ is the length of the bar parallel to the flow: 4.0 in or 0.333 ft. The kinematic viscosity of the water at 40°F is $1.67 \times 10^{-5}$ ft$^2$/s. Then

$$N_R = \frac{(4.00)(0.333)}{1.67 \times 10^{-5}} = 8.0 \times 10^4$$

Then, the drag coefficient $C_D = 2.05$. The maximum area perpendicular to the flow, $A$, can now be computed. $A$ can also be described as the projected area seen if you look directly at the bar. In this case, then, the bar is a rectangle 0.333 ft high and 6.00 ft long. That is,

$$A = (0.333 \text{ ft})(6.00 \text{ ft}) = 2.00 \text{ ft}^2$$

The density of the water is 1.94 slugs/ft$^3$. Equivalent units are 1.94 lb·s$^2$/ft$^4$. We can now compute the drag force:

$$F_D = (2.05)(\tfrac{1}{2})(1.94 \text{ lb·s}^2/\text{ft}^4)(4.00 \text{ ft/s})^2(2.00 \text{ ft}^2) = 63.6 \text{ lb}$$

∎

## 17.6 FRICTION DRAG ON SPHERES IN LAMINAR FLOW

A special method of analysis is used for computing friction drag for spheres moving at low velocities in a viscous fluid, which results in very low Reynolds numbers. An important application of this phenomenon is the *falling ball viscometer*, discussed in Chapter 2. As a sphere falls through a viscous fluid, no separation occurs, and the boundary layer remains attached to the entire surface. Therefore, virtually all the drag is due to friction rather than to pressure drag.

In Reference 8, George G. Stokes presents important research on spheres moving through viscous fluids. He found that for Reynolds numbers less than about 1.0, the relationship between the drag coefficient and Reynolds number is $C_D = 24/N_R$. Special forms of the drag force equation can then be developed. The general form of the drag force equation is

$$F_D = C_D\left(\frac{\rho v^2}{2}\right)A$$

Letting $C_D = 24/N_R$ and letting $N_R = vD\rho/\mu$, we get

$$C_D = \frac{24}{N_R} = \frac{24\mu}{vD\rho}$$

Then, the drag force becomes

$$F_D = \frac{24\mu}{vD\rho}\left(\frac{\rho v^2}{2}\right)A = \frac{12\mu vA}{D} \tag{17-6}$$

When computing friction drag, we use the surface area of the object. For a sphere, the surface area is $A = \pi D^2$. Then

DRAG ON A SPHERE RELATED TO SURFACE AREA

$$F_D = \frac{12\mu vA}{D} = \frac{12\mu v(\pi D^2)}{D} = 12\pi\mu vD \tag{17-7}$$

To correlate drag in the low Reynolds number range with that already presented in Section 17.5 dealing with pressure drag, we must redefine the area to be the maximum cross-sectional area of the sphere, $A = \pi D^2/4$. Equation (17–6) then becomes

STOKES'S LAW: DRAG ON A SPHERE RELATED TO CROSS SECTIONAL AREA

$$F_D = \frac{12\mu vA}{D} = \left(\frac{12\mu v}{D}\right)\left(\frac{\pi D^2}{4}\right) = 3\pi\mu vD \tag{17-8}$$

This form for the drag on a sphere in a viscous fluid is commonly called *Stoke's law*. As shown in Fig. 17.3, the relation $C_D = 24/N_R$ plots as a straight line for the low Reynolds numbers.

**17.7**
**VEHICLE DRAG**

Decreasing drag is a major goal in designing most kinds of vehicles, because a significant amount of energy is required to overcome drag as vehicles move through fluids. You are familiar with the streamlined shapes of aircraft bodies and the hulls of ships. Race cars and sports cars have long had the sleek styling characteristic of low aerodynamic drag. More recently, passenger cars and highway trucks have been redesigned to decrease drag.

Many factors affect the overall drag coefficient for vehicles, including the following:

1. The shape of the forward end, or *nose*, of the vehicle
2. The smoothness of the surfaces of the body
3. Such appendages as mirrors, door handles, antennas, and so forth
4. The shape of the tail section of the vehicle
5. The effect of nearby surfaces, such as the ground beneath an automobile
6. Discontinuities, such as wheels and wheel wells
7. The effect of other vehicles nearby
8. The direction of the vehicle with respect to prevailing winds
9. Air intakes to provide engine cooling or ventilation
10. The ultimate purpose of the vehicle (critical for commercial trucks)
11. The accommodation of passengers
12. Visibility afforded to operators and passengers
13. Stability and control of the vehicle
14. Aesthetics (the attractiveness of the design)

**17.7.1**
**Automobiles**

The overall drag coefficient, as defined in Eq. (17–1) based on the maximum projected frontal area, varies widely for passenger cars. Reference 5 lists a nominal mean value of 0.45, with a range of 0.30 to 0.60. Experimental shapes for cars have shown values as low as 0.175. An approximate value of 0.25 is practical for a "low drag" design.

The basic principles of drag reduction for automobiles include providing rounded, smooth contours for the forward part; elimination or streamlining of appendages; blending of changes in contour (such as at the hood/windshield interface); and rounding of rear corners.

---

☐ **EXAMPLE PROBLEM 17.2**

A prototype automobile has an overall drag coefficient of 0.35. Compute the total drag as it moves at 25 m/s through still air at 20°C. The maximum projected frontal area is 2.50 m$^2$.

**Solution**

We will use the drag force equation:

$$F_D = C_D \left( \frac{\rho v^2}{2} \right) A$$

From Appendix E, $\rho = 1.204$ kg/m$^3$. Then

$$F_D = 0.35 \left[ \frac{(1.204)(25)^2}{2} \right] (2.50) = 329 \text{ kg·m/s}^2 = 329 \text{ N}$$

■

**17.7.2**
**Power Required to**
**Overcome Drag**

*Power* is defined as the rate of doing work. When a force is continuously exerted on an object while the object is moving at a constant velocity, power equals force times velocity. Then, the power required to overcome drag is

$$P_D = F_D v$$

Using the data from Example Problem 17.2, we get

$$P_D = (329 \text{ N})(25 \text{ m/s}) = 8230 \text{ N·m/s} = 8230 \text{ W} = 8.23 \text{ kW}$$

In U.S. Customary System units, this would convert to 11.0 hp, a sizable power loss.

**17.7.3**
**Trucks**

The shapes commonly used for trucks fall into the category called *bluff bodies*. Reference 5 indicates that the approximate contribution of various parts of a truck to its total drag are

        70 percent—the design of the front

        20 percent—the design of the rear

        10 percent—friction drag on body surfaces

As with automobiles, rounded smooth contours offer large improvements. For trucks with box-shaped cargo containers, designing corners with a large radius can assist in keeping the boundary layer from separating at the corners, consequently reducing the size of the turbulent wake behind the vehicle and reducing drag. In theory, providing a long, streamlined tail similar to the shape of an aircraft fuselage will reduce drag. However, such a vehicle would be too long to be practical or useful. Newer large highway trucks have drag coefficients in the range from 0.55 to 0.75.

**17.7.4**
**Trains**

Early locomotives had drag coefficients in the range of 0.80 to 1.05 (Reference 1). High-speed, streamlined trains can have values of approximately 0.40. For long passenger and freight trains, skin friction can be significant.

**17.7.5**
**Aircraft**

As with automobiles, wide variations in the overall drag coefficients of aircraft are to be expected with changes in the size and shape to accommodate different uses. For subsonic aircraft, the typical rounded, fairly blunt-nosed design with smooth blends at wings and tail structures and a long-tapered tail section results in drag coefficients of approximately 0.12 to 0.22. At supersonic speeds, the nose is usually sharp to diminish the effect of the shock wave. Operating at much lower speeds, the airshift (dirigible or blimp) has a drag coefficient in the range of 0.04.

**17.7.6**
**Ships**

The total resistance to the motion of floating ships through water is due to skin friction, pressure or form drag, and wave-making resistance. Wave-making resistance, a large contributor to the total resistance, makes analyzing drag on ships quite different from analyzing drag on ground vehicles or aircraft. Reference 1 defines the *total ship resistance*, $R_{ts}$, as the force required to overcome all forms of drag. To normalize the value for different sizes of ships within a given class, values are reported as the ratio $R_{ts}/\Delta$, where $\Delta$ is the displacement of the ship. Representative values of $R_{ts}/\Delta$ are given in Table 17.2. The resistance values can be combined with the speed of the ship $(v)$ to compute the effective power required to propel it through the water:

$$P_E = R_{ts} v \tag{17–9}$$

TABLE 17.2   Resistance of ships

| Ship Type | $R_{ts}/\Delta$ |
|---|---|
| Ocean freighter | 0.001 |
| Passenger liner | 0.004 |
| Tugboat | 0.006 |
| Fast navy ship | 0.01 to 0.12 |

☐ **EXAMPLE PROBLEM 17.3**

Assume that a tugboat has a displacement of 625 long tons (1 long ton = 2240 lb) and is moving through water at 35 ft/s. Compute the total ship resistance and the total effective power required to drive the boat.

*Solution*

From Table 17.2, we find the specific resistance ratio to be $R_{st}/\Delta = 0.006$. Then, the total ship resistance is

$$\Delta = (625 \text{ tons})(2240 \text{ lb/ton}) = 1.4 \times 10^6 \text{ lb}$$
$$R_{ts} = (0.006)(\Delta) = (0.006)(1.4 \times 10^6 \text{ lb}) = 8400 \text{ lb}$$

The power required is

$$P_E = R_{ts}v = (8400 \text{ lb})(35 \text{ ft/s}) = 0.294 \times 10^6 \text{ lb-ft/s}$$

Using 550 lb-ft/s = 1.0 hp, we get

$$P_E = (0.294 \times 10^6)/550 = 535 \text{ hp}$$

∎

**17.7.7 Submarines**

A floating submarine's resistance can be computed in the same way as can a ship's. However, when completely submerged, none of the submarine's motion causes surface waves, and the computation of resistance is similar to that for an aircraft. The hull shape is similar to the shape of an aircraft fuselage, and skin friction plays a major role in the total resistance. Of course, the total *magnitude* of the drag for a submarine is significantly greater than that for an aircraft, because the density of water is far greater than that of air.

**17.8 COMPRESSIBILITY EFFECTS AND CAVITATION**

The results reported in Section 17.5 are for conditions in which the compressibility of the fluid (usually air) has little effect on the drag coefficient. These data are valid if the velocity of flow is less than about one-half the speed of sound in the fluid. Above that speed for air, the character of the flow changes and the drag coefficient increases rapidly.

When the fluid is a liquid such as water, we need not consider compressibility since liquids are very slightly compressible. However, we must consider another phenomenon called *cavitation*. As the liquid flows past a body, the static pressure decreases. If the pressure becomes sufficiently low, the liquid vaporizes, forming bubbles. Since the region of low pressure is generally small, the bubbles burst when they leave that region. When the collapsing of the vapor bubbles occurs near a surface of the body, rapid erosion or pitting results. Cavitation has other adverse effects when it occurs near control surfaces of boats or on propellers. The bubbles in the water decrease the forces exerted on rudders and control vanes and decrease thrust and performance of propellers.

**17.9**
**LIFT AND DRAG**
**ON AIRFOILS**

We define lift as a force acting on a body in a direction perpendicular to that of the flow of fluid. We will discuss the concepts concerning lift with reference to airfoils. The shape of the airfoil comprising the wings of an airplane determines its performance characteristics.

The manner in which an airfoil produces lift when placed in a moving air stream (or when moving in still air) is illustrated in Fig. 17.7. As the air flows over the airfoil, it achieves a high velocity on the top surface with a corresponding decrease in pressure. At the same time the pressure on the lower surface is increased. The net result is an upward force called *lift*. We express the *lift force* $F_L$ as a function of a lift coefficient $C_L$ in a manner similar to that presented for drag:

⇨ **LIFT FORCE**

$$F_L = C_L(\rho v^2/2)A \qquad (17\text{–}10)$$

**FIGURE 17.7** Pressure distribution on an airfoil.

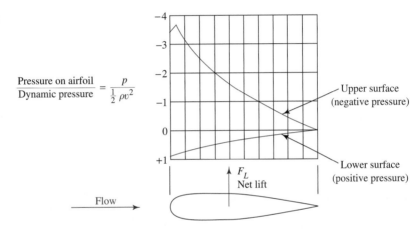

The velocity $v$ is the velocity of the free stream of fluid relative to the airfoil. In order to achieve uniformity in the comparison of one shape with another, we usually define the area $A$ as the product of the span of the wing and the length of the airfoil section called the *chord*. In Fig. 17.8, the span is $b$ and the chord length is $c$.

**FIGURE 17.8** Span and chord lengths for an airfoil.

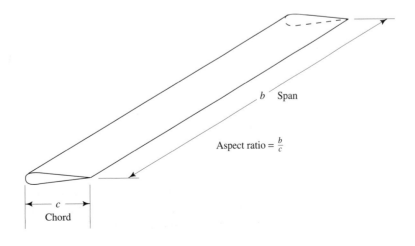

The value of the lift coefficient $C_L$ is dependent on the shape of the airfoil and also on the angle of attack. Figure 17.9 shows that the angle of attack is the angle between the chord line of the airfoil and the direction of the fluid velocity. Other factors affecting lift are the Reynolds number, the surface roughness, the turbulence of the air stream, the ratio of the velocity of the fluid stream to the speed of sound, and the aspect ratio. *Aspect ratio* is the name given to the ratio of the span $b$ of the wing to the chord length $c$. It is important because the characteristics of the flow at the wing tips are different from those toward the center of the span.

**FIGURE 17.9**   Induced drag.

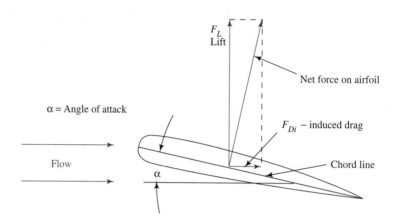

The total drag on an airfoil has three components. Friction drag and pressure drag occur as described before. The third component is called *induced drag*, which is a function of the lift produced by the airfoil. At a particular angle of attack, the net resultant force on the airfoil acts essentially perpendicular to the chord line of the section, as shown in Fig. 17.9. Resolving this force into vertical and horizontal components produces the true lift force $F_L$ and the induced drag $F_{Di}$. Expressing the induced drag as a function of a drag coefficient gives

$$F_{Di} = C_{Di}(\rho v^2/2)A \qquad (17\text{–}11)$$

It can be shown that $C_{Di}$ is related to $C_L$ by the relation

$$C_{Di} = \frac{C_L^2}{\pi(b/c)} \qquad (17\text{–}12)$$

The total drag is then

$$F_D = F_{Df} + F_{Dp} + F_{Di} \qquad (17\text{–}13)$$

Normally, it is the total drag which is of interest in design. We determine a single drag coefficient $C_D$ for the airfoil, from which the total drag can be calculated using the relation

$$F_D = C_D(\rho v^2/2)A \qquad (17\text{–}14)$$

As before, the area $A$ is the product of the span $b$ and the chord length $c$.

We use two methods to present the performance characteristics of airfoil profiles. In Fig. 17.10, the values of $C_L$, $C_D$, and the ratio of lift to drag $F_L/F_D$ are all plotted versus the angle of attack as the abscissa. Note that the scale factors are different for each variable. The airfoil to which the data apply has the designation

**FIGURE 17.10**   Airfoil performance curves.

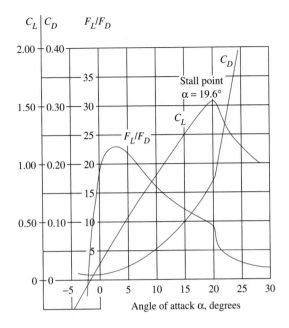

NACA 2409 according to a system established by the National Advisory Committee. NACA Technical Report 610 explains the code used to describe airfoil profiles. NACA Reports 586, 647, 669, 708, and 824 present the performance characteristics of several airfoil sections.

The second method of presenting data for airfoils is shown in Fig. 17.11. This is called the *polar diagram* and is constructed by plotting $C_L$ versus $C_D$ with the angle of attack indicated as points on the curve.

In both Fig. 17.10 and Fig. 17.11 it can be seen that the lift coefficient increases with increasing angle of attack up to a point where it abruptly begins to decrease. This point of maximum lift is called the *stall point*; at this angle of attack, the boundary layer of the air stream separates from the upper side of the airfoil. A large turbulent wake is created, greatly increasing drag and decreasing lift.

**FIGURE 17.11**   Airfoil polar diagram.

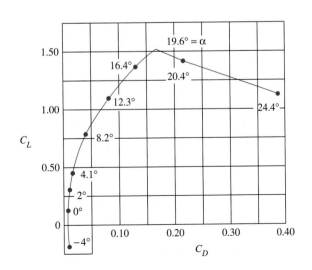

## REFERENCES

1. Avallone, Eugene A., and Theodore Baumeister III, eds. 1996. *Marks' Standard Handbook for Mechanical Engineers.* 10th ed. New York: McGraw-Hill.

2. Daugherty, R. L., and J. B. Franzini. 1985. *Fluid Mechanics With Engineering Applications.* 8th ed. New York: McGraw-Hill.

3. Dunlop Sports Company. 1982. *MAXFLI DDH: The Technical Story.* Greenville, SC: Dunlop Sports Company.

4. Lindsey, W. F., 1938. *Drag of Cylinders of Simple Shapes.* Report No. 619. National Advisory Committee for Aeronautics.

5. Morel, T., and C. Dalton, eds. 1979. *Aerodynamics of Transportation.* New York: The American Society of Mechanical Engineers.

6. Prandtl, L. 1952. *Essentials of Fluid Dynamics.* New York: Hafner Publishing Co.

7. SAE International. 1999. Vehicle Aerodynamics and Wind Noise. Warrendale, PA: SAE International.

8. Stokes, George G. 1901. *Mathematical and Physical Papers.* Vol. 3. London: Cambridge University Press.

9. Streeter, V. L. 1985. *Fluid Mechanics.* 8th ed. New York: McGraw-Hill.

10. von Mises, R. 1959. *Theory of Flight.* New York: Dover Publications. (First published in 1945 by the McGraw-Hill Book Co., New York.)

## PRACTICE PROBLEMS

**17.1M** *A cylinder 25 mm in diameter is placed perpendicular to a fluid stream with a velocity of 0.15 m/s. If the cylinder is 1 m long, calculate the total drag force if the fluid is (a) water at 15°C and (b) air at 10°C and atmospheric pressure.*

**17.2M** *As part of an advertising sign on the top of a tall building, a 2-m diameter sphere called a "weather ball" glows different colors if the temperature is predicted to drop, rise, or remain about the same. Calculate the force on the weather ball due to winds of 15, 30, 60, 120 and 160 km/h if the air is at 0°C.*

**17.3M** *Determine the terminal velocity (see Section 2.6.4) of a 75-mm diameter sphere made of solid aluminum (specific weight = 26.6 kN/m³) in free fall in (a) castor oil at 25°C, (b) water at 25°C, and (c) air at 20°C and standard atmospheric pressure. Consider the effect of buoyancy.*

**17.4M** *Calculate the moment at the base of a flagpole caused by a wind of 150 km/h. The pole is made of three sections, each 3 m long, of different size Schedule 80 steel pipe. The bottom section is 6-in, the middle is 5-in, and the top is 4-in. The air is at 0°C and standard atmospheric pressure.*

**17.5M** *A pitcher throws a baseball without spin with a velocity of 20 m/s. If the ball has a circumference of 225 mm, calculate the drag force on the ball in air at 30°C.*

**17.6M** *A parachute in the form of a hemispherical cup, 1.5 m in diameter, is deployed from a car trying for the land speed record. Determine the force exerted on the car if it is moving at 1100 km/h in air at atmospheric pressure and 20°C.*

**17.7M** *Calculate the required diameter of a parachute in the form of a hemispherical cup supporting a man weigh-* ing 800 N if the terminal velocity (see Section 2.6.4) in air at 40°C is to be 5 m/s.

**17.8M** *A ship tows an instrument in the form of a 30° cone, point first, at 7.5 m/s in sea water. If the base of the cone has a diameter of 2.20 m, calculate the force in the cable to which the cone is attached.*

**17.9M** *A highway sign is being designed to withstand winds of 125 km/h. Calculate the total force on a sign 4 m by 3 m if the wind is flowing perpendicular to the face of the sign. The air is at −10°C. Compare the force calculated for this problem with that for Problem 16.4. Discuss the reasons for the differences.*

**17.10M** *Assuming that a semitrailer behaves as a square cylinder, calculate the force exerted if a wind of 20 km/h strikes it broadside. The trailer is 2.5 m by 2.5 m by 12 m. The air is at 0°C and standard atmospheric pressure.*

**17.11M** *A type of level indicator incorporates four hemispherical cups with open fronts mounted as shown in Fig. 17.12. Each cup is 25 mm in diameter. A motor drives the cups at a constant rotational speed. Calculate the torque that the motor must produce to maintain the motion at 20 r/min when the cups are in (a) air at 30°C and (b) gasoline at 20°C.*

**17.12M** *Determine the wind velocity required to overturn the mobile home sketched in Fig. 17.13 if it is 10 m long and weighs 50 kN. Consider it to be a square cylinder. The air is at 0°C.*

**17.13M** *A bulk liquid transport truck incorporates a cylindrical tank 2 m in diameter and 8 m long. For the tank alone, calculate the pressure drag when the truck is traveling at 100 km/h in still air at 0°C.*

FIGURE 17.12   Problem 17.11.

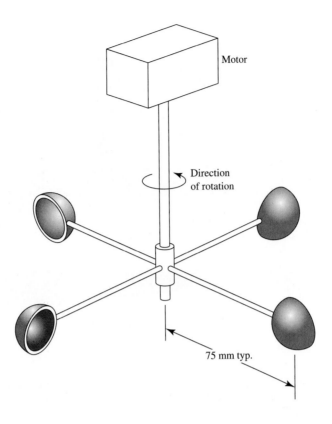

Motor

Direction
of rotation

75 mm typ.

FIGURE 17.13   Problem 17.12.

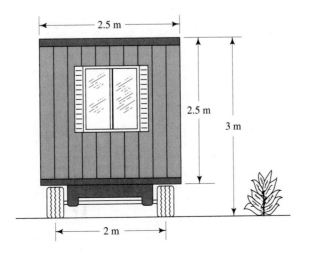

2.5 m

2.5 m

3 m

2 m

**FIGURE 17.14**   Problems 17.14 and 17.15.

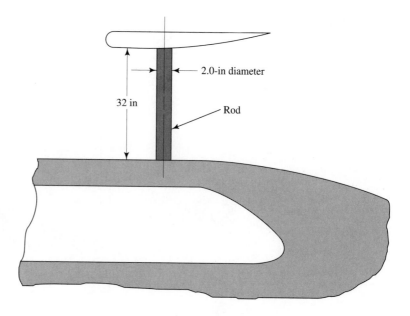

17.14E A wing on a race car is supported by two cylindrical rods, as shown in Fig. 17.14. Compute the drag force exerted on the car due to these rods when the car is traveling through still air at −20°F at a speed of 150 mph.

17.15E In an attempt to decrease the drag on the car shown in Fig. 17.14 and described in Problem 17.14, the cylindrical rods are replaced by elongated elliptical cylinders having a length to breadth ratio of 8:1. By how much will the drag be reduced? Repeat for the *Navy strut*.

17.16E The four designs shown in Fig. 17.15 for the cross section of an emergency flasher lighting system for police vehicles are being evaluated. Each has a length of 60 in and a width of 9.00 in. Compare the drag force exerted on each proposed design when the vehicle moves at 100 mph through still air at −20°F.

17.17E A four-wheel drive utility vehicle incorporates a roll bar that extends above the cab and is in the free stream of air. The bar, made from a 3-in Schedule 40 steel pipe, has a total length of 92 in exposed to the wind. Compute the drag exerted on the vehicle by the bar when the vehicle travels at 65 mph through still air at −20°F.

17.18E An advertising sign for the ABC Paper Company is shown in Fig. 17.16. It is made from three flat disks, each with a diameter of 56.0 in. The disks are joined by 4.50-in diameter tubes measuring 30 in between the disks. Compute the total force on the sign if it is faced into a 100-mph wind. The air is at −20°F.

17.19E An antenna in the shape of a cylindrical rod projects from the top of a locomotive. If the antenna is 42 in long

and 0.200 inches in diameter, compute the drag force on it when the locomotive is traveling at 160 mph in still air at −20°F.

17.20E A ship tows an instrument package in the form of a hemisphere with an open back at a velocity of 25.0 ft/s through seawater at 77°F. The diameter of the hemisphere is 7.25 ft. Compute the force in the cable to which the package is attached.

17.21E A flat rectangular plate, 8.50 × 11.00 in in size, is inserted into lake water at 60°F from a boat moving at 30 mph. What force is required to hold the plate steady relative to the boat with the flat face toward the water?

17.22E The windshield on an antique Stutz Bearcat automobile is a flat circular disk approximately 28 in in diameter. Compute the drag caused by the windshield when the car travels at 60 mph in still air at 40°F.

17.23E Assume that curve 2 in Fig. 17.4 is a true representation of the performance of a golf ball with a diameter of 1.25 in. If the Reynolds number is $1.5 \times 10^5$, compute the drag force on the golf ball and compare it to the drag force on a smooth sphere of the same diameter whose drag coefficient conforms to curve 1. The air is at 40°F.

17.24E In a falling ball viscometer, a steel sphere with a diameter of 1.200 in drops through a heavy syrup and travels 18.0 inches in 20.40 s at a constant speed. Compute the viscosity of the syrup. The syrup has a specific gravity of 1.18. Note that the free body diagram of the sphere should include its weight acting down and the buoyant

**FIGURE 17.15**   Problem 17.16.

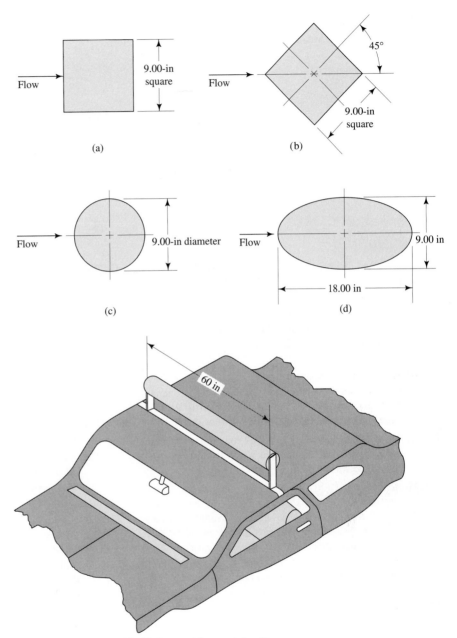

(a)

(b)

(c)

(d)

(e) Pictorial of light assembly mounted on the car

**FIGURE 17.16**    Problem 17.18

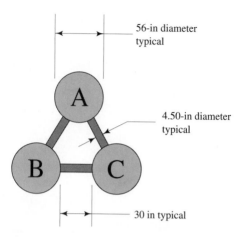

56-in diameter typical

4.50-in diameter typical

30 in typical

force and the drag force acting up. The steel has a specific gravity of 7.83. See also Chapter 2.

**17.25E** Compute the power required to overcome drag on a truck with a drag coefficient of 0.75 when the truck moves at 65 mph through still air at 40°F. The maximum cross section of the truck is a rectangle 8.00 ft wide and 12.00 ft high.

**17.26E** A small, fast boat has a specific resistance ratio of 0.06 (see Table 17.2) and displaces 125 long tons. Compute the total ship resistance and the power required to overcome drag when moving at 50 ft/s in seawater at 77°F.

**17.27E** A passenger liner displaces 8700 long tons. Compute the total ship resistance and the power required to overcome drag when moving at 30 ft/s in seawater at 77°F.

**17.28M** *Assume that Fig. 17.10 shows the performance of the wing on the race car shown in Fig. 17.14. Note that it is mounted in the inverted position, so the lift pushes down to aid in skid resistance. Compute the downward force exerted on the car by the wing and the drag when the angle of attack is set at 15° and the speed is 25 m/s. The chord length is 780 mm and the span is 1460 mm.*

**17.29M** *Calculate the total drag on an airfoil that has a chord length of 2 m and a span of 10 m. The airfoil is at 3000 m flying at (a) 600 km/h and (b) 150 km/h. Use Fig. 17.10 for $C_D$ and $\alpha = 15°$.*

**17.30M** *For the airfoil with the performance characteristics shown in Fig. 17.10, determine the lift and drag at an angle of attack of 10°. The airfoil has a chord length of 1.4 m and a span of 6.8 m. Perform the calculation at a speed of 200 km/h in the standard atmosphere at (a) 200 m and (b) 10 000 m.*

**17.31M** *Repeat Problem 17.30 if the angle of attack is the stall point, 19.6°.*

**17.32M** *For the airfoil in Problem 17.30, what load could be lifted from the ground at a takeoff speed of 125 km/h when the angle of attack is 15°? The air is at 30°C and standard atmosphere pressure.*

**17.33M** *Determine the required wing area for a 1350-kg airplane to cruise at 125 km/h if the airfoil is set at an angle of attack of 2.5°. The airfoil has the characteristics shown in Fig. 17.10. The cruise altitude is 5000 m in the standard atmosphere.*

# 18 Fans, Blowers, Compressors, and the Flow of Gases

## 18.1 The Big Picture

### Discussion Map

- Fans, blowers, and compressors are used to increase pressure and to cause the flow of air and other gases in ducts and piping systems.

- The compressibility of gases requires special methods of analysis of the performance of such devices.

- You should become familiar with recommended ways of evaluating the performance of systems carrying gas flow.

### Discover

- Where have you seen fans, blowers, or compressors used in your home?
- Check in your car.
- Look around stores, offices or a shopping mall.
- If you have access to a factory, look for places where moving air or air under pressure is used.

This chapter will describe some of the more common types of fans, blowers, and compressors. The techniques of design and analysis of selected kinds of gas flow systems will also be discussed.

Fans, blowers, and compressors are used to increase pressure and to cause the flow of air and other gases in a gas flow system. Their function is similar to that of pumps in a liquid flow system, as discussed in Chapter 13. Some of the principles already developed for the flow of liquids and the application of pumps can be applied also to the flow of gases. However, the compressibility of gases causes some important differences.

As you observed where fans and blowers are used in your home, an obvious example is when you use a fan to circulate air when it is too warm for comfort. The fan draws air from the ambient in the room, accelerates it by the action of the fan blades, and delivers it at a higher velocity. The moving air tends to create a cooling effect.

What other examples did you think of? How does your list compare with these?

- Does your home have a forced air heating and air conditioning system? Inspect the unit if it is accessible and discover what causes the flow of air through the ductwork. Perhaps it looks like those shown later in Figures 18.2 or 18.3. The air is drawn from a return air duct into the center of the bladed rotor. The rotation of the rotor imparts energy to the air, throws it outward along the spiral housing and delivers it at higher pressure and velocity to the discharge ductwork.

- For an air conditioner, look for the condensing unit outside the home. A large fan draws ambient air and forces it over the condensing coils to carry heat away from them. The refrigerant in the coils then condenses and passes on to the evaporator of the system. The evaporator fan performs the important function of enhancing the cooling effect of the air.

- Many household appliances incorporate fans and blowers: a hair dryer, a clothes dryer, your computer, most refrigerators (both inside the cabinet and in the machine compartment), the vacuum cleaner, and power drills and saws.

- How many fans did you find on your car? The engine cooling fan is under the hood, moving air over the engine to remove heat from the coolant through the radiator and by direct convection from the engine itself. The car's heater and air conditioning system uses a blower in a similar manner to the home's furnace, air conditioner, and refrigerator.
- Stores, offices, and shopping malls must also use heating systems and air conditioning units to provide a comfortable environment for the people using those spaces. But consider that the overall demands for heating, cooling, and ventilating air are much larger. Therefore the air handling blowers are much larger as well.
- Now consider the factory. In addition to the heating, ventilating, and air conditioning systems (HVAC), there may be a need for high pressure air to operate many of the processes in the plant. Compressed air is used to power screwdrivers, drills, air cylinders (also called linear pneumatic actuators), and other pneumatic devices. Larger plants typically employ central compressors to deliver a steady supply of air at approximately 100 psi (690 kPa) to all parts of the factory. Individual work cells can tap into the system as needed. Perhaps the central compressor looks like that in Figure 18.5.

Special techniques for the design of flow systems carrying gases, such as air, have been developed by professionals after years of experience. The detailed analysis of the phenomena involved requires knowledge of thermodynamics. Because knowledge of thermodynamics is not necessary for the readers of this book, some of the methods in this chapter are presented without extensive development. New terms or concepts required for understanding the methods are, of course, described.

## 18.2 OBJECTIVES

After completing this chapter, you should be able to:

1. Describe the general characteristics of fans, blowers, and compressors.
2. Describe propeller fans, duct fans, and centrifugal fans.
3. Describe blowers and compressors of the centrifugal, axial, vane-axial, reciprocating, lobed, vane, and screw types.
4. Specify suitable sizes for pipes carrying steam, air, and other gases at higher pressures.
5. Compute the flow rate of air and other gases through nozzles.

## 18.3 GAS FLOW RATES AND PRESSURES

When working in U.S. Customary System units, the flow rate of air or another gas is most frequently expressed in ft$^3$/min, abbreviated cfm. Velocities are typically reported in ft/min. Although these are not the standard units for the U.S. Customary System, they are convenient for the range of flows typically encountered in residential, commercial, or industrial applications.

In the SI system, m$^3$/s for flow rate and m/s for velocity are the units most often used. For systems carrying relatively low flow rates, the unit of L/s is sometimes used. Convenient conversions are listed below.

$$1.0 \text{ ft}^3/\text{s} = 60 \text{ ft}^3/\text{min} = 60 \text{ cfm}$$
$$1.0 \text{ m}^3/\text{s} = 2120 \text{ ft}^3/\text{min} = 2120 \text{ cfm}$$
$$1.0 \text{ ft/s} = 60 \text{ ft/min}$$
$$1.0 \text{ m/s} = 3.28 \text{ ft/s}$$
$$1.0 \text{ m/s} = 197 \text{ ft/min}$$

Pressures can be measured in lb/in$^2$ (abbreviated psi) in U.S. Customary System units when relatively large pressures are encountered. However, in most air handling systems, the pressures are small and are measured in *inches of water gage*, abbreviated as in $H_2O$. This unit is derived from the practice of using a pitot tube and water manometer to measure the pressure in ducts, as illustrated in Fig. 15.18. The equivalent pressure can be derived from the pressure-elevation relation, $\Delta p = \gamma h$. If we use $\gamma = 62.4$ lb/ft$^3$ for water, a pressure of 1.00 in $H_2O$ is equivalent to

$$\Delta p = \gamma h = \frac{62.4 \text{ lb}}{\text{ft}^3} \cdot 1.00 \text{ in} \cdot \frac{1 \text{ ft}^3}{1728 \text{ in}^3} = 0.0361 \text{ lb/in}^2$$

Stated differently, 1.0 psi = 27.7 in $H_2O$. In many air flow systems, the pressures involved are only a few inches of water or even a fraction of an inch.

The standard SI unit of Pascals (Pa) is itself quite small and is used directly when designing a system in SI units. Also used are bars, mm of $H_2O$, and mm of mercury. Some useful conversion factors are listed below.

$$1.0 \text{ bar} = 100 \text{ kPa}$$
$$1.0 \text{ psi} = 6895 \text{ Pa}$$
$$1.0 \text{ in } H_2O = 248.8 \text{ Pa}$$
$$1.0 \text{ mm } H_2O = 9.81 \text{ Pa}$$
$$1.0 \text{ mm of mercury} = 132.8 \text{ Pa}$$

**18.4**
**CLASSIFICATION OF FANS, BLOWERS, AND COMPRESSORS**

Fans, blowers, and compressors are all used to increase the pressure of and move air or other gases. The primary differences among them are their physical construction and the pressures that they are designed to develop. A *fan* is designed to operate against small static pressures, up to about 2.0 psi (13.8 kPa). But typical operating pressures for fans are from 0 to 6 in $H_2O$ (0.00 to 0.217 psi or 0.00 to 1500 Pa). At pressures from 2.0 psi up to approximately 10.0 psi (69.0 kPa), the gas mover is called a *blower*. To develop higher pressures, even as high as several thousand psi, *compressors* are used (see Reference 4).

Fans are used to circulate air within a space, to bring air into or exhaust it from a space, or to move air through ducts in ventilation, heating, or air conditioning systems. Types of fans include propeller fans, duct fans, and centrifugal fans.

*Propeller fans* operate at virtually zero static pressure and are composed of two to six blades with the appearance of aircraft propellers. Thus, they draw air in from one side and discharge it from the other side in an approximately axial direction. This type of fan is popular for circulating air in living or working spaces to improve comfort. When mounted in windows or other openings in the walls of a building, they deliver fresh outside air into the building or exhaust air from the building. When mounted in the ceiling or roof, they are often called *ventilators*.

Propeller fans are available from small sizes (a few inches in diameter, delivering a few hundred cfm) to 60 in or more in diameter, delivering over 50 000 cfm at zero static pressure. Operating speeds typically range from about 600 rpm to 1725 rpm. These fans are driven by electric motors, either directly or through belt drives.

*Duct fans* have a construction similar to that of propeller fans, except that the fan is mounted inside a cylindrical duct, as shown in Fig. 18.1. The duct could be a part of a larger duct system delivering air to or exhausting it from a remote area. Duct fans can operate against static pressures up to about 1.50 in $H_2O$ (375 Pa). Sizes range from very small, delivering a few hundred cfm, to about 36 in, delivering over 20 000 cfm.

Two examples of *centrifugal fans* or *centrifugal blowers*, along with their rotors, are shown in Figs. 18.2 and 18.3. Air enters at the center of the rotor, also

**FIGURE 18.1**    Duct fan. (Source: Hartzell Fan, Inc., Piqua, OH)

called the *impeller*, and is thrown outward by the rotating blades, thus adding kinetic energy. The high velocity gas is collected by the volute surrounding the rotor, where the kinetic energy is converted into an increased gas pressure for delivery through a duct system for ultimate use.

The construction of the rotor is typically one of four basic designs, as shown in Fig. 18.4. The *backward-inclined blade* is often made with simple flat plates. As the rotor rotates, the air tends to leave parallel to the blade along the vector called $v_b$ in the figure. However, that is added vectorially to the tangential velocity of the blade itself, $v_t$, which gives the resultant velocity shown as $v_R$. The *forward-curved blades* yield a generally higher resultant air velocity, because the two component vectors are more nearly in the same direction. For this reason, a rotor with forward

**FIGURE 18.2**    Centrifugal blower with straight, radial bladed rotor. (Source: Hartzell Fan, Inc., Piqua, OH)

**FIGURE 18.3** Centrifugal blower with backward inclined bladed rotor. (Source: Hartzell Fan, Inc., Piqua, OH)

**FIGURE 18.4** Four types of centrifugal fan rotors.

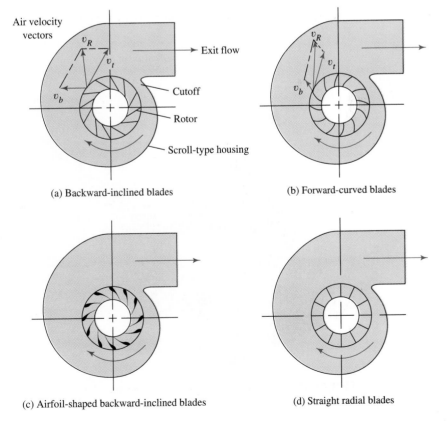

(a) Backward-inclined blades

(b) Forward-curved blades

(c) Airfoil-shaped backward-inclined blades

(d) Straight radial blades

curved blades will run at a slower speed than a similar sized backward-inclined bladed fan for the same air flow and pressure. However, the backward-inclined fan typically requires lower power for the same service (Reference 9). *Airfoil shaped, backward-inclined fan blades* operate more quietly and efficiently than flat, backward-inclined blades. All these types of fans are used for ventilation systems and some industrial process uses. *Radial-blade* fans have many applications in industry for supplying large volumes of air at moderate pressures for boilers, cooling towers, material dryers, and bulk material conveying.

*Centrifugal compressors* employ impellers similar to those in centrifugal pumps (Figs. 15.7 and 15.8). However, the specific geometry is tailored to handle gas rather than liquid. Figure 18.5 shows a large, single-stage, centrifugal compressor. When a single-rotor compressor cannot develop a sufficiently high pressure, a multistage compressor, as shown in Fig. 18.6, is used. Centrifugal compressors are used for flows from about 500 to 100 000 cfm (0.24 to 47 m³/s) at pressures as high as 8000 psi (55 MPa).

A multistage *axial compressor* is shown in Fig. 18.7. Only the lower half of the casing, the multistaged rotor, and the shaft assembly are shown. The gas is drawn into the large end, moved axially and compressed by the series of bladed rotors, and discharged from the small end. Axial compressors are employed to deliver large flow rates, approximately 8000 to 1.0 million cfm (3.8 to 470 m³/s) at a discharge pressure up to 100 psi (690 kPa).

*Vaneaxial blowers* are similar to duct fans, described earlier, except they typically have blades that are airfoil-shaped and include vanes within the cylindrical housing to redirect the flow axially within the following duct. This results in higher static pressure capability for the blower and reduces swirl of the air.

*Positive displacement blowers and compressors* come in a variety of designs:

- Reciprocating—single-acting or double-acting
- Rotary—lobe, vane, or screw

**FIGURE 18.5**   Single-stage centrifugal compressor. (Source: Dresser-Rand, Turbo Products Division, Olean, NY)

**FIGURE 18.6**   Lower case of a
horizontally split multistage centrifugal compressor with rotor installed.
(Source: Dresser-Rand, Turbo
Products Division, Olean, NY)

The construction of a reciprocating compressor looks similar to that of an engine.
A rotating crank and a connecting rod drive the piston. The piston reciprocates within
its cylinder, drawing in low pressure gas as it travels away from the cylinder head
and then compressing it within the cylinder as it travels toward the head. When the
pressure of the gas reaches the desired level, discharge valves open to deliver the
compressed gas to the piping system. Figure 13.5 shows the arrangement of both
single-acting and double-acting pistons. Small versions of such compressors are seen

**FIGURE 18.7**   Lower case of an
axial compressor with rotor installed.
(Source: Dresser-Rand, Turbo
Products Division, Olean, NY)

in small shops and service stations. However, for many industrial users, they can be very large, delivering up to 10 000 cfm (4.7 m³/s) at pressures up to 60 000 psi (413 MPa).

Lobe and vane compressors appear very similar to the pumps shown in Figs. 13.1 and 13.4. Lobe-type designs can develop up to approximately 15 psi (100 kPa) and are often called blowers. Vane type compressors are capable of developing several hundred psi and are often used in pneumatic fluid power systems.

*Screw compressors* are used in construction and industrial applications requiring compressed air up to 500 psi (3.4 MPa) with delivery up to 20 000 cfm (9.4 m³/s). In the single screw design, air is trapped between adjacent "threads" rotating inside a close-fitting housing. The axial progression of the threads delivers the air to the outlet. In some designs, the pitch of the threads decreases along the length of the screw, providing compression within the housing as well as delivering it against system resistance. Two or more screws in a mesh can also be employed. See Fig. 13.2.

### 18.5 FLOW OF COMPRESSED AIR AND OTHER GASES IN PIPES

Many industries use compressed air in fluid power systems to power production equipment, material handling devices, and automation machinery. A common operating pressure for such systems is in the range of 60 to 125 psig (414 to 862 kPa gage). Performance and productivity of the equipment are degraded if the pressure drops below the design pressure. Therefore, careful attention must be paid to pressure losses between the compressor and the point of use. A detailed analysis of the piping system should be completed, using the methods outlined in Chapters 6 to 12 of this book, modified for the compressibility of air.

When large changes in pressure or temperature of the compressed air occur along the length of a flow system, the corresponding changes in the specific weight of the air should be taken into account. However, if the change in pressure is less than about 10 percent of the inlet pressure, variations in specific weight will have negligible effect. When the pressure drop is between 10 percent and 40 percent of the inlet pressure, we can use the average of the specific weight for the inlet and outlet conditions to produce results with reasonable accuracy (Reference 5). When the predicted pressure change is greater than 40 percent, either redesign the system or consult other references.

### 18.5.1 Specific Weight for Air

Figure 18.8 shows the variation of the specific weight for air as a function of changes in pressure and temperature. Note the large magnitude of the changes, particularly as pressure changes. The specific weight for any conditions of pressure and temperature can be computed from the *ideal gas law* from thermodynamics that states

IDEAL GAS LAW

$$\frac{p}{\gamma T} = \text{constant} = R \tag{18–1}$$

where    $p$ = absolute pressure of the gas

$\gamma$ = specific weight of the gas

$T$ = absolute temperature of the gas, that is, the temperature above absolute zero

$R$ = gas constant for the gas being considered

Also, Eq. (18–1) can be solved for the specific weight:

$$\gamma = \frac{p}{RT} \tag{18–2}$$

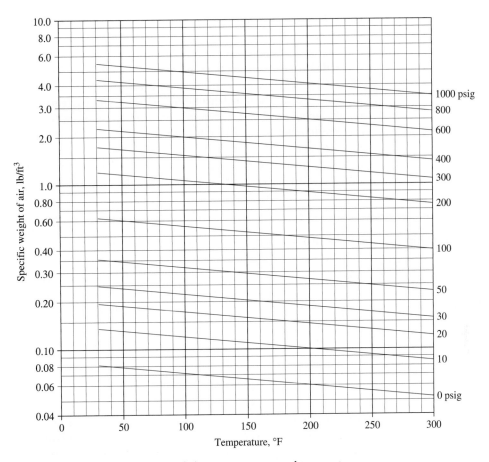

**FIGURE 18.8** Specific weight of air versus pressure and temperature.

The absolute temperature is found by adding a constant to the measured temperature. In U.S. Customary System units,

$$T = (t°F + 460) °R$$

where °R is degrees in Rankine, the standard unit for absolute temperature measured relative to absolute zero. In SI units,

$$T = (t°C + 273) K$$

where K (Kelvin) is the standard SI unit for absolute temperature.

As presented in Section 3.3 (Eq. 3–2), absolute pressure is found by adding the prevailing atmospheric pressure to the gage pressure reading. We use $p_{atm} = 14.7$ psia in U.S. Customary System units and $p_{atm} = 101.3$ kPa absolute in SI units unless the true local atmospheric pressure is known.

The value of the gas constant $R$ for air is 53.3 ft·lb/lb·°R in U.S. Customary System units. The numerator of the unit for $R$ is energy in foot pounds (ft·lb), and thus the pound (lb) unit indicates *force*. The corresponding SI unit for energy is the newton meter (N·m), where

$$1.0 \text{ ft·lb} = 1.356 \text{ N·m}$$

The pound unit in the denominator for $R$ is the weight of the air, also a force. To convert the U.S. Customary System unit °R to the SI unit K, we use 1.0 K = 1.8°R.

Using these conversions, we can show that the value of $R$ for air in the SI system is 29.2 N·m/N·K. Appendix N shows values for $R$ for other gases.

In choosing to express the values for the gas constant $R$ in terms of *weight*, we facilitate the computation of the *specific weight* $\gamma$, as shown in Example Problem 18.1. Later, we will use $\gamma$ in calculations for the *weight flow rate* of a gas through a nozzle. It should be noted that the use of the resulting relationships based on weight should be confined to applications near the earth's surface, where the value of the acceleration due to gravity, $g$, is fairly constant.

In other types of analysis, particularly those in the field of thermodynamics, $R$ is defined in terms of *mass* rather than weight. In aerospace environments, in which $g$ and the weight can approach zero, the mass basis should also be used.

☐ **EXAMPLE PROBLEM 18.1**　Compute the specific weight of air at 100 psig and 80°F.

**Solution**　Using Eq. 18–2, we find

$$p = p_{atm} + p_{gage} = 14.7 \text{ psia} + 100 \text{ psig} = 114.7 \text{ psia}$$
$$T = t + 460 = 80°F + 460 = 540°R$$

Then,

$$\gamma = \frac{p}{RT} = \frac{114.7 \text{ lb}}{\text{in}^2} \cdot \frac{\text{lb·°R}}{53.3 \text{ ft·lb}} \cdot \frac{1}{540°R} \cdot \frac{144 \text{ in}^2}{\text{ft}^2}$$
$$\gamma = 0.574 \text{ lb/ft}^3$$

Note that the quantities 53.3 and 144 will always be used for air in this type of calculation. Then, a convenient, unit-specific equation for specific weight of air can be derived as follows:

$$\gamma = 2.70 p / T \qquad (18\text{–}3)$$

This gives $\gamma$ for air directly in pounds per cubic foot (lb/ft³) when the absolute pressure is in psia and the absolute temperature is in °R.

☐ **EXAMPLE PROBLEM 18.2**　Compute the specific weight of air at 750 kPa gage and 30°C.

**Solution**　Using Eq. (18–2), we get

$$p = p_{atm} + p_{gage} = 101.3 \text{ kPa} + 750 \text{ kPa} = 851.3 \text{ kPa}$$
$$T = t + 273 = 30°C + 273 = 303 \text{ K}$$

Then,

$$\gamma = \frac{p}{RT} = \frac{851.3 \times 10^3 \text{ N}}{\text{m}^2} \cdot \frac{\text{N·K}}{29.2 \text{ N·m}} \cdot \frac{1}{303 \text{ K}} = 96.2 \text{ N/m}^3$$

■

**18.5.2**
**Flow Rates for**
**Compressed Air Lines**

Ratings for equipment using compressed air and for compressors delivering the air are given in terms of *free air*, sometimes called *free air delivery* (f.a.d.). This gives the quantity of air delivered per unit time assuming that the air is at standard atmospheric pressure (14.7 psia or 101.3 kPa absolute) and at the standard temperature of 60°F or 15°C (absolute temperatures of 520°R or 285 K). To determine the flow rate at other conditions, the following equation can be used:

$$Q_a = Q_s \cdot \frac{P_{\text{atm}-s}}{P_{\text{atm}} + p_a} \cdot \frac{T_a}{T_s}$$  (18–4)

where
$Q_a$ = volume flow rate at actual conditions
$Q_s$ = volume flow rate at standard conditions
$P_{\text{atm}-s}$ = standard absolute atmospheric pressure
$P_{\text{atm}}$ = actual absolute atmospheric pressure
$p_a$ = actual gage pressure
$T_a$ = actual absolute pressure
$T_s$ = standard absolute temperature = 520°R or 285 K

Using these values and those of the standard atmosphere, we can write Eq. (18–4) as follows.

In U.S. Customary System units:

$$Q_a = Q_s \cdot \frac{14.7 \text{ psia}}{P_{\text{atm}} + p_a} \cdot \frac{(t + 460)°R}{520°R}$$  (18–4a)

In SI units:

$$Q_a = Q_s \cdot \frac{101.3 \text{ kPa}}{P_{\text{atm}} + p_a} \cdot \frac{(t + 273)K}{285 \text{ K}}$$  (18–4b)

---

☐ **EXAMPLE PROBLEM 18.3**    An air compressor has a rating of 500 cfm free air. Compute the flow rate in a pipe line in which the pressure is 100 psig and the temperature is 80°F.

*Solution*    Using Eq. (18–4a) and assuming that the local atmospheric pressure is 14.7 psia, we get

$$Q_a = 500 \text{ cfm} \cdot \frac{14.7 \text{ psia}}{14.7 + 100} \cdot \frac{(80 + 460)}{520} = 66.5 \text{ cfm}$$

■

---

**18.5.3**
**Pipe Size Selection**

Many factors must be considered to specify a suitable pipe size for carrying compressed air in industrial plants. Some of those factors and the parameters involved follow.

- *Pressure drop.* Because friction losses are proportional to the square of the velocity of flow, it is desirable to use as large a pipe size as is feasible, to ensure adequate pressure at all points of use in a system.
- *Compressor power requirement.* The power required to drive the compressor increases as the pressure drop increases. Therefore, it is desirable to use large pipes to minimize pressure drop.
- *Cost of piping.* Large pipes cost more than small pipes, which makes the use of smaller pipes preferable.
- *Cost of the compressor.* In general, a compressor designed to operate at a higher pressure will cost more, which makes the use of larger pipes that minimize pressure drop preferable.

- *Installation cost.* Small pipes are easier to handle, but that is usually not a major factor.
- *Space required.* Small pipes take less space and provide less interference with other equipment or operations.
- *Future expansion.* To allow the addition of more air-using equipment in the future, large pipes are desirable.
- *Noise.* When air flows at a high velocity through pipes, valves, and fittings, it creates high noise level. It is desirable to use large pipes to permit lower velocities.

There is no clearly optimum pipe size for every installation, and the designer should evaluate the overall performance of several proposed sizes before making a final specification. To aid in beginning the process, Table 18.1 lists some suggested sizes.

As with other pipe line systems studied earlier, compressed air pipe systems typically contain valves and fittings to control the amount and direction of flow. We account for their effects by using the equivalent length technique, described in Section 10.10. Values for the $L_e/D$ ratio are listed in Table 10.4.

Figure 18.9 shows a sketch of a typical layout of a piping system serving an industrial operation. The following are its basic features.

- The compressor draws ambient air and increases its pressure for delivery to the system.
- The aftercooler conditions the air.
- The compressed air goes to a receiver that has a relatively large volume to ensure that an even supply of air is available to the system.
- A trap is installed before the receiver to remove moisture.
- Piping serving the factory systems is laid out in a loop arrangement.

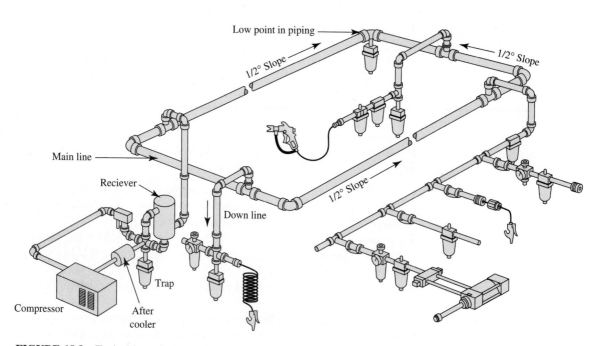

**FIGURE 18.9**   Typical layout of the piping for an industrial compressed air system.

- Connections to the loop are made at the top of the main loop piping to inhibit the delivery of moisture to branch lines and the tools used there.
- Piping in the loop system slopes away from the compressor with one or more traps to remove moisture at low points in the system.
- Branch lines are sized to carry their given flow rates with the same nominal velocity as in the loop system.
- Pressure regulators are installed in branch lines to enable the adjustment of pressure to suit the tools used in that line.

☐ **EXAMPLE PROBLEM 18.4**

Specify a suitable size pipe for the delivery of 500 cfm (free air) at 100 psig at 80°F to an automated machine. The total length of straight pipe required between the compressor and the machine is 140 ft. The line also contains two fully open gate valves, six standard elbows, and two standard tees, in which the flow is through the run of the tee. Then, analyze the pressure required at the compressor to ensure that the pressure at the machine is no less than 100 psig.

**Solution**

As a tentative choice, let's consult Table 18.1 and specify a $1^1/_2$-inch Schedule 40 steel pipe to carry the air. Then, from Appendix F we find $D = 0.1342$ ft and $A = 0.01414$ ft$^2$. We should now check to determine the actual pressure drop through the system and judge its acceptability. The solution procedure is similar to that used in Chapter 11. Special circumstances relative to air will be discussed.

*Step 1.* Write the energy equation between the outlet from the compressor and the inlet to the machine:

$$\frac{p_1}{\gamma_1} + z_1 + \frac{v_1^2}{2g} - h_L = \frac{p_2}{\gamma_2} + z_2 + \frac{v_2^2}{2g}$$

Note that the specific weight terms have been identified with subscripts for the reference points. Because air is compressible, there could be a significant change in specific weight.

**TABLE 18.1** Suggested compressed air system pipe sizes

| Maximum Flow Rate (cfm) | | Pipe Size (Schedule 40) |
|---|---|---|
| Free Air | Compressed Air (100 psig, 60°F) | |
| 4 | 0.513 | $^1/_8$ |
| 8 | 1.025 | $^1/_4$ |
| 20 | 2.563 | $^3/_8$ |
| 35 | 4.486 | $^1/_2$ |
| 80 | 10.25 | $^3/_4$ |
| 150 | 19.22 | 1 |
| 300 | 38.45 | $1^1/_4$ |
| 450 | 57.67 | $1^1/_2$ |
| 900 | 115.3 | 2 |
| 1400 | 179.4 | $2^1/_2$ |
| 2500 | 320.4 | 3 |
| 3500 | 448.6 | $3^1/_2$ |
| 5000 | 640.8 | 4 |

*Note:* The sizes listed are the smallest standard Schedule 40 steel pipes that will carry the given flow rate at a pressure of 100 psig (690 kPa) with no more than 5.0 psi (34.5 kPa) pressure drop in 100 ft (30.5 m) of pipe. See Appendix F for pipe dimensions.

However, it is our intention in this design to have a small change in pressure between points 1 and 2. If this is achieved, the change in specific weight can be ignored. Therefore, let $\gamma_1 \approx \gamma_2$. The conditions at point 2 are the same as those used in Example Problem 18.1. Then, we will use $\gamma = 0.574$ lb/ft$^3$.

No information was given about the elevations of the compressor and the machine. Because the specific weight of air and other gases is so small, it is permissible to ignore elevation differences when dealing with the flow of gases, unless these differences are very large. As indicated in Sections 3.4 and 3.5, the pressure change is directly proportional to the specific weight of the fluid and to the change in elevation. For $\gamma = 0.574$ lb/ft$^3$ for air in this problem, an elevation change of 100 ft (about the height of a 10-story building), would change the pressure only 0.40 psi.

The velocity at the two reference points will be the same, because we will use the same size pipe throughout. Then, the velocity head terms can be cancelled from the energy equation.

*Step 2.* Solve for the pressure at the compressor:

$$p_1 = p_2 + \gamma h_L$$

*Step 3.* Evaluate the energy loss $h_L$ by using Darcy's equation, and include the effects of minor losses:

$$h_L = f\left(\frac{L}{D}\right)\left(\frac{v^2}{2g}\right) + f_T\left(\frac{L_e}{D}\right)\left(\frac{v^2}{2g}\right)$$

The $L/D$ term is the actual pipe length to flow diameter ratio.

$$\text{Pipe: } L/D = (140 \text{ ft}/0.1342 \text{ ft}) = 1043$$

The equivalent $L_e/D$ for the valves and fittings are found in Table 10.4.

$$
\begin{array}{rl}
2 \text{ valves:} & L_e/D = 2(8) = 16 \\
6 \text{ elbows:} & L_e/D = 6(30) = 180 \\
2 \text{ tees:} & L_e/D = 2(20) = 40 \\
\text{Total:} & L_e/D = 236
\end{array}
$$

The flow velocity can be computed from the continuity equation. It was determined in Example Problem 18.3 that the flow rate of 500 cfm of free air at actual conditions of 100 psig and 80°F is 66.5 cfm. Then,

$$v = \frac{Q}{A} = \frac{66.5 \text{ ft}^3}{\text{min}} \cdot \frac{1}{0.01414 \text{ ft}^2} \cdot \frac{1 \text{ min}}{60 \text{ s}} = 78.4 \text{ ft/s}$$

The velocity head is

$$\frac{v^2}{2g} = \frac{(78.4)^2 \text{ ft}^2/\text{s}^2}{2(32.2 \text{ ft/s}^2)} = 95.44 \text{ ft}$$

To evaluate the friction factor $f$, the density and viscosity of the air is needed. Knowing the specific weight of the air, we can compute the density from

$$\rho = \frac{\gamma}{g} = \left(\frac{0.574 \text{ lb}}{\text{ft}^3}\right)\left(\frac{\text{s}^2}{32.2 \text{ ft}}\right) = \frac{0.0178 \text{ lb}\cdot\text{s}^2}{\text{ft}^4} = 0.0178 \text{ slug/ft}^3$$

The dynamic viscosity of a gas does not change much as pressure changes. So, we can use the data from Appendix E, even though they are for standard atmospheric pressure. The dynamic viscosity is found to be $\mu = 3.85 \times 10^{-7}$ lb·s/ft$^2$.

It would be incorrect to use the kinematic viscosity listed for air in Appendix E, because that value includes the density, which is dramatically different at 100 psig than it is at atmospheric pressure.

We can now compute the Reynolds number:

$$N_R = \frac{vD\rho}{\mu} = \frac{(78.4)(0.1342)(0.0178)}{3.85 \times 10^{-7}} = 4.86 \times 10^5$$

The relative roughness $D/\epsilon$ is

$$D/\epsilon = 0.1342/1.5 \times 10^{-4} = 895$$

Then, from the Moody diagram (Fig. 9.2), we read $f = 0.021$. The value for $f_T$ used for the valves and fittings can be found from Table 10.5 to be 0.021 for the $1\frac{1}{2}$ in Schedule 40 pipe. Because this is equal to the friction factor for the pipe itself, the $L/D$ for the pipe can be added to the total $L_e/D$ for the valves and fittings.

$$(L_e/D)_{\text{total}} = 1043 + 236 = 1279$$

And now the energy loss can be computed:

$$h_L = f_T\left(\frac{L_e}{D}\right)_{\text{total}}\left(\frac{v^2}{2g}\right) = (0.021)(1279)(95.44) = 2563 \text{ ft}$$

*Step 4.* Compute the pressure drop in the pipe line.

$$p_1 - p_2 = \gamma h_L = \frac{0.574 \text{ lb}}{\text{ft}^3} \cdot 2563 \text{ ft} \cdot \frac{1 \text{ ft}^2}{144 \text{ in}^2} = 10.22 \text{ psi}$$

*Step 5.* Compute the pressure at the compressor.

$$p_1 = p_2 + 10.22 \text{ psi} = 100 \text{ psig} + 10.22 \text{ psi} = 110.2 \text{ psig}$$

*Step 6.* Because the change in pressure is less than 10 percent, the assumption that the specific weight of the air is constant is satisfactory. If a larger pressure drop had occurred, we could either redesign the system with a larger pipe size or adjust the specific weight to the average of those at the beginning and end of the system. This system design appears to be satisfactory with regard to pressure drop.

■

**18.6
FLOW OF AIR AND
OTHER GASES
THROUGH NOZZLES**

The typical design of a nozzle is a converging section through which a fluid flows from a region of higher pressure to a region of lower pressure. Figure 18.10 shows a nozzle installed in the side of a relatively large tank with flow from the tank to the atmosphere. The nozzle shown converges smoothly and gradually, terminating at its smallest section, called the *throat*. Other designs for nozzles, including abrupt orifices and those attached to smaller pipes at the inlet, require special treatment, as we will discuss later.

Some concepts from the field of thermodynamics must now be discussed, along with some additional properties of gases.

When the flow of a gas occurs very slowly, heat from the surroundings can transfer to or from the gas to maintain its temperature constant. Such flow is called *isothermal*. However, when the flow occurs very rapidly or when the system is very well insulated, very little heat can be transferred to or from the gas. Under ideal conditions with *no* heat transferred, the flow is called *adiabatic*. Real systems

**FIGURE 18.10**   Discharge of a gas from a tank through a smooth convergent nozzle.

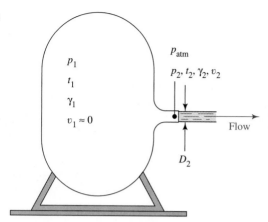

behave in some manner between isothermal and adiabatic. However, for rapid flow through a nozzle, we will assume that the flow is adiabatic.

**18.6.1**
**Nozzle Flow for**
**Adiabatic Process**

For an adiabatic process, the equation that describes the relationship between the absolute pressure and the specific weight of the gas is

$$\frac{p}{\gamma^k} = \text{constant} \qquad (18-5)$$

The exponent $k$ is called the *adiabatic exponent*, a dimensionless number, and its value for air is 1.40. Appendix N gives the values for $k$ for other gases.

Equation (18–5) can be used to compute the condition of a gas at a point of interest if the condition at some other point is known and if an adiabatic process occurs between the two points. That is,

$$\frac{p}{\gamma^k} = \text{constant} = \frac{p_1}{\gamma_1^k} = \frac{p_2}{\gamma_2^k} \qquad (18-6)$$

Stated differently,

$$\frac{p_2}{p_1} = \left(\frac{\gamma_2}{\gamma_1}\right)^k \qquad (18-7)$$

or,

$$\frac{\gamma_2}{\gamma_1} = \left(\frac{p_2}{p_1}\right)^{1/k} \qquad (18-8)$$

Here, $p_2$ is in the nozzle and $p_1$ is in the tank. The pressure outside the nozzle is $p_{\text{atm}}$.

The weight flow rate of gas exiting the tank through the nozzle in Fig. 18.10 is

$$W = \gamma_2 v_2 A_2 \qquad (18-9)$$

The principles of thermodynamics can be used to show that the velocity of the flow in the nozzle is

VELOCITY OF FLOW THROUGH A
NOZZLE

$$v_2 = \left\{ \left(\frac{2gp_1}{\gamma_1}\right)\left(\frac{k}{k-1}\right)\left[1 - \left(\frac{p_2}{p_1}\right)^{(k-1)/k}\right] \right\}^{1/2} \qquad (18-10)$$

Note that the pressures here are *absolute pressures*. Equations (18–6) through (18–10) can be combined to produce a convenient equation for the weight rate of flow from the tank in terms of the conditions of the gas in the tank and the pressure ratio $p_2/p_1$.

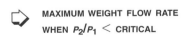

WEIGHT FLOW RATE WHEN
$P_2/P_1$ > CRITICAL RATIO

$$W = A_2 \sqrt{ \frac{2gk}{k-1}(p_1\gamma_1) \left[ \left( \frac{p_2}{p_1} \right)^{2/k} - \left( \frac{p_2}{p_1} \right)^{(k+1)/k} \right] }$$    (18–11)

Note that a *decreasing* pressure ratio $p_2/p_1$ actually indicates an *increasing* pressure difference $(p_1 - p_2)$ and that, therefore, it is expected that the weight flow rate $W$ will increase as the pressure ratio is decreased. This is true for the higher values of the pressure ratio. Under these conditions $p_2$ in the throat is equal to $p_{\text{atm}}$.

However, it can be shown that the flow rate reaches a maximum at a *critical pressure ratio*, defined as

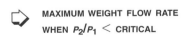

CRITICAL PRESSURE RATIO

$$\left( \frac{p_2'}{p_1} \right)_c = \left( \frac{2}{k+1} \right)^{k/(k-1)}$$    (18–12)

Because the value of the critical pressure ratio is a function only of the adiabatic exponent $k$, it is a constant for any particular gas.

When the critical pressure ratio is reached, the velocity of flow at the throat of the nozzle is equal to the speed of sound in the gas at the conditions that prevail there. *And this velocity of flow remains constant regardless of how much the downstream pressure is lowered.* The speed of sound in the gas is

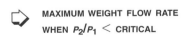

SONIC VELOCITY

$$c = \sqrt{ \frac{kgp_2'}{\gamma_2} }$$    (18–13)

Another name for $c$ is the *sonic velocity*, the velocity that a sound wave would travel in the gas. This is the maximum velocity of flow of a gas through a converging nozzle.

*Supersonic velocity*, velocity greater than the speed of sound, can be obtained only with a nozzle that first converges and then diverges. The analysis of such nozzles is not presented here.

The name *Mach number* is given to the ratio of the actual velocity of flow to the sonic velocity. That is,

$$N_M = v/c$$    (18–14)

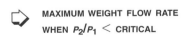

MACH NUMBER

Equation (18–11) must be used to compute the weight flow rate of gas from a tank though a converging nozzle for values of $N_M < 1.0$ for which the pressure ratio, $p_{\text{atm}}/p_1$, is greater than the critical pressure ratio. For $N_M = 1$, substituting the critical pressure ratio from Eq. (18–12) into Eq. (18–11) yields

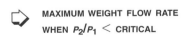

MAXIMUM WEIGHT FLOW RATE
WHEN $P_2/P_1$ < CRITICAL

$$W_{\text{max}} = A_2 \sqrt{ \left( \frac{2gk}{k+1} \right)(p_1\gamma_1) \left( \frac{2}{k+1} \right)^{2/(k-1)} }$$    (18–15)

This equation must be used when the pressure ratio $p_{\text{atm}}/p_1$ is less than the critical ratio.

Figure 18.11 shows the behavior of gas flow through a nozzle from a relatively large tank, according to Eqs. (18–11) and (18–15). The graph in (a) is for the case in which the pressure in the tank, $p_1$, is held constant and the pressure outside the nozzle, $p_{\text{atm}}$, is decreased. When $p_{\text{atm}} = p_1$, $p_{\text{atm}}/p_1 = 1.00$, and there is obviously no flow through the nozzle. As $p_{\text{atm}}$ is decreased, the relatively larger

**FIGURE 18.11**    Weight flow rate
of a gas through a nozzle.

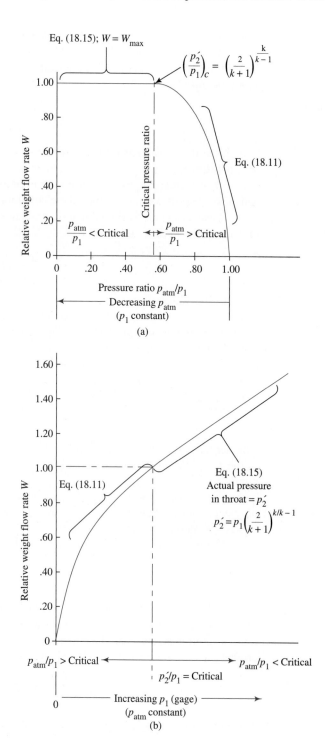

difference in pressure $(p_1 - p_{atm})$ causes an increase in the weight flow rate as computed from Eq. (18–11). However, when the critical pressure ratio, $(p_2'/p_1)_c$, is reached, the velocity in the throat reaches sonic velocity and the pressure remains at the critical pressure, $p_2'$, computed from Eq. (18–12). That is,

⇨ CRITICAL PRESSURE, $p_2'$

$$p_2 = p_2' = p_1 \left( \frac{2}{k+1} \right)^{k/(k-1)} \qquad \textbf{(18–16)}$$

Decreasing the pressure outside the nozzle any further would *not* increase the rate of flow from the tank.

Figure 18.11(b) shows a different interpretation of the variation of weight flow rate versus pressure ratio. In this case, the pressure outside the nozzle, $p_{atm}$, is held constant, while the pressure in the tank, $p_1$, is increased. Obviously, when the gage pressure in the tank is zero, no flow occurs, because there is no pressure differential. As $p_1$ is increased, the pressure ratio $p_{atm}/p_1$ is at first greater than the critical pressure ratio, and Eq. (18–11) applies. When the critical pressure ratio is reached or exceeded, the velocity in the throat will be at sonic velocity *for the condition of the gas in the throat.*

Note, however, that for any given value of $p_1$, the critical pressure in the throat, $p_2'$, is given by Eq. (18–16). But, because $p_1$ is increasing, $p_2'$ is also increasing. Still, because the pressure ratio between the tank and the throat is at the critical value, Eq. (18–15) must be used to compute the weight flow rate through the nozzle. The weight flow rate is now dependent on $p_1$ and $\gamma_1$. Note also that $\gamma_1$ is directly proportional to $p_1$, as can be seen from Eq. (18–2). Then, after the critical pressure ratio is reached, the weight flow rate increases linearly as the pressure in the tank increases.

If the nozzle has an abrupt reduction rather than the smooth shape shown in Fig. 18.10, the flow will be less than that predicted from Eq. (18–11) or (18–15). A discharge coefficient should be applied, similar to the discharge coefficients described in Chapter 15 for venturi meters, flow nozzles, and orifice meters. Also, if the upstream section is relatively small, as in a pipe, some correction for the velocity of approach should be applied (see Reference 7).

In summary, the following procedure can be used to compute the weight flow rate of a gas through a nozzle of the type shown in Fig. 18.10, assuming that the flow is adiabatic. Figure 18.12 charts the process.

### COMPUTATION OF ADIABATIC FLOW OF A GAS THROUGH A NOZZLE

1. Compute the actual pressure ratio between the pressure outside the nozzle and that in the tank, $p_{atm}/p_1$.

2. Compute the critical pressure ratio by using Eq. (18–12).

3a. If the actual pressure ratio is greater than the critical pressure ratio, use Eq. (18–11) to compute the weight flow rate through the nozzle with $p_2 = p_{atm}$. If desired, the velocity of flow can be computed by using Eq. (18–10).

3b. If the actual pressure ratio is less than the critical pressure ratio, use Eq. (18–15) to compute the weight flow rate through the nozzle. Also, recognize that the velocity of flow in the throat of the nozzle is equal to the sonic velocity, computed from Eq. (18–13), and that the pressure at the throat is that called $p_2'$ in Eq. (18–16). The gas then expands to $p_{atm}$ as it leaves the nozzle.

**FIGURE 18.12** Flow chart for computing weight flow rate of a gas from a nozzle.

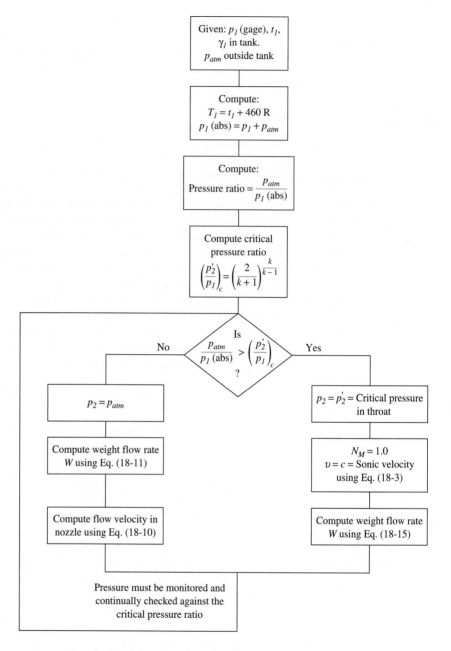

□ **EXAMPLE PROBLEM 18.5**    For the tank with a nozzle in its side, shown in Fig. 18.10, compute the weight flow rate of air leaving the tank for the following conditions:

$$p_1 = 10.0 \text{ psig} = \text{pressure in the tank}$$

$$p_{\text{atm}} = 14.2 \text{ psia} = \text{atmospheric pressure outside the tank}$$

$$t_1 = 80°F = \text{temperature of the air in the tank}$$

$$D_2 = 0.100 \text{ in} = \text{diameter of the nozzle at its outlet}$$

**Solution**     Using the procedure outlined above:

1. Actual pressure ratio:

$$\frac{p_{atm}}{p_1} = \frac{14.2 \text{ psia}}{(10.0 + 14.2) \text{ psia}} = 0.587$$

2. Determine the critical pressure ratio from Appendix N. For air, it is 0.528.
3a. Because the actual ratio is greater than the critical ratio, Eq. (18–11) is used for the weight flow rate. We must compute the nozzle throat area, $A_2$.

$$A_2 = \pi(D_2)^2/4 = \pi(0.100 \text{ in})^2/4 = 0.00785 \text{ in}^2$$

Converting to $ft^2$, we get

$$A_2 = 0.00785 \text{ in}^2 \,(1.0 \text{ ft}^2/144 \text{ in}^2) = 5.45 \times 10^{-5} \text{ ft}^2$$

Eq. (18–3) can be used to compute $\gamma_1$:

$$\gamma_1 = \frac{2.70 \, p_1}{T_1} = \frac{2.70(24.2 \text{ psia})}{(80 + 460)°R} = 0.121 \text{ lb/ft}^3$$

It is helpful to convert $p_1$ to the units of $lb/ft^2$.

$$p_1 = \frac{24.2 \text{ lb}}{\text{in}^2} \frac{144 \text{ in}^2}{\text{ft}^2} = 3485 \text{ lb/ft}^2$$

Then, using Eq. (18–11), with consistent U.S. Customary System units, we find the result for $W$ in lb/s. For these conditions, $p_2 = p_{atm}$.

$$W = A_2\sqrt{\frac{2gk}{k-1}(p_1\gamma_1)\left[\left(\frac{p_2}{p_1}\right)^{2/k} - \left(\frac{p_2}{p_1}\right)^{(k+1)/k}\right]} \tag{18–11}$$

$$W = (5.45 \times 10^{-5})\sqrt{\frac{2(32.2)(1.4)(3485)(0.121)}{(1.4-1)}[(0.587)^{2/1.4} - (0.587)^{2.4/1.4}]}$$

$$W = 4.32 \times 10^{-3} \text{ lb/s}$$

---

☐ **EXAMPLE PROBLEM 18.6**     For the conditions used in Example Problem 18.5, compute the velocity of flow in the throat of the nozzle and the Mach number for the flow.

**Solution**     Eq. (18–10) must be used to compute the velocity in the throat. For consistent U.S. Customary System units, the velocity will be in ft/s.

$$v_2 = \left\{\frac{2gp_1}{\gamma_1}\left(\frac{k}{k-1}\right)\left[1 - \left(\frac{p_2}{p_1}\right)^{(k-1)/k}\right]\right\}^{1/2} \tag{18–10}$$

$$v_2 = \left\{\left(\frac{2(32.2)(3485)}{0.121}\right)\left(\frac{1.40}{0.40}\right)(1 - (0.587)^{0.4/1.4})\right\}^{1/2}$$

$$v_2 = 957 \text{ ft/s}$$

To compute the Mach number, we need to compute the speed of sound in the air at the conditions existing in the throat by using Eq. (18–13):

$$c = \sqrt{\frac{kgp_2}{\gamma_2}} \tag{18–13}$$

The pressure $p_2 = p_{atm} = 14.2$ psia. Converting to lb/ft$^2$, we get

$$p_2 = \left(\frac{14.2\ \text{lb}}{\text{in}^2}\right)\left(\frac{144\ \text{in}^2}{1.0\ \text{ft}^2}\right) = 2045\ \text{lb/ft}^2$$

The specific weight, $\gamma_2$, can be computed from Eq. (18–8).

$$\frac{\gamma_2}{\gamma_1} = \left(\frac{p_2}{p_1}\right)^{1/k} \tag{18–8}$$

Knowing that $\gamma_1 = 0.121$ lb/ft$^3$, we get

$$\gamma_2 = \gamma_1\left(\frac{p_2}{p_1}\right)^{1/k}$$

$$\gamma_2 = (0.121)(0.587)^{1/1.4} = 0.0827\ \text{lb/ft}^3$$

Then the speed of sound is

$$c = \sqrt{\frac{kgp_2}{\gamma_2}} \tag{18–13}$$

$$c = \sqrt{\frac{(1.4)(32.2)(2045)}{0.0827}} = 1056\ \text{ft/s}$$

Now the Mach number can be computed.

$$N_M = \frac{v}{c} = \frac{957\ \text{ft/s}}{1056\ \text{ft/s}} = 0.906$$

---

☐ **EXAMPLE PROBLEM 18.7**

Compute the weight flow rate of air from the tank through the nozzle shown in Fig. 18.10 if the pressure in the tank is raised to 20.0 psig. All other conditions are the same as in Example Problem 18.5.

**Solution**

Use the same procedure outlined before:

1. Actual pressure ratio:

$$\frac{p_{atm}}{p_1} = \frac{14.2\ \text{psia}}{(20.0 + 14.2)\ \text{psia}} = 0.415$$

2. The critical pressure ratio is, again, 0.528 for air.
3.b Because the actual pressure ratio is less than the critical pressure ratio, Eq. (18–15) should be used.

$$W_{max} = A_2\sqrt{\frac{2gk}{k+1}(p_1\gamma_1)\left(\frac{2}{k+1}\right)^{2/(k-1)}} \tag{18–15}$$

Computing $\gamma_1$ for $p_1 = 34.2$ psia, we get

$$\gamma_1 = \frac{270p_1}{T_1} = \frac{(2.70)(34.2\ \text{psia})}{540°\text{R}} = 0.171\ \text{lb/ft}^3$$

We need the pressure $p_1$ in lb/ft$^2$.

$$p_1 = \left(\frac{34.2\ \text{lb}}{\text{in}^2}\right)\left(\frac{144\ \text{in}^2}{\text{ft}^2}\right) = 4925\ \text{lb/ft}^2$$

Then the weight flow rate is

$$W_{max} = (5.45 \times 10^{-5}) \sqrt{\frac{2(32.2)(1.4)(4925)(0.171)}{2.4}\left[\frac{2}{2.4}\right]^{2/0.4}}$$

$$W_{max} = 6.15 \times 10^{-3} \text{ lb/s}$$

The velocity of the air flow at the throat will be the speed of sound at the throat conditions. However, the pressure at the throat must be determined from the critical pressure ratio, Eq. (18–12).

$$\left(\frac{p'_2}{p_1}\right)_c = \left(\frac{2}{k+1}\right)^{k/(k-1)} = 0.528 \qquad (18\text{--}12)$$

$$p'_2 = p_1(0.528) = (4925 \text{ lb/ft}^2)(0.528) = 2600 \text{ lb/ft}^2$$

Knowing that $\gamma_1 = 0.171 \text{ lb/ft}^3$, we find

$$\gamma_2 = \gamma_1\left(\frac{p_2}{p_1}\right)^{1/k}$$

$$\gamma_2 = (0.171)(0.528)^{1/1.4} = 0.1084 \text{ lb/ft}^3$$

Then, the speed of sound and also the velocity in the throat is

$$c = \sqrt{\frac{kgp_2}{\gamma_2}} = \sqrt{\frac{(1.4)(32.2)(2600)}{0.1084}} = 1040 \text{ ft/s}$$

Of course, the Mach number in the throat is then 1.0.

■

## REFERENCES

1. Air Movement and Control Association International. 1999. *Fan Application Manual Publication B200-3*. Arlington Heights, Illinois: AMCA.
2. American Society of Mechanical Engineers. 1997. *Glossary of Terms Used in the Measurement of Fluid Flow in Pipes*. Standard ANSI/ASME MFC-1M. New York: Author.
3. Bleier, Frank P. 1997. *Fan Handbook*. New York: McGraw-Hill.
4. Chopey, Nicholas P., ed., and the staff of Chemical Engineering. 1994. *Fluid Movers: Pumps, Compressors, Fans and Blowers*. 2nd ed. New York: McGraw-Hill.
5. Crane Co. 1988. *Flow of Fluids Through Valves, Fittings, and Pipe*. Technical Paper 410. Joliet, Ill.: Author.
6. Hartzell Fan, Inc. 1999. *Hartzell Stock Fans and Blowers*. Piqua, Ohio: Author.
7. Idelchik, I. E. 1991. *Fluid Dynamics of Industrial Equipment*. Washington, D.C.: Hemisphere Publishing Corporation.
8. Osborne, W. C., 1977. *Fans*. 2nd ed. New York: Pergamon Press.
9. The Trane Company. 1996. *Trane Air Conditioning Manual*. La Crosse, Wisconsin: Author.

## PRACTICE PROBLEMS

### Units and Conversion Factors

**18.1E** A pipe in a compressed air system is carrying 2650 cfm. Compute the flow rate in ft³/s.

**18.2E** A duct in a heating system carries 8320 cfm. Compute the flow rate in ft³/s.

**18.3C** A pipe in a compressed air system carries 2650 cfm. Compute the flow rate in m³/s.

**18.4C** A duct in a heating system carries 8320 cfm. Compute the flow rate in m³/s.

**18.5C** The velocity of flow in a ventilation duct is 1140 ft/min. Compute the velocity in m/s.

**18.6C** The velocity of flow in an air conditioning system duct is 5.62 m/s. Compute the velocity in ft/s.

**18.7E** A measurement for the static pressure in a heating duct is 4.38 in H$_2$O. Express this pressure in psi.

**18.8C** A certain fan is rated to deliver 4760 cfm of air at a static pressure of 0.75 in $H_2O$. Express the flow rate in $m^3/s$ and the pressure in Pa.

**18.9C** The static pressure in a gas pipe is measured to be 925 Pa. Express this pressure in inches of $H_2O$.

**18.10C** Express the pressure of 925 Pa in psi.

## Fans, Blowers, and Compressors

**18.11** Describe a centrifugal fan with backward-inclined blades.

**18.12** Describe a centrifugal fan with forward-curved blades.

**18.13** Describe a duct fan.

**18.14** Describe a vaneaxial blower, and compare it with a duct fan.

**18.15** Name four types of positive displacement compressors.

**18.16** Name a type of compressor often used for pneumatic fluid power systems.

## Specific Weight of Air

**18.17E** Compute the specific weight of air at 80 psig and 75°F.

**18.18E** Compute the specific weight of air at 25 psig and 105°F.

**18.19E** Compute the specific weight of natural gas at 4.50 in $H_2O$ and 55°F.

**18.20E** Compute the specific weight of nitrogen at 32 psig and 120°F.

**18.21M** *Compute the specific weight of air at 1260 Pa(gage) and 25°C.*

**18.22E** Compute the specific weight of propane at 12.6 psig and 85°F.

## Flow of Compressed Air in Pipes

**18.23E** An air compressor delivers 820 cfm of free air. Compute the flow rate of air in a pipe in which the pressure is 80 psig and the temperature is 75°F.

**18.24E** An air compressor delivers 2880 cfm of free air. Compute the flow rate of air in a pipe in which the pressure is 65 psig and the temperature is 95°F.

**18.25E** Specify a size of Schedule 40 steel pipe suitable for carrying 750 cfm (free air delivery) at 100 psig with no more than 5.0 psi pressure drop in 100 ft of pipe.

**18.26E** Specify a size of Schedule 40 steel pipe suitable for carrying 165 cfm (free air delivery) at 100 psig with no more than 5.0 psi pressure drop in 100 ft of pipe.

**18.27E** Specify a size of Schedule 40 steel pipe suitable for carrying 800 cfm (free air) to a reactor vessel in a chemical processing plant in which the pressure must be at least 100 psig at 70°F. The total length of pipe from the compressor to the reactor vessel is 350 ft. The line contains eight standard elbows, two fully open gate valves, and one swing-type check valve. After completing the design, determine the required pressure at the compressor.

**18.28E** For an aeration process, a sewage treatment plant requires 3000 cfm of compressed air. The pressure must be 80 psig, and the temperature must be 120°F. The compressor is located in a utility building, and 180 ft of pipe is required. The line also contains one fully open butterfly valve, 12 elbows, four tees with the flow through the run, and one ball-type check valve. Specify a suitable size of Schedule 40 steel pipe, and determine the required pressure at the compressor.

## Gas Flow Through Nozzles

**18.29E** Air flows from a reservoir in which the pressure is 40.0 psig and the temperature is 80°F to a pipe in which the pressure is 20.0 psig. The flow is to be considered adiabatic. Compute the specific weight of the air both in the reservoir and in the pipe.

**18.30M** *Air flows from a reservoir in which the pressure is 275 kPa and the temperature is 25°C to a pipe in which the pressure is 140 kPa. The flow is to be considered adiabatic. Compute the specific weight of the air both in the reservoir and in the pipe.*

**18.31E** Refrigerant 12 expands adiabatically from 35.0 psig at a temperature of 60°F to 3.6 psig. Compute the specific weight of the refrigerant at both conditions.

**18.32E** Oxygen is discharged from a tank in which the pressure is 125 psig and the temperature is 75°F through a nozzle with a diameter of 0.120 in. The oxygen flows into the atmosphere, where the pressure is 14.40 psia. Compute the weight flow rate from the tank and the velocity of flow through the nozzle.

**18.33E** Repeat Problem 18.32, but change the pressure in the tank to 7.50 psig.

**18.34E** A high performance racing tire is charged with nitrogen at 50 psig and 70°F. At what rate would the nitrogen escape through a valve with a diameter of 0.062 in into the atmosphere at a pressure of 14.60 psia?

**18.35E** Repeat Problem 18.34 at internal pressures of 45 psig through 0 psig in 5.0 psig decrements. Plot a graph of weight flow rate versus the internal pressure in the tire.

**18.36E** Figure 18.13 shows a two-compartment vessel. The compartments are connected by a smooth convergent nozzle. The left compartment contains propane gas and is maintained at a constant 25.0 psig and 65°F. The right compartment starts at an equal 25.0 psig pressure, and then the pressure is decreased to 0.0 psig. The local atmospheric pressure is 14.28 psia. Compute the weight rate of flow of propane through the 0.5 in nozzle as the pressure decreases in 5.0 psi decrements. Plot the weight flow rate versus the pressure in the right compartment.

**FIGURE 18.13**    Vessel for Problem 18.36.

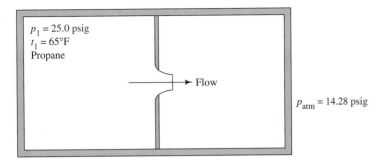

$p_1 = 25.0$ psig
$t_1 = 65°F$
Propane

→ Flow

$p_{atm} = 14.28$ psig

**18.37M**  *Air flows from a large tank through a smooth convergent nozzle into the atmosphere, where the pressure is 98.5 kPa absolute. The temperature in the tank is 95°C. Compute the minimum pressure in the tank required to produce sonic velocity in the nozzle.*

**18.38M**  *For the conditions of Problem 18.37, compute the magnitude of the sonic velocity in the nozzle.*

**18.39M**  *For the conditions of Problem 18.37, compute the weight flow rate of air from the tank if the nozzle diameter is 10.0 mm.*

**18.40M**  *A tank of Refrigerant 12 is at 150 kPa gage and 20°C. At what rate would the refrigerant flow from the tank into the atmosphere, where the pressure is 100.0 kPa absolute, through a smooth nozzle having a throat diameter of 8.0 mm?*

**18.41M**  *For the tank described in Problem 18.40, compute the weight flow rate through the nozzle for tank gage pressures of 125 kPa, 100 kPa, 75 kPa, 50 kPa, and 25 kPa. Assume that the temperature in the tank is 20°C for all cases. Plot the weight flow rate versus tank pressure.*

## COMPUTER PROGRAMMING ASSIGNMENTS

1. Write a program or spreadsheet to perform the calculations called for in Eqs. (18–2) and (18–3) for the specific weight of a gas and the correction of the volume flow rate for pressure and temperatures, different from standard free air conditions.

2. Write a program or spreadsheet for the analysis of the flow of compressed air in a pipe line system. The program should use a procedure similar to that outlined in Example Problem 18.4. Note that some of the features of the program are similar to those used in earlier chapters for the flow of liquids in pipe line systems.

3. Write a program or spreadsheet to compute the velocity and weight rate of flow of a gas from a tank through a smooth convergent nozzle. The program should use Eqs. (18–6) through (18–16), which involve critical pressure ratio and sonic velocity.

4. Use the program or spreadsheet from assignment 3 to solve problems 18.32 and 18.33, 18.35, 18.36, and 18.41. These problems call for analysis at several conditions.

# ■ ■ ■ ■ ■ 19 Flow of Air in Ducts

## 19.1 The Big Picture

### Discussion Map

☐ Ventilation, heating, and air-conditioning systems distribute air at relatively low pressure through ducts.

☐ The fans or blowers responsible for moving the air are generally considered to be high-volume, low pressure devices.

☐ Knowledge of the pressures in the duct system is required to properly match a fan to a given system, to ensure the delivery of an adequate amount of air, and to balance the flow to various parts of the system.

### Discover

■ Acquire data about a forced air heating, air-conditioning, or ventilation system that you can gain access to. It might be in your home, in your school, in a commercial building, or in an industrial plant.

■ Describe the system in as much detail as possible including the size and shape of the ducts, the kind of fan used, the location of the fan, and how the air is distributed to the conditioned space.

In this chapter, you will learn some basic methods for analyzing and designing ducts to carry air in heating, ventilation, and air-conditioning systems.

Here are some things you should look for when examining a duct system.

■ Where is the main fan or blower that forces the air through the system? Describe its physical size and configuration, referring to Figures 18.1 and 18.3 in Chapter 18 for examples. Can you find ratings for the fan such as its speed, delivery (volume flow rate), and pressure rating? The delivery may be reported in units such as *cfm* that is an abbreviation for *cubic feet per minute*. The pressure rating might be in psi or inches of water or some other unit.

■ How does the air get to the intake of the fan? Where does it come from?

■ Where does the air go directly from the discharge section of the fan? Is it a part of the burner of a furnace or a space heater? Or does it deliver air over the cooling coils of an air-conditioning system? Or is the flow delivered directly to ductwork for ventilation without affecting its temperature?

■ Follow the ductwork from the fan outlet to each of its discharge points. Try to get measurements for the dimensions of the duct. Is it round, square, or rectangular? Are there any bends, reducers, or expansions in the ductwork?

■ Are there any control devices such as dampers installed in the ducts that allow the air flow to be partially blocked? This allows the system operator to balance the flow to ensure that an adequate amount of conditioned air is delivered to each destination point.

■ Describe the grilles or registers that control the air delivery at each destination. What are their critical dimensions?

■ Are there provisions for the air to return back from the conditioned spaces to the fan system to encourage air circulation? If so, how is that accomplished?

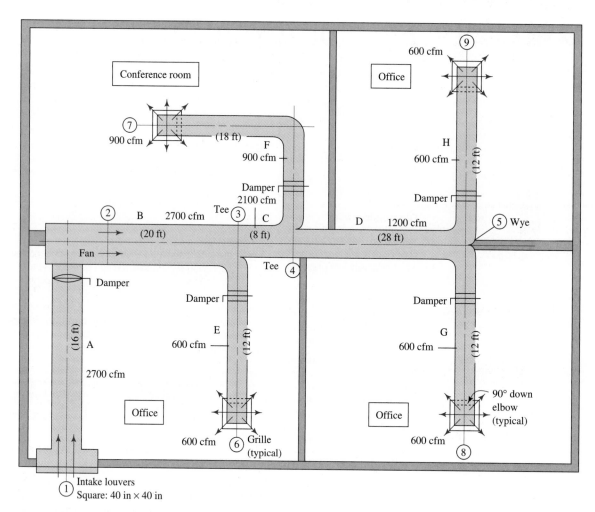

**FIGURE 19.1**   Air distribution system.

***An Example Air Distribution System.***   Figure 19.1 is a sketch of the layout of an air distribution system. Outside air enters the building at point 1 through louvers that protect the ductwork from wind and rain. The velocity of the air flow through the louvers should be relatively low, approximately 500 ft/min (2.5 m/s), to minimize entrainment of undesirable contaminants. The duct then reduces to a smaller size to deliver the air to the suction side of a fan. A sudden contraction of the duct is shown, although a more gradual reduction would have a lower pressure loss.

The damper in the intake duct can partially close off the duct to decrease the flow, if desired. The intake duct delivers the air to the fan inlet, where its pressure is increased by the action of the fan wheel.

The outlet from the fan is carried by a main duct from which four branches deliver the air to its points of use. Dampers, shown in each of the branches, permit balancing of the system while in operation. Grilles are used at each outlet to distribute the air to the conditioned spaces (in this example, three offices and a conference room).

The ductwork shown is mostly above the ceiling. Note that an elbow is placed above each outlet grille to direct the air flow downward through the ceiling system.

It is important to understand the basic operating parameters of such an air-handling system. Obviously, the air outside the building is at the prevailing atmospheric pressure. To cause flow into the duct through the intake louvers, the fan must create a pressure lower than atmospheric in the duct. This is a negative gage pressure. As the air flows through the duct, friction losses cause the pressure to decrease further. Also, any obstruction to the flow, such as a damper, and tees or wyes that direct the flow, cause pressure drops. The fan increases the pressure of the air and forces it through the supply ductwork to the outlet grilles.

The air inside the rooms of the building may be slightly above or below atmospheric pressure. Some air handling-system designers prefer to have a slight positive pressure in the building for better control and to eliminate drafts. However, when designing the ductwork, one usually considers the pressure inside the building to be the same as that outside the building.

**19.2**
**OBJECTIVES**

After completing this chapter, you should be able to:

1. Describe the basic elements of an air distribution system that may be used for heating, ventilation, or air conditioning.
2. Determine energy losses in ducts considering straight sections and fittings.
3. Determine the circular equivalent diameters of rectangular ducts.
4. Analyze and design ductwork to carry air to spaces needing conditioning and to achieve balance in the system.
5. Identify the fan selection requirements for the system.

**19.3**
**ENERGY LOSSES IN DUCTS**

Two kinds of energy losses in duct systems cause the pressure to drop along the flow path. Friction losses occur as the air flows through straight sections, while dynamic losses occur as the air flows through such fittings as tees and wyes and through flow control devices.

*Friction losses* can be estimated by using the Darcy equation introduced in Chapter 9 for flow of liquids. However, special charts have been prepared by the American Society of Heating, Refrigerating, and Air-Conditioning Engineers (ASHRAE) for the typical conditions found in duct design. Figures 19.2 and 19.3 show friction loss $h_L$ as a function of volume flow rate, with two sets of diagonal lines showing the diameter of circular ducts and the velocity of flow. The units used for the various quantities and the assumed conditions are summarized in Table 19.1. Reference 2 shows correction factors for other conditions. See Section 18.3 for information on air flow rates and pressures.

**TABLE 19.1** Units and assumed conditions for friction charts

| | **U.S. Customary System Units** | **SI Units** |
|---|---|---|
| Flow rate | ft³/min (cfm) | m³/s |
| Friction loss, $h_L$ | in of water per 100 ft (in H₂O/100 ft) | Pa/m |
| Velocity | ft/min | m/s |
| Duct diameter | in | mm |
| Specific weight of air | 0.075 lb/ft³ | 11.81 N/m³ |
| Duct surface roughness | $5 \times 10^{-4}$ ft | $1.5 \times 10^{-4}$ m |
| Condition of air | 14.7 psia; 68°F | 101.3 kPa; 20°C |

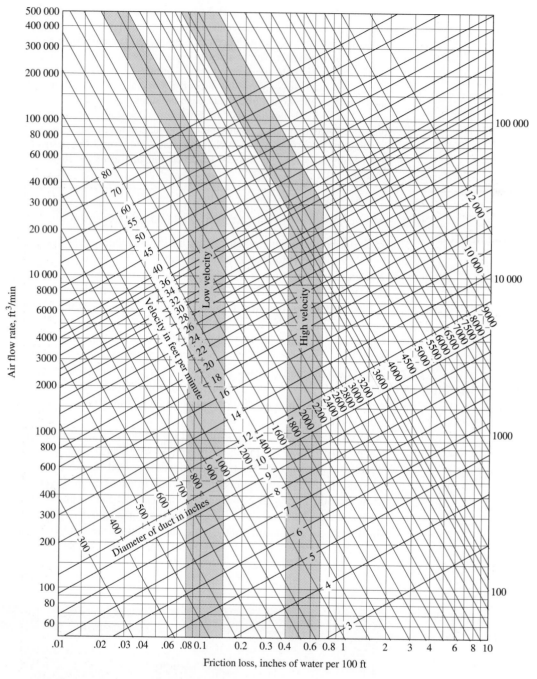

**FIGURE 19.2** Friction loss in ducts—U.S. Customary System units. (Reprinted by permission from *ASHRAE Handbook: 1981 Fundamentals*).

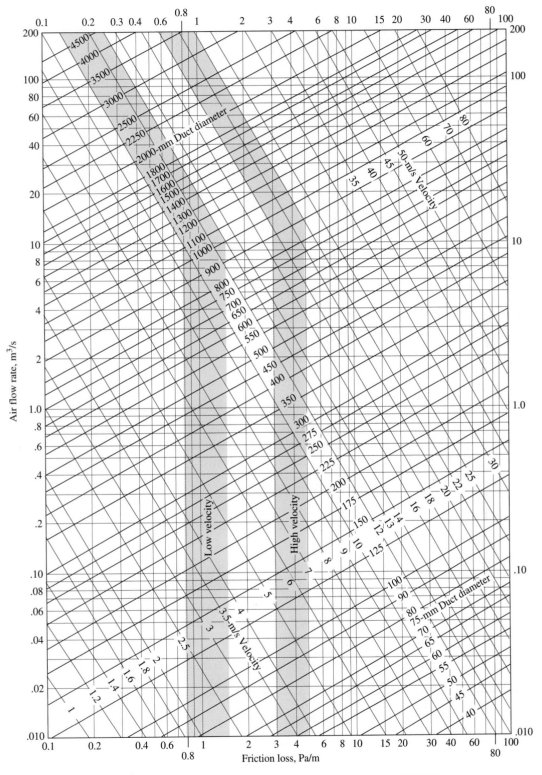

**FIGURE 19.3** Friction loss in ducts—SI units. (Reprinted by permission from *ASHRAE Handbook: 1981 Fundamentals*).

**TABLE 19.2**   Circular equivalent diameters of rectangular ducts

| Side $a$ (in) | Side $b$ (in) | | | | | | | | | | | | |
|---|---|---|---|---|---|---|---|---|---|---|---|---|---|
| | 6 | 8 | 10 | 12 | 14 | 16 | 18 | 20 | 22 | 24 | 26 | 28 | 30 |
| 6 | 6.6 | | | | | | | | | | | | |
| 8 | 7.6 | 8.7 | | | | | | | | | | | |
| 10 | 8.4 | 9.8 | 10.9 | | | | | | | | | | |
| 12 | 9.1 | 10.7 | 12.0 | 13.1 | | | | | | | | | |
| 14 | 9.8 | 11.5 | 12.9 | 14.2 | 15.3 | | | | | | | | |
| 16 | 10.4 | 12.2 | 13.7 | 15.1 | 16.4 | 17.5 | | | | | | | |
| 18 | 11.0 | 12.9 | 14.5 | 16.0 | 17.3 | 18.5 | 19.7 | | | | | | |
| 20 | 11.5 | 13.5 | 15.2 | 16.8 | 18.2 | 19.5 | 20.7 | 21.9 | | | | | |
| 22 | 12.0 | 14.1 | 15.9 | 17.6 | 19.1 | 20.4 | 21.7 | 22.9 | 24.0 | | | | |
| 24 | 12.4 | 14.6 | 16.5 | 18.3 | 19.9 | 21.3 | 22.7 | 23.9 | 25.1 | 26.2 | | | |
| 26 | 12.8 | 15.1 | 17.1 | 19.0 | 20.6 | 22.1 | 23.5 | 24.9 | 26.1 | 27.3 | 28.4 | | |
| 28 | 13.2 | 15.6 | 17.7 | 19.6 | 21.3 | 22.9 | 24.4 | 25.8 | 27.1 | 28.3 | 29.5 | 30.6 | |
| 30 | 13.6 | 16.1 | 18.3 | 20.7 | 22.0 | 23.7 | 25.2 | 26.6 | 28.0 | 29.3 | 30.5 | 31.7 | 32.8 |

We use the symbol $h_L$ to indicate the friction loss per 100 ft of duct read from Figure 19.2. Then the total energy loss for a given duct length $L$ is called $H_L$ and is found from

$$H_L = h_L(L/100)$$

Other energy losses will also be designated by the $H$ symbol with subscripts pertinent to the unit being analyzed. See Example Problems 19.1 to 19.4 that follow later in the chapter.

Although circular ducts are often used for distributing air through heating, ventilating, or air-conditioning systems, it is usually most convenient to use rectangular ducts because of space limitations, particularly above ceilings. The hydraulic radius of the rectangular duct can be used to characterize its size (as discussed in Section 8.7). When the necessary substitutions of the hydraulic radius for the diameter are made in relationships for velocity, Reynolds number, relative roughness, and the corresponding friction factor, we see that the *equivalent diameter* for a rectangular duct is

 **EQUIVALENT DIAMETER FOR A RECTANGULAR DUCT**

$$D_e = \frac{1.3(ab)^{5/8}}{(a + b)^{1/4}} \tag{19-1}$$

where $a$ and $b$ are sides of the rectangle.

This enables you to use the friction loss charts in Figs. 19.2 and 19.3 for rectangular as well as circular ducts. Table 19.2 shows some results computed using Eq. 19–1.

---

☐ **EXAMPLE PROBLEM 19.1**      Determine the velocity of flow and the amount of friction loss that would occur as air flows at 3000 cfm through 80 ft of circular duct having a diameter of 22 in.

*Solution*      We can use Fig. 19.2 to determine that the velocity is approximately 1150 ft/min and that the friction loss per 100 ft of duct $(h_L)$ is 0.082 in $H_2O$. Then, by proportion, the loss for 80 ft is

$$H_L = h_L(L/100) = (0.082 \text{ in } H_2O)(\tfrac{80}{100}) = 0.066 \text{ in } H_2O$$

☐ **EXAMPLE PROBLEM 19.2**
Specify the dimensions of a rectangular duct that would have the same friction loss as the circular duct described in Example Problem 19.1.

*Solution*
From Table 19.2 we can specify a 14-by-30-in rectangular duct that would have the same loss as the 22.0-in diameter circular duct. Others that will have approximately the same loss are 16-by-26-in, 18-by-22-in, and 20-by-20-in rectangular ducts. Such a list gives the designer many options when fitting a duct system into given spaces.

■

*Dynamic losses* can be estimated by using published data for loss coefficients for air flowing through certain fittings (see References 2 and 4). Also, manufacturers of special air-handling devices publish extensive data about expected pressure drops. Table 19.3 presents a few examples for use in problems in this book. Note that these data are highly simplified. For example, actual loss coefficients for tees depend on the size of the branches and on the amount of air flow in each branch. As with the minor losses discussed in Chapter 10, changes in flow area or direction of flow should be made as smooth as possible to minimize dynamic losses. The data for round, 90° elbows show the large variations possible.

The dynamic loss for a fitting is calculated from

$$H_L = C(H_v) \qquad\qquad (19\text{–}2)$$

where $C$ is the loss coefficient from Table 19.3 and $H_v$ is the *velocity pressure* or *velocity head*.

**TABLE 19.3** Examples of loss factors for duct fittings

| Dynamic Loss Coefficient, $C$ | | | | | |
|---|---|---|---|---|---|
| 90°elbows | | | | | |
| Smooth, round | 0.22 | | | | |
| 5-piece, round | 0.33 | | | | |
| 4-piece, round | 0.37 | | | | |
| 3-piece, round | 0.42 | | | | |
| Mitered, round | 1.20 | | | | |
| Smooth, rectangular | 0.18 | | | | |
| Tee, branch | 1.00 | | | | |
| Tee, flow through main | 0.10 | | | | |
| Symmetrical wye | 0.30 | | | | |

| Damper: | | | | | | |
|---|---|---|---|---|---|---|
| Position | 0° (wide open) | 10° | 20° | 30° | 40° | 50° |
| $C$ | 0.20 | 0.52 | 1.50 | 4.5 | 11.0 | 29 |

Outlet grille: Assume total pressure drop through grille is 0.06 in $H_2O$ (15 Pa).

Intake louvers: Assume total pressure drop across louver is 0.07 in $H_2O$ (17 Pa).

Dynamic loss for fittings = $C(H_v)$, where $H_v$ = velocity pressure upstream of the fitting.

Values shown are examples, for use only in solving problems in this book. Many factors affect the actual values for a given style of fitting. Refer to Reference 2 or manufacturers' catalogs for more complete data.

In U.S. Customary System units, pressure levels and losses are typically expressed in inches of water, which is actually a measure of pressure head. Then,

$$H_v = \frac{\gamma_a v^2}{2g\gamma_w} \qquad (19\text{--}3)$$

where $\gamma_a$ is the specific weight of air, $v$ is the flow velocity, and $\gamma_w$ is the specific weight of water. When the velocity is expressed in feet per minute and the conditions of standard air are used, Eq. (19–3) reduces to

**VELOCITY PRESSURE FOR AIR FLOW (U.S.)**

$$H_v = \left(\frac{v}{4005}\right)^2 \qquad (19\text{--}4)$$

When we use SI units, pressure levels and losses are measured in the pressure unit of Pa. Then,

$$H_v = \frac{\gamma_a v^2}{2g} \qquad (19\text{--}5)$$

When the velocity is expressed in m/s and the conditions of standard air are used, Eq. (19–5) reduces to

**VELOCITY PRESSURE FOR AIR FLOW (SI)**

$$H_v = \left(\frac{v}{1.289}\right)^2 \text{ Pa} \qquad (19\text{--}6)$$

---

☐ **EXAMPLE PROBLEM 19.3**   Estimate the pressure drop that occurs when 3000 cfm of air flows around a smooth, rectangular, 90° elbow with side dimensions of 14 × 24 in.

**Solution**   Use Table 19.2 to find that the equivalent diameter for the duct is 19.9 in. From Fig. 19.2 we find the velocity of flow to be 1400 ft/min. Then, using Eq. (19–4), we compute

$$H_v = \left(\frac{v}{4005}\right)^2 = \left(\frac{1400}{4005}\right)^2 = 0.122 \text{ in H}_2\text{O}$$

From Table 19.3, we find $C = 0.18$. Then the pressure drop is

$$H_L = C(H_v) = (0.18)(0.122) = 0.022 \text{ in H}_2\text{O}$$

■

---

**19.4**
**DUCT DESIGN EXAMPLE**

In Section 19.1, the general features of ducts for carrying the flow of air were described. Figure 19.1 shows a simple duct system, the operation of which has been described. In this section, we describe one method of designing such a duct system.

The goals of the design process are to specify reasonable dimensions for the various sections of the ductwork, to estimate the air pressure at key points, to determine the requirements to be met by the fan in the system, and to balance the system. Balance requires that the pressure drop from the fan outlet to each outlet grille is the same when the duct sections are carrying their design capacities.

The design method illustrated here uses Fig. 19.2 or 19.3 to specify a suitable velocity of flow in each duct section. The duct size, either circular or rectangular, is then decided. Then pressure drops are estimated for friction and for fittings. After all branch designs are tentatively completed, the total pressure drop to each grille is computed. An ideal design is one in which all such pressure drops are equal, and the system is *balanced*. If pressure drops are not equal, as is often the case, design changes are required. For small imbalances, such flow controls as dampers can be used to balance the system. When major deviations in pressure drops occur among the duct sections, the duct sizes or the design of fittings should be changed.

This procedure is sometimes referred to as the *equal friction method* or the *velocity reduction method*. The logic of these names is illustrated in Fig. 19.2. When we specify designs within the shaded part of the figure, the friction loss in each section will be nearly the same. Also, as the volume flow rate in each duct along a branch decreases, the recommended velocity of flow decreases. Note that the shaded part of the friction loss chart applies only to supply ducts and branches on the outlet side of fans. Proper flow of air into a fan requires a somewhat lower velocity, typically 700 to 1000 ft/min (3.5 to 5.0 m/s). Lower velocities are also used when low noise is desirable.

Other duct system design methods include the *static regain method* and the *balanced capacity method* (see References 2 and 3).

Most smaller systems for homes and light commercial applications are of the "low velocity" type, in which ductwork and fittings are relatively simple. Noise is usually not a major problem if the limits shown in Figs. 19.2 and 19.3 are not exceeded. However, the resulting sizes for ducts in a low velocity system are relatively large.

Space limitations in the design of large office buildings and certain industrial applications make "high velocity systems" attractive. The name comes from the practice of using smaller ducts to carry a given flow rate. However, several consequences arise:

**1.** Noise is usually a factor and special noise-attenuation devices must be employed.
**2.** Duct construction must be more substantial, and sealing is more critical.
**3.** Operating costs are generally higher because of greater pressure drops and higher fan total pressures.

High velocity systems can be justified when lower building costs result or when more efficient use of space can be achieved.

The following design example is for a low velocity system.

---

☐ **EXAMPLE PROBLEM 19.4**   The system shown in Fig. 19.1 is being designed for a small office building. The air is drawn from outside the building by a fan and delivered through four branches to three offices and a conference room. The air flows shown at each outlet grille have been determined by others to provide adequate ventilation to each area. Dampers in each branch permit final adjustment of the system.

Complete the design of the duct system, specifying the size of each section of the ductwork for a low velocity system. Compute the expected pressure drop for each section and at each fitting. Then, compute the total pressure drop along each branch from the fan to the four outlet grilles and check for system balance. If a major imbalance is predicted, redesign appropriate parts of the system to achieve a more nearly balanced system. Then, determine the total pressure required for the fan. Use Fig. 19.2 for estimating friction losses and Table 19.3 for dynamic loss coefficients.

*Solution*   First, treat each section of the duct and each fitting separately. Then, analyze the branches.

   **1.** Intake duct A: $Q = 2700$ cfm; $L = 16$ ft
   Let $v \approx 800$ ft/min.
   From Fig. 19.2, required $D = 25.0$ in.

$h_L = 0.035$ in $H_2O$/100 ft

$H_L = 0.035(16/100) = 0.0056$ in $H_2O$.

2. Damper in duct A: $C = 0.20$ (assume wide open) (Table 19.3)

   For 800 ft/min, $H_v = (800/4005)^2 = 0.040$ in $H_2O$

   $H_L = 0.20(0.040) = 0.0080$ in $H_2O$

3. Intake louvers: The 40-by-40-in size has been specified to give approximately 600 ft/min velocity through the open space of the louvers. Use $H_L = 0.070$ in $H_2O$ from Table 19.3.

4. Sudden contraction between louver housing and intake duct: From Fig. 10.7, we know that the resistance coefficient depends on the velocity of flow and the ratio $D_1/D_2$ for circular conduits. Because the louver housing is a 40-by-40-in square, we can compute its equivalent diameter from Eq. (19–1)

$$D_e = \frac{1.3(ab)^{5/8}}{(a + b)^{1/4}} = \frac{1.3(40 \times 40)^{5/8}}{(40 + 40)^{1/4}} = 43.7 \text{ in}$$

Then, in Fig. 10.7,

$$D_1/D_2 = 43.7/25 = 1.75$$

and $K = C = 0.31$. Then,

$$H_L = C(H_v) = 0.31(0.04) = 0.0124 \text{ in } H_2O$$

5. Total loss in intake system:

$$H_L = 0.0056 + 0.0080 + 0.07 + 0.0124 = 0.096 \text{ in } H_2O$$

Because the pressure outside the louvers is atmospheric, the pressure at the inlet to the fan is $-0.096$ in $H_2O$, a negative gage pressure. An additional loss may occur at the fan inlet if a geometry change is required to mate the intake duct with the fan. Knowledge of the fan design is required, and this potential loss is ignored in this example.

*Note:* All ducts on the fan outlet side are rectangular.

6. Fan outlet − duct B: $Q = 2700$ cfm; $L = 20$ ft

   Let $v \approx 1200$ ft/min; $h_L = 0.110$ in $H_2O$/100 ft.

   $D_e = 20.0$ in; use 12-by-30-in size to minimize overhead space required.

   $H_L = 0.110(20/100) = 0.0220$ in $H_2O$

   $H_v = (1200/4005)^2 = 0.090$ in $H_2O$

7. Duct E: $Q = 600$ cfm; $L = 12$ ft

   Let $v \approx 800$ ft/min; $h_L = 0.085$ in $H_2O$/100 ft.

   $D_e = 12.0$ in; use 12-by-10-in size.

   $H_L = 0.085(12/100) = 0.0102$ in $H_2O$

   $H_v = (800/4005)^2 = 0.040$ in $H_2O$

8. Damper in duct E: $C = 0.20$ (assume wide open)

   $H_L = 0.20(0.040) = 0.0080$ in $H_2O$

9. Elbow in duct E: Smooth rectangular elbow; $C = 0.18$

   $H_L = 0.18(0.040) = 0.0072$ in $H_2O$

10. Grille 6 for duct E: $H_L = 0.060$ in $H_2O$

11. Tee 3 from duct B to branch E, flow in branch: $C = 1.00$

    $H_L$ based on velocity ahead of tee in duct B

    $H_L = 1.00(0.090) = 0.090$ in $H_2O$

12. Duct C: $Q = 2100$ cfm; $L = 8$ ft

    Let $v \approx 1200$ ft/min; $h_L = 0.110$ in $H_2O$/100 ft.

$D_e = 18.5$ in; use 12-by-24-in size.

$H_L = 0.110(8/100) = 0.0088$ in $H_2O$

$H_v = (1200/4005)^2 = 0.090$ in $H_2O$

13. Tee 3 from duct B to duct C, flow through main: $C = 0.10$

    $H_L = 0.10(0.090) = 0.009$ in $H_2O$

14. Duct F: $Q = 900$ cfm; $L = 18$ ft

    Let $v \approx 800$ ft/min; $h_L = 0.068$ in $H_2O$/100 ft

    $D_e = 14.3$ in; use 12-by-14-in size

    $H_L = 0.068(18/100) = 0.0122$ in $H_2O$

    $H_v = (800/4005)^2 = 0.040$ in $H_2O$

15. Damper in duct F: $C = 0.20$ (assume wide open)

    $H_L = 0.20(0.040) = 0.0080$ in $H_2O$

16. Two elbows in duct F: Smooth rectangular elbow; $C = 0.18$

    $H_L = 2(0.18)(0.040) = 0.0144$ in $H_2O$

17. Grille 7 for duct F: $H_L = 0.060$ in $H_2O$

18. Tee 4 from duct C to branch F, flow in branch: $C = 1.00$

    $H_L$ based on velocity ahead of tee in duct C

    $H_L = 1.00(0.090) = 0.090$ in $H_2O$

19. Duct D: $Q = 1200$ cfm; $L = 28$ ft

    Let $v \approx 1000$ ft/min; $h_L = 0.100$ in $H_2O$/100 ft

    $D_e = 14.7$ in; use 12-by-16-in size

    Actual $D_e = 15.1$ in; new $h_L = 0.087$ in $H_2O$/100 ft

    $H_L = 0.087(28/100) = 0.0244$ in $H_2O$

    New $v = 960$ ft/min

    $H_v = (960/4005)^2 = 0.057$ in $H_2O$

20. Tee 4 from duct C to duct D, flow through main: $C = 0.10$

    $H_L = 0.10(0.090) = 0.009$ in $H_2O$

21. Wye 5 between duct D and ducts G and H: $C = 0.30$

    $H_L = 0.30(0.057) = 0.017$ in $H_2O$

    This loss applies to either duct G or duct H.

22. Ducts G and H are identical to duct E, and losses from steps 7, 8, 9, and 10 can be applied to these paths.

This completes the evaluation of the pressure drops through components in the system. Now we can sum the losses through any path from the fan outlet to the outlet grilles.

a. Path to grille 6 in duct E: Sum of losses from steps 6, 7, 8, 9, 10, and 11.

$$H_6 = 0.0220 + 0.0102 + 0.0080 + 0.0072 + 0.060 + 0.090$$
$$= 0.1974 \text{ in } H_2O$$

b. Path to grille 7 in duct F: Sum of losses from steps 6, 12, 13, 14, 15, 16, 17, and 18.

$$H_7 = 0.0220 + 0.0088 + 0.0090 + 0.0122 + 0.0080 + 0.0144 + 0.060 + 0.090$$
$$= 0.2244 \text{ in } H_2O$$

c. Path to either grille 8 in duct G or grille 9 in duct H: Sum of losses from steps 6, 12, 13, 19, 20, 21, 7, 8, 9, and 10.

$$H_8 = 0.0220 + 0.0088 + 0.0090 + 0.0244 + 0.0090 + 0.0170$$
$$+ 0.0102 + 0.008 + 0.0072 + 0.06$$
$$= 0.1756 \text{ in } H_2O$$

**Redesign to Achieve a Balanced System**

The ideal system design would be one in which the loss along any path, a, b, or c, is the same. Because that is not the case here, some redesign is called for. The loss in path b to grille 7 in duct F is much higher than the others. The component losses from steps 12, 14, 15, 16, and 18 affect this branch, and some reduction can be achieved by reducing the velocity of flow in ducts C and F.

**12a.** Duct C: $Q = 2100$ cfm; $L = 8$ ft
Let $v \approx 1000$ ft/min; $h_L = 0.073$ in $H_2O/100$ ft
$D_e = 19.6$ in; use 12-by-28-in size
$H_L = 0.073(8/100) = 0.0058$ in $H_2O$
$H_v = (1000/4005)^2 = 0.0623$ in $H_2O$

**14a.** Duct F: $Q = 900$ cfm; $L = 18$ ft
Let $v \approx 600$ ft/min; $h_L = 0.033$ in $H_2O/100$ ft
$D_e = 16.5$ in; use 12-by-18-in size; $D_e = 16.0$ in
Actual $v = 630$ ft/min; $h_L = 0.038$ in $H_2O/100$ ft
$H_L = 0.038(18/100) = 0.0068$ in $H_2O$
$H_v = (630/4005)^2 = 0.0247$ in $H_2O$

**15a.** Damper in duct F: $C = 0.20$ (assume wide open)
$H_L = 0.20(0.0247) = 0.0049$ in $H_2O$

**16a.** Two elbows in duct F: Smooth rectangular elbow; $C = 0.18$
$H_L = 2(0.18)(0.0247) = 0.0089$ in $H_2O$

**18a.** Tee 4 from duct C to branch F, flow in branch: $C = 1.00$
$H_L$ based on velocity ahead of tee in duct C
$H_L = 1.00(0.0623) = 0.0623$ in $H_2O$

Now we can recompute the total loss in path B to grille 7 in duct F. As before, this is the sum of the losses from steps 6, 12a, 13, 14a, 15a, 16a, 17, and 18a.

$$H_7 = 0.0220 + 0.0058 + 0.009 + 0.0068 + 0.0049 + 0.0089 + 0.06 + 0.0623$$
$$= 0.1797 \text{ in } H_2O$$

This is a significant reduction, which results in a total pressure drop less than that of path a. Therefore, let's see if we can reduce the loss in path a by also reducing the velocity of flow in duct E. Steps 7, 8, and 9 are affected.

**7a.** Duct E: $Q = 600$ cfm; $L = 12$ ft
Let $v \approx 600$ ft/min
$D_e = 13.8$ in; use 12-by-14-in size
Actual $D_e = 14.2$ in; $h_L = 0.032$ in $H_2O$; $v = 550$ ft/min
$H_L = 0.032(12/100) = 0.0038$ in $H_2O$
$H_v = (550/4005)^2 = 0.0189$ in $H_2O$

**8a.** Damper in duct E: $C = 0.20$ (assume wide open)
$H_L = 0.20(0.0189) = 0.0038$ in $H_2O$

**9a.** Elbow in duct E: Smooth rectangular elbow; $C = 0.18$
$H_L = 0.18(0.0189) = 0.0034$ in $H_2O$

Now, we can recompute the total loss in path a to grille 6 in duct E. As before, this is the sum of the losses from steps 6, 7a, 8a, 9a, 10, and 11.

$$H_6 = 0.0220 + 0.0038 + 0.0038 + 0.0034 + 0.060 + 0.090$$
$$= 0.1830 \text{ in } H_2O$$

This value is very close to that found for the redesigned path b, and the small difference can be adjusted with the dampers.

Now, note that path c to either grille 8 or grille 9 still has a lower total loss than either path a or path b. We could either use a slightly smaller duct size in branches G and H or depend on the adjustment of the dampers here also. To evaluate the suitability of using the dampers, let's estimate how much the dampers would have to be closed to increase the total loss to 0.1830 in $H_2O$ (to equal that in path a). The increased loss is

$$H_6 - H_8 = 0.1830 - 0.1756 \text{ in } H_2O = 0.0074 \text{ in } H_2O$$

With the damper wide open and with 600 cfm passing at a velocity of approximately 800 ft/min, the loss was 0.0080 in $H_2O$, as found in the original step 8. The loss now should be

$$H_L = 0.0080 + 0.0074 = 0.0154 \text{ in } H_2O$$

But, for the damper,

$$H_L = C(H_v)$$

Solving for $C$ gives

$$C = \frac{H_L}{H_v} = \frac{0.0154 \text{ in } H_2O}{0.040 \text{ in } H_2O} = 0.385$$

Referring to Table 19.3, you can see that a damper setting of less than 10° would produce this value of $C$, a very feasible setting. Thus, it appears that the duct system could be balanced as redesigned and that the total pressure drop from the fan outlet to any outlet grille will be approximately 0.1830 in $H_2O$. This is the pressure that the fan would have to develop.

### SUMMARY OF THE DUCT SYSTEM DESIGN

- Intake duct A: Round; $D = 25.0$ in
- Duct B: Rectangular; 12 in × 30 in
- Duct C: Rectangular; 12 in × 28 in
- Duct D: Rectangular; 12 in × 16 in

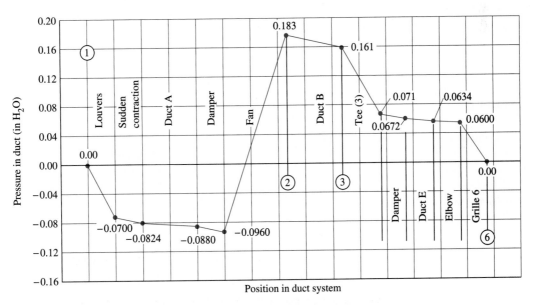

**FIGURE 19.4** Pressure in duct (in $H_2O$) versus position for system shown in Fig. 19.1.

- Duct E: Rectangular; 12 in × 14 in
- Duct F: Rectangular; 12 in × 18 in
- Duct G: Rectangular; 12 in × 10 in
- Duct H: Rectangular; 12 in × 10 in
- Pressure at fan inlet: −0.096 in $H_2O$
- Pressure at fan outlet: 0.1830 in $H_2O$
- Total pressure rise by the fan: 0.1830 + 0.096 = 0.279 in $H_2O$
- Total delivery by the fan: 2700 cfm

It is helpful to visualize the pressure changes that occur in the system. Figure 19.4 on page 545 shows a plot of air pressure versus position for the path from the intake louvers, through the fan, through ducts B and E, to outlet grille 6. Similar plots could be made for the other paths.

■

# REFERENCES

1. American Society of Heating, Refrigerating and Air-Conditioning Engineers (ASHRAE). 1996. *ASHRAE Handbook: HVAC Systems and Equipment.* Atlanta: Author.

2. ———. 1997. *ASHRAE Handbook: Fundamentals.* Atlanta: Author.

3. Haines, Roger W. and C. Lewis Wilson. 1998. *HVAC Systems Design Handbook, 3rd ed.* New York: McGraw-Hill.

4. Idelchik, I. E. 1991. *Fluid Dynamics of Industrial Equipment.* Washington, DC: Hemisphere Publishing Corporation.

5. Sheet Metal and Air-Conditioning Contractors National Association (SMACNA). 1995. *HVAC Duct Construction Standards—Metal and Flexible, 2nd ed.* Chantilly, VA: Author.

6. Sheet Metal and Air-Conditioning Contractors National Association (SMACNA). 1990. *HVAC Systems—Duct Design, 3rd ed.* Chantilly, VA: Author.

7. The Trane Company. 1996. *Air-Conditioning Manual.* La Crosse, WI: Author.

8. Sun, Tseng-Yao. 1994. *Air Handling Systems Design.* New York: McGraw-Hill.

# PRACTICE PROBLEMS

## Energy Losses in Straight Duct Sections

**19.1E** Determine the velocity of flow and the friction loss as 1000 cfm of air flows through 75 ft of 18-in diameter round duct.

**19.2E** Repeat Problem 19.1 for duct diameters of 16, 14, 12, and 10 in. Then, plot velocity and friction loss versus duct diameter.

**19.3E** Specify a diameter for a round duct suitable for carrying 1500 cfm of air with a maximum pressure drop of 0.10 in $H_2O$ per 100 ft of duct, rounding up to the next inch. For the actual size specified, give the friction loss per 100 ft of duct.

**19.4M** *Determine the velocity of flow and the friction loss as 3.0 $m^3/s$ of air flows through 25 m of 500 mm diameter round duct.*

**19.5M** *Repeat Problem 19.4 for duct diameters of 600, 700, 800, 900, and 1000 mm. Then, plot velocity and friction loss versus duct diameter.*

**19.6M** *Specify a diameter for a round duct suitable for carrying 0.40 $m^3/s$ of air with a maximum pressure drop of 1.00 Pa/m of duct, rounding up to the next 50-mm increment. For the actual size specified, give the friction loss in Pa/m.*

**19.7E** A heating duct for a forced air furnace measures 10 × 30 in. Compute the circular equivalent diameter. Then, determine the maximum flow rate of air that the duct could carry while limiting the friction loss to 0.10 in $H_2O$ per 100 ft.

**19.8E** A branch duct for a heating system measures 3 × 10 in. Compute the circular equivalent diameter. Then, determine the maximum flow rate of air that the duct could carry while limiting the friction loss to 0.10 in $H_2O$ per 100 ft.

**19.9E** A ventilation duct in a large industrial warehouse measures 42 × 60 in. Compute the circular equivalent diameter. Then, determine the maximum flow rate of air that the duct could carry while limiting the friction loss to 0.10 in $H_2O$ per 100 ft.

**19.10M** *A heating duct for a forced air furnace measures 250 × 500 mm. Compute the circular equivalent diam-*

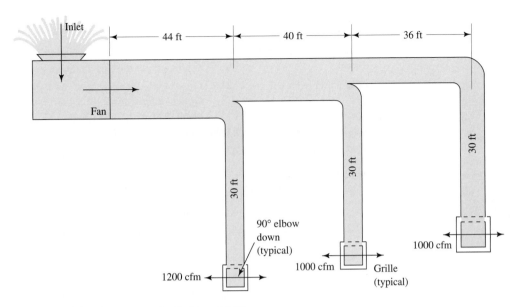

**FIGURE 19.5**  Duct system for Problem 19.27.

eter. Then, determine the maximum flow rate of air that the duct could carry while limiting the friction loss to 0.80 Pa/m.

**19.11M** *A branch duct for a heating system measures 75 × 250 mm. Compute the circular equivalent diameter. Then, determine the maximum flow rate of air that it could carry while limiting the friction loss to 0.80 Pa/m.*

**19.12E** Specify a size for a rectangular duct suitable for carrying 1500 cfm of air with a maximum pressure drop of

0.10 in $H_2O$ per 100 ft of duct. The maximum vertical height of the duct is 12.0 in.

**19.13E** Specify a size for a rectangular duct suitable for carrying 300 cfm of air with a maximum pressure drop of 0.10 in $H_2O$ per 100 ft of duct. The maximum vertical height of the duct is 6.0 in.

### Energy Losses in Ducts with Fittings

**19.14E** Compute the pressure drop as 650 cfm of air flows through a three-piece 90° elbow in a 12-in diameter round duct.

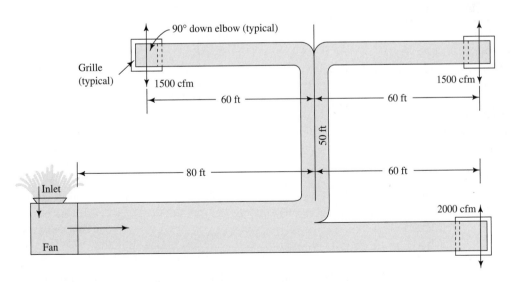

**FIGURE 19.6**  Duct system for Problem 19.28.

**19.15E** Repeat Problem 19.14, but use a 5-piece elbow.

**19.16E** Compute the pressure drop as 1500 cfm of air flows through a wide open damper installed in a 16-in diameter duct.

**19.17E** Repeat Problem 19.16 with the damper partially closed at 10°, 20°, and 30°.

**19.18E** A part of a duct system is a 10-by-22-in rectangular main duct carrying 1600 cfm of air. A tee to a branch duct, 10 × 10 in, draws 500 cfm from the main. The main duct remains the same size downstream from the branch. Determine the velocity of flow and the velocity pressure in all parts of the duct.

**19.19E** For the conditions of Problem 19.18, estimate the loss in pressure as the flow enters the branch duct through the tee.

**19.20E** For the conditions of Problem 19.18, estimate the loss in pressure for the flow in the main duct due to the tee.

**19.21M** *Compute the pressure drop as 0.20 m³/s of air flows through a three-piece 90° elbow in a 200-mm diameter round duct.*

**19.22M** *Repeat Problem 19.21, but use a mitered elbow.*

**19.23M** *Compute the pressure drop as 0.85 m³/s of air flows through a damper set at 30° installed in a 400-mm diameter duct.*

**19.24E** A section of duct system consists of 42 ft of straight 12-in diameter round duct, a wide open damper, two

**FIGURE 19.7**   Duct system for Problem 19.29.

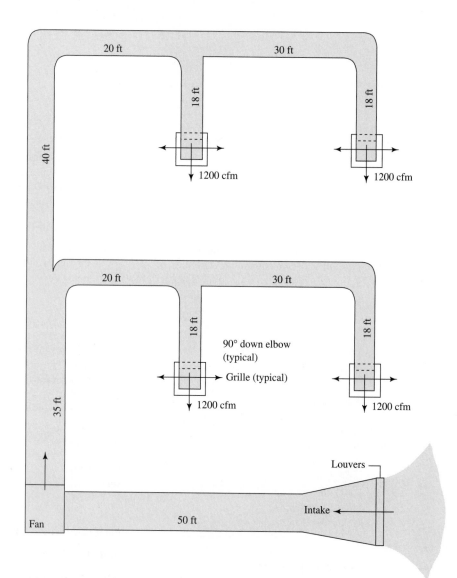

**FIGURE 19.8** Duct system for Problem 19.30.

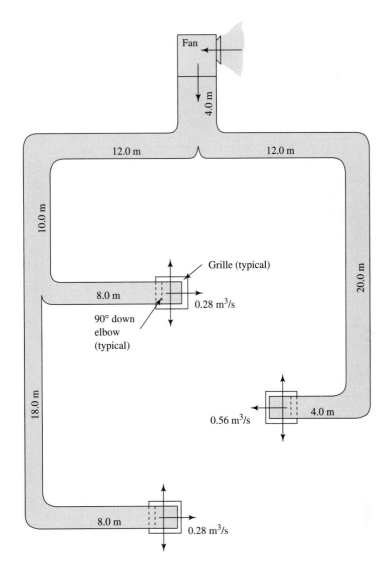

three-piece 90° elbows, and an outlet grille. Compute the pressure drop along this duct section for Q = 700 cfm.

**19.25E** A section of duct system consists of 38 ft of straight 12-by-20-in rectangular duct, a wide open damper, three smooth 90° elbows, and an outlet grille. Compute the pressure drop along this duct section for Q = 1500 cfm.

**19.26M** *The intake duct to a fan consists of intake louvers, 5.8 m of square duct (800 × 800 mm), a sudden contraction to a 400-mm diameter round duct, and 9.25 m of the round duct. Estimate the pressure at the fan inlet when the duct carries 0.80 m³/s of air.*

## Design of Ducts

For the conditions shown in Figs. 19.5 through 19.8, complete the design of the duct system by specifying the sizes for all duct sections necessary to achieve a system that is balanced when it carries the flow rates shown. Compute the pressure at the fan outlet, assuming that the final outlets from the duct system are to atmospheric pressure. When an inlet duct section is shown, also complete its design and compute the pressure at the fan inlet. Note that there is no single best solution to these problems, and several design decisions must be made. It may be desirable to change certain features of the suggested system design to improve its operation or to make it simpler to achieve a balanced system.

**19.27E** Use Fig. 19.5 on page 547.

**19.28E** Use Fig. 19.6 on page 547.

**19.29E** Use Fig. 19.7 on page 548.

**19.30M** *Use Fig. 19.8.*

# A Properties of Water

**TABLE A.1**  SI Units [101 kPa (abs)]

| Temperature ($°C$) | Specific Weight $\gamma$ ($kN/m^3$) | Density $\rho$ ($kg/m^3$) | Dynamic Viscosity $\mu$ ($Pa·s$) or ($N·s/m^2$) | Kinematic Viscosity $\nu$ ($m^2/s$) |
|---|---|---|---|---|
| 0 | 9.81 | 1000 | $1.75 \times 10^{-3}$ | $1.75 \times 10^{-6}$ |
| 5 | 9.81 | 1000 | $1.52 \times 10^{-3}$ | $1.52 \times 10^{-6}$ |
| 10 | 9.81 | 1000 | $1.30 \times 10^{-3}$ | $1.30 \times 10^{-6}$ |
| 15 | 9.81 | 1000 | $1.15 \times 10^{-3}$ | $1.15 \times 10^{-6}$ |
| 20 | 9.79 | 998 | $1.02 \times 10^{-3}$ | $1.02 \times 10^{-6}$ |
| 25 | 9.78 | 997 | $8.91 \times 10^{-4}$ | $8.94 \times 10^{-7}$ |
| 30 | 9.77 | 996 | $8.00 \times 10^{-4}$ | $8.03 \times 10^{-7}$ |
| 35 | 9.75 | 994 | $7.18 \times 10^{-4}$ | $7.22 \times 10^{-7}$ |
| 40 | 9.73 | 992 | $6.51 \times 10^{-4}$ | $6.56 \times 10^{-7}$ |
| 45 | 9.71 | 990 | $5.94 \times 10^{-4}$ | $6.00 \times 10^{-7}$ |
| 50 | 9.69 | 988 | $5.41 \times 10^{-4}$ | $5.48 \times 10^{-7}$ |
| 55 | 9.67 | 986 | $4.98 \times 10^{-4}$ | $5.05 \times 10^{-7}$ |
| 60 | 9.65 | 984 | $4.60 \times 10^{-4}$ | $4.67 \times 10^{-7}$ |
| 65 | 9.62 | 981 | $4.31 \times 10^{-4}$ | $4.39 \times 10^{-7}$ |
| 70 | 9.59 | 978 | $4.02 \times 10^{-4}$ | $4.11 \times 10^{-7}$ |
| 75 | 9.56 | 975 | $3.73 \times 10^{-4}$ | $3.83 \times 10^{-7}$ |
| 80 | 9.53 | 971 | $3.50 \times 10^{-4}$ | $3.60 \times 10^{-7}$ |
| 85 | 9.50 | 968 | $3.30 \times 10^{-4}$ | $3.41 \times 10^{-7}$ |
| 90 | 9.47 | 965 | $3.11 \times 10^{-4}$ | $3.22 \times 10^{-7}$ |
| 95 | 9.44 | 962 | $2.92 \times 10^{-4}$ | $3.04 \times 10^{-7}$ |
| 100 | 9.40 | 958 | $2.82 \times 10^{-4}$ | $2.94 \times 10^{-7}$ |

**TABLE A.2**  U.S. Customary
System Units (14.7 psia)

| Temperature (°F) | Specific Weight $\gamma$ (lb/ft$^3$) | Density $\rho$ (slugs/ft$^3$) | Dynamic Viscosity $\mu$ (lb-s/ft$^2$) | Kinematic Viscosity $\nu$ (ft$^2$/s) |
|---|---|---|---|---|
| 32 | 62.4 | 1.94 | $3.66 \times 10^{-5}$ | $1.89 \times 10^{-5}$ |
| 40 | 62.4 | 1.94 | $3.23 \times 10^{-5}$ | $1.67 \times 10^{-5}$ |
| 50 | 62.4 | 1.94 | $2.72 \times 10^{-5}$ | $1.40 \times 10^{-5}$ |
| 60 | 62.4 | 1.94 | $2.35 \times 10^{-5}$ | $1.21 \times 10^{-5}$ |
| 70 | 62.3 | 1.94 | $2.04 \times 10^{-5}$ | $1.05 \times 10^{-5}$ |
| 80 | 62.2 | 1.93 | $1.77 \times 10^{-5}$ | $9.15 \times 10^{-6}$ |
| 90 | 62.1 | 1.93 | $1.60 \times 10^{-5}$ | $8.29 \times 10^{-6}$ |
| 100 | 62.0 | 1.93 | $1.42 \times 10^{-5}$ | $7.37 \times 10^{-6}$ |
| 110 | 61.9 | 1.92 | $1.26 \times 10^{-5}$ | $6.55 \times 10^{-6}$ |
| 120 | 61.7 | 1.92 | $1.14 \times 10^{-5}$ | $5.94 \times 10^{-6}$ |
| 130 | 61.5 | 1.91 | $1.05 \times 10^{-5}$ | $5.49 \times 10^{-6}$ |
| 140 | 61.4 | 1.91 | $9.60 \times 10^{-6}$ | $5.03 \times 10^{-6}$ |
| 150 | 61.2 | 1.90 | $8.90 \times 10^{-6}$ | $4.68 \times 10^{-6}$ |
| 160 | 61.0 | 1.90 | $8.30 \times 10^{-6}$ | $4.38 \times 10^{-6}$ |
| 170 | 60.8 | 1.89 | $7.70 \times 10^{-6}$ | $4.07 \times 10^{-6}$ |
| 180 | 60.6 | 1.88 | $7.23 \times 10^{-6}$ | $3.84 \times 10^{-6}$ |
| 190 | 60.4 | 1.88 | $6.80 \times 10^{-6}$ | $3.62 \times 10^{-6}$ |
| 200 | 60.1 | 1.87 | $6.25 \times 10^{-6}$ | $3.35 \times 10^{-6}$ |
| 212 | 59.8 | 1.86 | $5.89 \times 10^{-6}$ | $3.17 \times 10^{-6}$ |

# B Properties of Common Liquids

**TABLE B.1** SI Units [101 kPa (abs) and 25°C]

| | Specific Gravity sg | Specific Weight $\gamma$ (kN/m$^3$) | Density $\rho$ (kg/m$^3$) | Dynamic Viscosity $\mu$ (Pa·s) or (N·s/m$^2$) |
|---|---|---|---|---|
| Acetone | 0.787 | 7.72 | 787 | $3.16 \times 10^{-4}$ |
| Alcohol, ethyl | 0.787 | 7.72 | 787 | $1.00 \times 10^{-3}$ |
| Alcohol, methyl | 0.789 | 7.74 | 789 | $5.60 \times 10^{-4}$ |
| Alcohol, propyl | 0.802 | 7.87 | 802 | $1.92 \times 10^{-3}$ |
| Ammonia | 0.826 | 8.10 | 826 | — |
| Benzene | 0.876 | 8.59 | 876 | $6.03 \times 10^{-4}$ |
| Carbon tetrachloride | 1.590 | 15.60 | 1 590 | $9.10 \times 10^{-4}$ |
| Castor oil | 0.960 | 9.42 | 960 | $6.51 \times 10^{-1}$ |
| Ethylene glycol | 1.100 | 10.79 | 1 100 | $1.62 \times 10^{-2}$ |
| Gasoline | 0.68 | 6.67 | 680 | $2.87 \times 10^{-4}$ |
| Glycerine | 1.258 | 12.34 | 1 258 | $9.60 \times 10^{-1}$ |
| Kerosene | 0.823 | 8.07 | 823 | $1.64 \times 10^{-3}$ |
| Linseed oil | 0.930 | 9.12 | 930 | $3.31 \times 10^{-2}$ |
| Mercury | 13.54 | 132.8 | 13 540 | $1.53 \times 10^{-3}$ |
| Propane | 0.495 | 4.86 | 495 | $1.10 \times 10^{-4}$ |
| Seawater | 1.030 | 10.10 | 1 030 | $1.03 \times 10^{-3}$ |
| Turpentine | 0.870 | 8.53 | 870 | $1.37 \times 10^{-3}$ |
| Fuel oil, medium | 0.852 | 8.36 | 852 | $2.99 \times 10^{-3}$ |
| Fuel oil, heavy | 0.906 | 8.89 | 906 | $1.07 \times 10^{-1}$ |

**TABLE B.2**   U.S. Customary System Units (14.7 psia and 77°F)

|  | Specific Gravity sg | Specific Weight $\gamma$ (lb/ft$^3$) | Density $\rho$ (slugs/ft$^3$) | Dynamic Viscosity $\mu$ (lb-s/ft$^2$) |
|---|---|---|---|---|
| Acetone | 0.787 | 48.98 | 1.53 | $6.60 \times 10^{-6}$ |
| Alcohol, ethyl | 0.787 | 49.01 | 1.53 | $2.10 \times 10^{-5}$ |
| Alcohol, methyl | 0.789 | 49.10 | 1.53 | $1.17 \times 10^{-5}$ |
| Alcohol, propyl | 0.802 | 49.94 | 1.56 | $4.01 \times 10^{-5}$ |
| Ammonia | 0.826 | 51.41 | 1.60 | — |
| Benzene | 0.876 | 54.55 | 1.70 | $1.26 \times 10^{-5}$ |
| Carbon tetrachloride | 1.590 | 98.91 | 3.08 | $1.90 \times 10^{-5}$ |
| Castor oil | 0.960 | 59.69 | 1.86 | $1.36 \times 10^{-2}$ |
| Ethylene glycol | 1.100 | 68.47 | 2.13 | $3.38 \times 10^{-4}$ |
| Gasoline | 0.68 | 42.40 | 1.32 | $6.00 \times 10^{-6}$ |
| Glycerine | 1.258 | 78.50 | 2.44 | $2.00 \times 10^{-2}$ |
| Kerosene | 0.823 | 51.20 | 1.60 | $3.43 \times 10^{-5}$ |
| Linseed oil | 0.930 | 58.00 | 1.80 | $6.91 \times 10^{-4}$ |
| Mercury | 13.54 | 844.9 | 26.26 | $3.20 \times 10^{-5}$ |
| Propane | 0.495 | 30.81 | 0.96 | $2.30 \times 10^{-6}$ |
| Seawater | 1.030 | 64.00 | 2.00 | $2.15 \times 10^{-5}$ |
| Turpentine | 0.870 | 54.20 | 1.69 | $2.87 \times 10^{-5}$ |
| Fuel oil, medium | 0.852 | 53.16 | 1.65 | $6.25 \times 10^{-5}$ |
| Fuel oil, heavy | 0.906 | 56.53 | 1.76 | $2.24 \times 10^{-3}$ |

# C Typical Properties of Petroleum Lubricating Oils

| Type | Specific Gravity | Kinematic Viscosity, $\nu$ | | | | Viscosity Index |
|------|---|---|---|---|---|---|
| | | at 40°C (104°F) | | at 100°C (212°F) | | |
| | | ($m^2$/s) | ($ft^2$/s) | ($m^2$/s) | ($ft^2$/s) | |
| Automotive hydraulic systems | 0.887 | $3.99 \times 10^{-5}$ | $4.30 \times 10^{-4}$ | $7.29 \times 10^{-6}$ | $7.85 \times 10^{-5}$ | 141 |
| Machine tool hydraulic systems | | | | | | |
|   Light | 0.887 | $3.20 \times 10^{-5}$ | $3.44 \times 10^{-4}$ | $4.79 \times 10^{-6}$ | $5.16 \times 10^{-5}$ | 64 |
|   Medium | 0.895 | $6.70 \times 10^{-5}$ | $7.21 \times 10^{-4}$ | $7.29 \times 10^{-6}$ | $7.85 \times 10^{-5}$ | 66 |
|   Heavy | 0.901 | $1.96 \times 10^{-4}$ | $2.11 \times 10^{-3}$ | $1.40 \times 10^{-5}$ | $1.51 \times 10^{-4}$ | 70 |
|   Low temperature | 0.844 | $1.40 \times 10^{-5}$ | $1.51 \times 10^{-4}$ | $5.20 \times 10^{-6}$ | $5.60 \times 10^{-5}$ | 226 |
| Machine tool lubricating oils | | | | | | |
|   Light | 0.881 | $2.20 \times 10^{-5}$ | $2.37 \times 10^{-4}$ | $3.90 \times 10^{-6}$ | $4.20 \times 10^{-5}$ | 80 |
|   Medium | 0.915 | $6.60 \times 10^{-5}$ | $7.10 \times 10^{-4}$ | $7.00 \times 10^{-6}$ | $7.53 \times 10^{-5}$ | 83 |
|   Heavy | 0.890 | $2.00 \times 10^{-4}$ | $2.15 \times 10^{-3}$ | $1.55 \times 10^{-5}$ | $1.67 \times 10^{-4}$ | 80 |

See also Tables 2.4 and 2.5 in Chapter 2 for properties of SAE grades of engine and transmission oils.

# D Variation of Viscosity with Temperature

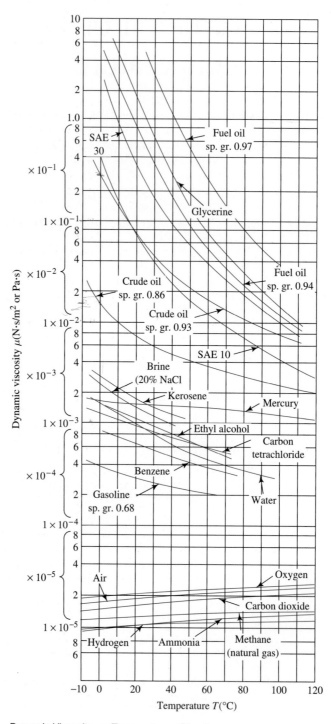

Dynamic Viscosity vs. Temperature—SI units

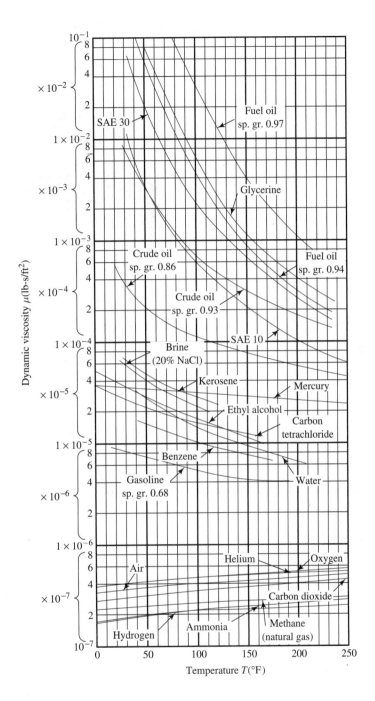

Dynamic Viscosity vs. Temperature—U.S. Customary Units

# E Properties of Air

**TABLE E.1** Properties of air versus temperature in SI units at standard atmospheric pressure

| Temperature $T$ (°C) | Density $\rho$ (kg/m$^3$) | Specific Weight $\gamma$ (N/m$^3$) | Dynamic Viscosity $\mu$ (Pa·s) | Kinematic Viscosity $\nu$ (m$^2$/s) |
|---|---|---|---|---|
| −40 | 1.514 | 14.85 | $1.51 \times 10^{-5}$ | $9.98 \times 10^{-6}$ |
| −30 | 1.452 | 14.24 | $1.56 \times 10^{-5}$ | $1.08 \times 10^{-5}$ |
| −20 | 1.394 | 13.67 | $1.62 \times 10^{-5}$ | $1.16 \times 10^{-5}$ |
| −10 | 1.341 | 13.15 | $1.67 \times 10^{-5}$ | $1.24 \times 10^{-5}$ |
| 0 | 1.292 | 12.67 | $1.72 \times 10^{-5}$ | $1.33 \times 10^{-5}$ |
| 10 | 1.247 | 12.23 | $1.77 \times 10^{-5}$ | $1.42 \times 10^{-5}$ |
| 20 | 1.204 | 11.81 | $1.81 \times 10^{-5}$ | $1.51 \times 10^{-5}$ |
| 30 | 1.164 | 11.42 | $1.86 \times 10^{-5}$ | $1.60 \times 10^{-5}$ |
| 40 | 1.127 | 11.05 | $1.91 \times 10^{-5}$ | $1.69 \times 10^{-5}$ |
| 50 | 1.092 | 10.71 | $1.95 \times 10^{-5}$ | $1.79 \times 10^{-5}$ |
| 60 | 1.060 | 10.39 | $1.99 \times 10^{-5}$ | $1.89 \times 10^{-5}$ |
| 70 | 1.029 | 10.09 | $2.04 \times 10^{-5}$ | $1.99 \times 10^{-5}$ |
| 80 | 0.9995 | 9.802 | $2.09 \times 10^{-5}$ | $2.09 \times 10^{-5}$ |
| 90 | 0.9720 | 9.532 | $2.13 \times 10^{-5}$ | $2.19 \times 10^{-5}$ |
| 100 | 0.9459 | 9.277 | $2.17 \times 10^{-5}$ | $2.30 \times 10^{-5}$ |
| 110 | 0.9213 | 9.034 | $2.22 \times 10^{-5}$ | $2.40 \times 10^{-5}$ |
| 120 | 0.8978 | 8.805 | $2.26 \times 10^{-5}$ | $2.51 \times 10^{-5}$ |

*Note:* Properties of air for standard conditions at sea level are

| | |
|---|---|
| Temperature | 15°C |
| Pressure | 101.325 kPa |
| Density | 1.225 kg/m$^3$ |
| Specific weight | 12.01 N/m$^3$ |
| Dynamic viscosity | $1.789 \times 10^{-5}$ Pa·s |
| Kinematic viscosity | $1.46 \times 10^{-5}$ m$^2$/s |

**TABLE E.2**   Properties of air versus temperature in U.S. customary units at standard atmospheric pressure

| Temperature (°F) | Density (slugs/ft³) | Specific Weight (lb/ft³) | Dynamic Viscosity (lb-s/ft²) | Kinematic Viscosity (ft²/s) |
|---|---|---|---|---|
| −40 | $2.94 \times 10^{-3}$ | 0.0946 | $3.15 \times 10^{-7}$ | $1.07 \times 10^{-4}$ |
| −20 | $2.80 \times 10^{-3}$ | 0.0903 | $3.27 \times 10^{-7}$ | $1.17 \times 10^{-4}$ |
| 0 | $2.68 \times 10^{-3}$ | 0.0864 | $3.41 \times 10^{-7}$ | $1.27 \times 10^{-4}$ |
| 20 | $2.57 \times 10^{-3}$ | 0.0828 | $3.52 \times 10^{-7}$ | $1.37 \times 10^{-4}$ |
| 40 | $2.47 \times 10^{-3}$ | 0.0795 | $3.64 \times 10^{-7}$ | $1.47 \times 10^{-4}$ |
| 60 | $2.37 \times 10^{-3}$ | 0.0764 | $3.74 \times 10^{-7}$ | $1.58 \times 10^{-4}$ |
| 80 | $2.28 \times 10^{-3}$ | 0.0736 | $3.85 \times 10^{-7}$ | $1.69 \times 10^{-4}$ |
| 100 | $2.20 \times 10^{-3}$ | 0.0709 | $3.97 \times 10^{-7}$ | $1.80 \times 10^{-4}$ |
| 120 | $2.13 \times 10^{-3}$ | 0.0685 | $4.06 \times 10^{-7}$ | $1.91 \times 10^{-4}$ |
| 140 | $2.06 \times 10^{-3}$ | 0.0662 | $4.16 \times 10^{-7}$ | $2.02 \times 10^{-4}$ |
| 160 | $1.99 \times 10^{-3}$ | 0.0641 | $4.27 \times 10^{-7}$ | $2.15 \times 10^{-4}$ |
| 180 | $1.93 \times 10^{-3}$ | 0.0621 | $4.38 \times 10^{-7}$ | $2.27 \times 10^{-4}$ |
| 200 | $1.87 \times 10^{-3}$ | 0.0602 | $4.48 \times 10^{-7}$ | $2.40 \times 10^{-4}$ |
| 220 | $1.81 \times 10^{-3}$ | 0.0584 | $4.58 \times 10^{-7}$ | $2.52 \times 10^{-4}$ |
| 240 | $1.76 \times 10^{-3}$ | 0.0567 | $4.68 \times 10^{-7}$ | $2.66 \times 10^{-4}$ |

**TABLE E.3**  Properties of the atmosphere

| SI Units | | | | U.S. Customary System Units | | | |
|---|---|---|---|---|---|---|---|
| Altitude (m) | Temperature (°C) | Pressure (kPa) | Density (kg/m$^3$) | Altitude (ft) | Temperature (°F) | Pressure (psi) | Density (slugs/ft$^3$) |
| 0 | 15.00 | 101.3 | 1.225 | 0 | 59.00 | 14.696 | $2.38 \times 10^{-3}$ |
| 200 | 13.70 | 98.9 | 1.202 | 500 | 57.22 | 14.433 | $2.34 \times 10^{-3}$ |
| 400 | 12.40 | 96.6 | 1.179 | 1000 | 55.43 | 14.173 | $2.25 \times 10^{-3}$ |
| 600 | 11.10 | 94.3 | 1.156 | 5000 | 41.17 | 12.227 | $2.05 \times 10^{-3}$ |
| 800 | 9.80 | 92.1 | 1.134 | 10000 | 23.34 | 10.106 | $1.76 \times 10^{-3}$ |
| 1000 | 8.50 | 89.9 | 1.112 | 15000 | 5.51 | 8.293 | $1.50 \times 10^{-3}$ |
| 2000 | 2.00 | 79.5 | 1.007 | 20000 | -12.62 | 6.753 | $1.27 \times 10^{-3}$ |
| 3000 | -4.49 | 70.1 | 0.9093 | 30000 | -47.99 | 4.365 | $8.89 \times 10^{-4}$ |
| 4000 | -10.98 | 61.7 | 0.8194 | 40000 | -69.70 | 2.720 | $5.85 \times 10^{-4}$ |
| 5000 | -17.47 | 54.0 | 0.7364 | 50000 | -69.70 | 1.683 | $3.62 \times 10^{-4}$ |
| 10000 | -49.90 | 26.5 | 0.4135 | 60000 | -69.70 | 1.040 | $2.24 \times 10^{-4}$ |
| 15000 | -56.50 | 12.11 | 0.1948 | 70000 | -67.30 | 0.644 | $1.38 \times 10^{-4}$ |
| 20000 | -56.50 | 5.53 | 0.0889 | 80000 | -61.81 | 0.400 | $8.45 \times 10^{-5}$ |
| 25000 | -51.60 | 2.55 | 0.0401 | 90000 | -56.32 | 0.251 | $5.22 \times 10^{-5}$ |
| 30000 | -46.64 | 1.20 | 0.0184 | 100000 | -50.84 | 0.158 | $3.25 \times 10^{-5}$ |

Data from *U.S. Standard Atmosphere, 1976,* NOAA-S/T76-1562. Washington, D.C.: National Oceanic and Atmospheric Administration.

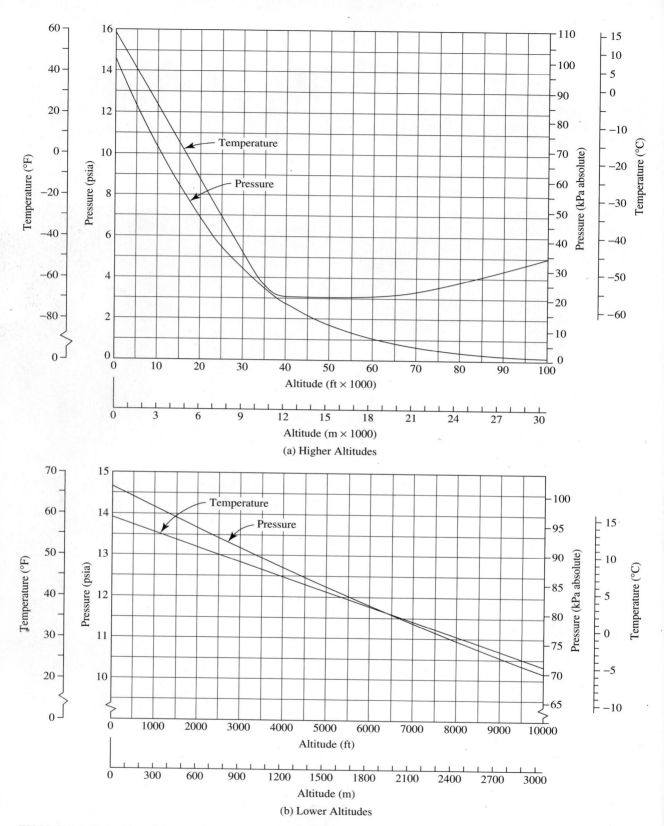

**FIGURE E.1** Properties of the standard atmosphere versus altitude

# F Dimensions of Steel Pipe

**TABLE F.1**  Schedule 40

| Nominal Pipe Size (in) | Outside Diameter (in) | Outside Diameter (mm) | Wall Thickness (in) | Wall Thickness (mm) | Inside Diameter (in) | Inside Diameter (ft) | Inside Diameter (mm) | Flow Area (ft²) | Flow Area (m²) |
|---|---|---|---|---|---|---|---|---|---|
| 1/8 | 0.405 | 10.3 | 0.068 | 1.73 | 0.269 | 0.0224 | 6.8 | 0.000 394 | $3.660 \times 10^{-5}$ |
| 1/4 | 0.540 | 13.7 | 0.088 | 2.24 | 0.364 | 0.0303 | 9.2 | 0.000 723 | $6.717 \times 10^{-5}$ |
| 3/8 | 0.675 | 17.1 | 0.091 | 2.31 | 0.493 | 0.0411 | 12.5 | 0.001 33 | $1.236 \times 10^{-4}$ |
| 1/2 | 0.840 | 21.3 | 0.109 | 2.77 | 0.622 | 0.0518 | 15.8 | 0.002 11 | $1.960 \times 10^{-4}$ |
| 3/4 | 1.050 | 26.7 | 0.113 | 2.87 | 0.824 | 0.0687 | 20.9 | 0.003 70 | $3.437 \times 10^{-4}$ |
| 1 | 1.315 | 33.4 | 0.133 | 3.38 | 1.049 | 0.0874 | 26.6 | 0.006 00 | $5.574 \times 10^{-4}$ |
| 1 1/4 | 1.660 | 42.2 | 0.140 | 3.56 | 1.380 | 0.1150 | 35.1 | 0.010 39 | $9.653 \times 10^{-4}$ |
| 1 1/2 | 1.900 | 48.3 | 0.145 | 3.68 | 1.610 | 0.1342 | 40.9 | 0.014 14 | $1.314 \times 10^{-3}$ |
| 2 | 2.375 | 60.3 | 0.154 | 3.91 | 2.067 | 0.1723 | 52.5 | 0.023 33 | $2.168 \times 10^{-3}$ |
| 2 1/2 | 2.875 | 73.0 | 0.203 | 5.16 | 2.469 | 0.2058 | 62.7 | 0.033 26 | $3.090 \times 10^{-3}$ |
| 3 | 3.500 | 88.9 | 0.216 | 5.49 | 3.068 | 0.2557 | 77.9 | 0.051 32 | $4.768 \times 10^{-3}$ |
| 3 1/2 | 4.000 | 101.6 | 0.226 | 5.74 | 3.548 | 0.2957 | 90.1 | 0.068 68 | $6.381 \times 10^{-3}$ |
| 4 | 4.500 | 114.3 | 0.237 | 6.02 | 4.026 | 0.3355 | 102.3 | 0.088 40 | $8.213 \times 10^{-3}$ |
| 5 | 5.563 | 141.3 | 0.258 | 6.55 | 5.047 | 0.4206 | 128.2 | 0.139 0 | $1.291 \times 10^{-2}$ |
| 6 | 6.625 | 168.3 | 0.280 | 7.11 | 6.065 | 0.5054 | 154.1 | 0.200 6 | $1.864 \times 10^{-2}$ |
| 8 | 8.625 | 219.1 | 0.322 | 8.18 | 7.981 | 0.6651 | 202.7 | 0.347 2 | $3.226 \times 10^{-2}$ |
| 10 | 10.750 | 273.1 | 0.365 | 9.27 | 10.020 | 0.8350 | 254.5 | 0.547 9 | $5.090 \times 10^{-2}$ |
| 12 | 12.750 | 323.9 | 0.406 | 10.31 | 11.938 | 0.9948 | 303.2 | 0.777 1 | $7.219 \times 10^{-2}$ |
| 14 | 14.000 | 355.6 | 0.437 | 11.10 | 13.126 | 1.094 | 333.4 | 0.939 6 | $8.729 \times 10^{-2}$ |
| 16 | 16.000 | 406.4 | 0.500 | 12.70 | 15.000 | 1.250 | 381.0 | 1.227 | 0.1140 |
| 18 | 18.000 | 457.2 | 0.562 | 14.27 | 16.876 | 1.406 | 428.7 | 1.553 | 0.1443 |
| 20 | 20.000 | 508.0 | 0.593 | 15.06 | 18.814 | 1.568 | 477.9 | 1.931 | 0.1794 |
| 24 | 24.000 | 609.6 | 0.687 | 17.45 | 22.626 | 1.886 | 574.7 | 2.792 | 0.2594 |

**TABLE F.2**    Schedule 80

| Nominal Pipe Size (in) | Outside Diameter | | Wall Thickness | | Inside Diameter | | | Flow Area | |
|---|---|---|---|---|---|---|---|---|---|
| | (in) | (mm) | (in) | (mm) | (in) | (ft) | (mm) | (ft$^2$) | (m$^2$) |
| 1/8 | 0.405 | 10.3 | 0.095 | 2.41 | 0.215 | 0.017 92 | 5.5 | 0.000 253 | $2.350 \times 10^{-5}$ |
| 1/4 | 0.540 | 13.7 | 0.119 | 3.02 | 0.302 | 0.025 17 | 7.7 | 0.000 497 | $4.617 \times 10^{-5}$ |
| 3/8 | 0.675 | 17.1 | 0.126 | 3.20 | 0.423 | 0.035 25 | 10.7 | 0.000 976 | $9.067 \times 10^{-5}$ |
| 1/2 | 0.840 | 21.3 | 0.147 | 3.73 | 0.546 | 0.045 50 | 13.9 | 0.001 625 | $1.510 \times 10^{-4}$ |
| 3/4 | 1.050 | 26.7 | 0.154 | 3.91 | 0.742 | 0.061 83 | 18.8 | 0.003 00 | $2.787 \times 10^{-4}$ |
| 1 | 1.315 | 33.4 | 0.179 | 4.55 | 0.957 | 0.079 75 | 24.3 | 0.004 99 | $4.636 \times 10^{-4}$ |
| 1 1/4 | 1.660 | 42.2 | 0.191 | 4.85 | 1.278 | 0.106 5 | 32.5 | 0.008 91 | $8.278 \times 10^{-4}$ |
| 1 1/2 | 1.900 | 48.3 | 0.200 | 5.08 | 1.500 | 0.125 0 | 38.1 | 0.012 27 | $1.140 \times 10^{-3}$ |
| 2 | 2.375 | 60.3 | 0.218 | 5.54 | 1.939 | 0.161 6 | 49.3 | 0.020 51 | $1.905 \times 10^{-3}$ |
| 2 1/2 | 2.875 | 73.0 | 0.276 | 7.01 | 2.323 | 0.193 6 | 59.0 | 0.029 44 | $2.735 \times 10^{-3}$ |
| 3 | 3.500 | 88.9 | 0.300 | 7.62 | 2.900 | 0.241 7 | 73.7 | 0.045 90 | $4.264 \times 10^{-3}$ |
| 3 1/2 | 4.000 | 101.6 | 0.318 | 8.08 | 3.364 | 0.280 3 | 85.4 | 0.061 74 | $5.736 \times 10^{-3}$ |
| 4 | 4.500 | 114.3 | 0.337 | 8.56 | 3.826 | 0.318 8 | 97.2 | 0.079 86 | $7.419 \times 10^{-3}$ |
| 5 | 5.563 | 141.3 | 0.375 | 9.53 | 4.813 | 0.401 1 | 122.3 | 0.126 3 | $1.173 \times 10^{-2}$ |
| 6 | 6.625 | 168.3 | 0.432 | 10.97 | 5.761 | 0.480 1 | 146.3 | 0.181 0 | $1.682 \times 10^{-2}$ |
| 8 | 8.625 | 219.1 | 0.500 | 12.70 | 7.625 | 0.635 4 | 193.7 | 0.317 4 | $2.949 \times 10^{-2}$ |
| 10 | 10.750 | 273.1 | 0.593 | 15.06 | 9.564 | 0.797 0 | 242.9 | 0.498 6 | $4.632 \times 10^{-2}$ |
| 12 | 12.750 | 323.9 | 0.687 | 17.45 | 11.376 | 0.948 0 | 289.0 | 0.705 6 | $6.555 \times 10^{-2}$ |
| 14 | 14.000 | 355.6 | 0.750 | 19.05 | 12.500 | 1.042 | 317.5 | 0.852 1 | $7.916 \times 10^{-2}$ |
| 16 | 16.000 | 406.4 | 0.842 | 21.39 | 14.314 | 1.193 | 363.6 | 1.117 | 0.1038 |
| 18 | 18.000 | 457.2 | 0.937 | 23.80 | 16.126 | 1.344 | 409.6 | 1.418 | 0.1317 |
| 20 | 20.000 | 508.0 | 1.031 | 26.19 | 17.938 | 1.495 | 455.6 | 1.755 | 0.1630 |
| 24 | 24.000 | 609.6 | 1.218 | 30.94 | 21.564 | 1.797 | 547.7 | 2.535 | 0.2344 |

# G Dimensions of Steel Tubing

| Outside Diameter | | Wall Thickness | | Inside Diameter | | | Flow Area | |
|---|---|---|---|---|---|---|---|---|
| (in) | (mm) | (in) | (mm) | (in) | (ft) | (mm) | (ft²) | (m²) |
| 1/8 | 3.18 | 0.032 | 0.813 | 0.061 | 0.00508 | 1.549 | $2.029 \times 10^{-5}$ | $1.885 \times 10^{-6}$ |
|  |  | 0.035 | 0.889 | 0.055 | 0.00458 | 1.397 | $1.650 \times 10^{-5}$ | $1.533 \times 10^{-6}$ |
| 3/16 | 4.76 | 0.032 | 0.813 | 0.124 | 0.01029 | 3.137 | $8.319 \times 10^{-5}$ | $7.728 \times 10^{-6}$ |
|  |  | 0.035 | 0.889 | 0.117 | 0.00979 | 2.985 | $7.530 \times 10^{-5}$ | $6.996 \times 10^{-6}$ |
| 1/4 | 6.35 | 0.035 | 0.889 | 0.180 | 0.01500 | 4.572 | $1.767 \times 10^{-4}$ | $1.642 \times 10^{-5}$ |
|  |  | 0.049 | 1.24 | 0.152 | 0.01267 | 3.861 | $1.260 \times 10^{-4}$ | $1.171 \times 10^{-5}$ |
| 5/16 | 7.94 | 0.035 | 0.889 | 0.243 | 0.02021 | 6.160 | $3.207 \times 10^{-4}$ | $2.980 \times 10^{-5}$ |
|  |  | 0.049 | 1.24 | 0.215 | 0.01788 | 5.448 | $2.509 \times 10^{-4}$ | $2.331 \times 10^{-5}$ |
| 3/8 | 9.53 | 0.035 | 0.889 | 0.305 | 0.02542 | 7.747 | $5.074 \times 10^{-4}$ | $4.714 \times 10^{-5}$ |
|  |  | 0.049 | 1.24 | 0.277 | 0.02308 | 7.036 | $4.185 \times 10^{-4}$ | $3.888 \times 10^{-5}$ |
| 1/2 | 12.70 | 0.049 | 1.24 | 0.402 | 0.03350 | 10.21 | $8.814 \times 10^{-4}$ | $8.189 \times 10^{-5}$ |
|  |  | 0.065 | 1.65 | 0.370 | 0.03083 | 9.40 | $7.467 \times 10^{-4}$ | $6.937 \times 10^{-5}$ |
| 5/8 | 15.88 | 0.049 | 1.24 | 0.527 | 0.04392 | 13.39 | $1.515 \times 10^{-3}$ | $1.407 \times 10^{-4}$ |
|  |  | 0.065 | 1.65 | 0.495 | 0.04125 | 12.57 | $1.336 \times 10^{-3}$ | $1.242 \times 10^{-4}$ |
| 3/4 | 19.05 | 0.049 | 1.24 | 0.652 | 0.05433 | 16.56 | $2.319 \times 10^{-3}$ | $2.154 \times 10^{-4}$ |
|  |  | 0.065 | 1.65 | 0.620 | 0.05167 | 15.75 | $2.097 \times 10^{-3}$ | $1.948 \times 10^{-4}$ |
| 7/8 | 22.23 | 0.049 | 1.24 | 0.777 | 0.06475 | 19.74 | $3.293 \times 10^{-3}$ | $3.059 \times 10^{-4}$ |
|  |  | 0.065 | 1.65 | 0.745 | 0.06208 | 18.92 | $3.027 \times 10^{-3}$ | $2.812 \times 10^{-4}$ |
| 1 | 25.40 | 0.065 | 1.65 | 0.870 | 0.07250 | 22.10 | $4.128 \times 10^{-3}$ | $3.835 \times 10^{-4}$ |
|  |  | 0.083 | 2.11 | 0.834 | 0.06950 | 21.18 | $3.794 \times 10^{-3}$ | $3.524 \times 10^{-4}$ |
| 1 1/4 | 31.75 | 0.065 | 1.65 | 1.120 | 0.09333 | 28.45 | $6.842 \times 10^{-3}$ | $6.356 \times 10^{-4}$ |
|  |  | 0.083 | 2.11 | 1.084 | 0.09033 | 27.53 | $6.409 \times 10^{-3}$ | $5.954 \times 10^{-4}$ |
| 1 1/2 | 38.10 | 0.065 | 1.65 | 1.370 | 0.1142 | 34.80 | $1.024 \times 10^{-2}$ | $9.510 \times 10^{-4}$ |
|  |  | 0.083 | 2.11 | 1.334 | 0.1112 | 33.88 | $9.706 \times 10^{-3}$ | $9.017 \times 10^{-4}$ |
| 1 3/4 | 44.45 | 0.065 | 1.65 | 1.620 | 0.1350 | 41.15 | $1.431 \times 10^{-2}$ | $1.330 \times 10^{-3}$ |
|  |  | 0.083 | 2.11 | 1.584 | 0.1320 | 40.23 | $1.368 \times 10^{-2}$ | $1.271 \times 10^{-3}$ |
| 2 | 50.80 | 0.065 | 1.65 | 1.870 | 0.1558 | 47.50 | $1.907 \times 10^{-2}$ | $1.772 \times 10^{-3}$ |
|  |  | 0.083 | 2.11 | 1.834 | 0.1528 | 46.58 | $1.835 \times 10^{-2}$ | $1.704 \times 10^{-3}$ |

# H Dimensions of Type K Copper Tubing

| Nominal Size | Outside Diameter | | Wall Thickness | | Inside Diameter | | | Flow Area | |
|---|---|---|---|---|---|---|---|---|---|
| (in) | (in) | (mm) | (in) | (mm) | (in) | (ft) | (mm) | (ft²) | (m²) |
| 1/8 | 0.250 | 6.35 | 0.035 | 0.889 | 0.180 | 0.0150 | 4.572 | $1.767 \times 10^{-4}$ | $1.642 \times 10^{-5}$ |
| 1/4 | 0.375 | 9.53 | 0.049 | 1.245 | 0.277 | 0.0231 | 7.036 | $4.185 \times 10^{-4}$ | $3.888 \times 10^{-5}$ |
| 3/8 | 0.500 | 12.70 | 0.049 | 1.245 | 0.402 | 0.0335 | 10.21 | $8.814 \times 10^{-4}$ | $8.189 \times 10^{-5}$ |
| 1/2 | 0.625 | 15.88 | 0.049 | 1.245 | 0.527 | 0.0439 | 13.39 | $1.515 \times 10^{-3}$ | $1.407 \times 10^{-4}$ |
| 5/8 | 0.750 | 19.05 | 0.049 | 1.245 | 0.652 | 0.0543 | 16.56 | $2.319 \times 10^{-3}$ | $2.154 \times 10^{-4}$ |
| 3/4 | 0.875 | 22.23 | 0.065 | 1.651 | 0.745 | 0.0621 | 18.92 | $3.027 \times 10^{-3}$ | $2.812 \times 10^{-4}$ |
| 1 | 1.125 | 28.58 | 0.065 | 1.651 | 0.995 | 0.0829 | 25.27 | $5.400 \times 10^{-3}$ | $5.017 \times 10^{-4}$ |
| 1 1/4 | 1.375 | 34.93 | 0.065 | 1.651 | 1.245 | 0.1037 | 31.62 | $8.454 \times 10^{-3}$ | $7.854 \times 10^{-4}$ |
| 1 1/2 | 1.625 | 41.28 | 0.072 | 1.829 | 1.481 | 0.1234 | 37.62 | $1.196 \times 10^{-2}$ | $1.111 \times 10^{-3}$ |
| 2 | 2.125 | 53.98 | 0.083 | 2.108 | 1.959 | 0.1632 | 49.76 | $2.093 \times 10^{-2}$ | $1.945 \times 10^{-3}$ |
| 2 1/2 | 2.625 | 66.68 | 0.095 | 2.413 | 2.435 | 0.2029 | 61.85 | $3.234 \times 10^{-2}$ | $3.004 \times 10^{-3}$ |
| 3 | 3.125 | 79.38 | 0.109 | 2.769 | 2.907 | 0.2423 | 73.84 | $4.609 \times 10^{-2}$ | $4.282 \times 10^{-3}$ |
| 3 1/2 | 3.625 | 92.08 | 0.120 | 3.048 | 3.385 | 0.2821 | 85.98 | $6.249 \times 10^{-2}$ | $5.806 \times 10^{-3}$ |
| 4 | 4.125 | 104.8 | 0.134 | 3.404 | 3.857 | 0.3214 | 97.97 | $8.114 \times 10^{-2}$ | $7.538 \times 10^{-3}$ |
| 5 | 5.125 | 130.2 | 0.160 | 4.064 | 4.805 | 0.4004 | 122.0 | $1.259 \times 10^{-1}$ | $1.170 \times 10^{-2}$ |
| 6 | 6.125 | 155.6 | 0.192 | 4.877 | 5.741 | 0.4784 | 145.8 | $1.798 \times 10^{-1}$ | $1.670 \times 10^{-2}$ |
| 8 | 8.125 | 206.4 | 0.271 | 6.883 | 7.583 | 0.6319 | 192.6 | $3.136 \times 10^{-1}$ | $2.914 \times 10^{-2}$ |
| 10 | 10.125 | 257.2 | 0.338 | 8.585 | 9.449 | 0.7874 | 240.0 | $4.870 \times 10^{-1}$ | $4.524 \times 10^{-2}$ |
| 12 | 12.125 | 308.0 | 0.405 | 10.287 | 11.315 | 0.9429 | 287.4 | $6.983 \times 10^{-1}$ | $6.487 \times 10^{-2}$ |

# I Dimensions of Ductile Iron Pipe

**TABLE I.1**  Class 150 for 150 psi (1.03 MPa) pressure service

| Nominal Pipe Size | Outside Diameter | | Wall Thickness | | Inside Diameter | | | Flow Area | |
|---|---|---|---|---|---|---|---|---|---|
| (in) | (in) | (mm) | (in) | (mm) | (in) | (ft) | (mm) | (ft$^2$) | (m$^2$) |
| 3 | 3.96 | 100.6 | 0.320 | 8.13 | 3.32 | 0.277 | 84.3 | 0.0601 | $5.585 \times 10^{-3}$ |
| 4 | 4.80 | 121.9 | 0.350 | 8.89 | 4.10 | 0.342 | 104.1 | 0.0917 | $8.518 \times 10^{-3}$ |
| 6 | 6.90 | 175.3 | 0.380 | 9.65 | 6.14 | 0.512 | 156.0 | 0.2056 | $1.910 \times 10^{-2}$ |
| 8 | 9.05 | 229.9 | 0.410 | 10.41 | 8.23 | 0.686 | 209.0 | 0.3694 | $3.432 \times 10^{-2}$ |
| 10 | 11.10 | 281.9 | 0.440 | 11.18 | 10.22 | 0.852 | 259.6 | 0.5697 | $5.292 \times 10^{-2}$ |
| 12 | 13.20 | 335.3 | 0.480 | 12.19 | 12.24 | 1.020 | 310.9 | 0.8171 | $7.591 \times 10^{-2}$ |
| 14 | 15.65 | 397.5 | 0.510 | 12.95 | 14.63 | 1.219 | 371.6 | 1.167 | 0.1085 |
| 16 | 17.80 | 452.1 | 0.540 | 13.72 | 16.72 | 1.393 | 424.7 | 1.525 | 0.1417 |
| 18 | 19.92 | 506.0 | 0.580 | 14.73 | 18.76 | 1.563 | 476.5 | 1.920 | 0.1783 |
| 20 | 22.06 | 560.3 | 0.620 | 15.75 | 20.82 | 1.735 | 528.8 | 2.364 | 0.2196 |
| 24 | 26.32 | 668.5 | 0.730 | 18.54 | 24.86 | 2.072 | 631.4 | 3.371 | 0.3132 |

# J Areas of Circles

■ ■ ■ ■

**TABLE J.1** U.S. Customary
System Units

| Diameter | | Area | |
|---|---|---|---|
| (in) | (ft) | (in$^2$) | (ft$^2$) |
| 0.25 | 0.0208 | 0.0491 | $3.409 \times 10^{-4}$ |
| 0.50 | 0.0417 | 0.1963 | $1.364 \times 10^{-3}$ |
| 0.75 | 0.0625 | 0.4418 | $3.068 \times 10^{-3}$ |
| 1.00 | 0.0833 | 0.7854 | $5.454 \times 10^{-3}$ |
| 1.25 | 0.1042 | 1.227 | $8.522 \times 10^{-3}$ |
| 1.50 | 0.1250 | 1.767 | $1.227 \times 10^{-2}$ |
| 1.75 | 0.1458 | 2.405 | $1.670 \times 10^{-2}$ |
| 2.00 | 0.1667 | 3.142 | $2.182 \times 10^{-2}$ |
| 2.50 | 0.2083 | 4.909 | $3.409 \times 10^{-2}$ |
| 3.00 | 0.2500 | 7.069 | $4.909 \times 10^{-2}$ |
| 3.50 | 0.2917 | 9.621 | $6.681 \times 10^{-2}$ |
| 4.00 | 0.3333 | 12.57 | $8.727 \times 10^{-2}$ |
| 4.50 | 0.3750 | 15.90 | 0.1104 |
| 5.00 | 0.4167 | 19.63 | 0.1364 |
| 6.00 | 0.5000 | 28.27 | 0.1963 |
| 7.00 | 0.5833 | 38.48 | 0.2673 |
| 8.00 | 0.6667 | 50.27 | 0.3491 |
| 9.00 | 0.7500 | 63.62 | 0.4418 |
| 10.00 | 0.8333 | 78.54 | 0.5454 |
| 12.00 | 1.00 | 113.1 | 0.7854 |
| 18.00 | 1.50 | 254.5 | 1.767 |
| 24.00 | 2.00 | 452.4 | 3.142 |

**TABLE J.2**   SI Units

| Diameter | | Area | |
|---|---|---|---|
| (mm) | (m) | (mm$^2$) | (m$^2$) |
| 6 | 0.006 | 28.27 | $2.827 \times 10^{-5}$ |
| 12 | 0.012 | 113.1 | $1.131 \times 10^{-4}$ |
| 18 | 0.018 | 254.5 | $2.545 \times 10^{-4}$ |
| 25 | 0.025 | 490.9 | $4.909 \times 10^{-4}$ |
| 32 | 0.032 | 804.2 | $8.042 \times 10^{-4}$ |
| 40 | 0.040 | 1257 | $1.257 \times 10^{-3}$ |
| 45 | 0.045 | 1590 | $1.590 \times 10^{-3}$ |
| 50 | 0.050 | 1963 | $1.963 \times 10^{-3}$ |
| 60 | 0.060 | 2827 | $2.827 \times 10^{-3}$ |
| 75 | 0.075 | 4418 | $4.418 \times 10^{-3}$ |
| 90 | 0.090 | 6362 | $6.362 \times 10^{-3}$ |
| 100 | 0.100 | 7854 | $7.854 \times 10^{-3}$ |
| 115 | 0.115 | $1.039 \times 10^4$ | $1.039 \times 10^{-2}$ |
| 125 | 0.125 | $1.227 \times 10^4$ | $1.227 \times 10^{-2}$ |
| 150 | 0.150 | $1.767 \times 10^4$ | $1.767 \times 10^{-2}$ |
| 175 | 0.175 | $2.405 \times 10^4$ | $2.405 \times 10^{-2}$ |
| 200 | 0.200 | $3.142 \times 10^4$ | $3.142 \times 10^{-2}$ |
| 225 | 0.225 | $3.976 \times 10^4$ | $3.976 \times 10^{-2}$ |
| 250 | 0.250 | $4.909 \times 10^4$ | $4.909 \times 10^{-2}$ |
| 300 | 0.300 | $7.069 \times 10^4$ | $7.069 \times 10^{-2}$ |
| 450 | 0.450 | $1.590 \times 10^5$ | $1.590 \times 10^{-1}$ |
| 600 | 0.600 | $2.827 \times 10^5$ | $2.827 \times 10^{-1}$ |

# ■ ■ ■ ■ ■ K Conversion Factors

**TABLE K.1** Conversion factors

**Mass**    Standard SI unit: Kilogram (kg). Equivalent unit: $N \cdot s^2/m$.

| $\dfrac{14.59 \text{ kg}}{\text{slug}}$ | $\dfrac{32.174 \text{ lb}_m}{\text{slug}}$ | $\dfrac{2.205 \text{ lb}_m}{\text{kg}}$ | $\dfrac{453.6 \text{ grams}}{\text{lb}_m}$ | $\dfrac{2000 \text{ lb}_m}{\text{ton}_m}$ | $\dfrac{1000 \text{ kg}}{\text{metric ton}_m}$ |
|---|---|---|---|---|---|

**Force**    Standard SI unit: Newton (N). Equivalent unit: $kg \cdot m/s^2$.

| $\dfrac{4.448 \text{ N}}{\text{lb}_f}$ | $\dfrac{10^5 \text{ dynes}}{\text{N}}$ | $\dfrac{4.448 \times 10^5 \text{ dynes}}{\text{lb}_f}$ | $\dfrac{224.8 \text{ lb}_f}{\text{kN}}$ |
|---|---|---|---|

**Length**

| $\dfrac{3.281 \text{ ft}}{\text{m}}$ | $\dfrac{39.37 \text{ in}}{\text{m}}$ | $\dfrac{12 \text{ in}}{\text{ft}}$ | $\dfrac{1.609 \text{ km}}{\text{mi}}$ | $\dfrac{5280 \text{ ft}}{\text{mi}}$ | $\dfrac{6080 \text{ ft}}{\text{nautical mile}}$ |
|---|---|---|---|---|---|

**Area**

| $\dfrac{144 \text{ in}^2}{\text{ft}^2}$ | $\dfrac{10.76 \text{ ft}^2}{\text{m}^2}$ | $\dfrac{645.2 \text{ mm}^2}{\text{in}^2}$ | $\dfrac{10^6 \text{ mm}^2}{\text{m}^2}$ | $\dfrac{43{,}560 \text{ ft}^2}{\text{acre}}$ | $\dfrac{10^4 \text{ m}^2}{\text{hectare}}$ |
|---|---|---|---|---|---|

**Volume**

| $\dfrac{1728 \text{ in}^3}{\text{ft}^3}$ | $\dfrac{231 \text{ in}^3}{\text{gal}}$ | $\dfrac{7.48 \text{ gal}}{\text{ft}^3}$ | $\dfrac{264 \text{ gal}}{\text{m}^3}$ | $\dfrac{3.785 \text{ L}}{\text{gal}}$ | $\dfrac{35.3 \text{ ft}^3}{\text{m}^3}$ |
|---|---|---|---|---|---|
| $\dfrac{28.32 \text{ L}}{\text{ft}^3}$ | $\dfrac{1000 \text{ L}}{\text{m}^3}$ | $\dfrac{61.03 \text{ in}^3}{\text{L}}$ | $\dfrac{1000 \text{ cm}^3}{\text{L}}$ | $\dfrac{1.201 \text{ U.S. gal}}{\text{Imperial gallon}}$ | |

**Volume Flow Rate**

| $\dfrac{449 \text{ gal/min}}{\text{ft}^3/\text{s}}$ | $\dfrac{35.3 \text{ ft}^3/\text{s}}{\text{m}^3/\text{s}}$ | $\dfrac{15{,}850 \text{ gal/min}}{\text{m}^3/\text{s}}$ | $\dfrac{3.785 \text{ L/min}}{\text{gal/min}}$ |
|---|---|---|---|
| $\dfrac{60{,}000 \text{ L/min}}{\text{m}^3/\text{s}}$ | $\dfrac{2120 \text{ ft}^3/\text{min}}{\text{m}^3/\text{s}}$ | $\dfrac{16.67 \text{ L/min}}{\text{m}^3/\text{hr}}$ | $\dfrac{101.9 \text{ m}^3/\text{hr}}{\text{ft}^3/\text{s}}$ |

**Density** (Mass/unit volume)

| $\dfrac{515.4 \text{ kg/m}^3}{\text{slug/ft}^3}$ | $\dfrac{1000 \text{ kg/m}^3}{\text{gram/cm}^3}$ | $\dfrac{32.17 \text{ lb}_m/\text{ft}^3}{\text{slug/ft}^3}$ | $\dfrac{16.018 \text{ kg/m}^3}{\text{lb}_m/\text{ft}^3}$ |
|---|---|---|---|

**Specific Weight** (Weight/unit volume)

| $\dfrac{157.1 \text{ N/m}^3}{\text{lb}_f/\text{ft}^3}$ | $\dfrac{1728 \text{ lb/ft}^3}{\text{lb/in}^3}$ |
|---|---|

**TABLE K.1**   Conversion factors
(*continued*)

***Pressure***   Standard SI unit: Pascal (Pa). Equivalent units: $N/m^2$ or $kg/m \cdot s^2$.

$$\frac{144 \text{ lb/ft}^2}{\text{lb/in}^2} \qquad \frac{47.88 \text{ Pa}}{\text{lb/ft}^2} \qquad \frac{6895 \text{ Pa}}{\text{lb/in}^2} \qquad \frac{1 \text{ Pa}}{\text{N/m}^2} \qquad \frac{100 \text{ kPa}}{\text{bar}} \qquad \frac{14.50 \text{ lb/in}^2}{\text{bar}}$$

$$\frac{27.7 \text{ in H}_2\text{O}}{\text{lb/in}^2} \qquad \frac{249 \text{ Pa}}{\text{in H}_2\text{O}} \qquad \frac{2.04 \text{ in Hg}}{\text{lb/in}^2} \qquad \frac{3386 \text{ Pa}}{\text{in Hg}} \qquad \frac{132.8 \text{ Pa}}{\text{mm Hg}} \qquad \frac{51.9 \text{ mm Hg}}{\text{lb/in}^2}$$

$$\frac{14.696 \text{ lb/in}^2}{\text{Std. Atmosphere}} \qquad \frac{101.325 \text{ Pa}}{\text{Std. Atmosphere}} \qquad \frac{29.92 \text{ in Hg}}{\text{Std. Atmosphere}} \qquad \frac{760 \text{ mm Hg}}{\text{Std. Atmosphere}}$$

***Energy***   Standard SI unit: Joule (J). Equivalent units: $N \cdot m$ or $kg \cdot m^2/s^2$.

$$\frac{1.356 \text{ J}}{\text{lb-ft}} \qquad \frac{1.0 \text{ J}}{\text{N} \cdot \text{m}} \qquad \frac{8.85 \text{ lb-in}}{\text{J}} \qquad \frac{1.055 \text{ kJ}}{\text{Btu}} \qquad \frac{3.600 \text{ kJ}}{\text{W} \cdot \text{hr}} \qquad \frac{778 \text{ ft-lb}}{\text{Btu}}$$

***Power***   Standard SI unit: Watt (W). Equivalent unit: J/s or $N \cdot m/s$.

$$\frac{745.7 \text{ W}}{\text{hp}} \qquad \frac{1.0 \text{ W}}{\text{N} \cdot \text{m/s}} \qquad \frac{550 \text{ lb-ft/s}}{\text{hp}} \qquad \frac{1.356 \text{ W}}{\text{lb-ft/s}} \qquad \frac{3.412 \text{ Btu/hr}}{\text{W}} \qquad \frac{1.341 \text{ hp}}{\text{kW}}$$

***General approach to application of conversion factors.*** Arrange the conversion factor from the table in such a manner that, when multiplied by the given quantity, the original units cancel out, leaving the desired units.

**Example 1**     Convert 0.24 $m^3$/s to the units of gal/min.

$$0.24 \text{ m}^3\text{/s} \; \frac{15{,}850 \text{ gal/min}}{\text{m}^3\text{/s}} = 3804 \text{ gal/min}$$

**Example 2**     Convert 150 gal/min to the units of $m^3$/s.

$$150 \text{ gal/min} \; \frac{1 \text{ m}^3\text{/s}}{15{,}850 \text{ gal/min}} = 9.46 \times 10^{-3} \text{ m}^3\text{/s}$$

**TABLE K.2** Units and conversion factors for dynamic viscosity, $\mu$ (Multiply number in table by viscosity in given unit to obtain viscosity in desired unit.)

| Given Unit ↓ / Desired Unit → | lb-s/ft$^2$ | Pa·s | Poise |
|---|---|---|---|
| lb-s/ft$^2$ | 1 | 47.88 | 478.8 |
| Pa·s* | $2.089 \times 10^{-2}$ | 1 | 10 |
| poise** | $2.089 \times 10^{-3}$ | 0.1 | 1 |
| centipoise | $2.089 \times 10^{-5}$ | 0.001 | 0.01 |

\* Standard SI units
   Equivalent unit: N·s/m$^2$

\*\* dyne·s/cm$^2$
   Equivalent unit: g/cm·s

Example: Given a viscosity measurement of 200 centipoises, the viscosity in Pa·s is

$$200 \text{ centipoises} \times \frac{0.001 \text{ Pa·s}}{\text{centipoise}} = 0.20 \text{ Pa·s}$$

**TABLE K.3** Units and conversion factors for kinematic viscosity, $\nu$ (Multiply number in table by viscosity in given unit to obtain viscosity in desired unit.)

| Given Unit ↓ / Desired Unit → | ft$^2$/s | SSU | m$^2$/s | Stoke |
|---|---|---|---|---|
| ft$^2$/s | 1 | $4.29 \times 10^5$ | $9.290 \times 10^{-2}$ | 929.0 |
| SSU* | $2.33 \times 10^{-6}$ | 1 | $2.17 \times 10^{-7}$ | $2.17 \times 10^{-3}$ |
| m$^2$/s† | 10.764 | $4.61 \times 10^6$ | 1 | $10^4$ |
| stoke‡ | $1.076 \times 10^{-3}$ | $4.61 \times 10^2$ | $10^{-4}$ | 1 |
| centistoke | $1.076 \times 10^{-5}$ | 4.61 | $10^{-6}$ | 0.01 |

\* Saybolt Seconds, Universal (conversions approximate for SSU > 100)
   For SSU < 100: $\nu = (0.226 \text{ SSU} - 195/\text{SSU})(10^{-6}) \text{ m}^2/\text{s}$

† Standard SI unit

‡ cm$^2$/s

Example: Given a viscosity measurement of 200 centistokes, the viscosity in m$^2$/s is

$$200 \text{ centistokes} \times \frac{10^{-6} \text{ m}^2/\text{s}}{\text{centistoke}} = 200 \times 10^{-6} \text{ m}^2/\text{s}$$

# ■ ■ ■ ■ L Properties of Areas

| Section | Area of Section, $A$ | Distance to Centroidal Axis, $\overline{y}$ | Moment of Inertia About Centroidal Axis, $I_c$ |
|---|---|---|---|
| Square | $H^2$ | $H/2$ | $H^4/12$ |
| Rectangle | $BH$ | $H/2$ | $BH^3/12$ |
| Triangle | $BH/2$ | $H/3$ | $BH^3/36$ |

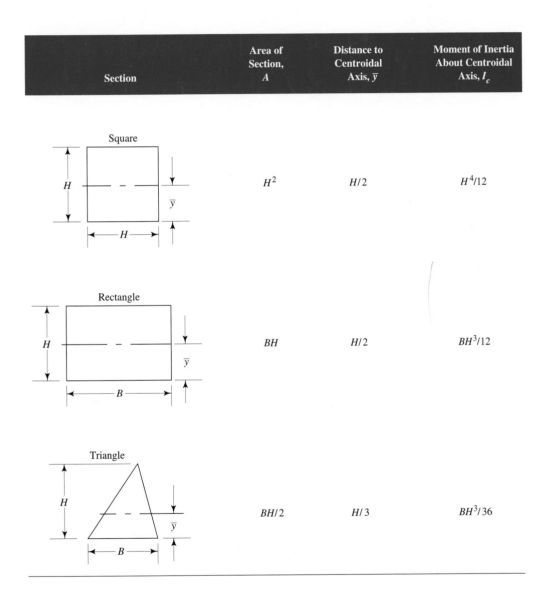

Properties of Areas (*continued*)

| Section | Area of Section, $A$ | Distance to Centroidal Axis, $\bar{y}$ | Moment of Inertia About Centroidal Axis, $I_c$ |
|---|---|---|---|

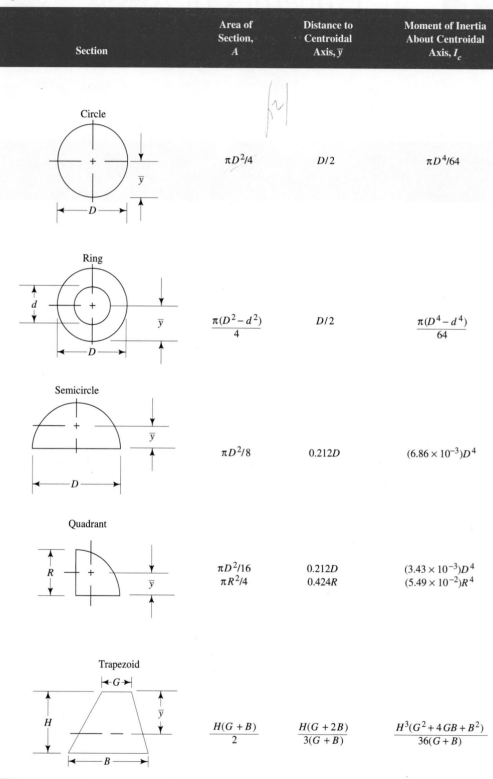

Circle

$\pi D^2/4$

$D/2$

$\pi D^4/64$

Ring

$\dfrac{\pi(D^2-d^2)}{4}$

$D/2$

$\dfrac{\pi(D^4-d^4)}{64}$

Semicircle

$\pi D^2/8$

$0.212D$

$(6.86 \times 10^{-3})D^4$

Quadrant

$\pi D^2/16$
$\pi R^2/4$

$0.212D$
$0.424R$

$(3.43 \times 10^{-3})D^4$
$(5.49 \times 10^{-2})R^4$

Trapezoid

$\dfrac{H(G+B)}{2}$

$\dfrac{H(G+2B)}{3(G+B)}$

$\dfrac{H^3(G^2+4GB+B^2)}{36(G+B)}$

# M  Properties of Solids

| Shape | Volume $V$ | Distance to Centroid, $\bar{y}$ |
|---|---|---|
| 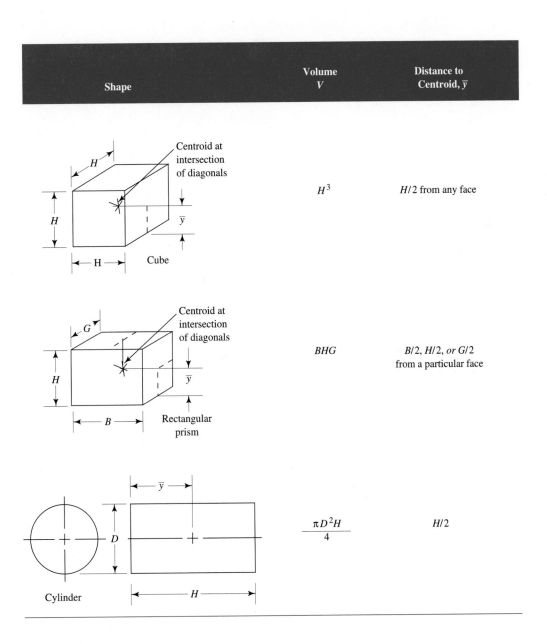 Cube | $H^3$ | $H/2$ from any face |
| Rectangular prism | $BHG$ | $B/2$, $H/2$, or $G/2$ from a particular face |
| Cylinder | $\dfrac{\pi D^2 H}{4}$ | $H/2$ |

Properties of Solids (*continued*)

| Shape | Volume $V$ | Distance to Centroid, $\overline{y}$ |
|---|---|---|

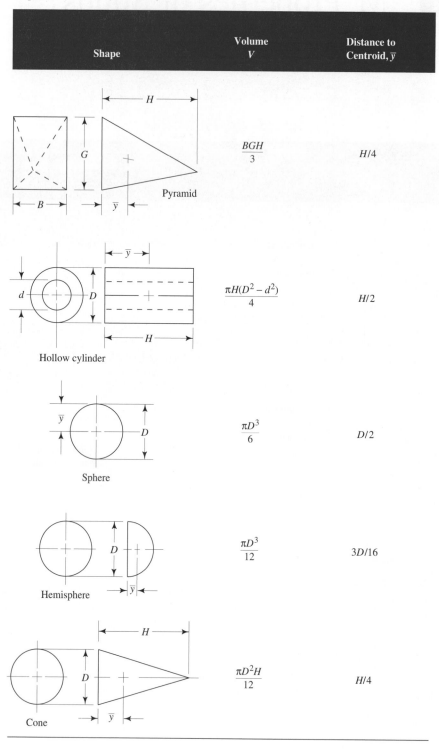

Pyramid $\qquad \dfrac{BGH}{3} \qquad H/4$

Hollow cylinder $\qquad \dfrac{\pi H(D^2 - d^2)}{4} \qquad H/2$

Sphere $\qquad \dfrac{\pi D^3}{6} \qquad D/2$

Hemisphere $\qquad \dfrac{\pi D^3}{12} \qquad 3D/16$

Cone $\qquad \dfrac{\pi D^2 H}{12} \qquad H/4$

# ■■■■ N Gas Constant, Adiabatic Exponent, and Critical Pressure Ratio for Selected Gases

| Gas | Gas Constant, $R$ | | $k$ | Critical Pressure Ratio |
|---|---|---|---|---|
| | $\dfrac{\text{ft·lb}}{\text{lb·°R}}$ | $\dfrac{\text{N·m}}{\text{N·K}}$ | | |
| Air | 53.3 | 29.2 | 1.40 | 0.528 |
| Ammonia | 91.0 | 49.9 | 1.32 | 0.542 |
| Carbon dioxide | 35.1 | 19.3 | 1.30 | 0.546 |
| Natural gas (typical; depends on gas) | 79.1 | 43.4 | 1.27 | 0.551 |
| Nitrogen | 55.2 | 30.3 | 1.41 | 0.527 |
| Oxygen | 48.3 | 26.5 | 1.40 | 0.528 |
| Propane | 35.0 | 19.2 | 1.15 | 0.574 |
| Refrigerant 12 | 12.6 | 6.91 | 1.13 | 0.578 |

# Answers to Selected Problems

## Chapter 1

**1.1**   1.25 m
**1.3**   $3.65 \times 10^{-6} \text{ m}^3$
**1.5**   $391 \times 10^6 \text{ mm}^3$
**1.7**   22.2 m/s
**1.9**   2993 m
**1.11**  786 m
**1.13**  $7.39 \times 10^{-3} \text{ m}^3$
**1.15**  1.83 m/s
**1.17**  47.2 m/s
**1.19**  48.7 mi/h
**1.21**  $8.05 \times 10^{-2} \text{ m/s}^2$
**1.23**  $0.264 \text{ ft/s}^2$
**1.25**  10.8 N-m
**1.27**  1.76 kN-m
**1.29**  37.4 g
**1.31**  1.56 m/s
**1.33**  26 700 ft-lb
**1.35**  6.20 slugs
**1.37**  4.63 ft/s
**1.39**  2.49 runs/game
**1.41**  129 innings
**1.43**  354 psi
**1.45**  2.72 MPa
**1.47**  119 psi
**1.49**  40.25 kN
**1.51**  2.26 in
**1.57**  1300 psi        8.96 MPa
**1.59**  1890 psi        13.03 MPa
**1.61**  −1.59%
**1.63**  884 lb/in
**1.65**  14 137 lb/in
**1.67**  62.2 kg
**1.69**  8093 N
**1.71**  0.242 slug
**1.73**  50.9 lb
**1.75**  $m = 4.97$ slugs
          $w = 712$ N
          $m = 72.5$ kg
**1.77**  9810 N
**1.81**  $1.225 \text{ kg/m}^3$
**1.83**  0.903 at 5°C
          0.865 at 50°C
**1.85**  Density = $883 \text{ kg/m}^3$
          Specific weight = $8.66 \text{ kN/m}^3$
          Specific gravity = 0.883
**1.87**  634 N
**1.89**  $2.72 \times 10^{-3} \text{ m}^3$
**1.91**  Density = $789 \text{ kg/m}^3$
          Specific weight = $7.74 \text{ kN/m}^3$
**1.93**  $w = 3.536$ MN        $m = 360.5$ Mg
**1.95**  3.23 N
**1.97**  $2.38 \times 10^{-3} \text{ slugs/ft}^3$
**1.99**  0.904 at 40°F        0.865 at 120°F
**1.101** Specific weight = $56.1 \text{ lb/ft}^3$
          Density = $1.74 \text{ slugs/ft}^3$
          Specific gravity = 0.899
**1.103** 142 lb
**1.105** $2745 \text{ cm}^3$
**1.107** $1.53 \text{ slugs/ft}^3$; $0.79 \text{ g/cm}^3$
**1.109** Volume = $1.16 \times 10^5$ gal
          Weight = $6.60 \times 10^5$ lb
**1.111** 0.724 lb

## Chapter 2

**2.19**  $1.5 \times 10^{-3}$ Pa-s
**2.23**  1.90 Pa-s
**2.25**  $8.9 \times 10^{-6} \text{ lb-s/ft}^2$
**2.29**  $4.1 \times 10^{-3} \text{ lb-s/ft}^2$
**2.31**  $2.8 \times 10^{-5} \text{ lb-s/ft}^2$
**2.33**  $9.5 \times 10^{-5} \text{ lb-s/ft}^2$
**2.35**  $2.2 \times 10^{-4} \text{ lb-s/ft}^2$
**2.55**  $\mu = 3\,500$ mPa·s: Cranking at −20°C
          $\mu = 60\,000$ mPa·s: Pumping at −30°C
          $\nu = 9.3 \times 10^{-6} \text{ m}^2/\text{s}$ at 100°C minimum
          $\nu = 12.5 \times 10^{-6} \text{ m}^2/\text{s}$ at 100°C maximum
          $\mu = 2.9$ mPa·s: at 150°C minimum
**2.57**  $5.60 \times 10^{-6} \text{ m}^2/\text{s}$        $6.03 \times 10^{-5} \text{ ft}^2/\text{s}$
**2.59**  $1.36 \times 10^{-4} \text{ lb-s/ft}^2$
**2.61**  0.402 Pa·s
**2.63**  $2.07 \times 10^{-4} \text{ lb-s/ft}^2$

## Chapter 3

**3.11**  12.7 psia
**3.13**  Zero gage pressure
**3.15**  56 kPa(gage)
**3.17**  −23 kPa(gage)
**3.19**  384 kPa(abs)

585

**3.21**  105.4 kPa(abs)
**3.23**  13 kPa(abs)
**3.25**  8.1 psig
**3.27**  −3.2 psig
**3.29**  55.7 psia
**3.31**  15.3 psia
**3.33**  1.9 psia
**3.35**  1.05
**3.37**  5.56 psig
**3.39**  32.37 kPa(gage)
**3.41**  177.9 psig
**3.43**  $p_{surface} = 24.77$ kPa(abs)
        $p_{bottom} = 67.93$ kPa(abs)
**3.45**  61.73 psig
**3.47**  13.36 ft
**3.49**  6.84 m
**3.51**  70.6 kPa(gage)
**3.53**  110 MPa
**3.55**  −22.47 kPa(gage)
**3.63**  $p_B - p_A = -0.258$ psi
**3.65**  $p_A - p_B = 96.03$ kPa
**3.67**  $p_A = 90.05$ kPa(gage)
**3.69**  $p_A - p_B = 2.73$ psi
**3.71**  $p_A = 0.254$ kPa(gage)
**3.77**  30.06 in
**3.81**  83.44 kPa
**3.83**  14.99 psia
**3.85**  98.94 kPa(abs)

## Chapter 4

**4.1**  1673 lb
**4.3**  125 lb
**4.5**  2.47 kN
**4.7**  22.0 kN
**4.9**  6.05 lb
**4.11**  137 kN
**4.13**  1.26 MN
**4.15**  $F_R = 126\ 300$ lb
        $d_p = 10.33$ ft vertical depth to center of pressure
        $L_p = 11.93$ ft
**4.17**  $F_R = 46.8$ kN
        $d_p = 0.933$ m vertical depth to center of pressure
        $L_p = 1.32$ m
**4.19**  $F_R = 1.09$ kN    $L_p = 966$ mm
        $L_p - L_c = 13.3$ mm
**4.21**  $F_R = 1787$ lb    $L_p = 13.51$ ft
        $L_p - L_c = 0.0136$ ft
**4.23**  $F_R = 1.213$ kN    $L_p = 1.122$ m
        $L_p - L_c = 5.98$ mm
**4.25**  $F_R = 5.79$ kN    $L_p = 1.372$ m
        $L_p - L_c = 0.0637$ m
**4.27**  $F_R = 11.97$ kN    $L_p = 1.693$ m
        $L_p - L_c = 0.0235$ m
**4.29**  $F_R = 329.6$ lb    $L_p = 47.81$ in
        $L_p - L_c = 0.469$ in

**4.31**  $F_R = 247$ N    $L_p = 196.5$ mm
        $L_p - L_c = 0.0465$ m
**4.33**  $F_R = 29\ 950$ lb    $L_p = 5.333$ ft
**4.35**  $F_R = 34\ 586$ lb    $L_p = 6.158$ ft
**4.37**  $F_R = 343$ kN    $L_p = 3.067$ m
**4.39**  $F_R = 11.92$ kN    $L_p = 1.00$ m
**4.41**  Force on hinge = 4.85 kN to left
        Force on stop = 2.95 kN to left
**4.43**  $F_R = 3.29$ kN
        $L_{pe} - L_{ce} = 4.42$ mm
**4.45**  $F_R = 1826$ lb
        $L_{pe} - L_{ce} = 1.885$ in
**4.47**  $F_R = 48.58$ kN    $F_V = 35.89$ kN
        $F_H = 32.74$ kN
**4.49**  $F_R = 120\ 550$ lb    $F_V = 99\ 925$ lb
        $F_H = 67\ 437$ lb
**4.51**  $F_R = 959.1$ kN    $F_V = 927.2$ kN
        $F_H = 245.3$ kN
**4.53**  $F_R = 80.7$ kN    $F_V = 54.0$ kN
        $F_H = 60.0$ kN
**4.55**  $F_R = 64.49$ kN    $F_V = 47.15$ kN
        $F_H = 44.00$ kN
**4.57**  70.1 lb downward
**4.59**  Zero
**4.61**  70.1 lb downward

## Chapter 5

**5.1**  Buoyant force = 814 N    Tension = 556 N
**5.3**  It will sink
**5.5**  234 mm
**5.7**  0.217 m$^3$
**5.9**  $7.515 \times 10^{-3}$ m$^3$
**5.11**  5.055 ft$^3$
**5.13**  0.0249 lb
**5.15**  1.041
**5.17**  1447 lb
**5.19**  283.6 m$^3$
**5.21**  7.95 kN/m$^3$
**5.23**  237 mm
**5.25**  29 mm
**5.27**  10.05 kN
**5.29**  135 mm
**5.31**  1681 lb
**5.33**  4.67 in
**5.35**  300 lb
**5.37**  14.39 lb
**5.39**  $y_{mc} = 0.4844$ m  (unstable)
**5.41**  $y_{mc} = 8.256$ ft    (stable)
**5.43**  $y_{mc} = 488.8$ mm (unstable)
**5.45**  $y_{mc} = 10.55$ in   (unstable)
**5.47**  32.50 ft
**5.49**  $y_{mc} = 436$ mm    (stable)
**5.51**  $y_{mc} = 90.2$ mm    (stable)
**5.53**  $y_{mc} = 410.3$ mm    (stable)
**5.55**  $y_{mc} = 54.18$ in    (stable)

**5.57** $y_{mc}$ = 13.29 ft    (stable)
**5.59** $y_{mc}$ = 467 mm    (stable)
**5.61** $y_{mc}$ = 1.288 m    (stable)
**5.63** (a) 17.09 kN    (b) 11.85 kN/m$^3$
    (c) Unstable; $y_{mc}$ = 0.822 m; $y_{cg}$ = 0.950 m

## Chapter 6

**6.1** $1.89 \times 10^{-4}$ m$^3$/s
**6.3** 0.550 m$^3$/s
**6.5** $2.08 \times 10^{-3}$ m$^3$/s
**6.7** 0.250 m$^3$/s
**6.9** $3.30 \times 10^4$ L/min
**6.11** $2.96 \times 10^{-7}$ m$^3$/s
**6.13** 215 L/min
**6.15** 1.02 ft$^3$/s
**6.17** 5.57 ft$^3$/s
**6.19** 561 gal/min
**6.21** 3368 gal/min
**6.23** $Q$ = 500 gal/min = 1.11 ft$^3$/s = $3.15 \times 10^{-2}$ m$^3$/s
    $Q$ = 2500 gal/min = 5.57 ft$^3$/s = 0.158 m$^3$/s
**6.25** $2.77 \times 10^{-2}$ ft$^3$/s
**6.27** $1.76 \times 10^{-5}$ ft$^3$/s
**6.29** $W$ = 736 N/s    $M$ = 75.0 kg/s
**6.31** $Q$ = $7.47 \times 10^{-7}$ m$^3$/s    $M$ = $8.07 \times 10^{-4}$ kg/s
**6.33** $M$ = $2.48 \times 10^{-2}$ slugs/s    $W$ = 2878 lb/h
**6.35** 5.38 ft$^3$/s
**6.37** 3.09 ft
**6.39** $v_1$ = 0.472 m/s    $v_2$ = 1.89 m/s
**6.41** 6.17 m/s
**6.43** $1.97 \times 10^5$ lb/h
**6.45** 1¼ × 0.065 steel tube for $v \geq$ 8.0 ft/s min
    ⅞ × 0.065 steel tube for $v \leq$ 25.0 ft/s max
**6.47** 6 in Sch 40 pipe for Q = 1800 L/min
    14 in Sch 40 pipe for 9500 L/min
**6.49** 3.075 m/s
**6.51** 10.08 ft/s
**6.53** ¾ × 0.065 steel tube
**6.55** $Q_{shell}/Q_{tube}$ = 2.19
**6.57** $Q_{tube}$ = 0.3535 ft$^3$/s    $Q_{shell}$ = 1.998 ft$^3$/s
**6.59** $v_{pipe}$ = 7.98 ft/s    $v_{nozzle}$ = 181.8 ft/s
**6.61** 34.9 kPa
**6.63** 25.1 psig
**6.65** $p_A$ = 58.1 kPa    Q = 0.0213 m$^3$/s
**6.67** 2.90 ft$^3$/s
**6.69** $Q$ = $4.66 \times 10^{-3}$ m$^3$/s    $P_A$ = −2.82 kPa
    $P_B$ = −11.65 kPa
**6.71** 1.42 m
**6.73** 35.6 ft/s
**6.75** $3.98 \times 10^{-3}$ m$^3$/s
**6.77** $1.48 \times 10^{-3}$ m$^3$/s
**6.79** 1.035 ft$^3$/s
**6.81** 31.94 psig
**6.86** $6.00 \times 10^{-3}$ m$^3$/s
**6.90** 1.28 ft
**6.93** 10.18 psig

**6.95** 296 s
**6.97** 556 s
**6.99** 504 s
**6.101** 1155 s
**6.103** 252 s
**6.105** 1.94 s

## Chapter 7

**7.1** 34.5 lb-ft/lb
**7.3** $3.33 \times 10^{-2}$ m$^3$/s
**7.5** 15.7 lb-ft/lb
**7.7** 72.7
**7.9** 16.2 kW
**7.11** 0.700 hp    70.0%
**7.13** $h_A$ = 37.46 m    $P_A$ = 0.390 kW
**7.15** (a) $p_B$ = 1.07 psig    (b) $p_C$ = 21.8 psig
    (c) $h_A$ = 48 ft    (d) $P_A$ = 10.9 hp
**7.17** $h_A$ = 4.68 m    $P_A$ = 43.6 W
**7.19** 2.84 hp
**7.21** $P_A$ = 1.25 W    $P_I$ = 2.08 W
**7.23** 9.15 kW
**7.25** $P_R$ = 16.79 kW    $P_O$ = 12.60 kW
**7.27** 2.00 hp
**7.29** 35.0 kPa
**7.31** 12.8 ft
**7.33** 6.88 m
**7.35** 21.16 hp
**7.37** 219.1 psig
**7.39** 1.01 psig
**7.41** 5.76 psig
**7.43** 4.28 hp
**7.45** 1.26 hp

## Chapter 8

**8.1** 249
**8.3** $7.02 \times 10^{-3}$ m$^3$/s
**8.5** (a) 3 in Type K copper tube    (b) 5 in tube
    (c) ¾ in tube    (d) ⅛ in tube
**8.9** $4.76 \times 10^4$
**8.11** $9.59 \times 10^5$
**8.13** 33.4
**8.15** $5.61 \times 10^3$
**8.17** 2237 (Critical zone)
**8.19** 1105
**8.21** $2.12 \times 10^4$
**8.23** $Q_1$ = 0.1681 gal/min
    $Q_2$ = 0.3362 gal/min
**8.25** $1.06 \times 10^4$
**8.27** 32.4 mm
**8.31** $3.04 \times 10^4$
**8.33** $3.808 \times 10^5$
**8.35** In pipe: $N_R$ = $6.08 \times 10^4$
    In duct: Q = $1.39 \times 10^{-2}$ m$^3$/s
**8.37** $1.58 \times 10^5$
**8.39** $5.90 \times 10^3$

**8.41** $1.61 \times 10^7$
**8.43** 0.224 ft³/s
**8.45** $Q_{total} = 2.40 \times 10^{-3}$ m³/s    $N_R = 6.40 \times 10^3$
**8.47** Top channel: $N_R = 9430$ Turbulent
Both side channels: $N_R = 7127$ Turbulent

## Chapter 9

**9.1** $p_1 - p_2 = -471$ kPa
**9.3** 1.20 lb-ft/lb
**9.5** $p_1 - p_2 = 25.2$ kPa
**9.7** 45.7 ft
**9.9** (a) 12.60 ft    (b) 113.8 hp
**9.11** 46.9 psi
**9.13** (a) 853 kPa    (b) 17.1 kW
**9.15** 89.9 kPa
**9.17** $p_1 - p_2 = 39.6$ psi
**9.19** $p_1 - p_2 = 411$ psi
**9.21** 151 hp
**9.23** 2.64 hp
**9.25** $p_1 - p_2 = 110$ kPa
**9.27** 0.0273
**9.29** 0.0155
**9.31** 0.0213
**9.33** 0.0206
**9.35** 0.0175
**9.37** 0.713 psi
**9.39** 92.0 Pa
**9.41** 111 kPa
**9.43** 3.02 kPa
**9.45** 3.72 psi
**9.47** $v = 23.05$ ft/s    $Q = 187$ gal/min    $h_L = 51.9$ ft
**9.49** $7.36 \times 10^4$
**9.51** 1.84 m/s
**9.53** Selected values:

| y (mm) | U (m/s) |
|---|---|
| 10 | 0.530 |
| 30 | 0.628 |
| 50 | 0.674 |
| 100 | 0.735 |
| 300 | 0.833 |
| 600 | $0.895 = U_{max}$ |

**9.55** At $y = 2.44$ in, $U = v_{avg} = 6.00$ ft/s
At $y_1 = 2.94$ in, $U_1 = 6.12$ ft/s
At $y_2 = 1.94$ in, $U_2 = 5.85$ ft/s

**9.57**

| $N_R$ | f | $v/U_{max}$ |
|---|---|---|
| $4 \times 10^3$ | 0.041 | 0.775 |
| $1 \times 10^4$ | 0.032 | 0.796 |
| $1 \times 10^5$ | 0.021 | 0.828 |
| $1 \times 10^6$ | 0.0185 | 0.837 |

**9.59** Selected values: $v = 10.08$ ft/s

| y (in) | U (ft/s) |
|---|---|
| 0.05 | 5.83 |
| 0.15 | 7.98 |
| 0.50 | 10.35 |
| 1.00 | 11.71 |
| 1.50 | 12.51 |
| 2.013 | $13.09 = U_{max}$ |

**9.61** $h_L = 15.2$ ft
**9.63** $h_L = 28.5$ ft
**9.65** $h_L = 3.56$ m
**9.67** (a) $h_L = 61.4$ ft    (b) $h_L = 28.3$ ft
**9.69** $h_L = 14.7$ ft

## Chapter 10

**10.1** 0.239 m
**10.3** 4.55 ft
**10.5** $p_1 - p_2 = -0.0891$ psi
**10.7** 0.326 m
**10.13** 503.7 kPa
**10.15** 0.235 m
**10.17** 1.35 ft
**10.19** False
**10.21** 0.224 m
**10.23** 4.32 ft
**10.27** $K = 0.255$
**10.29** (a) 0.459 m    (b) 0.229 m    (c) 0.115 m
(d) 0.018 m
**10.31** 2.04 m
**10.33** 1.58 psi
**10.35** 7.36 kPa
**10.37** 9.87 psi
**10.39** 0.340 ft
**10.41** 1.29 m
**10.43** $h_{L_1} = 0.85$    $h_{L_2} = 1.11$
**10.45** 0.432 m
**10.47** 0.849 ft
**10.49** $K = 9.15$    $h_L = 15.5$ ft
**10.51** $K = 0.731$    $h_L = 1.25$ ft
**10.53** 175 psi
**10.55** $K = 143$
**10.57** $C_v = 0.645$

## Chapter 11

**11.1** 85.1 kPa
**11.3** 212.8 psig
**11.5** 12.74 MPa
**11.7** 24.3 kPa
**11.9** 3.16 m/s
**11.11** $1.64 \times 10^{-3}$ m³/s
**11.13** (a) 49.6 ft/s    (b) 129 ft/s
**11.15** $1.95 \times 10^{-2}$ m³/s
**11.17** 5 in Schedule 80 pipe
**11.19** 2.12 ft
**11.21** 3.35 m

**11.23** $h_A = 24.24$ ft $\quad P_A = 0.169$ hp
**11.25** 6.68 ft/s
**11.27** $-49.0$ kPa
**11.29** $h_A = 276.9$ ft $\quad P_A = 16.1$ hp $\quad P_I = 21.2$ hp
**11.31** 321.1 kPa
**11.33** 204 L/min
**11.35** 250 gal/min
**11.37** 270 gal/min
**11.39** $4.08 \times 10^{-2}$ m$^3$/s
**11.41** $7.32 \times 10^{-2}$ m$^3$/s
**11.43** 4-in Schedule 40 steel pipe
**11.45** 1$^1$/$_2$-in Schedule 40 plastic pipe
**11.47** 1$^1$/$_4$-in steel tube; $t = 0.065$ in
**11.49** $Q = 136.6$ L/min

## Chapter 12

**12.1** 0.0602 m$^3$/s
**12.3** (a) $Q_a = 518$ L/min (Upper pipe)
$\quad\quad Q_b = 332$ L/min (Lower pipe)
$\quad$ (b) 95.0 kPa
**12.5** $K = 160$
**12.6** (a) 1.841 ft$^3$/s $\quad$ (b) 1.385 ft$^3$/s $\quad$ (c) 0.456 ft$^3$/s
**12.7** $Q_6 = 2.805$ ft$^3$/s in 6-in pipe
$\quad Q_6 = 0.205$ ft$^3$/s in 2-in pipe
**12.11** Flow rates after 6 iterations
$\quad \Delta Q < 0.25\%$ in any pipe
$\quad Q_1 = 7.504$ ft$^3$/s $\quad Q_2 = 4.204$ ft$^3$/s
$\quad Q_3 = 7.996$ ft$^3$/s $\quad Q_4 = 7.504$ ft$^3$/s
$\quad Q_5 = 2.704$ ft$^3$/s $\quad Q_6 = 3.637$ ft$^3$/s
$\quad Q_7 = 2.678$ ft$^3$/s $\quad Q_8 = 4.359$ ft$^3$/s
$\quad Q_9 = 3.259$ ft$^3$/s $\quad Q_{10} = 1.382$ ft$^3$/s
$\quad Q_{11} = 1.359$ ft$^3$/s $\quad Q_{12} = 1.618$ ft$^3$/s

## Chapter 13

**13.16** Capacity is cut in half
**13.17** Reduced by a factor of 4
**13.18** Reduced by a factor of 8
**13.19** Reduced by 25%
**13.20** Reduced by 44%
**13.21** Reduced by 58%
**13.23** 1-$^1$/$_2$ $\times$ 3 $-$ 10
**13.25** $Q = 280$ gal/min $\quad P = 26$ hp
$\quad$ Efficiency $= 53\%$ $\quad NSPH_R = 10.9$ ft
**13.26** Head $= 250$ ft $\quad Q = 220$ gal/min
$\quad P = 24.0$ hp $\quad$ Efficiency $= 56\%$
$\quad NPSH_R = 8.0$ ft
**13.35** $Q = 390$ gal/min $\quad N_s = 619$ $\quad D_s = 2.94$
**13.37** 795 rpm
**13.39** 2659
**13.41** 2475
**13.51** 4.97 ft
**13.53** 20.70 ft
**13.55** 3.43 m

## Chapter 14

**14.1** 75 mm
**14.3** 0.940 ft
**14.5** 40.3 mm
**14.7** 1.606 in
**14.9** 0.909 m
**14.11** 0.295 ft$^3$/s
**14.13** 0.0125
**14.15** (a) 34.7 ft$^3$/s $\quad$ (b) 141.1 ft$^3$/s
**14.17** 1.69 m
**14.19** $Q = 15.89$ m$^3$/s
$\quad N_F = 0.629$ for depth $= 1.50$ m
$\quad y_c = 1.16$ m
**14.21** 1.29 ft
**14.23** $y = 0.833$ m $\quad R = 0.270$ m
**14.24** $Q = 0.0116$ m$^3$
**14.25 and 14.26**
Selected values:

| Width (m) | Depth (m) | $R$ (m) | $S$ |
|---|---|---|---|
| 0.50 | 1.333 | 0.2105 | 0.0162 |
| 1.00 | 0.667 | 0.2857 | 0.0108 |
| 1.50 | 0.444 | 0.2791 | 0.0111 |
| 2.00 | 0.333 | 0.2500 | 0.0129 |

**14.27** $A = 7.50$ ft$^2$ $\quad R = 0.936$ ft
**14.28** $Q = 44.49$ ft$^3$/s
**14.29** $Q = 75.63$ ft$^3$/s
**14.30 and 14.31**
Selected values:

| $y$ (in) | $A$ (ft$^2$) | $R$ (ft) | $Q$ (ft$^3$/s) |
|---|---|---|---|
| 6.00 | 1.375 | 0.3616 | 4.33 |
| 10.00 | 2.708 | 0.5412 | 11.15 |
| 18.00 | 6.375 | 0.8605 | 35.75 |
| 24.00 | 10.00 | 1.086 | 65.47 |

**14.33** $A = 0.0358$ m$^2$ $\quad R = 0.0742$ m
**14.35** $Q = 0.0168$ m$^3$/s
**14.37**

| | $S$ |
|---|---|
| Rectangle | 0.00519 |
| Triangle | 0.00518 |
| Trapezoid | 0.00471 |
| Semicircle | 0.00441 |

**14.39** (a) $y_c = 0.917$ m $\quad$ (b) $E = 1.38$ m
$\quad$ (d) $E = 2.04$ m $\quad y = 1.90$ m
$\quad$ (e) For $y = 0.50$ m:
$\quad\quad v = 5.5$ m/s $\quad N_F = 2.48$
$\quad\quad$ For $y = 1.90$ m:
$\quad\quad v = 1.45$ m/s $\quad N_F = 0.34$
$\quad$ (f) For $y = 0.50$ m, $S = 0.0378$
$\quad\quad$ For $y = 1.90$ m, $S = 0.0011$

## Chapter 15

**15.1** $2.12 \times 10^{-2}$ m$^3$/s
**15.3** 0.0336 ft$^3$/s
**15.5** 8.45 m/s

**15.7**  5.11 m/s

**15.9**  2.66 ft

**15.12**  (a) 18.8 ft$^3$/s       (b) 16.95 ft$^3$/s       (c) 6.84 ft$^3$/s

**15.14**  0.10 ft minimum
2.01 ft maximum

**15.16**  $H$ = 1.687 ft for a 4-ft flume
$H$ = 1.155 ft for a 10-ft flume

## Chapter 16

**16.1**  2.76 kN

**16.3**  $R_x = R_y$ = 39.7 lb       Resultant = 56.1 lb       45°

**16.5**  $R_x$ = 10.13 lb to the right       $R_x$ = 37.79 lb up

**16.7**  $R_x$ = 873 N to the left       $R_x$ = 1512 N up

**16.9**  25.2 m/s

**16.11**  Spring force = 32.0 lb

**16.13**  368 lb

**16.15**  2676 lb

**16.17**  $R_x = R_y$ = 20.41 kN       Resultant = 28.9 kN       45°

**16.19**  $R_x$ = 10.17 kN       $R_y$ = 4.18 kN

Resultant = 11.0 kN       22.3°

**16.21**  $v$ = 45.6 m/s

**16.23**  2.72 × 10$^{-7}$ N

**16.25**  Moment = 0.336 lb-in

**16.27**  0.0307 lb

**16.29**  $R_s$ = 41.0 N; $R_y$ = 19.1 N

## Chapter 17

**17.1**  (a) 0.253 N       (b) 4.56 × 10$^{-4}$ N

**17.3**  1.50 m/s       2.05 m/s       105 m/s

**17.5**  0.42 N

**17.7**  7.32 m

**17.9**  11.2 kN

**17.11**  (a) 2.85 × 10$^{-6}$ N·m       (b) 1.67 × 10$^{-3}$ N·m

**17.13**  1364 N

**17.15**  $F_D$ = 12.05 lb

**17.17**  $F_D$ = 31.3 lb

**17.19**  $F_D$ = 5.86 lb

**17.21**  $F_D$ = 1414 lb

**17.23**  $F_D$ = 0.080 lb on golf ball
$F_D$ = 0.207 lb on smooth sphere

**17.25**  $P_D$ = 140 hp

**17.27**  $P_E$ = 4252 hp

**17.29**  (a) $F_D$ = 26.5 kN       (b) $F_D$ = 1.66 kN

**17.31**  (a) $F_L$ = 26.8 kN       $F_D$ = 2.83 kN
(b) $F_L$ = 9.24 kN       $F_D$ = 972 N

**17.33**  $A$ = 90.4 m$^2$

## Chapter 18

**18.1**  44.17 ft$^3$/s

**18.3**  1.25 m$^3$/s

**18.5**  5.79 m/s

**18.7**  0.158 psi

**18.9**  3.72 in H$_2$O

**18.17**  0.478 lb/ft$^3$

**18.19**  0.0525 lb/ft$^3$

**18.21**  11.79 N/m$^3$

**18.23**  131 cfm

**18.25**  2-in Schedule 40 pipe

**18.27**  2$^1$/$_2$-in Schedule 40 pipe, $p$ = 107 psig

**18.29**  0.2735 lb/ft$^3$ in the reservoir
0.198 lb/ft$^3$ in the pipe

**18.31**  1.092 lb/ft$^3$ at 35.0 psig
0.506 lb/ft$^3$ at 3.6 psig

**18.33**  $W$ = 5.76 × 10$^{-3}$ lb/s
$v$ = 811 ft/s

**18.34**  4.44 × 10$^{-3}$ lb/s

**18.37**  186.6 kPa

**18.39**  9.58 × 10$^{-3}$ N/s

**18.40 and 18.41**

| $p_1$ (kPa gage) | $W$ (N/s) |
| --- | --- |
| 150 | 0.555 |
| 125 | 0.500 |
| 100 | 0.444 |
| 75 | 0.389 |
| 50 | 0.326 |
| 25 | 0.238 |

## Chapter 19

**19.1**  $v$ = 570 ft/min       $h_L$ = 0.0203 in H$_2$O

**19.3**  $D$ = 17.0 in       $h_L$ = 0.078 in H$_2$O

**19.6**  $D$ = 350 mm       $h_L$ = 0.58 Pa/m

**19.8**  $D_e$ = 5.74 in       $Q$ = 95 cfm

**19.10**  $D_e$ = 381 mm       $Q$ = 0.60 m$^3$/s

**19.12**  10 × 24 or 12 × 20

**19.14**  0.0180 in H$_2$O

**19.16**  0.0145 in H$_2$O

**19.18**  Main duct; 1600 cfm:
$v$ = 1160 ft/min; $H_v$ = 0.0839 in H$_2$O
Main duct; 1100 cfm:
$v$ = 800 ft/min; $H_v$ = 0.0399 in H$_2$O
Branch; 500 cfm:
$v$ = 720 ft/min; $H_v$ = 0.0323 in H$_2$O

**19.20**  $H_L$ = 0.00839 in H$_2$O

**19.22**  $H_L$ = 29.6 Pa

**19.24**  $H_L$ = 0.1629 in H$_2$O

**19.26**  $p_{fan}$ = −27.6 Pa

# Index

# KEY EQUATIONS

FRICTION FACTOR FOR TURBULENT FLOW

$$f = \frac{0.25}{\left[ \log \left( \dfrac{1}{3.7 \, (D/\epsilon)} + \dfrac{5.74}{N_R^{0.9}} \right) \right]^2}$$

(9–5)

REYNOLDS NUMBER FOR NONCIRCULAR SECTIONS

$$N_R = \frac{v(4R)\rho}{\mu} = \frac{v(4R)}{\nu}$$

(9–6)

DARCY'S EQUATION FOR NONCIRCULAR SECTIONS

$$h_L = f \frac{L}{4R} \frac{v^2}{2g}$$

(9–7)

HAZEN-WILLIAMS FORMULA U.S. CUSTOMARY UNITS

$$v = 1.32 \, C_h R^{0.63} s^{0.54}$$

(9–11)

HAZEN-WILLIAMS FORMULA SI UNITS

$$v = 0.85 \, C_h R^{0.63} s^{0.54}$$

(9–12)

HYDRAULIC RADIUS

$$R = \frac{A}{WP} = \frac{\text{area}}{\text{wetted perimeter}}$$

(14–1)

REYNOLDS NUMBER FOR OPEN CHANNELS

$$N_R = \frac{vR}{\nu}$$

(14–3)

FROUDE NUMBER

$$N_F = \frac{v}{\sqrt{gy_h}}$$

(14–4)

HYDRAULIC DEPTH

$$y_h = A/T$$

(14–5)

MANNING'S EQUATION—SI UNITS

$$v = \frac{1.00}{n} R^{2/3} S^{1/2}$$

(14–6)

NORMAL DISCHARGE—SI UNITS

$$Q = \left( \frac{1.00}{n} \right) AR^{2/3} S^{1/2}$$

(14–8)

MANNING'S EQUATION—U.S. CUSTOMARY UNITS

$$v = \frac{1.49}{n} R^{2/3} S^{1/2}$$

(14–10)

NORMAL DISCHARGE—U.S. CUSTOMARY UNITS

$$Q = AV = \left( \frac{1.49}{n} \right) AR^{2/3} S^{1/2}$$

(14–11)

GENERAL FORM OF FORCE EQUATION

$$F = (m/\Delta t)\Delta v = M \, \Delta v = \rho Q \, \Delta v$$

(16–4)